G Protein-Coupled Receptors
From Structure to Function

RSC Drug Discovery Series

Titles in the Series:
1: Metabolism, Pharmacokinetics and Toxicity of Functional Groups: Impact of Chemical Building Blocks on ADMET
2: Emerging Drugs and Targets for Alzheimer's Disease; Volume 1: Beta-Amyloid, Tau Protein and Glucose Metabolism
3: Emerging Drugs and Targets for Alzheimer's Disease; Volume 2: Neuronal Plasticity, Neuronal Protection and Other Miscellaneous Strategies
4: Accounts in Drug Discovery: Case Studies in Medicinal Chemistry
5: New Frontiers in Chemical Biology: Enabling Drug Discovery
6: Animal Models for Neurodegenerative Disease
7: Neurodegeneration: Metallostasis and Proteostasis
8: G Protein-Coupled Receptors: From Structure to Function

How to obtain future titles on publication:
A standing order plan is available for this series. A standing order will bring delivery of each new volume immediately on publication.

For further information please contact:
Book Sales Department, Royal Society of Chemistry, Thomas Graham House, Science Park, Milton Road, Cambridge, CB4 0WF, UK
Telephone: +44 (0)1223 420066, Fax: +44 (0)1223 420247, Email: books@rsc.org
Visit our website at http://www.rsc.org/Shop/Books/

G Protein-Coupled Receptors
From Structure to Function

Edited by

Jesús Giraldo
Institut de Neurosciències and Unitat de Bioestadística, Universitat Autònoma de Barcelona, Bellaterra, Spain

Jean-Philippe Pin
Departement de Pharmacologie Moleculaire, Universités Montpellier I and II, Montpellier, France

RSC Publishing

RSC Drug Discovery Series No. 8

ISBN: 978-1-84973-183-6
ISSN: 2041-3203

A catalogue record for this book is available from the British Library

Published by The Royal Society of Chemistry,
Thomas Graham House, Science Park, Milton Road,
Cambridge CB4 0WF, UK

Registered Charity Number 207890

For further information see our web site at www.rsc.org

Preface

G protein-coupled receptors (GPCRs) are membrane proteins that share a common structure consisting of seven transmembrane helices connected by extracellular and intracellular loops. GPCRs are at the centre of current pharmacological research, from both an academic and an industrial side. The reasons for this arise from the ubiquitous presence of GPCRs in live systems and the many cellular functions they regulate. These receptors are encoded by the largest gene family in mammalian genomes, representing more than 3% of genes. Not surprisingly, these proteins have evolved to recognize a wide variety of signals, from photons to large proteins, through various types of neurotransmitters and hormones. Not only are GPCRs expressed in every cell, but each cell type is likely to express a defined subset of these receptors, thus offering ways to precisely target specific cell types with synthetic compounds acting on these receptors. It is therefore not surprising that GPCRs represent 30% of all identified drug targets and remain major targets for drug development programmes.

Over the last decade and particularly its second half, the GPCR field has experienced an explosion of knowledge from many of the currently available technologies. From crystallography, several structures of GPCRs either free or in complex with antagonists and agonists have been solved, providing researchers with a reliable basis for the construction of mechanistic hypothesis for the functional differentiation between ligands. Crystallography is fundamental in assessing the structural aspects of ligand–protein interactions and it is expected that the current partial unblocking of the technical obstacles that have made GPCR crystallization a titanic task in the recent past will lead to the progressive filling of the ligand–receptor structural space. Yet, the information that crystallography renders is framed in a static picture which cannot

RSC Drug Discovery Series No. 8
G Protein-Coupled Receptors: From Structure to Function
Edited by Jesús Giraldo and Jean-Philippe Pin

Published by the Royal Society of Chemistry, www.rsc.org

encompass the many conformations receptors can sample and, importantly, the process of receptor activation.

To obtain a dynamic view of proteins, time needs to be included. Experimental and theoretical studies provide complementary information. Nuclear magnetic resonance (NMR) spectroscopy—particularly solid-state NMR—is of key importance in exploring the receptor conformational landscape and the changes that the receptor experiences after activation. These molecular details are now closer to researchers' eyes thanks to the enormous advances in computer hardware and software which have led to longer and more accurate molecular dynamics trajectories.

GPCRs are not dancing alone and many protein–protein interactions have already been identified. Of special interest are those including GPCR kinases and arrestins. These proteins were initially associated strictly with desensitization processes. However, it is now known that they can have specific signalling effects. GPCRs not only interact with other proteins but also between them. An important flow of structural and functional knowledge has come from fluorescence- and luminescence-based approaches which have provided insights into receptor homo- and hetero-oligomerization and the concurrent receptor inter-subunits crosstalk. The complexity of GPCR organization and function is modulated by the membrane environment whose influence is being revealed to be more and more decisive.

All these investigations have led to a change of paradigm for GPCR signal transduction. The classical mechanism of activation of GPCRs involves the binding of an agonist to the extracellular side of the receptor coupled with the activation of a specific heterotrimeric G protein at the receptor intracellular region which, in turn, is connected with the activation of an effector system. This simple picture—one receptor, one G protein, one effector system—is no longer valid. The involvement of a particular receptor in multiple ligand-modulated signalling pathways, either through different G proteins or other accessory proteins, together with the existence of receptor constitutive activity, receptor oligomerization and allosteric binding sites has opened up the traditional drug discovery space, impelling collaborative efforts between researchers from all the disciplines involved—both experimental and computational.

GPCR research is evolving progressively towards systemic and transversal approaches. With these ideas in mind, we have sought to gather in this book the best current research in this exciting and dynamic field. We are very grateful to the authors who agreed to participate in this project and especially to the anonymous referees who reviewed the manuscripts. We hope you enjoy reading the book.

Jesús Giraldo and Jean-Philippe Pin

Contents

RSC Drug Discovery Series No. 8
G Protein-Coupled Receptors: From Structure to Function
Edited by Jesús Giraldo and Jean-Philippe Pin

Published by the Royal Society of Chemistry, www.rsc.org

HISTORICAL PERSPECTIVE

From Receptors to G Protein-coupled Receptors to Seven Transmembrane Receptors: A Journey of Discovery

ROBERT J. LEFKOWITZ

Department of Medicine and Biochemistry, Howard Hughes Medical Institute, Duke University Medical Center, Durham, NC 27710, USA

Early Ideas—Lingering Scepticism

Although some trace the notion of cellular receptors back to the work of Ehrlich in the late 19th century on the interactions of antigens with cells,[1] it was the British pharmacologist J. N. Langley[2] who first clearly and explicitly articulated the idea of 'receptors' and 'effectors' based on his work on the actions of adrenaline and cholinergic agents on skeletal muscle and salivary glands. Thus, in 1905, he wrote:

'So we may suppose that in all cells two constituents at least are to be distinguished, a chief substance, which is concerned with the chief function of the cell as contraction and secretion, and receptive substances which are acted upon by chemical bodies and in certain cases by nervous stimuli. The receptive substance affects or is capable of affecting the metabolism of the chief substance'.[2]

RSC Drug Discovery Series No. 8
G Protein-Coupled Receptors: From Structure to Function
Edited by Jesús Giraldo and Jean-Philippe Pin

Published by the Royal Society of Chemistry, www.rsc.org

Rather than being greeted with great acceptance, however, the concept was often either ignored or derided. For example, in 1943, Langley's former student H. H. Dale (a Nobelist for his work on cholinergic neurotransmission) wrote somewhat contemptuously of Langley's idea:

'It is a mere statement of fact to say that the action of adrenaline picks out certain such effector-cells and leaves others unaffected; it is a simple deduction that the affected cells have a special affinity of some kind for adrenaline; but I doubt whether the attribution to such cells of 'adrenaline-receptors' does more than re-state this deduction in another form'.[3]

Even more ironically, as late as 1973, Raymond Ahlquist, who in 1948 had proposed the idea of distinct α and β-adrenergic receptors, wrote of the notion of receptors:

'This would be true if I were so presumptuous as to believe that α and β receptors really did exist. There are those that think so and even propose to describe their intimate structure. To me they are an abstract concept conceived to explain observed responses of tissues produced by chemicals of various structure'.[4]

Despite this evident skepticism about the physicochemical existence of specific cellular receptor structures, the years between 1920 and 1970 witnessed the development of now classical receptor theory by pioneers such as Ariens, Clark, Stephenson, Black and Furchgott (reviewed in ref. 5).

A turning point occurred in the 1960s, largely due to the work of Earl Sutherland. He demonstrated the generation of a 'second messenger', cAMP, as a consequence of the activation of receptors such as those for adrenaline and glucagon and then many others.[6] This marked a transition in the field and signalled an emerging merger between the disciplines of pharmacology and biochemistry. For the first time the properties of putative receptors could be inferred not from some very downstream physiological readout, but rather from a much more proximate biochemical signal, in this case activation of the enzyme adenylate cyclase.

However, the receptors themselves remained elusive and controversial. As a biochemist, Sutherland was more comfortable with the idea of receptors as molecular entities, but he was still not willing to accord them an independent existence. Thus in 1969 he wrote:

'It seems likely that in most and perhaps all tissues the β receptor and adenyl cyclase are the same. The results of many previous studies have pointed to this conclusion, and we feel that the studies with the perfused rat heart have added further to its possible validity'.[6]

In 1971 Martin Rodbell discovered that the actions of hormones (initially glucagon) to stimulate adenylate cyclase required GTP and postulated the existence of a guanine nucleotide regulatory protein, distinct from the enzyme, which would mediate this effect.[7] The existence of these proteins was proven, and they were purified by Gilman and co-workers in the mid-1980s.[8]

The Molecular Era of Receptor Research—Radioligand Binding

As the 1970s dawned, the physical existence of discrete receptors coupled to second messenger generating enzymes remained unproven and controversial. To those of us who believed they did in fact exist, it was clear that the way forward would require a whole set of new techniques which did not then exist. The most essential of these would be some means of labelling the receptors for study, *i.e.* radioligand binding methods. For several reasons, I chose the adrenergic receptors and in particular the β2-adrenergic receptor (β2AR) as my primary models, and in the early 1970s developed radioligand binding techniques for studying both the α- and β-adrenergic receptors.[9] During this period, Soloman Snyder's lab amongst others was quite active in developing ligands for a number of neurotransmitter receptors beginning with the opiate receptors.[10] For the first time these methods permitted direct identification of the receptors and analysis of variations in their properties under a variety of physiological and pathophysiological circumstances. They also provided the means to begin the solubilization and purification of the receptors, and permitted detailed analysis of the binding of drugs and neurotransmitters to these receptors. This led to the elucidation of general paradigms for understanding the interactions of ligands, receptors and G proteins. Notable during this period was the discovery of high and low affinity agonist binding states of receptors[11] and their interconversion by guanine nucleotides and the ternary complex model (TCM).[12] Radioligand binding also led to the discovery of a number of new receptor subtypes and greatly facilitated the process of drug discovery and screening.

Purification and Reconstitution of Receptors

Efforts to prove the existence of receptors ultimately came down to the quest to isolate a molecule that would display the properties expected of a specific receptor:

- ability to bind ligands with the specificity and stereospecificity expected based on pharmacological experiments; and
- ability to convey to non-responsive systems responsiveness to an agonist ligand which binds to the isolated receptor.

This quest occupied my own laboratory for more than a decade. The work required developing procedures for the solubilization of active receptors—we found that only the plant glycoside digitonin gave favourable results—and the means for their purification and biospecific affinity chromatography matrices. The goal was daunting given the miniscule levels at which the receptors are expressed, requiring 1–300 000 fold purification. However, by the early 1980s we had succeeded in purifying the β2-adrenergic receptor to homogeneity from several sources, and within a few years had succeeded with all four subtypes of adrenergic receptors then known. In each case the isolated receptors consisted of a single subunit—a glycoprotein with molecular weight of approximately 60 000 daltons. In each case the isolated molecules bound ligands with all the appropriate specificity characteristics[13–15] (reviewed in ref. 16).

The isolated β2ARs were initially reconstituted, some by liposome fusion, into *Xenopus laevus* erythrocytes which contain adenylate cyclase and prostaglandin receptors but not β2-ARs; hence they do not respond to catecholamines. The incorporated receptors conveyed β-adrenergic responsiveness on the erythrocyte adenylate cyclase.[17] The following year (1984) in collaboration with Lutz Birnbaumer and Eva Neer, we were able to assemble, for the first time, a hormone-responsive adenylate cyclase from its three isolated and purified components—the β2ARs, heterotrimeric Gs, and the catalytic moiety of the cyclase. These experiments constituted the ultimate proof that a GPCR had in fact been isolated in a discrete and functional form.[18]

Receptor Cloning—Convergence with Rhodopsin

With *bone fide* validated receptors in hand, attention focused on learning about the primary structure of the receptors by molecular cloning. Since no immunological reagents were available to facilitate this effort, some amino acid sequence information was required. Using microsequencing techniques, and in collaboration with a group at Merck, we obtained five peptide sequences from the mammalian β2AR, totalling 80 residues. Nonetheless, cDNA cloning proved quite difficult because of the rarity of receptor mRNA. Ultimately we succeeded by first cloning the receptor gene from a genomic library, a task fortuitously facilitated by the fact that the β2AR is encoded by a single exon.[19] The deduced sequence highlighted all the cardinal features that today are recognized as canonical for GPCRs:

- seven hydrophobic membrane spanning domains;
- connecting extra and intracellular loops: conserved sites for glycosylation on the N terminus;
- sites for regulatory phosphorylation on intracellular domains especially the C-terminal tail.

And stunningly, and unexpectedly at the time, sequence homology with the visual pigment rhodopsin.

Twenty-five years later it seems hard to understand how this finding could have initially come as such a shock. After all, by 1986 (the year the β2AR was cloned), it was generally appreciated that visual signalling involved a photon capture molecule (rhodopsin), a G protein (transducin) and an effector enzyme (cGMP phosphodiesterase) all in obvious analogy with the β2AR and other Gs and adenylate cyclase coupled receptor systems. Nonetheless, no one at the time predicted this result.

The complete amino acid sequence of rhodopsin had been determined in 1982/3 by conventional protein sequencing (Edman degradation) and found to contain seven hydrophobic domains.[20,21] At the time the only seven membrane spanning protein known was bacteriorhodopsin, a light-sensitive proton pump found in Archaebacteria. Since both rhodopsin and bacteriorhodopsin are both light-sensitive proteins it was speculated that the seven membrane spanning arrangement might be a signature feature of light-sensitive proteins.[20,21] Only with the cloning of the β2AR did it begin to emerge that, in fact, it was instead the common feature of GPCRs.[19] This was rapidly confirmed by the cloning of seven more adrenergic receptors in my lab together with several serotonin receptors and muscarinic cholinergic receptors in other labs over the next several years (reviewed in ref. 16).

Orphan Receptors and Expansion of the Family

Once the first few GPCRs were cloned the floodgates were opened and the GPCR family expanded rapidly. Whereas the β- and α-adrenergic receptors and M1 and M2 muscarinic receptors were cloned based on peptide sequences, thereafter GPCR cloning was based almost exclusively on homology approaches and, in a few instances, expression cloning. Thus the long and difficult path to purifying the adrenergic receptors would ultimately pay rich dividends, serving as a Rosetta stone of sorts for homology cloning of other members of the receptor family.[16]

As the number of receptors grew, reports accumulated of cDNA sequences encoding what were clearly homologous members of the family, but for which neither the ligand nor the function were known—so called ‘orphan receptors’. We reported the first of these in 1987, a clone obtained by low stringency hybridization of genomic DNA with the β2AR cDNA.[22] Hopes that it was the β1AR were dashed by expression studies and only a year later would we ‘de-orphanize’ it as the serotonin, 5HT1A receptor.[23]

The largest family of ‘orphans’ was cloned by Buck and Axel in 1991—the olfactory receptors, numbering several hundred.[24] Many orphan receptors remain today, though gradually their ranks are shrinking as they are paired with ligands and functions. These receptors provide a potentially rich source of novel therapeutic targets.

Structure–Function Studies

By the late 1980s the field had focused its attention on trying to understand how the highly conserved structure of the receptors served to carry out the two core functions of binding specific ligands and activating specific G proteins. Approaches involved site-directed mutagenesis, deletion mutants and construction of chimeric receptors (reviewed in ref. 25). By the early 1990s, these approaches had established the major principles:

- binding of ligands to determinants in the extracellular regions and within the outer regions of the transmembrane helices; and

- coupling of G protein to regions on the cytoplasmic surface of the receptors, especially those in closest apposition to the plasma membrane.

Early work from the Merck group assigned specific residues within the membrane core of the β2AR to binding of specific functionalities on natural catecholamine agonists.[26] The chimeric receptor approach, in which regions of different receptors are stitched together to create hybrid molecules became quite popular, with more than 100 such studies published.[27] We had initially used this technique to transfer the ability to activate Gs to the normally Gi coupled α2AR by inserting the third cytoplasmic loop and flanking sequences from the Gs coupled β2AR.[28]

Constitutively Active Mutant Receptors (CAMs) and Inverse Agonism

In the course of studying mutations in the third cytoplasmic loop of the α1AR which were expected to reduce coupling to its G protein (Gq), we serendipitously discovered that some of these quite unexpectedly led to markedly enhanced constitutive activity of the receptor, *i.e.* activity in the absence of ligand.[29] The constitutively active mutations were in the distal portion of the third intracellular loop. Subsequently we found that analogous mutations in the β2AR also led to constitutive activation.[30] These studies, some 20 years ago, presaged more recent structural insights about the receptors, and in hindsight also had several other implications for subsequent work.

Nature of Receptor Activation Process

Although we did not at the time appreciate why our mutations led to constitutive activity, we speculated that they must abrogate certain intramolecular constraints that restrain the receptor in its inactive conformation.[30] Subsequently, Javitch and co-workers performed further mutagenesis and defined a proposed 'ionic lock' between R131 and E268 and located respectively at the cytoplasmic ends of TM3 and 6 in the β2AR which comprises just such a restraining force.[31] More recent crystal structures of both rhodopsin and the β2AR have confirmed this ionic lock, which is broken in the activated state.[32,33]

Extended Ternary Complex Model

While the experimental findings with the CAM β2AR could be accommodated within the framework of the TCM (see above) which postulates that receptor activation requires the agonist-promoted formation of an active 'ternary complex' of agonist, receptors and G protein, this required addition of an explicit isomerization step of the receptor (R) to an active state (R*). This rationalized findings such as, for example, the higher affinity of agonists and, to a lesser extent, partial agonists for the CAM receptor.[30]

Inverse Agonists

Prior to the late 1980s, ligands were viewed as having 'efficacies' varying from 0–1, *i.e.* they either did nothing, or something, to receptor activity. Conceptualized in the classical simplified framework of a two state receptor model ($R \rightleftarrows R^*$), they either stabilized R* to varying extents (agonists and partial agonists) or they did not perturb the equilibrium ('neutral antagonists'). The discovery of CAM receptors (and to a lesser extent the overexpression of wild-type receptors which permitted observation of their lower levels of constitutive activity) revealed the ability of many antagonists to actually inhibit 'basal' activity, *i.e.* to stabilize R. These ligands, so called 'inverse agonists', are of great utility as tool compounds in a wide variety of studies, including structural work, and may also have novel and potentially therapeutically useful properties.

CAM Receptor Diseases

Discovery of artificially created CAM receptors led to a search for spontaneously occurring CAMs in human diseases which was quickly rewarded. First reported were mutations in the distal portion of the third intracellular loop of the thyrotropin (TSH) receptor which lead to functioning thyroid adenomas (so called 'hot nodules').[34] Certain forms of male precocious puberty were found to be due to CAMs in luteinizing hormone (LH) receptors and more than a dozen such diseases have now been described (reviewed in ref. 35).

A Universal Mechanism Regulates Receptor Function

Throughout my career I have been fascinated by the phenomenon of desensitization, the complex set of processes by which cells and tissues dampen their responsiveness to stimulation which is persistent. Contemporaneous with work on the nature of GPCRs and their signalling through G proteins, a large body of work developed in the 1970s, 1980s and 1990s on the mechanisms of such desensitization (reviewed in ref. 36). Many cellular processes contribute to this, playing out over time frames from seconds to days, and involving mechanisms as diverse as regulation of receptor trafficking, degradation, gene transcription and translation. However, one mechanism appears to be quite universal and that is the two-step process by which activated receptors are phosphorylated by a G protein-coupled receptor kinase (GRK) and then bind a member of the arrestin family which sterically blocks further activation of the G protein.

Discovery of this two-step mechanism of desensitization again represented a convergence of contemporaneous work on the rhodopsin and β2AR systems. Both molecules were found to be hyperphosphorylated after stimulation. Isolation of the kinases responsible revealed that they were novel second messenger independent enzymes.[37,38] Purification and cloning of the enzymes [initially named β-adrenergic receptor kinase (BARK) and rhodopsin kinase, respectively] revealed that they were the founding members of a new kinase subfamily (the GRKs).[39,40] Today we know that there are seven members of this family—rhodopsin kinase being GRK1 and BARK being GRK2. All show a conserved three domain organization with a central catalytic domain flanked by highly variable regulatory regions. Only GRKs 2,3,5 and 6 are ubiquitously expressed whereas GRKs1 and 7 are largely confined to the retina.[41,42]

Visual arrestin (aka arrestin1) had been known for years as 'S-antigen or '48K protein', a highly abundant, highly antigenic retinal protein. In 1986, Kuhn discovered that its light-dependent translocation to the rod outer segment membrane was necessary for the full 'desensitizing' or turn off of visual signalling by rhodopsin kinase.[43] It was subsequently renamed arrestin. At this time we were struggling with the observation that, as we purified BARK to homogeneity, it progressively lost its ability to desensitize β2AR activation of Gs in a reconstituted assay system, even as its kinase specific activity increased. Reasoning that we were losing some necessary cofactor or component analogous to arrestin, we obtained some of the retinal protein from Kuhn. Indeed it did restore the βAR desensitization, albeit at high concentrations.[44] When a year later Shinohara cloned visual arrestin,[45] we were able to use it as a probe to clone β-arrestin1 (aka arrestin2)[46] and then β-arrestin2 (arrestin3).[47] With authentic recombinant proteins in hand we could demonstrate the marked specificity of the two β-arrestins for the β2AR *versus* rhodopsin and *vice versa* for visual arrestin.[47] In subsequent years the non-visual arrestins and the GRKs were found to act quite generally on activated GPCRs to cause their desensitization. There are four members of the arrestin gene family, but only β-arrestins 1 and 2 are ubiquitous whereas arrestins 1 and 4 are largely confined to the retina.[48]

In addition to the GRKs, a number of other kinase families have been found to phosphorylate and regulate the function of GPCRs, most notably the second messenger kinases, protein kinase A (PKA) and protein kinase C (PKC).[36] Since the GRKs require the activated form of the receptor as their substrate, the desensitization they mediate is often of the 'homologous' type, *i.e.* limited to the receptor type that has been stimulated. In contrast, PKA or PKC, activated downstream of one receptor can readily phosphorylate other types of receptors. This can lead to 'heterologous' desensitization of one receptor by stimulation of another. Unlike GRK-mediated receptor desensitization, PKA and PKC mediated desensitization does not require arrestins. Rather this appears to be due to conformational changes in the phosphorylated cytoplasmic regions of the receptors. Often PKA and PKC consensus sites are found localized to regions of the receptor involved in coupling to G proteins.[36]

Expanding Roles of β-Arrestins and GRKs

Although the GRK/β-arrestin system was discovered in the context of GPCR desensitization, over the past 10–15 years our appreciation of its cellular functions has been transformed. Benovic initially discovered that β-arrestins could serve as clathrin adaptors, thus facilitating endocytosis of the receptors.[49] Subsequently they have been found to bind a number of other elements of the clathrin-coated pit endocytic machinery including, for example, AP2, ARF6 and various E3 ubiquitin ligases (reviewed in ref. 50). Not only have the arrestins turned up as required adaptors for the endocytosis of all sorts of membrane proteins, not simply GPCRs, but they have emerged as adaptors for a growing list of E3 ligases and their substrates as well.[51]

Next it was discovered that arrestins could serve as signal transducers in their own right, generally by acting as scaffolds to assemble multiprotein signalling complexes. This was first documented for β2AR stimulation of the Src tyrosine kinase[52] and for the ERK Map kinases,[53,54] and then many other signalling systems (reviewed in refs. 55 and 56). Recently, powerful global proteomics approaches have documented that β-arrestins interact with dozens of protein partners,[57] and mediate signalling through numerous complex phosphorylation/dephosphorylation networks involved in disparate cellular processes.[58,59] In fact, the diversity and complexity of this signalling is reminiscent of that mediated by heterotrimeric G proteins. An interesting feature of this β-arrestin mediated signalling is that many of the pathways activated are also activated (or inhibited) through G protein dependent mechanisms. However the kinetic, cellular localization and physiological outcomes of these two types of signalling are generally quite distinct.[55,56]

Thus there are at least three types of functions carried out by the β-arrestins—desensitization, endocytosis and trafficking, and signalling. An interesting aspect of this topic which has recently begun to come into focus is how the different functions of β-arrestins are coordinated and engaged. The current thinking is that different GRKs phosphorylate distinct sets of sites on a receptor, thus setting up a 'bar code' of sorts which then 'instructs' the bound β-arrestins as to what conformation to assume and thus what functionalities to express.[60,61] In fact multiple conformations of β-arrestins have been demonstrated by several techniques.[62,63] For example in several systems, GRKs 5/6 seem to be more involved in mediating activation of ERK Map kinases whereas GRK2/3 oppose this. GRK2/3 on the other hand may be more involved in receptor desensitization.[60,61,64] This is currently a very active area of research.

A point worth emphasizing is that the roles of β-arrestins in cellular physiology are not at all limited to GPCRs. Rather they are now known to regulate a wide variety of other types of cellular receptors, ion channels and receptor tyrosine kinases.[65]

Biased Signalling

A particularly interesting phenomenon which has received greatly increased attention of late, and which has emerged from the discovery of β-arrestin mediated signalling, is 'biased agonism'. A biased agonist is a ligand which more effectively stimulates some responses than others through the same receptor.[66,67] Thus, for example, a series of agonists may show a different rank order of potency for one response than another. Such situations are formally incompatible with simple two state receptor models, in which all signalling is a function of the concentrations of the unitary receptor active state R*. In fact, biased signalling requires the existence of multiple active conformations of the receptor.

In the case of seven-transmembrane receptors (7TMRs), ligands may be biased toward a particular G protein or β-arrestin. Mutated receptors can also be biased.[66] Endogenous ligands can be biased, *e.g.* secondry lymphoid-organ chemokine (SLC) (G protein biased) and EBI-1 ligand chemokine (ELC) (unbiased) for the CCR7 chemokine receptor.[68,69] Endogenous receptors may also be biased. A case in point is the so-called 'decoy' chemokine receptor CXCR7. When it binds its ligands such as SDF1 or ITAC it does not activate G proteins, but does activate β-arrestins through which it can both internalize and signal.[70]

However, perhaps the most significant implication of such biased signalling is in the therapeutic area where it may be possible to design novel drugs, which by virtue of G protein or β-arrestin bias, have enhanced therapeutic efficacy and reduced side effects.[71] Such an outcome might be anticipated when desired therapeutic effects are mediated by one arm of this dual signalling paradigm whereas major side effects are mediated by the other. As an example, the powerful antinociceptive affects of opiates are mediated through activation of Gi protein signalling *via* the μ-opioid receptors. However, unwanted tolerance, constipation and respiratory depression appear to be a consequence primarily of β-arrestin2 mediated signaling.[72] Thus a G protein biased agonist for the μ-opioid receptor might have a desirable therapeutic profile. Recent reviews highlight numerous other potential examples.[71] Only time will tell whether such 'biased' drugs will succeed in the clinic.

Allosterism

An emerging concept over the past few years has been that of ligands interacting with allosteric rather than orthosteric sites, which has been described for a growing list of 7TMRs.[73] These may stimulate or inhibit signalling on their own, or act as allosteric modifiers which potentiate or inhibit the actions of drugs acting at the orthosteric binding site. Moreover, such allosterically acting ligands may also display the same sorts of bias as do orthosteric ligands. Such allosteric ligands have great potential both as novel therapeutics and as tool compounds to help elucidate the complex conformational dynamics of the receptors.

Receptor Dimerization

No introductory chapter would be complete without at least a mention of GPCR dimerization—a theme which has attracted greater and greater attention in recent years. Both hetero- and homodimerization have been described for numerous receptors using a wide variety of techniques, most involving receptor overexpression (reviewed in ref. 74). Both the existence and significance of such oligomeric receptor assembles has been controversial at times. However, an emerging consensus is that the phenomenology is in fact widespread and may create novel functional properties especially in the case of heteroligomerization. For homodimerization the issue of its functional significance remains to be determined. For both rhodopsin and the β2AR in reconstituted systems it has been demonstrated that monomeric receptor is sufficient for efficient signalling.[75,76]

Biophysics and Structural Biology of Receptors

A very recent and exciting development has been the advent of the first crystal structures of GPCRs. While the first structure of inactive rhodopsin appeared in 2000,[77] the first structures of a ligand binding receptor, the β2AR in an inactive conformation, did not follow until 2007.[78,79] Since then a spate of structures of class A receptors have been published, all of which show remarkable conservation of many features, albeit with some specializations especially in the case of rhodopsin.[80]

Even more recently the first active structures of opsin bound to a short transducin peptide, and the agonist occupied β2AR stabilized by a nanobody, have appeared.[32,33] These are remarkably similar [root mean square deviation (RMSD) < 1.5 Å] featuring disruption of the 'ionic lock', large outward movements of transmembrane 6 as well as rearrangements of transmembrane helices 3, 5 and 7.

As with other advances in this field, development of novel technical approaches was required such as substitution of T4 lysozyme for the very flexible third cytoplasmic loop of the β2AR, thermostabilizing mutations in the β1AR and the use of special techniques for membrane protein crystallization including lipidic cubic phase, bicelles, novel detergents, micro-focused X-ray beams for data collection and others. The next set of targets will obviously include class B and C receptors, as well as the receptors in complex with their downstream signalling and regulatory partners—the G proteins, arrestins and GRKs.

While crystallography will reveal much about receptor structure and the mechanisms of 7TMR ligand binding and activation it still provides, as many have noted, only a static snapshot of the receptor. Accumulating evidence strongly supports the conformational heterogeneity and complexity of the receptors especially when bound by functionally diverse ligands. Clearly a variety of other biophysical approaches aiming at dynamic information such as NMR, electron paramagnetic resonance (EPR), site-directed chemical modification, *etc.* will need to be used in conjunction with crystallographic studies to elucidate this complexity which undoubtedly underlies phenomena such as biased signalling.

Currently much excitement is being generated by the possible implications of this new structural information about the receptor for the design of new drugs using computational approaches. Only time will tell if this optimism is justified.

Conclusions

Forty years ago, when I began my own research career, receptors were a mere idea, a concept used to organize and 'explain' observed actions of drugs and hormones on tissues and physiological processes. Since then, steady progress has brought us to our current rich understanding of their structure, signalling properties and regulatory mechanisms. At every phase of this research, slow and painstaking development of novel technical approaches has been followed by periods of rapid progress. Despite obvious specializations in different members and classes of the large 7TMR superfamily, in my view, the most striking theme is the remarkable conservation in the structure, function and regulatory mechanisms of the receptors. Moreover, the 'druggability' of so many members of the GPCR family makes this information of potentially great value for the design of new drugs for the treatment of human illness.

References

1. P. Erhlich, *Lancet*, 1913, **2**, 445–451.
2. J. N. Langley, *J. Physiol.*, 1905, **33**, 374–413.
3. H. H. Dale, *Trans. Faraday Soc.*, 1943, **39**, 319–322.
4. R. Ahlquist, *Perspect. Biol. Med.*, 1973, **17**, 119–122.
5. L. E. Limbird, *Cell Surface Receptors: A Short Course on Theory and Methods*, Kluwer Academic, Boston, MA, 1996, pp. 1–26.
6. E. W. Sutherland, *Ann. N. Y. Acad. Sci.*, 1967, **139**, 703–723.
7. M. Rodbell, L. Birnbaumer, S. L. Pohl and H. M. J. Krans, *J. Biol. Chem.*, 1971, **246**, 1877–1882.
8. E. M. Ross and A. G. Gilman, *J. Biol. Chem.*, 1977, **252**, 6966–6969.
9. B. B. Hoffman and R. J. Lefkowitz, *Annu. Rev. Pharmacol. Toxicol.*, 1980, **20**, 581–608.
10. C. B. Pert and S. H. Snyder, *Science*, 1973, **179**, 1011–1014.
11. R. S. Kent, A. De Lean and R. J. Lefkowitz, *Mol. Pharmacol.*, 1980, **17**, 14–23.
12. A. De Lean, J. M. Stadel and R. J. Lefkowitz, *J. Biol. Chem.*, 1980, **255**, 7108–7117.
13. J. L. Benovic, R. G. L. Shorr, M. G. Caron and R. J. Lefkowitz, *Biochemistry*, 1984, **23**, 4510–4518.
14. J. W. Regan, H. Nakata, R. M. DeMarinis, M. G. Caron and R. J. Lefkowitz, *J. Biol. Chem.*, 1986, **261**, 3894–3900.
15. J. W. Lomasney, L. M. F. Leeb-Lundberg, S. Cotecchia, J. W. Regan, J. F. DeBernardis, M. G. Caron and R. J. Lefkowitz, *J. Biol. Chem.*, 1986, **261**, 7710–7716.
16. H. G. Dohlman, J. Thorner, M. G. Caron and R. J. Lefkowitz, *Annu. Rev. Biochem.*, **60**, 653–688.
17. R. A. Cerione, B. Strulovici, J. L. Benovic, C. D. Strader, M. G. Caron and R. J. Lefkowitz, *Proc. Natl. Acad. Sci. U. S. A.*, 1983, **80**, 4899–4903.
18. R. A. Cerione, D. R. Sibley, J. Codina, J. L. Benovic, J. Winslow, E. J. Neer, L. Birnbaumer, M. G. Caron and R. J. Lefkowitz, *J. Biol. Chem.*, 1984, **59**, 9979–9982.

19. R. A. F. Dixon, B. K. Kobilka, C. Strader, J. L. Benovic, H. G. Dohlman, T. Frielle, M. A. Bolanowski, C. D. Bennett, E. Rands, R. E. Diehl, R. A. Mumford, E. E. Slater, I. Sigal, M. G. Caron, R. J. Lefkowitz and C. D. Strader, *Nature*, 1986, **321**, 75–79.
20. Y. A. Ovchinnikov, *FEBS Lett.*, 1982, **148**, 179–191.
21. P. A. Hargrave, J. H. McDowell, D. R. Curtis, J. K. Wang, E. Juszczak, S.-L. Fong, J. K. M. Rao and P. Argos, *Biophys. Struct. Mech.*, 1983, **9**, 235–244.
22. B. K. Kobilka, T. Frielle, S. Collins, T. Yang-Feng, T. S. Kobilka, U. Francke, R. J. Lefkowitz and M. G. Caron, *Nature*, 1987, **329**, 75–79.
23. A. Fargin, R. J. Raymond, M. J. Lohse, B. K. Kobilka, M. G. Caron and R. J. Lefkowitz, *Nature*, 1988, **335**, 358–360.
24. L. Buck and R. Axel, *Cell*, 1991, **65**, 175–187.
25. J. Ostrowski, M. A. Kjelsberg, M. G. Caron and R. J. Lefkowitz, *Annu. Rev. Pharmacol. Toxicol.*, 1992, **32**, 167–183.
26. C. D. Strader, M. R. Candelore, W. S. Hill, I. Sigal and R. A. F. Dixon, *J. Biol. Chem.*, 1989, **264**, 13572–13578.
27. D. Yin, S. Gavi, H. Wang and C. C. Malbon, *Mol. Pharmacol.*, 2004, **65**, 1323–1332.
28. B. K. Kobilka, T. S. Kobilka, K. Daniel, J. W. Regan, M. G. Caron and R. J. Lefkowitz, *Science*, 1988, **240**, 1310–1316.
29. S. Cotecchia, J. Ostrowski, M. A. Kjelsberg, M. G. Caron and R. J. Lefkowitz, *J. Biol. Chem.*, 1992, **268**, 1633–1639.
30. P. Samama, S. Cotecchia, T. Costa and R. J. Lefkowitz, *J. Biol. Chem.*, 1993, **268**, 4625–4636.
31. J. A. Ballesteros, A. D. Jensen, G. Liapakis, S. G. F. Rasmussen, L. Shi, U. Gether and J. A. Javitch, *J. Biol. Chem.*, 2001, **276**, 29171–29177.
32. P. Scheerer, J. H. Park, P. W. Hildebrand, Y. J. Kim, N. Krauss, H. W. Choe, K. P. Hofmann and O. P. Ernst, *Nature*, 2008, **455**, 497–502.
33. S. G. F. Rasmussen, H.-J. Choi, J. J. Fung, E. Pardon, P. Casarosa, P. S. Chae, B. T. DeVree, D. M. Rosenbaum, F. S. Thian, T. S. Kobilka, A. Schnapp, I. Konetzki, R. K. Sunahara, S. H. Gellman, A. Pautsch, J. Steyaert, W. I. Weis and B. K. Kobilka, *Nature*, 2011, **469**, 175–181.
34. J. Parma, L. Duprez, J. Van Sande, P. Cochaux, C. Gervy, J. Mockel, J. Dumont and G. Vassart, *Nature*, 1993, **365**, 649–651.
35. A. M. Spiegel, in *G Proteins, Receptors and Disease*, ed. A. M. Spiegel, Humana Press, Totowa, NJ, 1998, pp. 1–21.
36. W. P. Hausdorff, M. G. Caron and R. J. Lefkowitz, *FASEB J.*, 1990, **4**, 2881–2889.
37. U. Widen and H. Kuhn, *Biochemistry*, 1982, **21**, 3014–3022.
38. J. L. Benovic, R. H. Strasser, M. G. Caron and R. J. Lefkowitz, *Proc. Natl. Acad. Sci. U. S. A.*, 1986, **83**, 2797–2801.
39. J. L. Benovic, A. DeBlasi, W. C. Stone, M. G. Caron and R. J. Lefkowitz, *Science*, 1989, **246**, 235–240.
40. W. Lorenz, J. Inglese, K. Palczewski, J. J. Onorato, M. G. Caron and R. J. Lefkowitz, *Proc. Natl. Acad. Sci. U.S.A.*, 1991, **88**, 8715–8719.

41. J. A. Pitcher, N. J. Freedman and R. J. Lefkowitz, *Annu. Rev. Biochem.*, 1998, **67**, 653–692.
42. D. T. Lodowski, J. A. Pitcher, W. D. Capel, R. J. Lefkowitz and J. J. G. Tesmer, *Science*, 2003, **300**, 1256–1262.
43. U. Wilden, S. W. Hall and H. Kuhn, *Proc. Natl. Acad. Sci. U. S. A.*, 1986, **83**, 1174–1178.
44. J. L. Benovic, H. Kuhn, I. Weyand, J. Codina, M. G. Caron and R. J. Lefkowitz, *Proc. Natl. Acad. Sci. U.S.A.*, 1987, **84**, 8879–8882.
45. T. Shinohara, B. Dietzschold, C. M. Craft, G. Wistow, J. J. Early, L. A. Donoso, J. Horwitz and R. Tao, *Proc. Natl. Acad. Sci. U.S.A.*, 1987, **84**, 6975–6979.
46. M. J. Lohse, J. L. Benovic, J. Codina, M. G. Caron and R. J. Lefkowitz, *Science*, 1987, **248**, 1547–1550.
47. H. Attramadal, J. L. Arriza, C. Aoki, T. M. Dawson, J. Codina, M. Kwatra, S. H. Snyder, M. G. Caron and R. J. Lefkowitz, *J. Biol. Chem.*, 1992, **267**, 17882–17890.
48. V. V. Gurevich and E. V. Gurecich, *Pharmacol. Ther.*, 2006, **110**, 465–502.
49. O. B. Goodman, Jr, J. G. Krupnick, F. Santini, V. V. Gurevich, R. B. Penn, A. W. Gagnon, J. H. Keen and J. L. Benovic, *Nature*, 1996, **383**, 447–450.
50. A. Claing, S. A. Laporte, M. G. Caron and R. J. Lefkowitz, *Prog. Neurobiol.*, 2002, **66**, 61–79.
51. S. K. Shenoy, *Circ. Res.*, 2007, **100**, 1142–1154.
52. L. M. Luttrell, S. S. G. Ferguson, Y. Daaka, W. E. Miller, S. Maudsley, G. J. Della Rocca, F.-T. Lin, H. Kawakatsu, K. Owada, D. K. Luttrell, M. G. Caron and R. J. Lefkowitz, *Science*, 1999, **283**, 655–661.
53. K. A. DeFea, J. Zalevsky, M. S. Thomas, O. Dery, R. D. Mullins and N. W. Bunnett, *J. Cell. Biol.*, 2000, **20**, 1267–1281.
54. L. M. Luttrell, F. L. Roudabush, E. W. Choy, W. E. Miller, M. E. Field, K. L. Pierce and R. J. Lefkowitz, *Proc. Natl. Acad. Sci., U. S. A.*, 2001, **98**, 2449–2454.
55. R. J. Lefkowitz and S. K. Shenoy, *Science*, 2005, **308**, 512–517.
56. S. M. DeWire, S. Ahn, R. J. Lefkowitz and S. K. Shenoy, *Annu. Rev. Physiology*, 2007, **69**, 483–510.
57. K. Xiao, D. B. McClatchy, A. K. Shukla, Y. Zhao, M. Chen, S. K. Shenoy, J. R. Yate and R. J. Lefkowitz, *Proc. Natl. Acad. Sci. U. S. A.*, 2007, **104**, 12011–12016.
58. K. Xiao, J. Sun, J. Kim, S. Rajagopal, B. Zhai, J. Villen, W. Haas, J. J. Kovacs, A. K. Shukla, M. R. Hara, M. Hernandez, A. Lachmann, S. Zhao, Y. Lin, Y. Cheng, K. Mizuno, A. Ma'ayan, S. P. Gygi and R. J. Lefkowitz, *Proc. Natl. Acad. Sci. U. S. A.*, 2010, **107**, 15299–15304.
59. G. L. Christensen, C. D. Kelstrup, C. Lyngso, U. Sarward, R. Bogebo, S. P. Sheikh, S. Gammeltoft, J. V. Olsen and J. L. Hansen, *Mol. Cell. Proteomics*, 2010, **7**, 1540–1553.
60. J. Kim, S. Ahn, X.-R. Ren, E. Whalen, E. Reiter, H. Wei and R. J. Lefkowitz, *Proc. Natl. Acad. Sci. U. S. A.*, 2005, **102**, 1442–1447.

61. X.-R. Ren, E. Reiter, S. Ahn, J. Kim, W. Chen and R. J. Lefkowitz, *Proc. Natl. Acad. Sci. U. S. A.*, 2005, **102**, 1448–1453.
62. K. Xiao, S. K. Shenoy, K. Nobles and R. J. Lefkowitz, *J. Biol. Chem.*, 2004, **279**, 55744–55753.
63. A. K. Shukla, J. D. Violin, E. J. Whalen, D. Gesty-Palmer, S. K. Shenoy and R. J. Lefkowitz, *Proc. Natl. Acad. Sci. U. S. A.*, 2008, **105**, 9988–999.
64 J. M. Busillo, S. Armando, R. Sengupta, O. Meucci, M. Bouvier and J. L. Benovic, *J. Biol. Chem.*, 2010, **285**, 7805–7817.
65. R. J. Lefkowitz, K. Rajagopal and E. J. Whalen, *Mol. Cell*, 2006, **24**, 643–652.
66. J. D. Violin and R. J Lefkowitz, *Trends Pharmacol. Sci.*, 2007, **28**, 416–422.
67. T. Kenakin, *J. Pharmacol. Exp. Ther.*, 2011, **336**, 296–302.
68. D. A. Zidar, J. D. Violin, E. J. Whalen and R. J. Lefkowitz, *Proc. Natl. Acad. Sci. U. S. A.*, 2009, **106**, 9649–9654.
69. T. A. Kohout, S. L. Nicholas, S. J. Perry, G. Reinhart, S. Junger and R. S. Struthers, *J. Biol. Chem.*, 2004, **279**, 23214–23222.
70. S. Rajagopal, J. Kim, S. Ahn, S. Craig, C. M. Lam, N. P. Gerard, C. Gerard and R. J. Lefkowitz, *Proc. Natl. Acad. Sci. U. S. A.*, 2010, **107**, 628–632.
71. S. Rajagopal, K. Rajagopal and R. J. Lefkowitz, *Nat. Rev. Drug Discov.*, 2010, **9**, 1–12.
72. K. M. Raehal, J. K. Walker and L. M. Bohn, *J. Pharmacol. Exp. Ther.*, 2005, **314**, 1195–1201.
73. L. Wang, B. Martin, R. Brenneman, L. M. Luttrell and S. J. Maudsley, *Pharmacol. Exp. Ther.*, 2009, **331**, 340–348.
74. M. J. Lohse, *Curr. Opin. Pharmacol.*, 2010, **10**, 53–58.
75. M. R. Whorton, M. P. Bokoch, S. G. Rasmussen, B. Huang, R. N. Zare, B. Kobilka and R. K. Sunahara, *Proc. Natl. Acad. Sci. U. S. A.*, 2007, **104**, 7682–7687.
76. M. R. Whorton, B. Jastrzebska, P. S. Park, D. Fotiadis, A. Engel, K. Palcewski and R. J. Sunahara, *J. Biol. Chem.*, 2008, **283**, 4387–4394.
77. K. Palczewski, T. Kumasaka, T. Hori, C. A. Behnke, H. Motoshima, B. A. Fox, I. LeTrong, D. C. Teller, T. Okada, R. E. Stenkamp, M. Yamamoto and M. Miyano, *Science*, 2000, **289**, 739–745.
78. S. G. Rasmussen, H. J. Choi, D. M. Rosenbaum, T. S. Kobilka, F. S. Thian, P. C. Edwards, M. Burghammer, V. R. Ratnala, R. Sanishvili, R. F. Fischetti, G. F. X. Schertler, W. I. Weis and B. K. Kobilka, *Nature*, 2007, **450**, 383–387.
79. V. Cherezov, D. M. Rosenbaum, M. A. Hanson, S. G. Rasmussen, F. S. Thian, T. S. Kobilka, H. J. Choi, P. Kuhn, W. I. Weis, B. K. Kobilka and R. C. Stevens, *Science*, 2007, **318**, 1258–1265.
80. M. A. Hanson and R. C. Stevens, *Structure*, 2009, **17**, 8–14.

Section I
G Protein-coupled Receptors: Membrane Proteins with Privileged Structures

CHAPTER 1

The Impact of G Protein-coupled Receptor (GPCR) Structures on Understanding Signal Transduction

DAVID T. LODOWSKI AND KRZYSZTOF PALCZEWSKI

Department of Pharmacology, School of Medicine, Case Western Reserve University, Cleveland, Ohio 44106-4965, USA

1.1 Introduction

Recent advances in the structural study of G protein-coupled receptors (GPCRs) have significantly enriched our understanding of the process of G protein signalling,[1–13] providing a structural framework to understand the huge volume of biochemical and biophysical studies on GPCRs performed and published over the past 50 years.[14–16] Determination of X-ray crystal structures provides the most direct methodology for examining the structural aspects of GPCR signalling and careful analysis of these structures reveals mechanistic details shared by all GPCRs (Table 1.1).

1.2 Early Approaches to Analysing GPCR Structure

Prior to the determination of the rhodopsin structure, similarities were noted between the light responsive photopigment, bacteriorhodopsin, and purified bovine rhodopsin, including their Schiff base linkage of chromophore,

RSC Drug Discovery Series No. 8
G Protein-Coupled Receptors: From Structure to Function
Edited by Jesús Giraldo and Jean-Philippe Pin

Published by the Royal Society of Chemistry, www.rsc.org

Table 1.1 Current list of all GPCR structures and their structural derivatives.[a]

Protein	*PDB ID*	*Derivative work*	*Comments*	*Reference*
Bovine rhodopsin	1F88		2.8 Å structure of bovine rhodopsin; first crystal structure of any GPCR Protein from native source Space-group $P4_1$	8
		1HZX		7
		1L9H	2.6 Å structure; more complete model with internal water molecules	50
		1U19	2.2 Å structure; highest resolution GPCR structure to date. Only complete native GPCR structure	11
		2HPY	2.8 Å structure of the lumirhodopsin photointermediate	73
		2G87	2.6 Å structure of the bathorhodopsin photointermediate	70
		2PED	2.95 Å structure of 9-*cis*-rhodopsin (isorhodopsin)	72
	1GZM		2.6 Å structure of bovine rhodopsin; solved in trigonal space-group $P3_1$	12
		2J4Y	3.4 Å structure, heterologously expressed thermostable rhodopsin mutant	118
		3C9L	2.6 Å structure of 1GZM reinterpreted in the higher symmetry hexagonal space-group $P6_5$	9
		3C9M	3.4 Å structure of 2J4Y reinterpreted in the higher symmetry hexagonal space-group $P6_5$	9
	2I36		4.1 Å structure of ground state bovine rhodopsin. space-group $P3_112$	13
		2I37	4.15 Å structure; photoactivated rhodopsin. Exhibits spectroscopic hallmarks of the meta II activated state.	13
	2I35		3.8 Å structure of bovine rhodopsin. space-group H32	13
	3CAP		2.9 Å structure of the inactive end-product of the phototransduction cascade, opsin. Space-group H3	2
		3DQB	3.2 Å structure of opsin bound to a peptide derived from $G_{t\alpha}$. space-group H32	1
Squid rhodopsin	2Z73		2.5 Å structure of squid rhodopsin; proteolysis necessary for crystallization. Space-group $P6_2$	10
	2ZIY		3.7 Å structure of squid rhodopsin; proteolysis necessary for crystallization. Space-group $C222_1$	77

Bovine β_2-adrenergic receptor	2RH1		2.4 Å structure of heterologously expressed β_2-adrenergic receptor-T4 lysozyme fusion bound to inverse agonist. Space-group C2	82
		3D4S	2.8 Å structure of β_2-adrenergic receptor T4 lysozyme fusion bound to inverse agonist and cholesterol. Space-group $P2_12_12_1$	119
	2R4R		3.4 Å structure of heterologously expressed β_2-adrenergic receptor bound to antagonist in complex with a Fab fragment. Space-group C2	78
		2R4S	3.4 Å structure of heterologously expressed β_2-adrenergic receptor in complex with a Fab fragment.	78
		3KJ6	3.4 Å structure of reductively methylated β_2-adrenergic receptor in complex with Fab fragment.	80
Turkey β_1-adrenergic receptor	2VT4		2.7 Å structure of heterologously expressed thermally stable mutant of the turkey β_1-adrenergic receptor. Space-group P1	6
A_{2A}-adenosine receptor	3EML		2.6 Å structure of A_{2A}-adenosine receptor t4 lysozyme fusion bound to inverse agonist. Space-group $P2_1$	5

[a]Rhodopsin was the first GPCR to have its X-ray structure determined. Improvements in resolution and model completeness followed, allowing complete tracing of the sequence, further solidification of the topology, location and orientation of the chromophore and location of important structural and functional motifs. Later derivative structures captured the early photointermediates, lumi- and bathorhodopsin, confirming that there are no significant structural changes in these intermediate states. Further work by Schertler and colleagues confirmed the structure determined by Palczewski *et al.* Careful analysis of the Schertler groups' structures revealed that crystals were mis-indexed in a lower symmetry trigonal space group, $P3_1$; re-indexing and refining in the higher symmetry hexagonal space group, $P6_5$, facilitated more precise determination of atomic positions and improved the model. The photoactivated structure of rhodopsin revealed that the changes in structure upon photoactivation were likely to be smaller than the large-scale rigid body movements which had been postulated to occur. This crystal structure of the photoactivated Rho exhibited the characteristic absorbance of meta II rhodopsin, contained only all-*trans*-retinal, and was fully capable of activating G protein when dissolved in detergent solution and had its structural changes restricted to the cytoplasmic loops that contact transducin. The crystal structures of opsin revealed the structure of the apo-protein. Although it was proposed that the opsin G_t peptide structure might resemble an activated state of a GPCR, there is little evidence for this supposition. Opsin is orders of magnitude less effective than meta II rhodopsin in activating G protein.

photoactivation and seven-transmembrane architecture (Figure 1.1A). After structure determination of bacteriorhodopsin *via* electron crystallography,[17] (reviewed in ref. 18), it was assumed that these two proteins shared a similar architecture.[19–21] Protein and DNA sequencing of rhodopsin[22–25] and the β_2-adrenergic receptor,[14] coupled with hydropat hy plots, allowed the construction of two-dimensional (2-D) topology models (Figure 1.1B and 1.1C).[14,22,23,26] Further clarification provided by electron crystallographic studies of 2-D crystals of bovine (Figure 1.1D) and invertebrate rhodopsin suggested that the arrangement of helices in bacteriorhodopsin *versus* rhodopsin and other

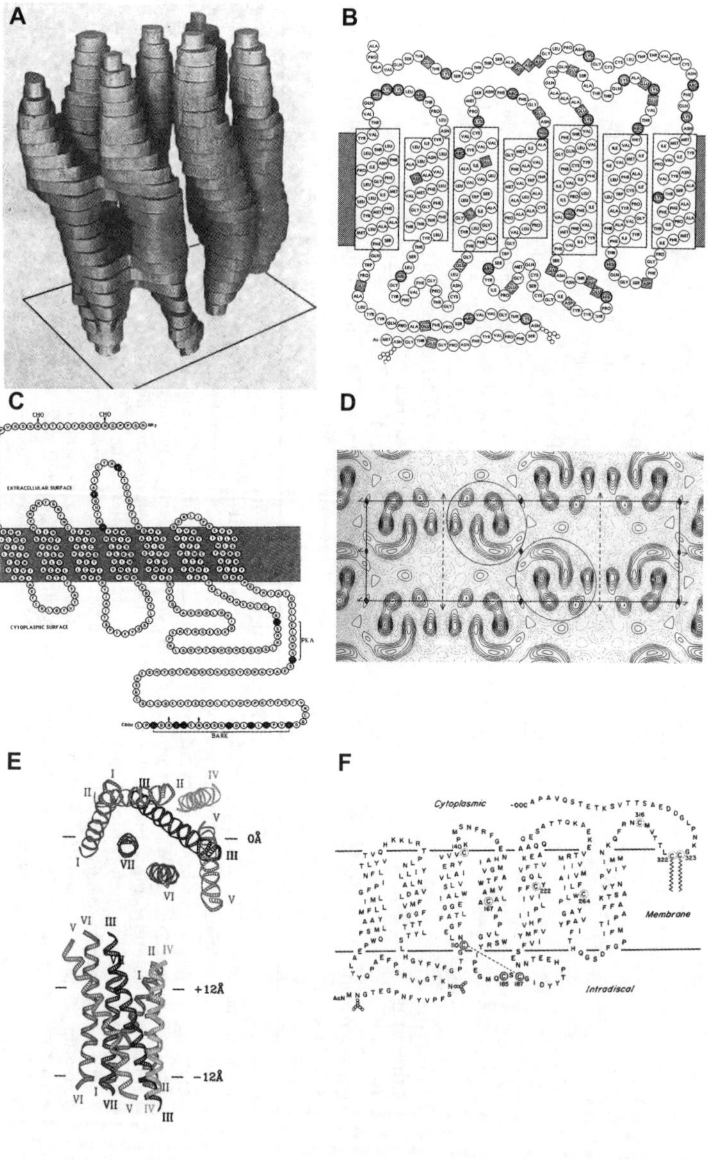

GPCRs were distinct.[27–29] These studies also improved predictions of the helical arrangement within the transmembrane bundle (Figure 1.1E).[30–33] Mutagenesis and biochemical studies allowed location assignments of disulfide bonds,[34,35] palmitoylation,[36,37] and phosphorylation sites[38–41] as well as further clarification of the border between loop and transmembrane regions (reviewed in ref. 42–46) (Figure 1.1F).

1.3 The Crystal Structure of Rhodopsin

While several laboratories were able to obtain crystals of rhodopsin,[47] these studies did not result in the determination of a crystal structure. A postdoctoral fellow and a technician in our laboratory, Dr. Tetsuji Okada and Preston Van Hoosier, were able to improve the purification of rhodopsin from a very high quality preparation of bovine rod outer segment membranes.[48] This provided homogenous bovine rhodopsin at the purity, quantity and concentration necessary to enable growth of diffraction quality crystals. These crystals led to the determination of the first crystal structure of rhodopsin[8] and subsequent rhodopsin structures extended the resolution to 2.2 Å,[11] facilitating

Figure 1.1 GPCR structures in the pre-atomic structural age. **A**. Bacteriorhodopsin structure. Prior to structural analysis of rhodopsin, bacteriorhodopsin was erroneously assumed to have a structure homologous to vertebrate rhodopsin due to its possession of a retinal chromophore, seven transmembrane helices, light-dependent protonation changes and shifts in absorbance spectra.[17] **B**. Two-dimensional secondary structure and sequence of bovine rhodopsin.[22,26] Advances in protein and gene sequencing, as well as the development of methods for predicting transmembrane segments, enabled reporting of the complete sequence of bovine rhodopsin as well as prediction of the seven transmembrane helices and sites of some post-translational modifications. **C**. Two-dimensional secondary structure and sequence of the β_2-adrenergic receptor (β_2-AR). Parallel work on the β_2-AR performed in the Strader and Lefkowitz laboratories enabled the determination of its complete sequence as well as its topology.[14] Due to their observed sequence and topological similarities, it was recognized that rhodopsin and β_2-AR were structurally related. **D**. Projection structure of bovine rhodopsin. This structure determined by electron crystallography revealed distinct differences in the topology of rhodopsin and bacteriorhodopsin, providing evidence that although both proteins are responsive to light, each has a very different function.[27] **E**. Proposed shared topology of GPCRs. Work by Baldwin postulated a similar topology for more than 200 GPCRs. Use of the projection structures of rhodopsin and gene sequences for the 204 GPCRs known at the time, coupled with hydropathy plots and mutational data, enabled a structural model template to be calculated for each GPCR.[32] **F**. Locations of disulfide and palmitoylation post-translational modifications. Khorana and colleagues through an exhaustive series of mutational experiments, predicted the locations of the disulfide, and the counter ion of the protonated Schiff base and confirmed the palmitoylation sites in rhodopsin.[42] These mutational experiments further solidified topological predictions. All figures are used with permission from their respective publications.

subsequent structural studies of additional GPCRs obtained by similar purification methods,[7,8,11,49,50] as well as those obtained by an alternative purification method.[12] These structures supplied the first true atomistic view of any GPCR, fully defining its transmembrane region and topology, including interhelical contacts. This detailed resolution cemented our knowledge as to the position of the chromophore and its contacts within the transmembrane (TM) region (Figure 1.2A). Using these structures, later biophysical studies probed the mechanisms by which conserved motifs act together to both maintain the GPCR in its inactive state and release these constraints upon activation.[13] The rhodopsin crystal structure has continued to provide a high-resolution template for the development of homology models of other GPCRs,[51] which has even been extended to virtual screening of ligands with these models.[52–56] The initial rhodopsin X-ray structure also allowed further revision of the 2D model introduced by Hargarve and Argos (Figure 1.2B).[22,26]

1.4 Conformational Intermediates of Rhodopsin

Because rhodopsin is relatively easily purified in its native state from bovine retina, a considerable amount of biochemical data exists pertaining to all aspects of the rhodopsin activation pathway.[57–59] Through the use of specific temperatures, wavelengths of light, chemicals and pH, it is possible to isolate various photointermediate states which display characteristic absorbance maxima (Figure 1.3).[60–62] Research by indirect methods predating the structural determination of rhodopsin relied on this photointermediate cascade to define functional intermediates which could be characterized (Figure 1.3). However, some of these photointermediates were identified under low temperature conditions which might trap a molecular species that does not occur physiologically, rendering the functional relevance of these photointermediate states open to question. Use of electron paramagnetic resonance (EPR) measurements upon attainment of the meta II state provided evidence for a longstanding proposed mechanism for the activation of GPCRs, which portrayed large-scale rigid body movements of helices to occur upon activation.[63] However, later refinements of these initial studies reduced the scale of these proposed movements to more thermodynamically feasible scales.[42,64–69]

Structures of several photointermediate states of rhodopsin determined by X-ray crystallography as well as by electron crystallography[1,2,13,70–74] have revealed changes in both the conformation of the chromophore upon photoactivation as well as the scale of changes accompanying photoactivation. Structures of early photointermediates of rhodopsin, batho- and lumirhodopsin, do not exhibit any significant changes within their protein backbones; only changes in the conformation of the chromophore are observed (Figure 1.4).[70,73] Determination of a low resolution structure of a photoactivated rhodopsin containing the characteristic deprotonated Schiff base linkage of the meta II activated state by our laboratory revealed that only small-scale changes were needed to achieve this state.[13] More recent work has revealed the conformation of the inactive apo-protein, opsin, in complex with a peptide which is

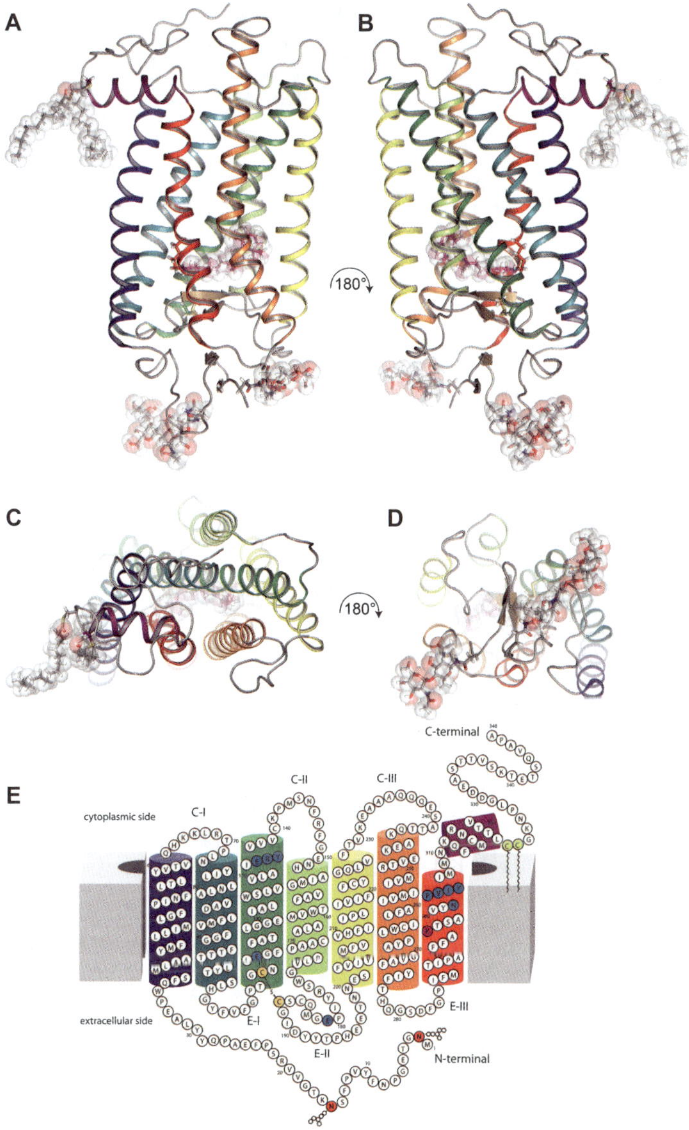

Figure 1.2 The crystal structure of rhodopsin provides the first true 3-D GPCR structural template. The initial crystal structure of rhodopsin revealed its topology, chromophore location, and contacts and geometries of important functional regions such as the D(E)RY and NPxxYx$_{(5,6)}$F motifs (see Figures 1.5 and 1.6). **A** and **B**. Views in the plane of the membrane. Ranging from blue to red, helices are coloured according to their locations in the primary sequence. The chromophore and post-translational modifications are shown as transparent spheres. **C** and **D**. Views from the extracellular and cytoplasmic face of rhodopsin, respectively. **E**. Revised 2-D representation of rhodopsin based upon crystal structures (adapted from ref. 8). The original rhodopsin crystal structure enabled better definition of the boundaries of individual helices, and absolute definition and confirmation of the topology.

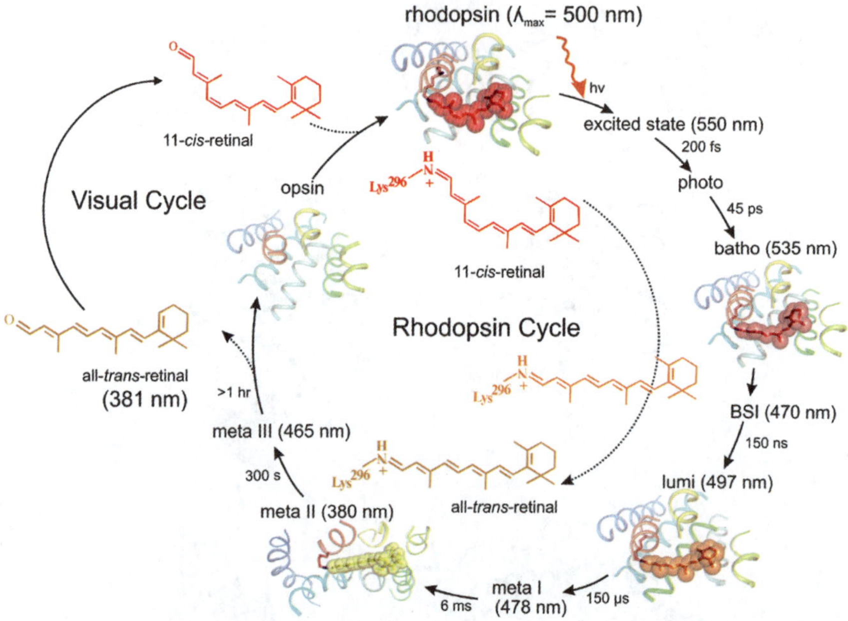

Figure 1.3 Schematic illustration of the rhodopsin cycle. Rhodopsin consists of an apoprotein, opsin, together with the covalently bound chromophore, 11-*cis*-retinylidiene, which conveys its red colour. In the dark (ground) state, the chromophore is attached through a protonated Schiff base linkage to Lys^{296}. Upon absorption of a photon, the chromophore undergoes a *cis* to *trans* isomerization which drives the activation of the protein through a progressive series of photointermediates (photo-, batho-, lumi-, BSI, meta I, meta II and meta III) rhodopsin. Once the meta I state is attained, deprotonation of the Schiff base occurs, yielding the active signalling meta II state—the only photointermediate capable of activating G protein efficiently. The meta II state decays to form the apo-protein, opsin, through the hydrolysis of the chromophore releasing all-*trans*-retinal. The 11-*cis*-retinal is regenerated through a series of enzymatic reactions, ultimately recombining with opsin, to re-form rhodopsin. Characteristic absorption wavelengths are shown in nm along with the timescale for each photointermediate. Figure reproduced with permission from *Encyclopedia of the Eye* (Elsevier).[117]

proposed to induce a conformation resembling the activated state.[1] Structural superposition reveals that the photoactivated meta II state falls midway between ground state rhodopsin and opsin.

In analysing structural information, it is important to consider that opsin is very inefficient at activating G protein compared with meta II Rho and that GPCR peptides are not equivalent to the proteins from which they are derived. While it is tempting to state that the peptide induces a conformation akin to the meta II active state, there is a fundamental disconnect; contacts that rhodopsin makes with transducin are much more extensive than this isolated helix, and docking of G_t using the position of this peptide would result in a collision with the

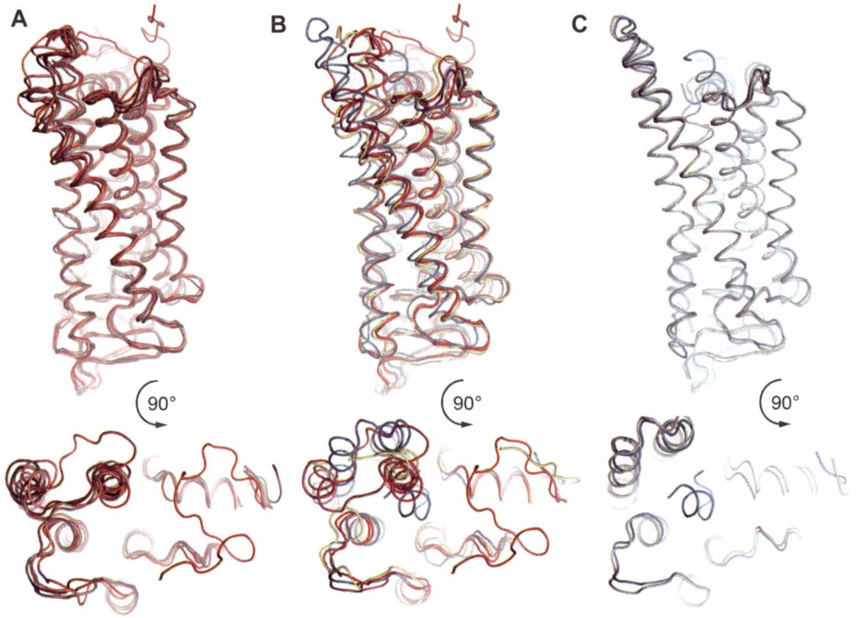

Figure 1.4 Photoactivated states of rhodopsin. **A**. Structural superposition of ground state rhodopsin (PDB IDs: 1U19 and 3C9L, red and magenta, respectively) and the early photointermediate states, bathorhodopsin and lumirhodopsin (PDB ID: 2G87 and 2HPY, dark pink and light pink, respectively). The slight changes in structure observed upon photoactivation to these early intermediate states are within the limits of uncertainty expected for atomic positions determined at this resolution and are of the same scale as observed between the different ground state structures of Rho. **B**. Structural superpositions of the entire rhodopsin photocycle from ground state (PDB IDs: 1U19 and 3C9L) to the meta II-like photoactivated activated structure to the inactive opsin. Yellow: photoactivated rhodopsin (PDB ID: 2I37) and light blue 'opsin*' structure (PDB ID: 3DQB). The large rigid body movements originally predicted to occur upon photoactivation are not evident. Furthermore, the position of the C-III loop and helices 5 and 6 in the photoactivated structure falls midway between ground state rhodopsin and the inactive apo-opsin. Because the EPR studies used to predict the structural changes accompanying attainment of meta II relied on single pointwise distance measurements in heterologously expressed mutant proteins, a low resolution crystal structure provides considerably better and more reliable evidence for the scale of structural movements. Crystallographic structures, rather than single pointwise measurements that have a certain error level associated with their measurements, provide positions of all atoms with a greater degree of certainty. The superpositioning was performed with the entire structures minus their helix V and VI and the connecting C-III loop to avoid obscuring any large scale movements of these helices/loops. **C**. Opsin and 'opsin*' structural superposition reveals that binding of the $G_{t\alpha}$ peptide does not result in any large-scale displacement of helices.

lipid bilayer. The wasp venom derived mastoparan peptide also very efficiently activates G protein subunits,[75,76] but it is unlikely that a co-crystal structure of mastoparan and G protein would reveal novel aspects of G protein activation which are germane to the GPCR–G protein interaction and nucleotide exchange.

1.5 Crystal Structures of Other GPCRs

More recent groundbreaking work has employed heterologously expressed GPCRs engineered to stabilize their conformations to maximize crystallization success. Squid rhodopsin also required modification for successful crystallization, even though it was obtained from a native source (Figure 1.5A).

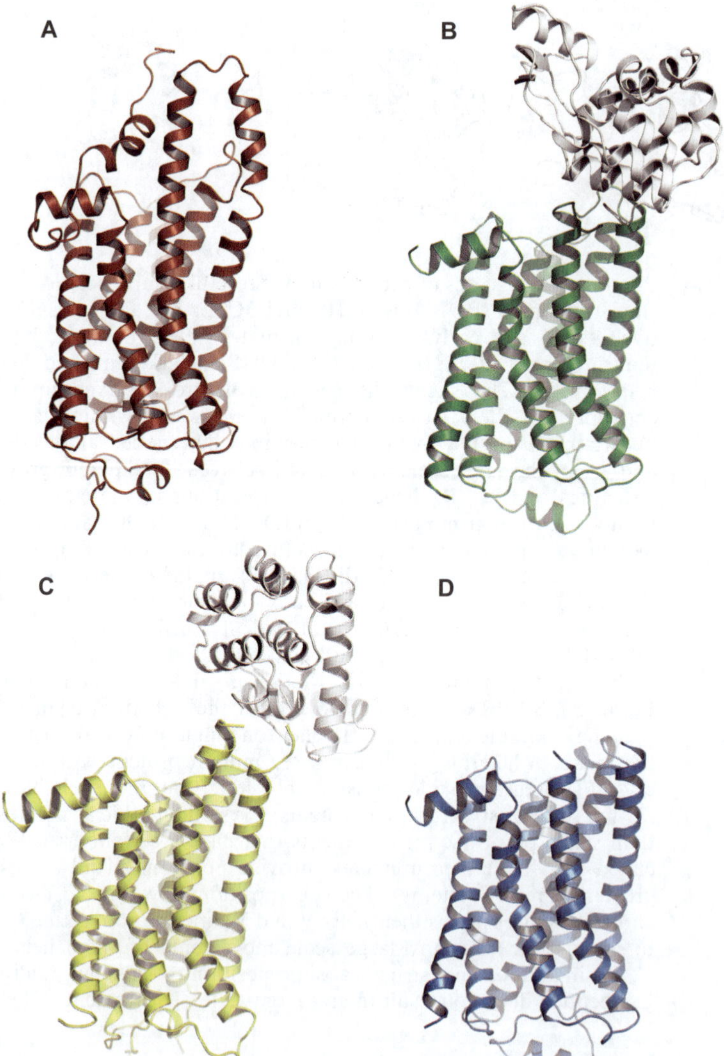

Proteolysis to remove a large cytoplasmic domain allowed crystallization and consequent structural determination.[10,77] Due to limited success with Fab fragment complexes in stabilizing the β_2-adrenergic receptor for crystallization,[78,79] it was proposed that mutations which confer stability or reduce conformational variability might assist in obtaining diffraction quality crystals.[4,80]

In the case of the β_2-adrenergic and A_{2A}-adenosine receptors,[4,5,81,82] heterologous expression was carried out to obtain the respective proteins with particular mutations, including the critical insertion of a T4-lysozyme into the corresponding C-III loops (Figure 1.5B and 1.5C, respectively); further stabilization was achieved with the addition of an antagonist to potentiate the inactive states of these receptors during crystallization. Crystallization of the β_1-adrenergic receptor also employed an antagonist for stabilization,[6] but also relied on mutations which were presumed to stabilize the structures based on positive increases in thermal stability (Figure 1.5D).[83,84] It is readily apparent that all these structures share the same global fold; they differ significantly only in their cytoplasmic and extracellular loops and ligand binding sites (Figure 1.5A–D).

Figure 1.5 Similarity of GPCR structures. While initial studies prior to any GPCR structural determination predicted a similar topology and fold for all GPCRs, determination of additional GPCR structures revealed the high level of structural similarity among family A GPCRs. Even though each GPCR is exquisitely sensitive to specific agonist/antagonist compounds, and the overall sequence homology is weak with the exception of several important functional motifs, all determined GPCR structures are structurally similar globally. Structural similarities are greater in the transmembrane segments compared with the cytoplasmic and extracellular loops and termini. All GPCRs exhibit a low level of sequence homology, yet they are recognized by a small set of G proteins, G protein receptor kinases and arrestins. Therefore, structural rather than sequence homology in GPCR regions that bind these proteins must account for binding specificity. **A**. The structure of squid rhodopsin. Proteolysis of natively expressed protein to remove a large cytoplasmic domain was required to grow crystals (PDB ID: 2Z73). **B**. Structure of the β_2-adrenergic receptor. Extensive mutation, truncation and insertion of a T4 lysozyme into cytoplasmic loop III allowed crystallization. Adaptation of the lipid cubic phase methodology as well a co-crystallization with the inverse agonists, carazolol or timolol, were also instrumental for this structural determination (PDB IDs: 2RH1 and 3D4S). **C**. Structure of the A_{2A}-adenosine receptor was determined using a strategy similar to that for the β_2-AR. Insertion of a T4 lysozyme domain into the third intracellular loop and the antagonist ZM241385 were needed to stabilize the protein for crystallization (PDB ID: 3EML). **D**. Structure of the β_1-adrenergic receptor (β_2-AR). Extensive protein engineering of heterologously expressed turkey adrenergic receptor and use of a novel thermostability assay in addition to microfocus X-ray sources enabled this structural determination. Inclusion of the inverse agonist, cyanopindolol, was also required for crystallization (PDB ID: 2VT4).

1.6 Sequence Similarities and Conserved Motifs within GPCRs

The structural similarity shared by all GPCR structures determined to date is quite high and is expected to extend to other GPCRs whose structures have not yet been determined.[85–87] However, the sequence homology shared by over 900 GPCRs is limited to a number of conserved motifs that are likely to play similar functional roles in most, if not all, GPCRs (Figure 1.6).[88] When all the reported structures of GPCRs are superposed, and the locations of crystallographically observed solvent (water) along with the side chains to which they are hydrogen bonded are examined, it is readily apparent that these waters occupy similar positions within the structures and that the sequence homology of the side chains with which they interact is high (Figure 1.7).[89,90] Because of this high degree of 'water homology', these waters and the residues to which they are hydrogen bonded most likely comprise a network along which the activation signal is passed upon agonist binding or photoactivation.[89,91,92] Radiolytic

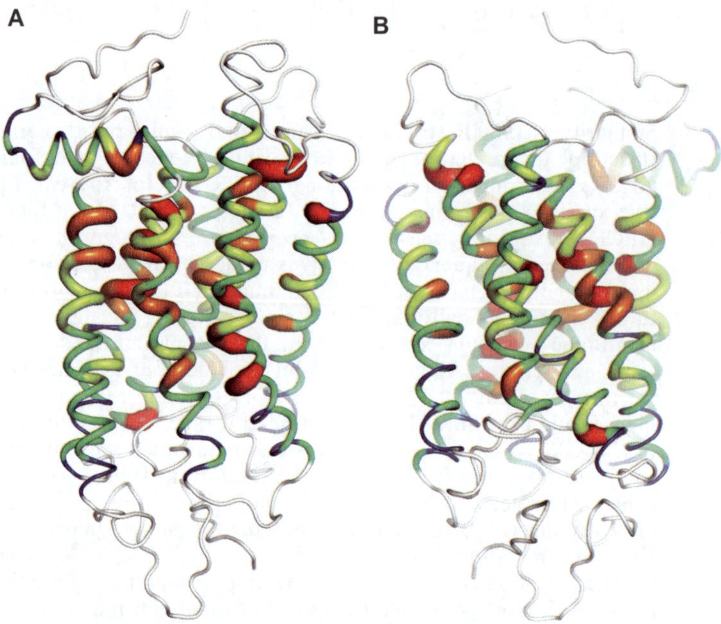

Figure 1.6 Amino acid homology within the transmembrane region of GPCRs. While overall sequence homology between GPCRs is low, many amino acid positions within the transmembrane region are highly conserved. **A** and **B**. Conserved sequences in all GPCRs were mapped onto the rhodopsin structure, ranging from red to yellow to blue to indicate high to low homology at individual positions. For clarity, the thickness of the trace also corresponds to the degree of conservation. Data adapted from ref. 88. Non-transmembrane regions vary too much in sequence and length to meaningfully interpret their homology across all GPCR sequences and are therefore depicted in light grey.

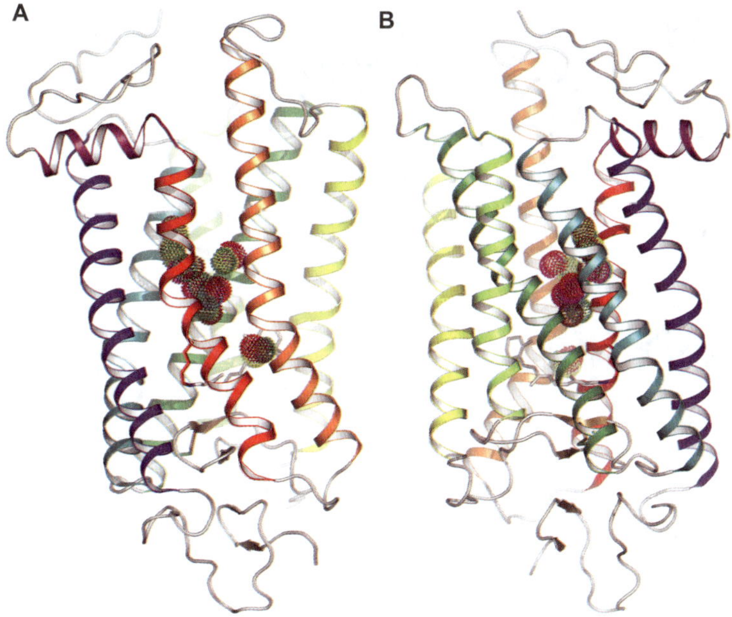

Figure 1.7 Homology of water positions within the transmembrane region of GPCRs. **A** and **B**. Conservation of the positions of crystallographically observed water molecules in all high resolution GPCR structures. Residues that directly bond with these waters also correspond to conserved residues in many cases, supporting a functional role for these water molecules. Crystallographically observed waters are depicted in red (PDB ID: 1U19) and magenta (PDB ID: 3C9L), in yellow for the A_{2A}-adenosine receptor (PDB ID: 3EML) and green for the β_2-adrenergic receptor (PDB ID: 2RH1). Superposition was performed only using the TM regions. Bovine rhodopsin (PDB ID: 1U19) is depicted in cartoon representation to show the relative positions of these waters within the TM region.

footprinting studies performed by our laboratory, in conjunction with the Chance laboratory, indicate that such waters do not freely exchange with bulk solvent, further supporting their roles as non-covalently bound prosthetic groups.[90,91] Meta-analysis of all GPCR structures links these clusters with conserved sequences such as the D(E)RY and NPxxYx$_{(5,6)}$F motifs that are thought to play crucial roles in receptor activation.[88]

1.7 The D(E)RY Motif and GPCR Activation

The ionic lock or D(E)RY region represents an energetic barrier which must be broken in order to reach the activated state.[93–96] Most but not all GPCRs contain this motif, suggesting that it plays an important but not wholly indispensible role in the activation process.[97] A conserved glutamic or aspartic acid residue located in helix-VI (E^{247} in Rho) makes a hydrogen bonding interaction with a conserved Arg (R^{135} in Rho) residue within helix-III (Figure 1.8).

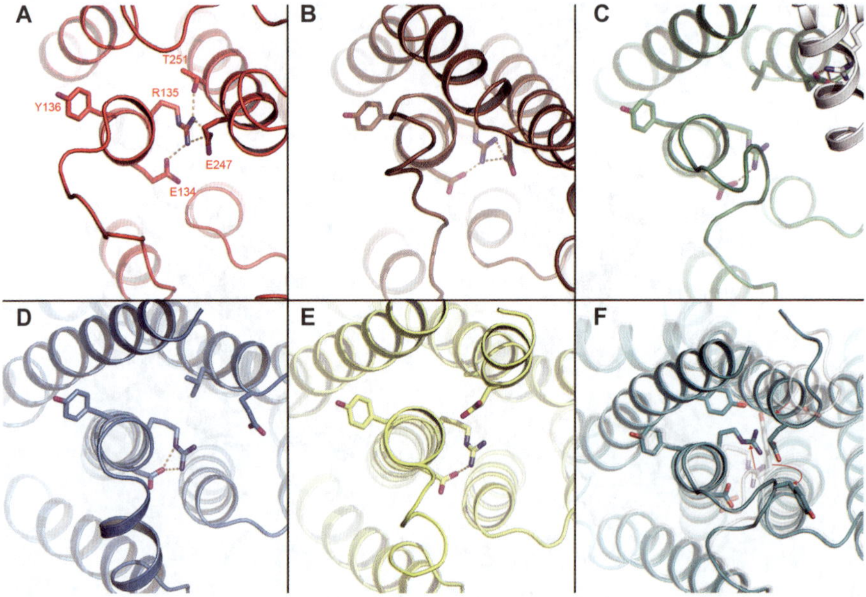

Figure 1.8 D(E)RY region of GPCRs. This region, termed the 'ionic lock', is highly conserved in family A GPCRs and critical for G protein activation. This hydrogen bond between a Glu (E^{247} in Rho) or Asp residue (in most other GPCRs) with an Arg (R^{135} in Rho) residue is thought to restrict the GPCR in an inactive conformation. **A**. D(E)RY region in rhodopsin. **B**. Squid rhodopsin D(E)RY region. **C**. β_2-adrenergic receptor D(E)RY region. **D**. β_1-Adrenergic D(E)RY region. **E**. A_{2A}-adenosine receptor D(E)RY region. While both squid and bovine rhodopsin in the ground state form the expected Glu–Arg hydrogen bonding interaction, this interaction is not observed in all the non-rhodopsin GPCR structures solved to date. This may be due to the mutations/protein engineering introduced to facilitate crystallization, or the GPCR may simply be trapped in this conformation due to the constraints of the crystal lattice or T4 lysozyme fusion. Disruption of this interaction also might be a manifestation of the basal G protein activation found for many GPCRs. **F**. Long timescale molecular dynamics simulations revealed that the ionic lock re-forms in the absence of the constraints imposed by the T4 lysozyme, strongly suggesting that the rigidity of the T4 lysozyme necessary for crystallization, or the subtle changes imposed in the TM region by the T4 lysozyme, forces the disruption of the ionic hydrogen bonding interaction that retains this motif in an inactive orientation.[101] The teal coloured cartoon indicates the small-scale movements/rotations needed to re-form these interactions and transparent grey coloured structures represent their original positions in the β_2-adrenergic receptor crystal structure (PDB ID: 2RH1).

Disruption of this bond is largely considered a hallmark of progression from the meta I to the meta II state. Protonation of the acidic residue within this motif is thought to accompany activation.[98–100] This hydrogen bond was disrupted in all recently determined structures of non-rhodopsin GPCRs (although bound to inverse agonists/antagonists), leading to much speculation that this lock was unique to rhodopsin and that, while the sequence motif was present in most GPCRs, it might not actually act as a functional 'lock'.[78,82] This supposition prompted a series of molecular dynamics simulations that revealed that, for each of these receptors, the ionic lock—while disrupted in the crystal structure reforms when the restraints imposed by the T4-lysozyme fusion or crystal lattice are removed (Figure 1.8).[101–104] However, partial occupancy of this broken ionic lock state might be an explanation for the agonist independent activation observed in these receptors.

1.8 The NPxxYx$_{(5,6)}$F Motif within GPCRs

The NPxxYx$_{(5,6)}$F motif is also highly conserved among members of class A GPCRs (Figure 1.9). Mutational analysis indicates that this region is important for receptor endocytosis and transport, interaction with small GPCR interacting proteins such as ADP ribosylation factors (ARFs) and Rho A, as well as for G protein binding.[95,105–108] The hydrogen bonding interaction between this motif's Tyr residue with a conserved Asn is postulated to directly affect G protein coupling.[109,110] Furthermore, hydrophobic interactions between the Tyr and Phe residues link helix-8 to the end of helix-VII, allowing changes in

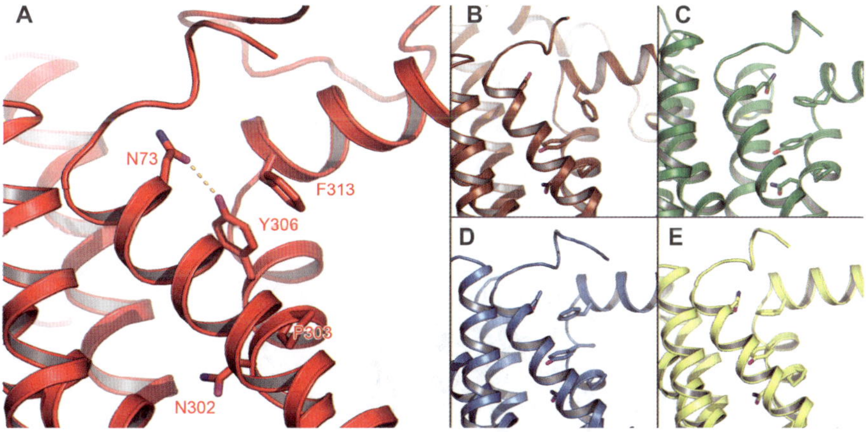

Figure 1.9 The NPxxYx$_{(5,6)}$F region in rhodopsin and other GPCRs. This highly conserved region was determined to be critical for retaining the receptor in an inactive conformation involved in receptor transport in addition to participating in photoactivation to the meta II state. This helix-turn-helix, as well as its sequence, is universally conserved in all family A GPCRs determined to date; however, the interaction between Y^{306} and N^{73} in bovine rhodopsin is not observed in any other GPCR structure. **A**. The NPxxYx$_{(5,6)}$F region in rhodopsin. **B**. Squid rhodopsin NPxxYx$_{(5,6)}$F region. **C**. β_2-Adrenergic receptor NPxxYx$_{(5,6)}$F region. **D**. β_1-Adrenergic NPxxYx$_{(5,6)}$F region. **E**. A_{2A}-Adenosine receptor NPxxYx$_{(5,6)}$F region.

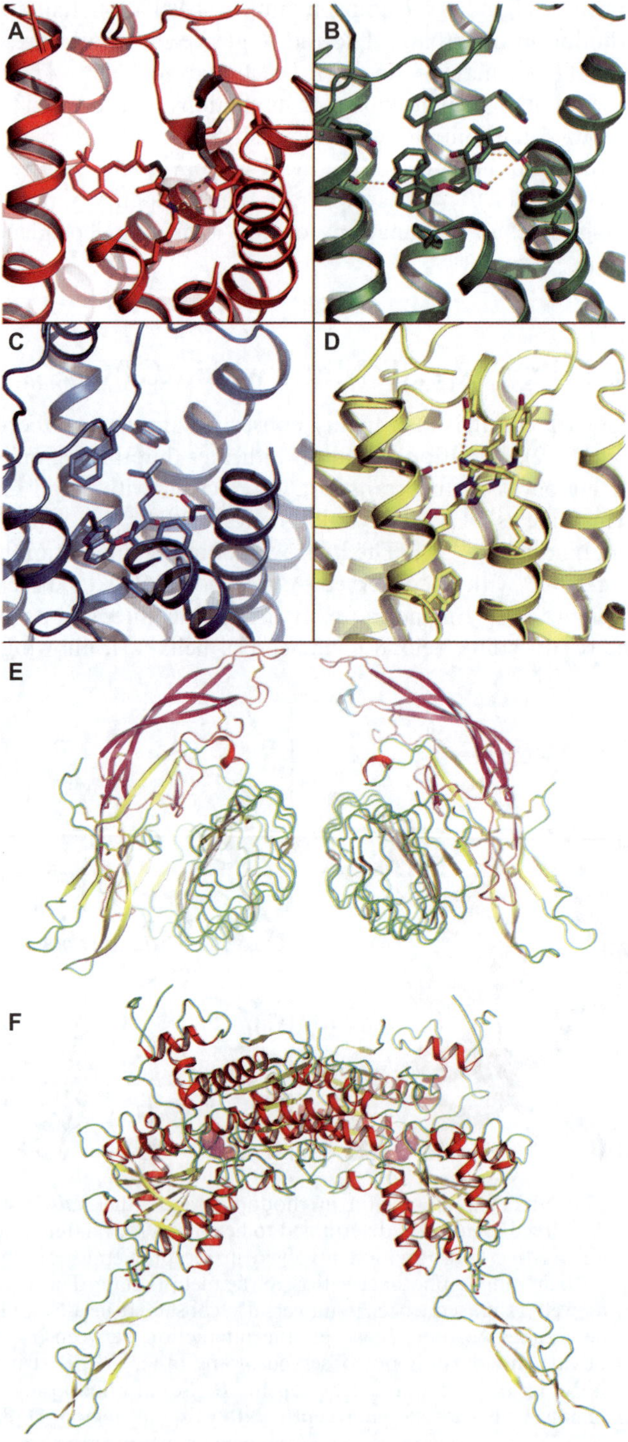
A
B
C
D
E
F

position of helix-VII to induce movements of helix-8 upon activation.[95] In higher resolution structures of rhodopsin, a cluster of three waters appear to link helices H-I, H-II and H-VI.[50,111] Further in-depth examination of bound water interactions in this vicinity reveals that similar networks of waters interact with the Asn residue (Figure 1.9). However, this interaction between N^{73} and Y^{306} observed in rhodopsin crystal structures is not seen in any other GPCR structure.

1.9 Ligand Binding Domains of GPCRs

In the human body, GPCRs respond to many diverse agents. Activation results from absorption of a single photon in the case of rhodopsin, binding of a small hormone molecule (*e.g.* β_1/β_2-adrenergic receptors), proteolysis of a tethered ligand (*e.g.* thrombin receptor), binding of a peptide (*e.g.* vasopressin receptor) or even binding of an entire protein (*e.g.* follicle-stimulating hormone receptor). It has been estimated that upwards of 60% of all current therapeutics act on or modulate GPCR-mediated signalling events, emphasizing the medical importance of this group of proteins.[112] Given the high diversity of GPCR activating compounds, it is hardly surprising that there are multiple sites for agonist binding. In many cases (including all structurally determined GPCRs), the ligand binding site is located within the transmembrane region (Figure 1.10A–D). However, for peptide and protein hormones the ligand binding site often is either on the extracellular face or extends to entire domains attached to the transmembrane region (Figure 1.10E and 1.10F). Limited structural data are available for a few of these extracellular domains, examples of which are shown in Figure 1.10.

Figure 1.10 Diversity of GPCR ligand binding sites. GPCRs feature a diverse set of activating agents, ranging from a photon to small hormones to proteolytic products and to even entire proteins. The ligand binding site within all full length GPCR structures determined thus far resides within the transmembrane region. However, larger and more hydrophilic agonist compounds, and the binding region of entire proteins, must access the corresponding ligand binding sites of other GPCRs, so these must be located outside the transmembrane region. **A–D**. Chromophore/ligand binding sites of rhodopsin, β_2-adrenergic receptor, β_1-adrenergic receptor and A_{2A}-adenosine receptors, respectively, are found in a similar transmembrane location and employ both hydrophilic and hydrophobic interactions to bind agonists/antagonists. **E**. Extracellular domain of the follicle stimulating hormone receptor in complex with FSH (PDB ID: 1XWD). **F**. Extracellular domain of the glutamate receptor (PDB ID: R2E4) in complex with glutamate (magenta spheres). Both FSH and glutamate receptors crystallized as dimers, and the glutamate receptor is known to function constitutively as a dimer. Regardless of the ligand binding site location, binding of an agonist must propagate a signal through the transmembrane region to cause changes on the cytoplasmic (G protein interacting) face of the GPCR.

Panels A–D in Figure 1.10 illustrate the chromophore/ligand binding sites of rhodopsin, β_2-adrenergic receptor, β_1-adrenergic receptor and A_{2A}-adenosine receptor, respectively. For other GPCRs, however, larger and more hydrophilic agonists and entire proteins must be able to access their respective ligand binding sites, so these sites are located outside the transmembrane region where they are more readily accessible. Examples of such domains are presented in Figure 1.10E and 1.10F. Because of the hydrophilic nature of ligands for these receptors, the location of the ligand binding site on the corresponding GPCR is outside the cell in the extracellular space. Both follicle-stimulating hormone (FSH) receptor[113] and glutamate receptor[114,115] crystallized as dimers, and the glutamate receptor is known to function as a dimer. Regardless of the locations of the ligand binding sites, agonist stimulation must propagate through the TM region, generating similar changes on the cytoplasmic (G protein interacting) face of the GPCR.

1.10 Problems with Alignment of GPCR Structures

The importance of a 'frame of reference" for comparing the various photostates of rhodopsin, as well as interpreting the state of activation of all determined GPCR structures, cannot be underestimated. Taking the opsin and rhodopsin structures as examples, different interpretations can result depending on which portions of these molecules are used for structural comparison (Figure 1.11). A global superposition minimizes observed differences in individual regions, whereas omitting such regions that differ substantially provides a much better indication of the magnitude of more subtle differences within regions of global structural homology. For example, the original opsin structure would be better solved in the higher symmetry space group R32 rather than in the R3 space group used, and the published structure contains temperature factors which do not conform to the distributions normally observed within X-ray crystal structures, complicating analysis. Uncertainties due to inadequate resolution, lack of data completeness, over-refinement/under-refinement and twinning must all be taken into consideration when comparing structures, as each of these criteria can lead to marked decreases in atomic certainty. Comparisons among determined GPCR structures also require special attention as to which regions to include; choosing sequence homology or identity as a criterion for superposition yields different results than a simple structural superposition of the transmembrane region.

1.11 Addressing Shortcomings in our Knowledge of GPCR and G Protein Activation

While our structural knowledge of GPCRs has expanded greatly over the last ten years, there remain some significant shortcomings in the current literature. In addition to obtaining structures of more GPCRs to increase the diversity of GPCR templates available for developing more effective homology models,

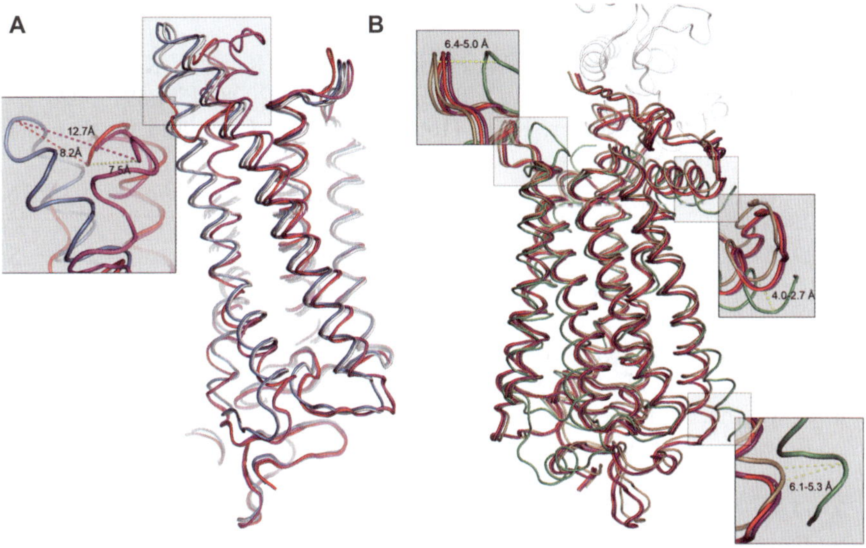

Figure 1.11 The problem with GPCR structure alignments. **A**. Opsin structure is superimposed upon the rhodopsin structure using two different frames of reference. Red and magenta (PDB IDs: 1U19 and 3C9L) backbone traces of ground state rhodopsin and opsin, respectively. Blue is a superposition in which H-5 and H-6 were left out of the alignment. Light grey is a global superposition. Superposition of only the TM regions yields a result very similar to these two alignments. **B**. Superposition of rhodopsin (PDB ID: 1U19) upon the β_2-adrenergic receptor structure (PDB ID: 2RH1) based on a variety of criteria results in 1–2 Å variations in distances, revealing difficulties in aligning structurally homologous proteins that share only weak sequence homology. Green: β_2-adrenergic receptor. Tan: Rho alignment based solely on Cα positions of TM helices and neglecting sequence. Red: Rho global sequence-based alignment. Purple: Rho superposition based solely on homologous residues within the TM region; similar superposition is observed when identity is used as a criterion.

a high-resolution structure of an activated GPCR is still required. Furthermore, while an agonist-bound GPCR structure might reveal the identity and scale of structural changes upon GPCR activation, this would only answer some of the questions pertaining to this process. Determination of the structures of important intermediate states of the G protein–GPCR complex[116] will ultimately reveal the important structural determinants of G protein activation by GPCRs and place the activated state of a GPCR in its cellular context.

Acknowledgements

We thank Drs. Wolfgang Baehr, Brian M. Kevany, John J. Mieyal, John C. Saari and Leslie T. Webster, Jr. for valuable comments on the manuscript. This work was supported in part by NIH grants EY0008061, EY019718 and GM079191. Krzysztof Palczewski is the John H. Hord Professor of Pharmacology.

References

1. P. Scheerer, J. H. Park, P. W. Hildebrand, Y. J. Kim, N. Krauss, H. W. Choe, K. P. Hofmann and O. P. Ernst, *Nature*, 2008, **455**, 497–502.
2. J. H. Park, P. Scheerer, K. P. Hofmann, H. W. Choe and O. P. Ernst, *Nature*, 2008, **454**, 183–187.
3. C. B. Roth, M. A. Hanson and R. C. Stevens, *J. Mol. Biol.*, 2008, **376**, 1305–1319.
4. D. M. Rosenbaum, V. Cherezov, M. A. Hanson, S. G. Rasmussen, F. S. Thian, T. S. Kobilka, H. J. Choi, X. J. Yao, W. I. Weis, R. C. Stevens and B. K. Kobilka, *Science*, 2007, **318**, 1266–1273.
5. V. P. Jaakola, M. T. Griffith, M. A. Hanson, V. Cherezov, E. Y. Chien, J. R. Lane, A. P. Ijzerman and R. C. Stevens, *Science*, 2008, **322**, 1211–1217.
6. T. Warne, M. J. Serrano-Vega, J. G. Baker, R. Moukhametzianov, P. C. Edwards, R. Henderson, A. G. Leslie, C. G. Tate and G. F. Schertler, *Nature*, 2008, **454**, 486–491.
7. D. C. Teller, T. Okada, C. A. Behnke, K. Palczewski and R. E. Stenkamp, *Biochemistry*, 2001, **40**, 7761–7772.
8. K. Palczewski, T. Kumasaka, T. Hori, C. A. Behnke, H. Motoshima, B. A. Fox, I. Le Trong, D. C. Teller, T. Okada, R. E. Stenkamp, M. Yamamoto and M. Miyano, *Science*, 2000, **289**, 739–745.
9. R. E. Stenkamp, *Acta Crystallogr. Sect. D: Biol. Crystallogr.*, 2008, **D64**, 902–904.
10. M. Murakami and T. Kouyama, *Nature*, 2008, **453**, 363–367.
11. T. Okada, M. Sugihara, A. N. Bondar, M. Elstner, P. Entel and V. Buss, *J. Mol. Biol.*, 2004, **342**, 571–583.
12. J. Li, P. C. Edwards, M. Burghammer, C. Villa and G. F. Schertler, *J. Mol. Biol.*, 2004, **343**, 1409–1438.
13. D. Salom, D. T. Lodowski, R. E. Stenkamp, I. Le Trong, M. Golczak, B. Jastrzebska, T. Harris, J. A. Ballesteros and K. Palczewski, *Proc. Natl. Acad. Sci. U. S. A.*, 2006, **103**, 16123–16128.
14. R. A. Dixon, B. K. Kobilka, D. J. Strader, J. L. Benovic, H. G. Dohlman, T. Frielle, M. A. Bolanowski, C. D. Bennett, E. Rands, R. E. Diehl, R. A. Mumford, E. E. Slater, I. S. Sigal, M. G. Caron, R. J. Lefkowitz and C. D. Strader, *Nature*, 1986, **321**, 75–79.
15. A. G. Gilman, *Annu. Rev. Biochem.*, 1987, **56**, 615–649.
16. G. A. Robison, R. W. Butcher and E. W. Sutherland, *Annu. Rev. Biochem.*, 1968, **37**, 149–174.
17. R. Henderson and P. N. Unwin, *Nature*, 1975, **257**, 28–32.
18. N. Unwin and R. Henderson, *Sci. Am.*, 1984, **250**, 78–94.
19. H. G. Khorana, *Ann. N. Y. Acad. Sci.*, 1986, **471**, 272–288.
20. M. Braiman, J. Bubis, T. Doi, H. B. Chen, S. L. Flitsch, R. R. Franke, M. A. Gilles-Gonzalez, R. M. Graham, S. S. Karnik, H. G. Khorana and *et al.*, *Cold Spring Harb. Symp. Quant. Biol.*, 1988, **53 Pt 1**, 355–364.
21. A. Ovchinnikov Yu, *FEBS Lett.*, 1982, **148**, 179–191.

22. P. A. Hargrave, J. H. McDowell, D. R. Curtis, J. K. Wang, E. Juszczak, S. L. Fong, J. K. Rao and P. Argos, *Biophys. Struct. Mech.*, 1983, **9**, 235–244.
23. A. Ovchinnikov Iu, N. G. Abdulaev, M. Feigina, I. D. Artamonov and A. S. Bogachuk, *Bioorg. Khim.*, 1983, **9**, 1331–1340.
24. J. Nathans and D. S. Hogness, *Cell*, 1983, **34**, 807–814.
25. J. Nathans and D. S. Hogness, *Proc. Natl. Acad. Sci. U. S. A.*, 1984, **81**, 4851–4855.
26. P. Argos, J. K. Rao and P. A. Hargrave, *Eur. J. Biochem.*, 1982, **128**, 565–575.
27. G. F. Schertler, C. Villa and R. Henderson, *Nature*, 1993, **362**, 770–772.
28. A. Krebs, C. Villa, P. C. Edwards and G. F. Schertler, *J. Mol. Biol.*, 1998, **282**, 991–1003.
29. A. Davies, B. E. Gowen, A. M. Krebs, G. F. Schertler and H. R. Saibil, *J. Mol. Biol.*, 2001, **314**, 455–463.
30. V. M. Unger, P. A. Hargrave, J. M. Baldwin and G. F. Schertler, *Nature*, 1997, **389**, 203–206.
31. J. M. Baldwin, *Curr. Opin. Cell Biol.*, 1994, **6**, 180–190.
32. J. M. Baldwin, *EMBO J.*, 1993, **12**, 1693–1703.
33. J. M. Baldwin, G. F. Schertler and V. M. Unger, *J. Mol. Biol.*, 1997, **272**, 144–164.
34. S. S. Karnik and H. G. Khorana, *J. Biol. Chem.*, 1990, **265**, 17520–17524.
35. S. S. Karnik, T. P. Sakmar, H. B. Chen and H. G. Khorana, *Proc. Natl. Acad. Sci. U. S. A.*, 1988, **85**, 8459–8463.
36. A. Ovchinnikov Yu, N. G. Abdulaev and A. S. Bogachuk, *FEBS Lett.*, 1988, **230**, 1–5.
37. S. S. Karnik, K. D. Ridge, S. Bhattacharya and H. G. Khorana, *Proc. Natl. Acad. Sci. U. S. A.*, 1993, **90**, 40–44.
38. K. Palczewski, J. Buczylko, M. W. Kaplan, A. S. Polans and J. W. Crabb, *J. Biol. Chem.*, 1991, **266**, 12949–12955.
39. H. Ohguro, K. Palczewski, L. H. Ericsson, K. A. Walsh and R. S. Johnson, *Biochemistry*, 1993, **32**, 5718–5724.
40. H. Ohguro, M. Rudnicka-Nawrot, J. Buczylko, X. Zhao, J. A. Taylor, K. A. Walsh and K. Palczewski, *J. Biol. Chem.*, 1996, **271**, 5215–5224.
41. H. Ohguro, N. Yoshida, H. Shindou, J. W. Crabb, K. Palczewski and M. Tsuda, *Photochem. Photobiol.*, 1998, **68**, 824–828.
42. W. L. Hubbell, C. Altenbach, C. M. Hubbell and H. G. Khorana, *Adv. Protein Chem.*, 2003, **63**, 243–290.
43. S. Filipek, R. E. Stenkamp, D. C. Teller and K. Palczewski, *Annu. Rev. Physiol.*, 2003, **65**, 851–879.
44. K. Palczewski, *Annu. Rev. Biochem.*, 2006, **75**, 743–767.
45. T. P. Sakmar, S. T. Menon, E. P. Marin and E. S. Awad, *Annu. Rev. Biophys. Biomol. Struct.*, 2002, **31**, 443–484.
46. S. T. Menon, M. Han and T. P. Sakmar, *Physiol. Rev.*, 2001, **81**, 1659–1688.

47. E. V. Yurkova, V. V. Demin and N. G. Abdulaev, *Biomed. Sci.*, 1990, **1**, 585–590.
48. T. Okada, I. Le Trong, B. A. Fox, C. A. Behnke, R. E. Stenkamp and K. Palczewski, *J. Struct. Biol.*, 2000, **130**, 73–80.
49. T. Okada and K. Palczewski, *Curr. Opin. Struct. Biol.*, 2001, **11**, 420–426.
50. T. Okada, Y. Fujiyoshi, M. Silow, J. Navarro, E. M. Landau and Y. Shichida, *Proc. Natl. Acad. Sci. U. S. A.*, 2002, **99**, 5982–5987.
51. S. Filipek, D. C. Teller, K. Palczewski and R. Stenkamp, *Annu. Rev. Biophys. Biomol. Struct.*, 2003, **32**, 375–397.
52. C. de Graaf, N. Foata, O. Engkvist and D. Rognan, *Proteins*, 2008, **71**, 599–620.
53. J. C. Mobarec, R. Sanchez and M. Filizola, *J. Med. Chem.*, 2009.
54. C. de Graaf and D. Rognan, *Curr. Pharm. Des.*, 2009, **15**, 4026–4048.
55. K. A. Reynolds, V. Katritch and R. Abagyan, *J. Comput. Aided Mol. Des.*, 2009, **23**, 273–288.
56. V. I. Perez-Nueno, D. W. Ritchie, O. Rabal, R. Pascual, J. I. Borrell and J. Teixido, *J. Chem. Inf. Model.*, 2008, **48**, 509–533.
57. Y. Shichida and H. Imai, *Cell. Mol. Life Sci.*, 1998, **54**, 1299–1315.
58. K. D. Ridge and K. Palczewski, *J. Biol. Chem.*, 2007, **282**, 9297–9301.
59. T. Okada, O. P. Ernst, K. Palczewski and K. P. Hofmann, *Trends Biochem. Sci.*, 2001, **26**, 318–324.
60. R. G. Matthews, R. Hubbard, P. K. Brown and G. Wald, *J. Gen. Physiol.*, 1963, **47**, 215–240.
61. T. Yoshizawa and G. Wald, *Nature*, 1964, **201**, 340–345.
62. T. E. Thorgeirsson, J. W. Lewis, S. E. Wallace-Williams and D. S. Kliger, *Biochemistry*, 1993, **32**, 13861–13872.
63. D. L. Farrens, C. Altenbach, K. Yang, W. L. Hubbell and H. G. Khorana, *Science*, 1996, **274**, 768–770.
64. C. Altenbach, A. K. Kusnetzow, O. P. Ernst, K. P. Hofmann and W. L. Hubbell, *Proc. Natl. Acad. Sci. U. S. A.*, 2008, **105**, 7439–7444.
65. B. Knierim, K. P. Hofmann, W. Gartner, W. L. Hubbell and O. P. Ernst, *J. Biol. Chem.*, 2008, **283**, 4967–4974.
66. B. Knierim, K. P. Hofmann, O. P. Ernst and W. L. Hubbell, *Proc. Natl. Acad. Sci. U. S. A.*, 2007, **104**, 20290–20295.
67. C. J. Lopez, M. R. Fleissner, Z. Guo, A. K. Kusnetzow and W. L. Hubbell, *Protein Sci.*, 2009, **18**, 1637–1652.
68. Z. T. Farahbakhsh, K. Hideg and W. L. Hubbell, *Science*, 1993, **262**, 1416–1419.
69. Z. T. Farahbakhsh, K. D. Ridge, H. G. Khorana and W. L. Hubbell, *Biochemistry*, 1995, **34**, 8812–8819.
70. H. Nakamichi and T. Okada, *Angew. Chem., Int. Ed. Engl.*, 2006, **45**, 4270–4273.
71. T. Morizumi, H. Imai and Y. Shichida, *J. Biochem.*, 2003, **134**, 259–267.

72. H. Nakamichi, V. Buss and T. Okada, *Biophys. J.*, 2007, **92**, L106–108.
73. H. Nakamichi and T. Okada, *Proc. Natl. Acad. Sci. U. S. A.*, 2006, **103**, 12729–12734.
74. T. Okada, *Biochem. Soc. Trans.*, 2004, **32**, 738–741.
75. S. Jones and J. Howl, *Curr. Protein Pept. Sci.*, 2006, **7**, 501–508.
76. J. Howl, U. Langel, S. R. Hawtin, A. Valkna, N. J. Yarwood, K. Saar and M. Wheatley, *FASEB J.*, 1997, **11**, 582–590.
77. T. Shimamura, K. Hiraki, N. Takahashi, T. Hori, H. Ago, K. Masuda, K. Takio, M. Ishiguro and M. Miyano, *J. Biol. Chem.*, 2008, **283**, 17753–17756.
78. S. G. Rasmussen, H. J. Choi, D. M. Rosenbaum, T. S. Kobilka, F. S. Thian, P. C. Edwards, M. Burghammer, V. R. Ratnala, R. Sanishvili, R. F. Fischetti, G. F. Schertler, W. I. Weis and B. K. Kobilka, *Nature*, 2007, **450**, 383–387.
79. P. W. Day, S. G. Rasmussen, C. Parnot, J. J. Fung, A. Masood, T. S. Kobilka, X. J. Yao, H. J. Choi, W. I. Weis, D. K. Rohrer and B. K. Kobilka, *Nat. Methods*, 2007, **4**, 927–929.
80. M. P. Bokoch, Y. Zou, S. G. Rasmussen, C. W. Liu, R. Nygaard, D. M. Rosenbaum, J. J. Fung, H. J. Choi, F. S. Thian, T. S. Kobilka, J. D. Puglisi, W. I. Weis, L. Pardo, R. S. Prosser, L. Mueller and B. K. Kobilka, *Nature*, 2010, **463**, 108–112.
81. M. A. Hanson and R. C. Stevens, *Structure*, 2009, **17**, 8–14.
82. V. Cherezov, D. M. Rosenbaum, M. A. Hanson, S. G. Rasmussen, F. S. Thian, T. S. Kobilka, H. J. Choi, P. Kuhn, W. I. Weis, B. K. Kobilka and R. C. Stevens, *Science*, 2007, **318**, 1258–1265.
83. T. Warne, M. J. Serrano-Vega, C. G. Tate and G. F. Schertler, *Protein Expr. Purif.*, 2009, **65**, 204–213.
84. C. G. Tate and G. F. Schertler, *Curr. Opin. Struct. Biol.*, 2009, **19**, 386–395.
85. D. Mustafi and K. Palczewski, *Mol. Pharmacol*, 2009, **75**, 1–12.
86. D. T. Lodowski and K. Palczewski, *Curr. Opin. HIV AIDS*, 2009, **4**, 88–95.
87. D. T. Lodowski, T. E. Angel and K. Palczewski, *Photochem. Photobiol.*, 2009, **85**, 425–430.
88. T. Mirzadegan, G. Benko, S. Filipek and K. Palczewski, *Biochemistry*, 2003, **42**, 2759–2767.
89. T. E. Angel, M. R. Chance and K. Palczewski, *Proc. Natl. Acad. Sci. U. S. A.*, 2009, **106**, 8555–8560.
90. T. Orban, S. Gupta, K. Palczewski and M. R. Chance, *Biochemistry*, 2010, **49**, 827–834.
91. T. E. Angel, S. Gupta, B. Jastrzebska, K. Palczewski and M. R. Chance, *Proc. Natl. Acad. Sci. U. S. A.*, 2009, **106**, 14367–14372.
92. E. Jardon-Valadez, A. N. Bondar and D. J. Tobias, *Biophys. J.*, 2009, **96**, 2572–2576.

93. E. H. Schneider, D. Schnell, A. Strasser, S. Dove and R. Seifert, *J. Pharmacol. Exp. Ther.*, 2010, **333**, 382–392.
94. A. E. Alewijnse, H. Timmerman, E. H. Jacobs, M. J. Smit, E. Roovers, S. Cotecchia and R. Leurs, *Mol. Pharmacol.*, 2000, **57**, 890–898.
95. O. Fritze, S. Filipek, V. Kuksa, K. Palczewski, K. P. Hofmann and O. P. Ernst, *Proc. Natl. Acad. Sci. U. S. A.*, 2003, **100**, 2290–2295.
96. M. Mahalingam, K. Martinez-Mayorga, M. F. Brown and R. Vogel, *Proc. Natl. Acad. Sci. U. S. A.*, 2008, **105**, 17795–17800.
97. C. A. Flanagan, *Mol. Pharmacol.*, 2005, **68**, 1–3.
98. S. Ye, T. Huber, R. Vogel and T. P. Sakmar, *Nat. Chem. Biol.*, 2009, **5**, 397–399.
99. S. Ludeke, M. Mahalingam and R. Vogel, *Photochem. Photobiol.*, 2009, **85**, 437–441.
100. R. Vogel, M. Mahalingam, S. Ludeke, T. Huber, F. Siebert and T. P. Sakmar, *J. Mol. Biol.*, 2008, **380**, 648–655.
101. R. O. Dror, D. H. Arlow, D. W. Borhani, M. O. Jensen, S. Piana and D. E. Shaw, *Proc. Natl. Acad. Sci. U. S. A.*, 2009, **106**, 4689–4694.
102. T. D. Romo, A. Grossfield and M. C. Pitman, *Biophys. J.*, 2010, **98**, 76–84.
103. E. Lyman, C. Higgs, B. Kim, D. Lupyan, J. C. Shelley, R. Farid and G. A. Voth, *Structure*, 2009, **17**, 1660–1668.
104. S. Vanni, M. Neri, I. Tavernelli and U. Rothlisberger, *Biochemistry*, 2009, **48**, 4789–4797.
105. L. Malaga-Dieguez, Q. Yang, J. Bauer, H. Pankevych, M. Freissmuth and C. Nanoff, *Mol. Pharmacol.*, **77**, 940–952.
106. I. Kalatskaya, S. Schussler, A. Blaukat, W. Muller-Esterl, M. Jochum, D. Proud and A. Faussner, *J. Biol. Chem.*, 2004, **279**, 31268–31276.
107. R. He, D. D. Browning and R. D. Ye, *J. Immunol.*, 2001, **166**, 4099–4105.
108. C. Gales, A. Kowalski-Chauvel, M. N. Dufour, C. Seva, L. Moroder, L. Pradayrol, N. Vaysse, D. Fourmy and S. Silvente-Poirot, *J. Biol. Chem.*, 2000, **275**, 17321–17327.
109. H. M. Miettinen, J. S. Mills, J. M. Gripentrog, E. A. Dratz, B. L. Granger and A. J. Jesaitis, *J. Immunol.*, 1997, **159**, 4045–4054.
110. J. S. Mills, H. M. Miettinen, D. Cummings and A. J. Jesaitis, *J. Biol. Chem.*, 2000, **275**, 39012–39017.
111. A. Faussner, A. Bauer, I. Kalatskaya, S. Schussler, C. Seidl, D. Proud and M. Jochum, *FEBS J.*, 2005, **272**, 129–140.
112. T. Klabunde and G. Hessler, *Chembiochem*, 2002, **3**, 928–944.
113. Q. R. Fan and W. A. Hendrickson, *Nature*, 2005, **433**, 269–277.
114. T. Muto, D. Tsuchiya, K. Morikawa and H. Jingami, *Proc. Natl. Acad. Sci. U. S. A.*, 2007, **104**, 3759–3764.
115. N. Kunishima, Y. Shimada, Y. Tsuji, T. Sato, M. Yamamoto, T. Kumasaka, S. Nakanishi, H. Jingami and K. Morikawa, *Nature*, 2000, **407**, 971–977.

116. B. Jastrzebska, Y. Tsybovsky and K. Palczewski, *Biochem. J.*, 2010, **428**, 1–10.
117. L. Pulagam and K. Palczewski, in *Encyclopedia of the Eye*, ed. D. A. Dartt, Elsevier, Oxford, English edn, 2010, vol. 3, pp. 403–417.
118. J. Standfuss, G. Xie, P. C. Edwards, M. Burghammer, D. D. Oprian and G. F. X. Schertler, *J. Mol. Biol.*, 2007, **372**, 1179–1188.
119. M. A. Hanson, V. Cherezov, M. T. Griffith, C. B. Roth, V. P. Jaakola, E. Y. Chien, J. Velasquez, P. Kuhn and R. C. Stevens, *Structure*, 2008, **16**, 897–905.

CHAPTER 2

Insights into GPCR Activation from NMR Spectroscopy

MARKUS EILERS AND STEVEN O. SMITH*

Department of Biochemistry and Cell Biology, Center for Structural Biology, Stony Brook University, Stony Brook, NY 11794-5215, USA

2.1 Introduction

G protein-coupled receptors (GPCRs) have arguably been one of the most challenging systems to understand in terms of structure–function in the past 20 years. Unlike soluble enzymes, they do not have a single active site, but rather have a dynamic structure where ligand binding on the extracellular side of the receptor induces conformational changes on the intracellular side. The relative simple architecture is deceiving since this framework has evolved to recognize literally thousands of different types of signals ranging from light to protein ligands. The crystal structure of rhodopsin solved in 2000 by Palczewski and Okada was a turning point in the structural biology of G protein-coupled receptors (Figure 2.1).[1] The structure revealed the position and interactions of conserved residues in the transmembrane helices and also revealed several features that were completely unexpected. The latter included the position of the second extracellular loop over the retinal binding site and the presence of a short amphipathic helix (H8) oriented parallel to the membrane surface. Importantly, the structure was of the dark, inactive conformation of the receptor and did not directly address the mechanism of receptor activation. In the past decade, there has been a concerted effort to obtain structures of ligand-activated GPCRs, and importantly to capture any GPCR in an active

RSC Drug Discovery Series No. 8
G Protein-Coupled Receptors: From Structure to Function
Edited by Jesús Giraldo and Jean-Philippe Pin

Published by the Royal Society of Chemistry, www.rsc.org

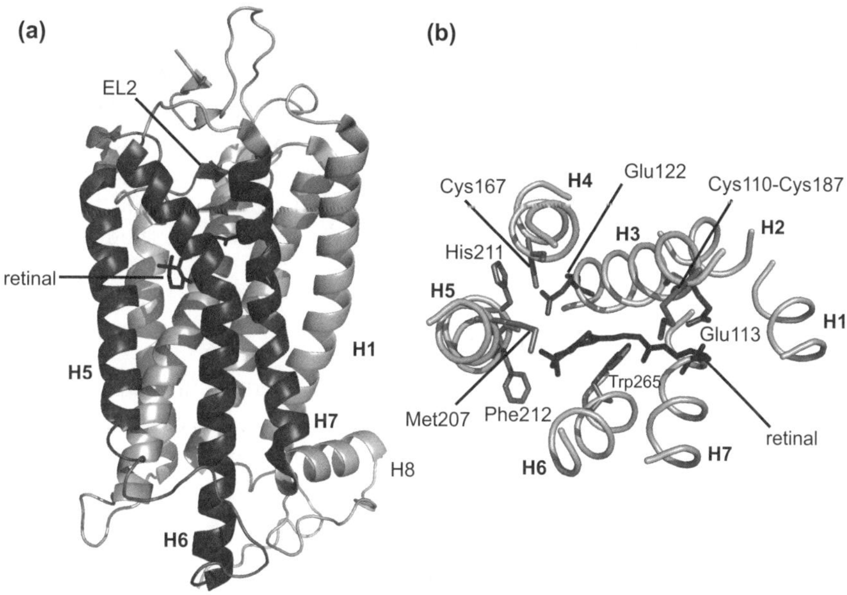

Figure 2.1 Crystal structure of rhodopsin. The first crystal structure of a GPCR was obtained in 2000 by Palczewski, Okada and colleagues.[1] Views of rhodopsin spanning the membrane bilayer (a) and from the extracellular (or intradiscal) surface of the membrane (b) show the seven-transmembrane α-helix architecture and the position of the 11-*cis* retinal chromophore. Light-induced isomerization of the retinal to the all-*trans* configuration converts the chromophore to a full agonist. Motion of helices H5–H7 upon activation opens up the G-protein binding site on the intracellular side of the receptor. Helices H1–H4 form the tightly packed core of the receptor.[143]

conformation. With the ligand-activated GPCRs, progress has been made using receptors that are engineered to favour or stabilize the inactive state.[2–5] However, obtaining a structure of an active GPCR has been more elusive. The crystal structures of opsin provide the highest resolution available of a GPCR with an active state conformation,[6,7] but it has not been possible to infer a detailed mechanism of activation since the structure lacks the agonist, all-*trans* retinal.

In this chapter, we summarize the progress made using NMR spectroscopy to probe the structure and activation mechanism of GPCRs. We focus on the visual receptor rhodopsin for which NMR measurements have been made on both the inactive and active conformations of the receptor. NMR spectroscopy offers an advantage over protein crystallography in that high-resolution distance measurements can be made in membrane environments on wild-type receptors without the use of mutations or protein insertions. Additionally the composition of the membrane can be changed to investigate its role in activation. NMR studies reported to date have been complementary to pioneering studies using electron paramagnetic resonance (EPR) and Fourier transform infrared (FTIR) spectroscopy. The EPR measurements have largely targeted the intracellular loops that can be readily spin labelled using site directed mutagenesis and thio-reactive spin labels.[8–12] In contrast, NMR measurements have targeted the transmembrane (TM) helix bundle and extracellular loops, particularly the second extracellular loop, which plays a key role in activation. FTIR spectroscopy, in combination with site directed mutagenesis, has been used extensively to identify changes in the protonation states and hydrogen-bonding strengths of amino acids in the activation pathway of rhodopsin.[13–21]

2.2 Experimental Approaches

In this section, we describe the approaches used to obtain milligram amounts of isotopically labelled GPCRs and the NMR techniques applied to acquire high-resolution spectra. A short overview of expression systems is presented along with an introduction to high-resolution solid state NMR methods.

2.2.1 Expression and Isotopic Labelling

The ability to generate sufficient amounts of purified and functional protein is essential for structural studies. Detailed NMR measurements further require samples to be labelled with stable isotopes such as ^{13}C and ^{15}N. Heterologous overexpression of GPCRs is needed to obtain the necessary milligram quantities of isotopically labelled receptors. Although GPCRs have been successfully expressed in *Escherichia coli*, yeast, insect, mammalian cells and cell-free translation systems, sample generation remains a challenging task.[22,23]

Expression of GPCRs in *E. coli* has the advantage of low cost, short generation time, and no post-translational modifications. The lack of post-translational modifications offers the advantage of homogenous receptors. On the other hand, glycosylation and palmitoylation of GPCRs are often needed for proper folding and receptor function. For example, ligand binding assays have demonstrated that glycosylation plays a role in the ligand binding of several GPCRs.[24–26] Furthermore, the carboxy terminal palmitoylation sites of rhodopsin have been shown to regulate downstream signalling targets.[27,28] Different strategies have been adopted to further circumvent the caveats of bacterial expression of GPCRs. Receptors are generally expressed in protease-deficient expression strains.[29] Auxiliary plasmids are also used for encoding rare codon tRNAs or for promoting disulfide bond formation. These strategies have been shown to increase the expression level of eukaryotic proteins.[30] Furthermore, two *E. coli* cell lines, C41 and C43, have been used to express membrane and toxic proteins.[31] Fusion with the *N*-terminal fragments of maltose binding protein (MBP),[29] Mistic[32] or OmpF[32] has also successfully been used to express large amounts of GPCRs.

Yeast expression systems allow mammalian-like expression and trafficking in compartmentalized organelles combined with short generation times in a genetically well-characterized organism. Yeast cells are capable of performing post-translational modifications, although the type and extent of the modifications can differ from mammalian systems. Recently, the yeast *Pichia pastoris* has been used to express the bradykinin receptor with significant post-translational modifications.[33] In *Saccharomyces cerevisiae*, only 2–4% of the opsin biosynthesized could be generated with retinal to rhodopsin,[34] whereas high yields of functional human adenosine A_2a receptor could be purified.[35] An advantage of yeast as an expression system is that functional analyses of expressed GPCRs can be performed prior to purification.[36,37]

Insect cell cultures infected with a GPCR-containing recombinant DNA virus allow the production of foreign proteins.[38,39] This eukaryotic expression system offers the advantage that it can be readily adapted to high-density suspension culture for large-scale expression and allows many of the post-translational modifications present in mammalian systems. The Sf9 insect system has also been used for expressing isotope-labelled rhodopsin and CCR5.[40,41]

Mammalian cells have the membrane composition and contain all the components needed to provide the highest level of post-translational modifications for the correct folding of functional GPCRs. However, these advantages come at the cost of the most demanding culturing conditions. HEK293S cells have been used successfully for incorporating isotope labels into several different GPCRs.[42,43] The expressing cells are cultured in media containing specifically labelled amino acids.[44] Selective incorporation has been attained for most the amino acids (see also next paragraph) [42,45–47]

Amino acids pathways in living cells can interfere with selective isotope labelling. For example, scrambling of an amino acid label into other amino acids is possible. The interconversion of 1-^{13}C glycine into 1-^{13}C serine through the enzymatic activity of serine hydroxymethyl transferase in mammalian cells was found in an HCNO experiment on rhodopsin grown in 2-^{15}N-lysine and 1-^{13}C-glycine enriched media.[48] Similarly, selectively labelled alanine, asparagine, aspartic acid, proline, glutamic acid and glutamine will result in isotope enrichment of other residues, and possibly reduced enrichment of the target residue.

Cell-free expression systems can produce large amounts of membrane proteins without the limitations of living organisms, *e.g.* label scrambling or protein toxicity. Further, the addition of detergents and other compounds to the reaction mixture may help functional folding and prevent aggregation. Several GPCRs have been successfully expressed with bacterial and wheat germ extracts.[49–53]

Receptors are generally solubilized in detergents and/or chaotropic agents for purification.[54] *E. coli* expressed membrane proteins, if not directed to the periplasmic membrane,[29,32] typically form inclusion bodies and require re-folding into a functional conformation.[55] The solubilized receptors can be purified by size exclusion, ion exchange and/or affinity chromatography. Affinity chromatography using immobilized ligands provides a method of purification of functional receptor. Ligand binding and G protein activation can also be used to estimate the amount of purified functional receptor. There are several structural genomics networks on membrane proteins that track GPCR expression levels in different organisms.[56]

An advantage of NMR methods over protein crystallography is the ability to study GPCRs reconstituted into membrane environments for structural measurements. The composition of the membrane for visual receptors has been shown to modulate the activity of the receptor.[57,58] The development of solubilized membrane patches, such as bicelles and nanodiscs, has opened up new approaches for NMR measurements in physiologically relevant membrane environments. These patches can be made by using short chain lipids (*i.e.* bicelles) or by using apo-lipoproteins (*i.e.* nanodiscs).[59,60] Nanodiscs have been used to solubilize a number of large membrane proteins, including the β_2 adrenergic receptor.[61] Rienstra and colleagues have shown that membrane proteins reconstituted in this manner can be precipitated[62] or lyophilized and rehydrated[63] and are subsequently suitable for characterization *via* solid-state NMR spectroscopy.

2.2.2 Solid-state NMR Spectroscopy

Structural studies rely on chemical shift and dipolar coupling measurements in both solution and solid-state NMR of proteins. In solution, rapid Brownian motion of the molecules averages the chemical shift interaction to its isotropic value and averages the dipolar interaction to zero. As a result, the solution NMR spectra of small proteins typically exhibit sharp, well-resolved resonances. Larger molecules tumble at a slower rate. The slower overall motion leads to a broadening of the NMR resonances as the chemical shift and dipolar interactions are no longer completely averaged. GPCRs have molecular weights that are often >40 kDa and must be solubilized in detergents or membrane vesicles. The large size of the solubilized receptor is the major drawback for multi-dimensional solution NMR experiments.

Solid-state NMR spectroscopy provides an alternative to solution state approaches, though the same problems with broadening of the NMR resonances due to the chemical shift and dipolar interactions are found in solid-state NMR (see Figure 2.2a). However, there are a number of different strategies possible with solid-state samples that can be applied to narrow the NMR lines and observe individual resonances.

The first is the use of magic angle spinning (MAS).[64–67] Both the chemical shift and the dipolar interaction have a $3\cos^2\theta - 1$ dependence on orientation with respect to the external magnetic field. Rapid mechanical rotation of the sample at the magic angle (the angle such that $3\cos^2\theta - 1 = 0$, *i.e.* $\theta = 54.7°$) averages the chemical shift interaction to its isotropic value and the dipolar interaction to zero. Figures 2.2a–c show the effect of MAS in narrowing the ^{13}C resonances of glycine. Under conditions of slow MAS (Figure 2.2b), the broad lineshapes are broken down into a resonance at the isotropic chemical shift value, flanked by additional lines with a spacing equal to the spinning frequency named the spinning sidebands. As the number of spins observed is

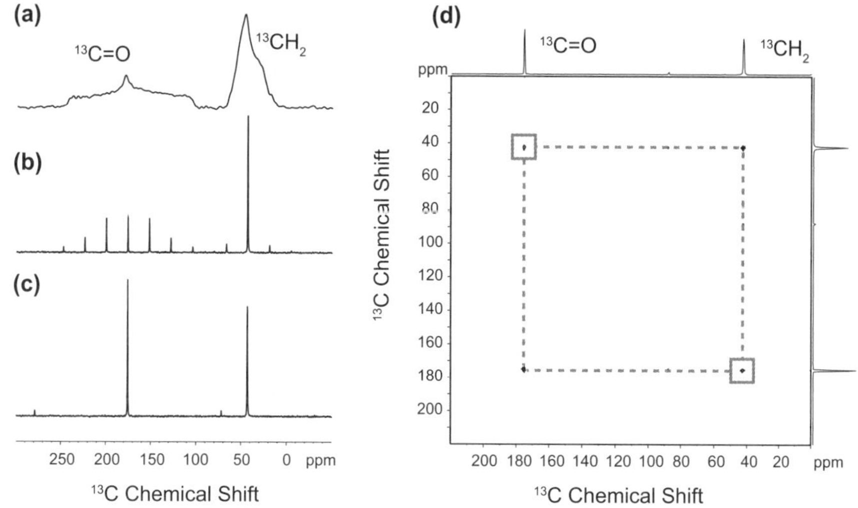

Figure 2.2 Solid-state magic angle spinning NMR spectroscopy. Solid-state ^{13}C NMR spectra of glycine illustrating the broad NMR resonances in static samples (a) and the effects of magic angle spinning at 3.0 kHz (b) and 13.0 kHz (c). All spectra were obtained using ^{1}H–^{13}C cross polarization and a ^{1}H decoupling field strength of 87 kHz was applied. MAS collapses the broad lineshapes into sharp centre bands at the isotropic chemical shifts and rotational sidebands spaced at the spinning frequency. (d) Two-dimensional ^{13}C dipolar assisted a rotational resonance (DARR)[82] experiment of glycine. A mixing time of 600 ms was used. The ^{1}H radio-frequency field strength during mixing was matched to the MAS speed to satisfy the condition $n = 1$.

reflected by the integral of their resonance, the height of the spinning sideband pattern is higher than the broad resonance from which they originate. When the spinning speed exceeds the width of the static NMR resonance, the full intensity shifts into the isotropic resonance. As a result, MAS improves both the resolution and the sensitivity of solid-state NMR spectra and has become a widely used technique for the study of interactions in the solid state.

Another strategy is to align the individual receptors relative to the magnetic field by layering lipid bilayers on glass slides or by incorporating receptors into membrane bicelles that have the ability to spontaneously orient in an external magnetic field.[68] Restricting the orientation of the receptor relative to the magnetic field narrows the frequency range of the chemical shift and dipolar interactions of a specific ^{13}C or ^{15}N nucleus, and consequently narrows the NMR resonance.

One-dimensional (1-D) MAS NMR difference spectra provide a method to easily isolate the resonances of amino acids whose chemical shift changes upon receptor activation. Figure 2.3 presents the ^{13}C NMR difference spectrum produced by subtracting the 1-D MAS NMR spectrum of the active state of

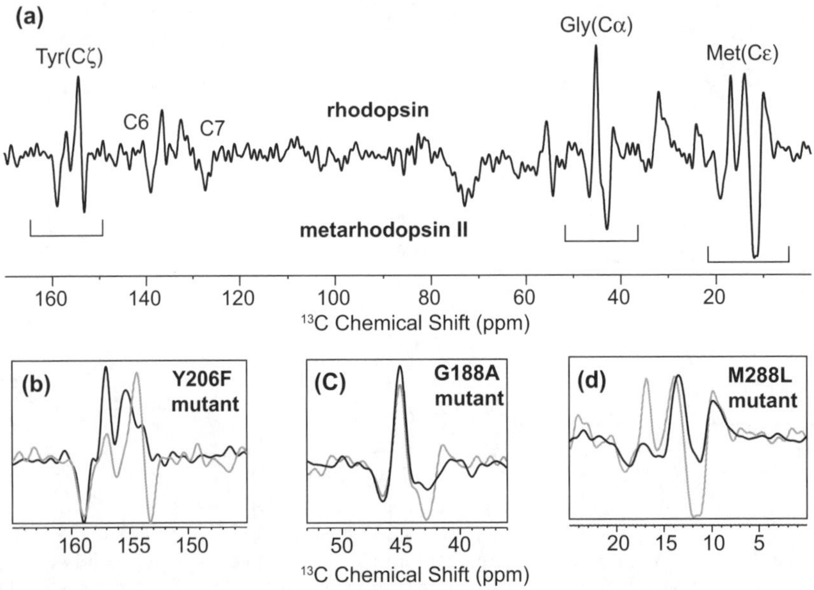

Figure 2.3 Solid-state NMR measurements of chemical shifts in rhodopsin. (**a**) The 150 MHz ^{13}C MAS difference spectrum is obtained by subtracting the spectrum of the active meta II intermediate from the inactive dark state of rhodopsin. By correlating the difference spectrum of the wild-type receptor (grey) with difference spectra of receptors with site-specific mutations (black), residues undergoing changes in chemical shift can be assigned (**b–d**).

rhodopsin from that of the inactive state. In this example, the receptor contains ^{13}C-labeled tyrosine, glycine and methionine. The chemical shifts for each type of amino acid change upon activation. These resonances can be assigned by comparison of difference spectra of the wild-type receptor and site-specific mutants. For example, there are two distinct negative peaks in the region of the ^{13}Cζ-resonances of tyrosine indicating that at least two tyrosines in the active meta II intermediate have changed chemical environment. The ^{13}Cζ-tyrosine chemical shift is sensitive to hydrogen bonding of the Cζ–OH hydroxyl group, with a downfield shift reflecting an increase in hydrogen bonding.[69] In Figure 2.3b, the loss of the upfield tyrosine resonance in meta II upon mutation of Tyr206$^{5.41}$ to phenylalanine allows this resonance to be assigned a specific amino acid. The upfield position of the chemical shift indicates that this tyrosine becomes more weakly hydrogen bonded upon activation.[46]

Measurements of chemical shifts can be used in several different ways. First, backbone ^{13}C and ^{1}H chemical shift measurements can be used to estimate protein secondary structure using TALOS, a database of chemical shifts compiled from proteins of known structure.[70] Secondly, backbone and side chain chemical shift (and relaxation) measurements provide information on protein mobility. Thirdly, chemical shifts can be used to characterize differences in the binding of antagonists and agonists, as well as in hydrogen bonding networks.

Dipolar couplings represent the interaction between the dipole moments associated with individual nuclear spins. The dipolar coupling has the functional form, $D = k\,(3\cos^2\theta - 1)\,\gamma_1\gamma_2/r^3$, where k is a constant, γ_1 and γ_2 are the gyromagnetic ratios of the two nuclei that are coupled, r is the distance between the coupled nuclei, and θ is the angle between r and the direction of the magnetic field. The dependence of the hetero- and homonuclear dipole–dipole interactions on the internuclear distance between two spins and orientation of the interacting spins with respect to the external magnetic field can be exploited to derive useful structural constraints on GPCRs.

In solution, rapid isotropic motion averages the dipolar interactions to zero. This improves the spectral resolution as the NMR resonances are narrowed, but the information on the distance-dependent dipolar couplings is lost. Nevertheless, it is possible to extract information on the dipolar interaction in solution NMR experiments. One well-established method is to measure the dipolar coupling indirectly through relaxation *via* the nuclear Overhauser effect (NOE). Another, more recent, technique is to partially orient the sample by restricting isotropic tumbling with filamentous Pf1 bacteriophages,[71] dilute lipid bicelles,[72] purple membrane fragments[73] or lamellar liquid-crystalline phases[74,75] that orient in a magnetic field. The residual dipolar couplings that result from partial orientation can be measured directly and used as constraints on the relative orientation of protein domains.

In solid-state NMR under MAS conditions, the dipolar couplings with magnitudes on the order of or less than the MAS frequency are averaged to zero. In the past 20 years, many methods have been developed to reintroduce dipolar couplings under MAS conditions in order to measure internuclear distances.[76–82] Since residual motion can average the dipolar couplings even in spectra obtained using solid-state NMR methods, the NMR data are often collected at low temperature. Figure 2.2d shows a two-dimensional MAS NMR experiment for measuring the $^{13}C...^{13}C$ dipolar couplings between the directly bonded carbons of glycine.

An example of a two-dimensional NMR experiment for measuring $^{13}C...^{13}C$ dipolar couplings in rhodopsin is shown in Figure 2.4. In this case, the method for reintroducing the dipolar couplings is referred to as dipolar-assisted rotational resonance (DARR).[82] The spectrum shown is of rhodopsin containing the same ^{13}C-labeled amino acids as in Figure 2.3. The resonances along the diagonal correspond to those observed in the 1-D spectrum. Off-diagonal cross peaks are observed between the diagonal resonances when the corresponding ^{13}C nuclei are separated by less than ~6 Å. The intensity of the cross peaks for unique sites can be related to the internuclear distance. In this example, the $^{13}C6$, $^{13}C7$ resonances are from unique sites on the retinal chromophore that are in close proximity to at least one methionine.

2.3 Retinal Conformation and Environment

The chromophore retinal in rhodopsin has two functions. The 11-*cis* isomer bound as a protonated Schiff base to Lys296$^{7.43}$ within the interior of the

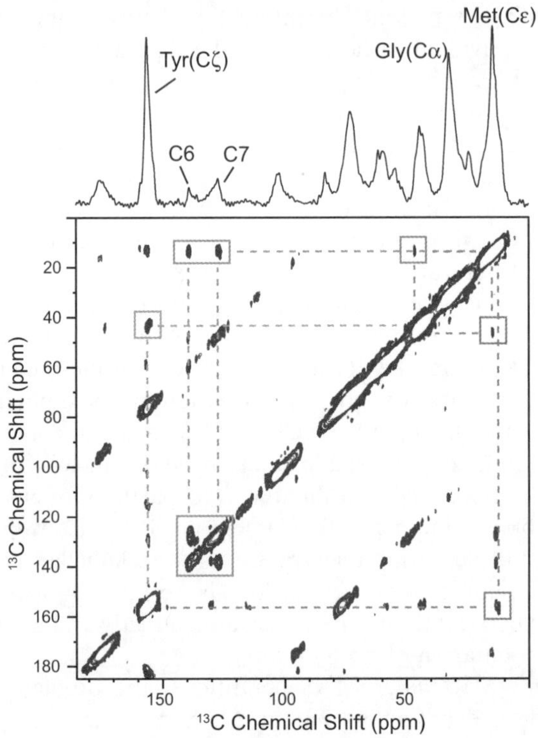

Figure 2.4 Solid-state NMR measurements of dipolar couplings in rhodopsin. The 150 MHz ^{13}C 2D DARR NMR spectrum is shown of rhodopsin specifically labelled with ^{13}Cα-glycine, ^{13}Cζ-tyrosine and ^{13}Cε-methionine and regenerated with 11-*cis* retinal ^{13}C-labeled at the C6 and C7 positions on the retinal polyene chain. DARR is used to reintroduce dipolar couplings between specific ^{13}C labelled sites while retaining the high resolution afforded by MAS.[82] The ^{13}C resonances introduced by specific labelling of the protein and retinal are observed along the diagonal. Off-diagonal cross peaks correspond to ^{13}C nuclei that are sufficiently close in space to transfer magnetization *via* the dipole–dipole interaction. By correlating the intensity of the observed cross peaks with internuclear distances obtained from the crystal structure of rhodopsin, the detection limit for the 2-D DARR NMR experiment is ~6–6.5 Å.

protein acts as an inverse agonist. After light-induced isomerization of the retinal, the all-*trans* isomer bound as an unprotonated Schiff base functions as a full agonist. These functions parallel those of small molecule ligands in other GPCRs. In this section, we first describe several of the unique properties of the retinal chromophore in visual pigments. We then describe how NMR spectroscopy has provided insights into the conformation of the retinal, its location within the seven-transmembrane helix bundle and some of the protein–retinal interactions that guide the efficient conversion of the inverse agonist to the full agonist and trigger receptor activation.

2.3.1 Retinal—the Photoreactive Trigger for Activation

The photoreactive chromophore in all visual receptors is the 11-*cis* isomer of the vitamin A aldehyde retinal (Figure 2.5a). Both the retinal polyene chain and its associated methyl groups contribute to the retinal's ability to trigger activation. The retinal is positively charged due to the protonation of the Schiff base bond; the positive charge is delocalized along the conjugated retinylidene chain and shifts the absorption band of the chromophore from the ultraviolet wavelength range into the middle of the visible spectrum. The protein counterion to the positive charge is $Glu113^{3.28}$ on transmembrane helix H3.[83,84] The protein environment of the apoprotein opsin shifts the chromophore absorption maximum from 460 nm for the protonated 11-*cis* retinylidene–butylimide model compound in chloroform[42] to 500 nm in rhodopsin. The $Glu113^{3.28}$ interaction with the protonated Schiff base is essential not only for spectral tuning, but also for maintaining the receptor in an inactive conformation. Comparative studies on the counterion in vertebrate and non-vertebrate visual pigments have shown that the counterion location can modulate the efficiency of G protein activation.[85]

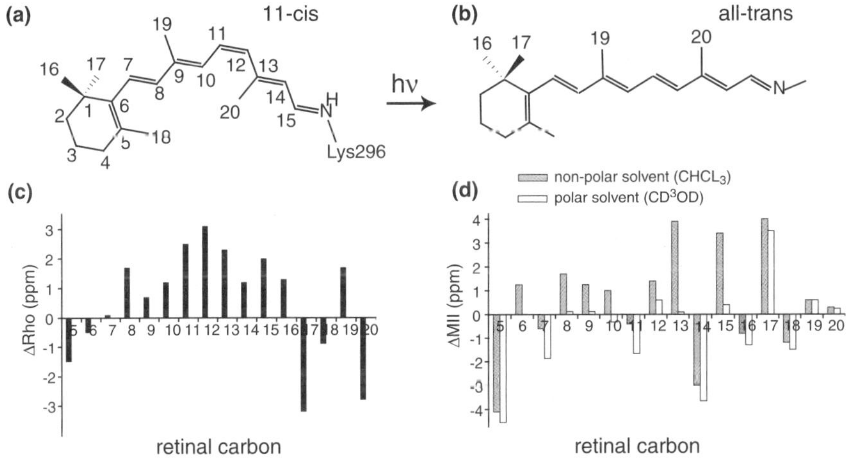

Figure 2.5 Comparison of retinal protonated Schiff base chemical shifts. (**a,b**) Structures of the 11-*cis* and all-*trans* retinal chromophores in the visual pigment rhodopsin. (**c**) Retinal ^{13}C chemical shift differences (ΔRho) are plotted between rhodopsin and the 11-*cis* protonated Schiff base model compound in solution ($CDCl_3$). The differences in chemical shift provide a pharmacophore map of retinal–protein interactions. For example, the positive ΔRho values between C8 and C15 are caused by close interaction of the retinal with a negatively charged glutamate residue (Glu181) in the retinal binding pocket. (**d**) Retinal ^{13}C chemical shift differences (ΔMII) are plotted between meta II and the all-*trans* retinal Schiff base model compound in solution (CD_3OD). The largest differences in chemical shift are observed in the C13=C14–C15 region of the retinal. These highly polarized bonds likely facilitate Schiff base hydrolysis in the conversion of meta II to opsin. The figure is adapted from ref. 102 with permission from the American Chemical Society.

Absorption of light by the retinal chromophore induces isomerization of the C11=C12 bond from *cis* to *trans*. The fast photochemistry[86] and high quantum yield[87] are due to the longitudinal restriction of the retinal within the binding site that imparts a twist about the C11=C12 double bond.[88–90] In addition, the C12–C13 single bond is twisted s-*cis* due to a steric contact between the C20 methyl group and the proton on C10. Together, these twists 'prime' the C20 methyl group to rotate in a clockwise direction (when viewed from the Schiff's base end of the retinal) as the retinal isomerizes.

Photoisomerization results in a highly strained all-*trans* retinal chromophore in the first reaction intermediates, photorhodopsin and bathorhodopsin, because the timescale is much too rapid for amino acid side-chains to rearrange in the tight protein binding site (Figure 2.6). Calorimetric studies show that bathorhodopsin has stored ~35 kcal/mol of the absorbed light energy.[91] The energy is released as the protein–retinal complex decays thermally through a series of distinct, spectrally defined intermediates. Crystal structures have been determined of the bathorhodopsin[92,93], lumirhodopsin[94] and metarhodopsin I intermediates.[95] There are no large scale changes observed in these transitions.

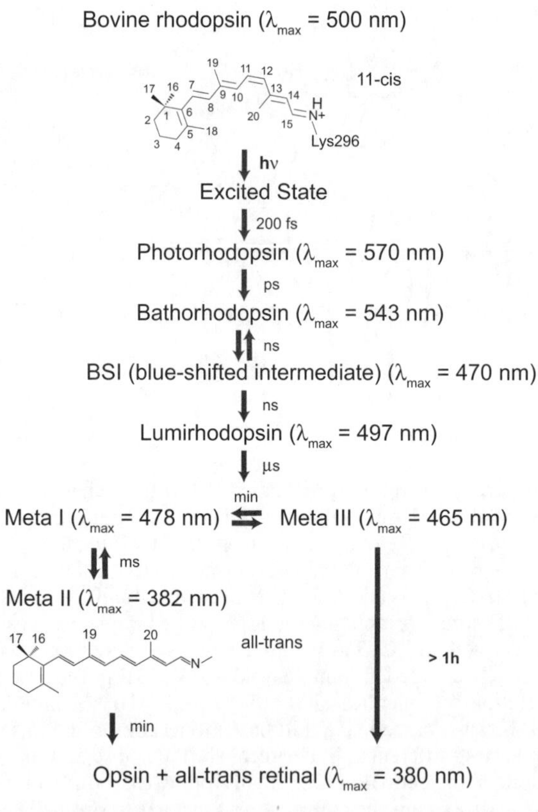

Figure 2.6 Photoreaction pathway of the retinal chromophore in rhodopsin.

In contrast, a large structural transition takes place in the conversion of metarhodopsin I (meta I) to metarhodopsin II (meta II). In the transition to the active meta II state, the Schiff base proton is transferred to Glu113$^{3.28}$. Meta II corresponds to the active state of rhodopsin, and the all-*trans* retinal chromophore with a deprotonated Schiff base is characterized by a 382 nm visible absorption maximum.

2.3.2 Retinal Conformation in the Visual Receptor Rhodopsin

The retinal chromophore functions as a light-activated ligand and offers two significant advantages for structural studies compared with other GPCRs. First, the retinal is covalently bound to the receptor. The covalent linkage through a protonated Schiff base to Lys296$^{7.43}$ insures 100% ligand occupancy in a defined manner. Secondly, the retinal functions as a pharmacological inverse agonist in the 11-*cis* configuration; binding of 11-*cis* retinal to the apoprotein opsin lowers the basal activity of the dark, inactive receptor to undetectable levels.[96,97] Light absorption rapidly converts the inverse agonist to a full agonist. Hence, for rhodopsin it is possible to isolate two well-defined, fully occupied states corresponding to the inactive and active receptor.

NMR measurements on the conformation of the 11-*cis* and all-*trans* chromophores in dark rhodopsin (inactive state) and the meta II intermediate (active state), respectively, have provided structural insights into how the retinal functions as a light-activated ligand. In addition, the chemical shifts of the ^{13}C atoms along the retinal provide information on the environment of its receptor binding site. Figure 2.5b presents a comparison of the chemical shifts of the 11-*cis* retinal protonated Schiff base in rhodopsin and an 11-*cis* protonated Schiff base model compound in solution. Significant differences in chemical shift are observed for the C8 to C13 positions.[98,99] The direction of the chemical shift differences reflects an increase in partial positive charge on these carbons. The most unusual difference is the chemical shift of C12 since charge delocalization along the conjugated polyene chain of the retinal typically leads to an alternation of charge where the odd-numbered carbons (C5, C7, C9, C11, C13 and C15) stabilize partial positive charge and the even-numbered carbons (C6, C8, C10, C12, C14) stabilize partial negative charge. When this chemical shift pattern was first observed in the early 1990s, a negative charge near C12 was proposed.[100,101] Indeed, the crystal structure of rhodopsin revealed that the carboxyl group of Glu181^{EL2} is positioned 4 Å from the retinal, with C12 being the closest point of contact.[1] In addition to the chemical shift differences along the conjugated chain, significant chemical shift differences are observed in the C16, C17, C18, C19 and C20 methyl groups, which are attributed to steric contacts in the retinal binding site compared to solution.[47,102]

In Figure 2.5c, the pattern of retinal chemical shifts in the all-*trans* chromophore of the active meta II intermediate changes considerably from the pattern of chemical shifts observed in the 11-*cis* chromophore of rhodopsin. The largest changes are observed at the two ends of the retinal, namely the β-ionone ring (C5, C17) and in the C13–C14–C15 region of the retinal–lysine Schiff base linkage. The large changes at C5 and C17 are due to structural changes in the β-ionone ring resulting from retinal isomerization within a tight receptor binding pocket, while the changes at C13–C15 are attributed to electrostatic interactions with the Glu113$^{3.28}$ carboxylic acid side chain, which becomes protonated upon receptor activation.[47,102]

Direct ^{15}N NMR measurements of the Schiff base nitrogen, which links the retinal to Lys296$^{7.43}$ on TM helix H7, have provided additional structural insights into how the retinal functions as a light-activated ligand. The interaction of the protonated Schiff base with its protein counterion, Glu113$^{3.28}$, is responsible for keeping the receptor off in the dark. Chemical shift measurements have been made on the protonated Schiff base by incorporating ^{15}N-labeled lysine into the protein.[42] The observed ^{15}N chemical shift of the protonated Schiff base in rhodopsin is upfield of that in retinal model compounds and suggestive of a weaker counterion interaction.[42] This observation is consistent with an extensive hydrogen-bonding network that connects Glu113$^{3.28}$ to polar residues in EL2 and reduces its effective charge.[103] More recent FTIR[104] and ^{2}H NMR data[105,106] provide further evidence for a complex counterion involving both Glu113$^{3.28}$ and Glu181^{EL2}. The complex counterion facilitates proton transfer from the Schiff base to Glu113$^{3.28}$ without a change in pH,[85] resulting in efficient activation of the G protein.[8,107] Indeed, Shichida and co-workers have shown that vertebrate rhodopsins activate G protein more efficiently than non-vertebrate rhodopsins and the muscarinic acetylcholine receptor.[85] A complex counterion structure was also observed for the retinal protonated Schiff base in bacteriorhodopsin[108] and may be responsible for the high Schiff base pKa needed to maintain the inactive state of the receptor.[109] ^{15}N chemical shift measurements of the meta II intermediate show a deprotonated Schiff base with a chemical shift significantly upfield of that of model compounds and bacteriorhodopsin.[102] This unusual ^{15}N chemical shift and a large downfield ^{13}C chemical shift at the adjacent C15 position (Figure 2.5c) indicate that the C15=N Schiff base bond is highly polarized with a significant partial positive charge localized on the C15 carbon.[102] This polarized structure is attributed to interaction with the protonated Glu113$^{3.28}$ side chain and may be responsible for the rapid hydrolysis of the retinal following activation and decay of the active meta II intermediate.[102]

2.3.3 Location and Environment of the Retinal in Activated Rhodopsin

As mentioned above, the retinal chromophore exists in a highly twisted 11-*cis* conformation in the ground state of rhodopsin.[110–115] After photoisomerization, the highly constrained chromophore relaxes on the timescale of milliseconds *via* several still inactive intermediates (bathorhodopsin, blue-shifted intermediate, lumirhodopsin,[94] metarhodopsin I[116,117]) to the active receptor conformation, metarhodopsin II. NMR measurements have been used to

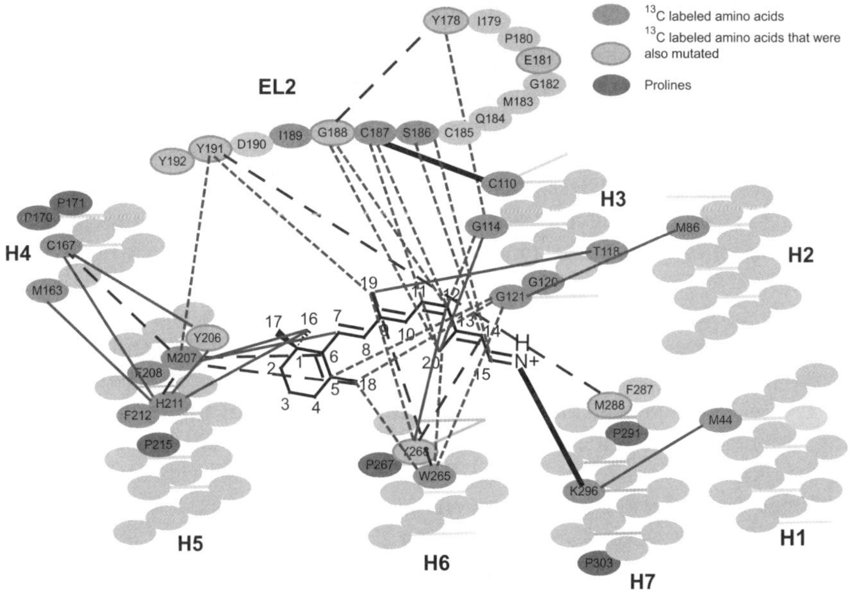

Figure 2.7 Two-dimensional residue map of close (<6 Å) interactions near the retinal-binding site determined by 2-D DARR NMR distance measurements in rhodopsin and meta II. Three main types interactions are shown: (1) retinal–protein interactions observed in both rhodopsin and meta II (solid lines); (2) interactions that are observed in rhodopsin and are lost upon conversion to meta II (dotted lines); and (3) interactions that are observed only in meta II (broken lines).

establish the location of the retinal following retinal isomerization and the protein conformational changes that accompany the dissipation of absorbed light energy.

Figure 2.7 presents a two-dimensional (2-D) residue map of close (<6 Å) interactions near the retinal-binding site determined by 2-D DARR NMR distance measurements in rhodopsin and meta II. EL2 forms a lid over the retinal-binding cavity, and the amino acids on the β4 strand of EL2 (Ser186, Cys187, Gly188 and Ile189) are interacting closely with the retinal polyene chain in rhodopsin. Upon activation, these distances increase and the chromophore moves toward H5.[46,47]

2.4 Receptor Structure and Conformational Changes Associated with Activation

In this section, we describe how the changes in the position of the retinal are coupled to the motion of EL2 and helices H5–H7. The consequences of the induced conformational changes that expose the G protein binding site are discussed.

2.4.1 Coupling of Retinal Isomerization to Helix Motion

Isomerization of the retinal chromophore in the visual GPCRs or binding of a signalling ligand in a ligand-activated GPCR causes a number of conformational changes on the extracellular side of the receptor that induce structural changes on the intracellular side of the receptor. The amino acid conservation within the TM core of Class A GPCRs argues strongly that there is a common mechanism for relaying the signal from extracellular ligand binding to the formation of a G protein binding site on the intracellular surface of the receptor.

Retinal isomerization leads to strong steric interactions within the protein binding site and deprotonation of the Schiff's base, which together serve to trigger the motion of helices H5, H6 and H7. Motion of the extracellular end of H5 results from displacement of EL2 and translation of the retinal β-ionone ring in meta II.[46,47] When viewed from the extracellular surface of rhodopsin, the β-ionone ring is packed against Glu122$^{3.37}$. Retinal movement disrupts the hydrogen bond between the main chain carbonyl of His211$^{5.38}$ and the side chain of Glu122$^{3.37}$, and a new hydrogen bond forms between Glu122$^{3.37}$ and the imidazole δ-nitrogen of His211$^{5.38}$. These interactions explain the requirement for an intact retinal β-ionone ring for rhodopsin activation[118,119] and the role of the His211$^{5.38}$ side-chain in meta II stability and activation.[120]

The switch for H6 motion has long been associated with a conserved aromatic cluster on the extracellular end of the helix.[121] In rhodopsin, the aromatic cluster is formed by three residues with relatively high identity across the class A GPCRs: Trp265$^{6.48}$, Phe261$^{6.44}$ and Tyr268$^{6.51}$. Trp265$^{6.48}$ lies within the arc formed by the retinal polyene chain and the Lys296$^{7.43}$ side chain, and is packed between Gly121$^{3.36}$ and Ala295$^{7.42}$. The retinal appears to function as a clamp to prevent motion of Trp265$^{6.48}$ and H6. Isomerization of the retinal and motion of the β-ionone ring toward H5 appear to facilitate motion of Trp265$^{6.48}$ toward the extracellular surface.[45]

Figure 2.8 shows the signalling pathway from Trp265$^{6.48}$ through Asn302$^{7.49}$ to Tyr306$^{7.53}$. Asn302$^{7.49}$ is not directly hydrogen-bonded to polar residues within the protein interior; rather a shell of water molecules surrounds the Asn302$^{7.49}$ side chain. Trp265$^{6.48}$ is connected to Asn302$^{7.49}$ through water-mediated hydrogen bonds. Asn302$^{7.49}$ is part of the conserved transmembrane core within GPCRs that appears to direct the outward rotation of H6. The sequence of events following retinal isomerization is likely to be very simple. Motion of Trp265$^{6.48}$ disrupts the hydrogen bonding contacts with Asn302$^{7.49}$.[122] One can speculate that the Asn302$^{7.49}$ side chain then rotates toward the conserved Asp83$^{2.54}$ on H2, since FTIR studies reveal a characteristic increase in hydrogen bonding of the Asp83$^{2.54}$ side chain in the meta I to meta II transition.[13] Motion of Asn302$^{7.49}$ then sets the stage for the outward rotation of the cytoplasmic end of H6 (see below).

Helices H5–H7 each contain a highly conserved proline (Pro215$^{5.42}$, Pro267$^{6.50}$ and Pro303$^{7.50}$) in the middle of the transmembrane sequence which are thought to be essential for mediating helix motion. There are three

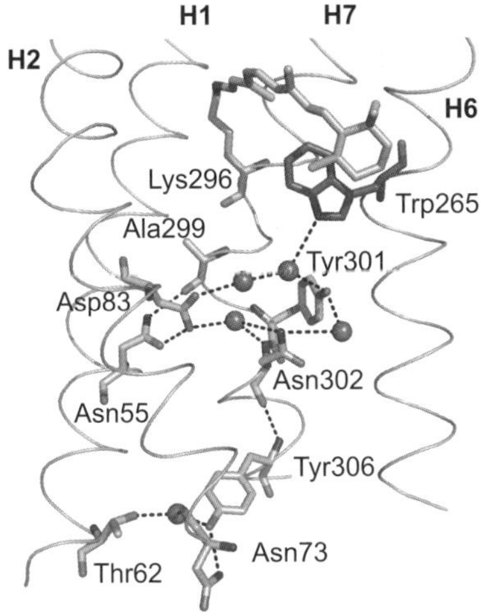

Figure 2.8 Signal transduction pathway through the conserved transmembrane core of rhodopsin. View of the rhodopsin crystal structure[144] shows the hydrogen-bonding networks between Trp265 of the retinal-binding pocket with Asn302 of the NPxxY sequence at the cytoplasmic surface. Asn302 and Tyr306 are signature residues on H7 that form stabilizing interactions with subfamily-specific residues on H6 and H2, respectively.[145] Tyr306 rotates into the space occupied by Met257 upon activation.[6,7,124]

additional prolines (Pro170$^{4.59}$, Pro171$^{4.60}$ and Pro291$^{7.38}$) with high subfamily specific conservation. The defining feature of a proline in transmembrane helices is that it is not able to form backbone hydrogen bond to the carbonyl group one helical turn away. In the case of Pro171$^{4.60}$, Pro215$^{5.42}$ and Pro303$^{7.50}$, the free backbone carbonyl makes key hydrogen bonding contacts. However, the free backbone carbonyls associated with Pro267$^{6.50}$ and Pro291$^{7.38}$ are oriented toward the lipids and in the crystal structures of rhodopsin and opsin do not form hydrogen bonds, suggesting that they allow the helical segments to swivel easily. These observations raise the possibility that the extracellular helix–loop–helix segment stretching from Pro267$^{6.50}$ to Pro291$^{7.38}$ pivots upon retinal isomerization. Coordinated motion of the extracellular ends of H6 and H7 is part of a global toggle switch mechanism proposed for GPCR activation.[123]

Solid-state NMR measurements have previously suggested that the displacement of EL2 upon activation is coupled to rotation of H5.[46] The coupled motion of EL2 and H5 was based on mutational experiments where substitutions in EL2 resulted in structural changes in H5.[46] More recently, we have shown that mutations in H5 (*i.e.* Y223F) result in structural changes in EL2 (*i.e.* Cys187^{EL2}).[124] In agreement with previous studies on Cys187^{EL2},[125] mutation of Tyr223$^{5.58}$ does not affect the wild-type properties of rhodopsin in the dark but leads to a less stable meta II intermediate upon activation. In wild-type rhodopsin at neutral pH, hydrolysis of the all-*trans* retinal Schiff base and loss of the retinal chromophore in the meta II to opsin transition shifts the receptor to an inactive conformation.[97] Inactive opsin is stabilized, at least in part, by electrostatic interactions involving Glu113$^{3.28}$ and Lys296$^{7.43}$, since mutation of either residue to a neutral amino acid increases constitutive receptor activity.[126] Interestingly, in opsin,[6] these residues do not interact directly but rather interact most closely with Cys187^{EL2} and Glu181^{EL2}, respectively, on EL2. Coupling of the position of EL2 to the orientation of H5 in the meta II to opsin transition would suggest there is a role for Tyr223$^{5.58}$ in both the opening and closing of the Arg135$^{3.50}$–Glu134$^{3.49}$ ionic lock.

Studies on rhodopsin mutants incapable of covalently binding retinal show that the covalent bond is required for formation of a stable active state. The K296G and K296A rhodopsin mutants cannot covalently bind retinal and have significant basal activity compared with the wild-type protein.[127,128] The absorption maxima for the K296G and K296A pigments are blue-shifted relative to wild-type rhodopsin. Oprian and coworkers suggested that the Schiff base nitrogen is less restricted in the non-covalently bound chromophore and can therefore approach the counterion Glu113$^{3.28}$ more closely.[127] The non-covalently bound pigments have a lower absorption coefficient than wild-type rhodopsin[127] and the activation mechanism triggered by photo-isomerization of the retinal is severely affected.[128] G-protein activation assays of K296G rhodopsin (K296G/nPrSB) clearly showed that an active state similar to the meta II intermediate of wild-type rhodopsin did not form in the bleaching process of this mutant, although the mutant exhibited an apparent light-dependent G-protein activity.[128] Shichida and coworkers further showed that the basal opsin activity and the light-dependent activity of K296G are very similar and caused by disrupting the opsin lock.[128]

FTIR measurements have shown that incorporating retinal analogs that lack the β-ionone ring result in an inability to adopt an active conformation.[118] Further experiments show that retinal analogs in which the conformation of the β-ionone ring is restricted also fail to achieve an active conformation.[129] Vogel and coworkers have described how these retinal–protein interactions control two protonation switches that mediate the transition to the active receptor conformation.[20,130]

2.4.2 Disruption of the Ionic Lock and G-protein Binding

The conserved Asp/Glu$^{3.49}$, Arg$^{3.50}$ and β-branched amino acid$^{3.54}$ motif at the cytoplasmic site of H3 is required for efficient signal transduction. This conserved hydrophobic cage motif was first described for the gonadotropin-releasing hormone receptor and called the ionic lock.[131] Ballesteros *et al.* proposed that a salt bridge between Asp$^{3.49}$ and Arg$^{3.50}$ stabilizes the inactive receptor and that upon activation Asp$^{3.49}$ becomes protonated with the charged Arg$^{3.50}$ side chain being prevented from orienting toward the cytoplasmic surface by Ile$^{3.54}$.[131] A more complex ionic lock involving the interaction of Arg$^{3.50}$ with both Asp$^{3.49}$ on H3 and Glu$^{6.30}$ on H6 was suggested, based on the results that both D3.49N and E6.30Q β_2 adrenergic receptor mutants show increased basal activity.[132] In the past few years, however, the crystal structures of the adenosine A2a,[5] and β_1[4] and β_2[133] adrenergic receptors have been determined and, in contrast to the crystal structure of rhodopsin, show no direct interaction between Arg$^{3.50}$ and Glu$^{6.30}$, although the Asp$^{3.49}$–Arg$^{3.50}$ salt bridge is retained.

Recent results on the interactions of Arg135$^{3.50}$ in meta II, along with amino acid conservation, provide insights into the nature of the closed and open states of the ionic lock.[124] Arginine (97%) at position 3.50 and Glu/Asp (90%) at position 3.49 show a high conservation, while lysine (0.9%) and histidine (0.2%), and Asn (2.4%) and Gln (1.7%) only rarely occur at positions 3.50 and 3.49, respectively. In contrast, Glu247$^{6.30}$ is not well conserved across the class A GPCRs, implying that there are various mechanisms for stabilizing the inactive conformations of GPCRs in the region of the ionic lock. In the β_1 and β_2 receptors, the position of Arg$^{3.50}$ is stabilized by a tyrosine on CL2, whose conservation in the amine subfamily (85%) is as high as the conservation of Glu$^{6.30}$ (82%). The conservation of two different residues that may stabilize Arg$^{3.50}$ in the amine receptors suggests that there may be multiple inactive states[134,135] or that these residues have other functions in the regulation of receptor activity.[136]

The motion of helices H5, H6 and H7 converge on the Arg135$^{3.50}$–Glu247$^{6.30}$ ionic lock on the intracellular side of rhodopsin. Three residues directly contribute to disrupting the ionic lock upon rhodopsin activation: Tyr223$^{5.50}$, Met257$^{6.40}$ and Tyr306$^{7.53}$. Tyr223$^{5.50}$ faces the surrounding lipids in rhodopsin and rotates toward Arg135$^{3.50}$ in meta II. In the opsin crystal structure, the Tyr223$^{5.50}$ side chain is tightly packed against Ala132$^{3.47}$, a group-conserved residue. Meta II decay measurements on Ala132$^{3.47}$ mutants suggest that a larger side chain at position 132$^{3.47}$ does not allow Tyr223$^{5.58}$ to form a stabilizing interaction with Arg135$^{3.50}$, indicating that the small side chain at this position creates a notch for orienting the tyrosine side-chain.[124]

Met257$^{6.40}$ on H6 changes its location next to Asn302$^{7.49}$ in the transmembrane core to a position close to Arg135$^{3.50}$ as H6 rotates outward (Figure 2.9). This motion facilitates the formation of two salt bridges on the intracellular surface of the receptor: Glu247$^{6.30}$–Lys231$^{5.58}$ and Glu249$^{6.28}$–Lys311^{CT}. In rhodopsin, the position of Tyr306$^{7.53}$ is constrained by hydrogen bonding with Asn73$^{2.45}$ (Figure 2.8) and Phe313^{H8}, both subfamily conserved residues. Tyr306$^{7.53}$ moves into the hydrophobic pocket vacated by Met257$^{6.40}$ when H6 rotates outward.

The ability of the receptor structure to bind the *C*-terminus of the α-subunit of transducin was shown by the opsin structure in complex with a synthetic peptide corresponding to the last 11 *C*-terminal amino acids of the α subunit of transducin (Gα-11mer).[6] Transducin is the G protein that interacts with the visual receptors. The Gα-11mer binds to opsin in an α-helical conformation to a site opened by the outward rotation of H6 and its structure is in good agreement with the structure of the Gα-11mer peptide bound to meta II obtained by transferred NOE measurements using solution NMR.[137] The peptide is disordered in solution, and binds weakly to rhodopsin upon light

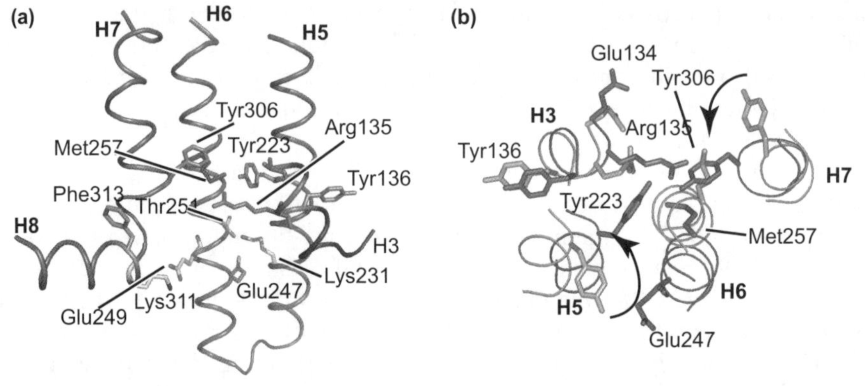

Figure 2.9 Intracellular ionic lock in rhodopsin. Views from the cytosolic surface of the opsin (PDB code 3CAP) (**a,b**) and rhodopsin (PDB code 1U19) (**b**) reveal the disruption of a salt bridge between Arg135 of the conserved E/DRY sequence and a glutamate side chain on H6 at position 247 upon activation. In concert with activation, the side chains of Tyr223 and Tyr306 on helices H5 and H7, respectively, rotate inward into close proximity to the guanidinium side chain of Arg135 as indicated in (**b**).

activation. The bound peptide has a helical turn followed by an open reverse turn centered at Gly348. By using a high magnetic field to align rhodopsin-containing unilamellar disks prepared from rod outer segments, Bax and coworkers were able to determine the orientation of two peptide NH groups in an 11-residue Gα peptide.[138,139] Peptides selectively labelled with ^{15}N at Leu5 and Gly9 exhibited residual dipolar couplings that allowed the NH vectors of Leu5 and Gly9 to be oriented with respect to the disk normal.[138,139]

The primary function of activated GPCRs is to catalyze the exchange of GTP for GDP in an intracellular heterotrimeric G protein. Receptor activation opens up a binding site on the intracellular surface of the receptor. However, there is only limited structural data on how the G protein complex binds. The most detailed data come from the experiments mentioned in the previous paragraph. The important question how receptor binding propagates a signal from the receptor surface to the GDP binding site of the G protein to facilitate GDP $\Leftrightarrow$ GTP exchange remains unanswered.

2.5 Conclusions

In this review we highlighted the progress that has been made to date on rhodopsin using NMR to correlate structure with function and outlined some of the unique possibilities NMR offers for structural studies on ligand-activated GPCRs. The diversity of GPCRs and their essential roles in cell biology and diseases make these receptors important structural targets. Further developments in expression and purification approaches, as well as NMR methodology, will move structural NMR studies forward. Dynamic nuclear polarization

(DNP) is one such application. DNP can increase the sensitivity of solid-state NMR experiments by using microwave irradiation to polarize unpaired electrons. For example, signal enhancements on the order of 40-fold have been obtained with DNP in ^{15}N MAS NMR studies of bacteriorhodopsin.[140] In solid-state MAS NMR measurements of ^{13}C-labeled rhodopsin, DNP has provided >20-fold increases in sensitivity.[141] As a result, the method opens up the possibility of structural studies on <1 mg of expressed and functional GPCRs. Furthermore, with the success of using engineered and constitutively active mutants for crystallization, there promises to be a number of crystal structures available that will provide the starting points for NMR studies in membrane environments.[142]

Acknowledgements

This work was supported by the National Institutes of Health through a grant (GM 41412) to S.O.S.

References

1. K. Palczewski, T. Kumasaka, T. Hori, C. A. Behnke, H. Motoshima, B. A. Fox, I. Le Trong, D. C. Teller, T. Okada, R. E. Stenkamp, M. Yamamoto and M. Miyano, *Science*, 2000, **289**, 739.
2. S. G. F. Rasmussen, H. J. Choi, D. M. Rosenbaum, T. S. Kobilka, F. S. Thian, P. C. Edwards, M. Burghammer, V. R. P. Ratnala, R. Sanishvili, R. F. Fischetti, G. F. X. Schertler, W. I. Weis and B. K. Kobilka, *Nature*, 2007, **450**, 383.
3. V. Cherezov, D. M. Rosenbaum, M. A. Hanson, S. G. F. Rasmussen, F. S. Thian, T. S. Kobilka, H. J. Choi, P. Kuhn, W. I. Weis, B. K. Kobilka and R. C. Stevens, *Science*, 2007, **318**, 1258.
4. T. Warne, M. J. Serrano-Vega, J. G. Baker, R. Moukhametzianov, P. C. Edwards, R. Henderson, A. G. Leslie, C. G. Tate and G. F. X. Schertler, *Nature*, 2008, **454**, 486.
5. V. P. Jaakola, M. T. Griffith, M. A. Hanson, V. Cherezov, E. Y. T. Chien, J. R. Lane, A. P. Ijzerman and R. C. Stevens, *Science*, 2008, **322**, 1211.
6. P. Scheerer, J. H. Park, P. W. Hildebrand, Y. J. Kim, N. Krauss, H. W. Choe, K. P. Hofmann and O. P. Ernst, *Nature*, 2008, **455**, 497.
7. J. H. Park, P. Scheerer, K. P. Hofmann, H. W. Choe and O. P. Ernst, *Nature*, 2008, **454**, 183.
8. D. L. Farrens, C. Altenbach, K. Yang, W. L. Hubbell and H. G. Khorana, *Science*, 1996, **274**, 768.
9. C. Altenbach, J. Klein-Seetharaman, K. W. Cai, H. G. Khorana and W. L. Hubbell, *Biochemistry*, 2001, **40**, 15493.
10. C. Altenbach, K. W. Cai, J. Klein-Seetharaman, F. G. Khorana and W. L. Hubbell, *Biochemistry*, 2001, **40**, 15483.
11. B. Knierim, K. P. Hofmann, O. P. Ernst and W. L. Hubbell, *Proc. Natl. Acad. Sci. U. S. A.*, 2007, **104**, 20290.

12. C. Altenbach, A. K. Kusnetzow, O. P. Ernst, K. P. Hofmann and W. L. Hubbell, *Proc. Natl. Acad. Sci. U. S. A.*, 2008, **105**, 7439.
13. K. Fahmy, F. Jäger, M. Beck, T. A. Zvyaga, T. P. Sakmar and F. Siebert, *Proc. Natl. Acad. Sci. U. S. A.*, 1993, **90**, 10206.
14. P. Rath, L. L. J. DeCaluwe, P. H. M. Bovee-Geurts, W. J. Degrip and K. J. Rothschild, *Biochemistry*, 1993, **32**, 10277.
15. S. Arnis, K. Fahmy, K. P. Hofmann and T. P. Sakmar, *J. Biol. Chem.*, 1994, **269**, 23879.
16. K. Fahmy, T. P. Sakmar and F. Siebert, *Biochemistry*, 2000, **39**, 10607.
17. R. Vogel, F. Siebert, G. Mathias, P. Tavan, G. B. Fan and M. Sheves, *Biochemistry*, 2003, **42**, 9863.
18. N. Lehmann, U. Alexiev and K. Fahmy, *J. Mol. Biol.*, 2007, **366**, 1129.
19. R. Vogel, M. Mahalingam, S. Lüdeke, T. Huber, F. Siebert and T. P. Sakmar, *J. Mol. Biol.*, 2008, **380**, 648.
20. M. Mahalingam, K. Martinez-Mayorga, M. F. Brown and R. Vogel, *Proc. Natl. Acad. Sci. U. S. A.*, 2008, **105**, 17795.
21. S. Lüdeke, M. Mahalingam and R. Vogel, *Photochem. Photobiol.*, 2009, **85**, 437.
22. V. Sarramegna, R. Talmont, P. Demange and A. Milon, *Cell. Mol. Life. Sci.*, 2003, **60**, 1529.
23. E. C. McCusker, S. E. Bane, M. A. O'Malley and A. S. Robinson, *Biotechnol. Prog.*, 2007, **23**, 540.
24. S. Rensdomiano and T. Reisine, *J. Biol. Chem.*, 1991, **266**, 20094.
25. D. Russo, G. D. Chazenbalk, Y. Nagayama, H. L. Wadsworth and B. Rapoport, *Mol. Endocrinol.*, 1991, **5**, 29.
26. T. Kusui, R. V. Benya, J. F. Battey and R. T. Jensen, *Biochemistry*, 1994, **33**, 12968.
27. Z. Y. Wang, X. H. Wen, Z. Ablonczy, R. K. Crouch, C. L. Makino and J. Lem, *J. Biol. Chem.*, 2005, **280**, 24293.
28. P. S. H. Park, K. T. Sapra, B. Jastrzebska, T. Maeda, A. Maeda, W. Pulawski, M. Kono, J. Lem, R. K. Crouch, S. Filipek, D. J. Muller and K. Palczewski, *Biochemistry*, 2009, **48**, 4294.
29. H. M. Weiss and R. Grisshammer, *Eur. J. Biochem.*, 2002, **269**, 82.
30. W. A. Prinz, F. Aslund, A. Holmgren and J. Beckwith, *J. Biol. Chem.*, 1997, **272**, 15661.
31. B. Miroux and J. E. Walker, *J. Mol. Biol.*, 1996, **260**, 289.
32. L. E. Petrovskaya, A. A. Shulga, O. V. Bocharova, Y. S. Ermolyuk, E. A. Kryukova, V. V. Chupin, M. J. J. Blommers, A. S. Arseniev and M. P. Kirpichnikov, *Biochemistry (Moscow)*, 2010, **75**, 881.
33. A. K. Shukla, W. Haase, C. Reinhart and H. Michel, *Protein Expr. Purif.*, 2007, **55**, 1.
34. R. Mollaaghababa, F. F. Davidson, C. Kaiser and H. G. Khorana, *Proc. Natl. Acad. Sci. U. S. A.*, 1996, **93**, 11482.
35. M. A. O'Malley, T. Lazarova, Z. T. Britton and A. S. Robinson, *J. Struct. Biol.*, 2007, **159**, 166.
36. S. J. Dowell and A. J. Brown, *Recept. Channels*, 2002, **8**, 343.

37. G. Ladds, A. Goddard and J. Davey, *Trends Biotechnol.*, 2005, **23**, 367.
38. M. J. Fraser, *Curr. Top. Microbiol. Immunol.*, 1992, **158**, 131.
39. T. A. Kost, J. P. Condreay and D. L. Jarvis, *Nat. Biotechnol.*, 2005, **23**, 567.
40. A. F. L. Creemers, C. H. W. Klaassen, P. H. M. Bovee-Geurts, R. Kelle, U. Kragl, J. Raap, W. J. De Grip, J. Lugtenburg and H. J. M. de Groot, *Biochemistry*, 1999, **38**, 7195.
41. L. Nisius, M. Rogowski, L. Vangelista and S. Grzesiek, *Protein Expr. Purif.*, 2008, **61**, 155.
42. M. Eilers, P. J. Reeves, W. W. Ying, H. G. Khorana and S. O. Smith, *Proc. Natl. Acad. Sci. U. S. A.*, 1999, **96**, 487.
43. S. J. Allen, S. Ribeiro, R. Horuk and T. M. Handel, *Protein Expr. Purif.*, 2009, **66**, 73.
44. M. Eilers, W. W. Ying, P. J. Reeves, H. G. Khorana and S. O. Smith, *Meth. Enzymol.*, 2002, **343**, 212.
45. E. Crocker, M. Eilers, S. Ahuja, V. Hornak, A. Hirshfeld, M. Sheves and S. O. Smith, *J. Mol. Biol.*, 2006, **357**, 163.
46. S. Ahuja, V. Hornak, E. C. Y. Yan, N. Syrett, J. A. Goncalves, A. Hirshfeld, M. Ziliox, T. P. Sakmar, M. Sheves, P. J. Reeves, S. O. Smith and M. Eilers, *Nat. Struct. Mol. Biol.*, 2009, **16**, 168.
47. S. Ahuja, E. Crocker, M. Eilers, V. Hornak, A. Hirshfeld, M. Ziliox, N. Syrett, P. J. Reeves, H. G. Khorana, M. Sheves and S. O. Smith, *J. Biol. Chem.*, 2009, **284**, 10190.
48. J. Klein-Seetharaman, P. J. Reeves, M. C. Loewen, E. V. Getmanova, L. Chung, H. Schwalbe, P. E. Wright and H. G. Khorana, *Proc. Natl. Acad. Sci. U. S. A.*, 2002, **99**, 3452.
49. C. Klammt, F. Löhr, B. Schäfer, W. Haase, V. Dötsch, H. Rüterjans, C. Glaubitz and F. Bernhard, *Eur. J. Biochem.*, 2004, **271**, 568.
50. G. Ishihara, M. Goto, M. Saeki, K. Ito, T. Hori, T. Kigawa, M. Shirouzu and S. Yokoyama, *Protein Expr. Purif.*, 2005, **41**, 27.
51. C. Klammt, D. Schwarz, N. Eifler, A. Engel, J. Piehler, W. Haase, S. Hahn, V. Dötsch and F. Bernhard, *J. Struct. Biol.*, 2007, **158**, 482.
52. L. Kaiser, J. Graveland-Bikker, D. Steuerwald, M. Vanberghem, K. Herlihy and S. G. Zhang, *Proc. Natl. Acad. Sci. U.S.A.*, 2008, **105**, 15726.
53. K. Sansuk, C. I. A. Balog, A. M. van der Does, R. Booth, W. J. de Grip, A. M. Deelder, R. A. Bakker, R. Leurs and P. J. Hensbergen, *J. Proteome Res.*, 2008, **7**, 621.
54. V. Sarramegna, I. Muller, A. Milon and F. Talmont, *Cell. Mol. Life. Sci.*, 2006, **63**, 1149.
55. H. Kiefer, *Biochim. Biophys. Acta Biomembr.*, 2003, **1610**, 57.
56. K. Lundstrom, *Bioorg. Med. Chem. Lett.*, 2005, **15**, 3654.
57. M. F. Brown, *Chem. Phys. Lipids*, 1994, **73**, 159.
58. E. Ramon, J. Marron, L. del Valle, L. Bosch, A. Andres, J. Manyosa and P. Garriga, *Vision Res.*, 2003, **43**, 3055.
59. C. R. Sanders and G. C. Landis, *Biochemistry*, 1995, **34**, 4030.
60. Y. Li, A. Z. Kijac, S. G. Sligar and C. M. Rienstra, *Biophys. J.*, 2006, **91**, 3819.

61. A. J. Leitz, T. H. Bayburt, A. N. Barnakov, B. A. Springer and S. G. Sligar, *Biotechniques*, 2006, **40**, 601.
62. A. Z. Kijac, Y. Li, S. G. Sligar and C. M. Rienstra, *Biochemistry*, 2007, **46**, 13696.
63. A. Kijac, A. Y. Shih, A. J. Nieuwkoop, K. Schulten, S. G. Sligar and C. M. Rienstra, *Biochemistry*, 2010, **49**, 9190.
64. I. J. Lowe, *Phys. Rev. Lett.*, 1959, **2**, 285.
65. E. R. Andrew, A. Bradbury and R. G. Eades, *Nature*, 1958, **182**, 1659.
66. J. Schaefer and E. O. Stejskal, *J. Am. Chem. Soc.*, 1976, **98**, 1031.
67. M. M. Maricq and J. S. Waugh, *J. Chem. Phys.*, 1979, **70**, 3300.
68. A. A. De Angelis, D. H. Jones, C. V. Grant, S. H. Park, M. F. Mesleh and S. J. Opella, *Meth. Enzymol.*, 2005, **394**, 350.
69. J. Herzfeld, S. K. Das Gupta, M. R. Farrar, G. S. Harbison, A. E. McDermott, S. L. Pelletier, D. P. Raleigh, S. O. Smith, C. Winkel, J. Lugtenburg and R. G. Griffin, *Biochemistry*, 1990, **29**, 5567.
70. G. Cornilescu, F. Delaglio and A. Bax, *J. Biomol. NMR*, 1999, **13**, 289.
71. M. R. Hansen, L. Mueller and A. Pardi, *Nat. Struct. Biol.*, 1998, **5**, 1065.
72. C. R. Sanders and J. P. Schwonek, *Biochemistry*, 1992, **31**, 8898.
73. B. W. Koenig, J. S. Hu, M. Ottiger, S. Bose, R. W. Hendler and A. Bax, *J. Am. Chem. Soc.*, 1999, **121**, 1385.
74. R. S. Prosser, J. A. Losonczi and I. V. Shiyanovskaya, *J. Am. Chem. Soc.*, 1998, **120**, 11010.
75. M. Rückert and G. Otting, *J. Am. Chem. Soc.*, 2000, **122**, 7793.
76. D. P. Raleigh, M. H. Levitt and R. G. Griffin, *Chem. Phys. Lett.*, 1988, **146**, 71.
77. T. Gullion and J. Schaefer, in *Advances in Magnetic Resonance*, ed. W. S. Warren, Academic Press, San Diego and London, 1989, vol. 13, 57–84.
78. A. E. Bennett, J. H. Ok, R. G. Griffin and S. Vega, *J. Chem. Phys.*, 1992, **96**, 8624.
79. D. K. Sodickson, M. H. Levitt, S. Vega and R. G. Griffin, *J. Chem. Phys.*, 1993, **98**, 6742.
80. Y. K. Lee, N. D. Kurur, M. Helmle, O. G. Johannessen, N. C. Nielsen and M. H. Levitt, *Chem. Phys. Lett.*, 1995, **242**, 304.
81. C. M. Rienstra, M. E. Hatcher, L. J. Mueller, B. Q. Sun, S. W. Fesik and R. G. Griffin, *J. Am. Chem. Soc.*, 1998, **120**, 10602.
82. K. Takegoshi, S. Nakamura and T. Terao, *Chem. Phys. Lett.*, 2001, **344**, 631.
83. T. P. Sakmar, R. R. Franke and H. G. Khorana, *Proc. Natl. Acad. Sci. U. S. A.*, 1989, **86**, 8309.
84. E. A. Zhukovsky and D. D. Oprian, *Science*, 1989, **246**, 928.
85. A. Terakita, M. Koyanagi, H. Tsukamoto, T. Yamashita, T. Miyata and Y. Shichida, *Nat. Struct. Mol. Biol.*, 2004, **11**, 284.
86. R. W. Schoenlein, L. A. Peteanu, R. A. Mathies and C. V. Shank, *Science*, 1991, **254**, 412.
87. H. J. A. Dartnall, *Vision Res.*, 1968, **8**, 339.

88. A. V. Struts, G. F. J. Salgad, K. Tanaka, S. Krane, K. Nakanishi and M. F. Brown, *J. Mol. Biol.*, 2007, **372**, 50.
89. T. Okada, M. Sugihara, A. N. Bondar, M. Elstner, P. Entel and V. Buss, *J. Mol. Biol.*, 2004, **342**, 571.
90. M. Sugihara, J. Hufen and V. Buss, *Biochemistry*, 2006, **45**, 801.
91. A. Cooper, *Nature*, 1979, **282**, 531.
92. H. Nakamichi and T. Okada, *Angew. Chem., Int. Ed. Engl.*, 2006, **45**, 4270.
93. M. Schreiber, M. Sugihara, T. Okada and V. Buss, *Angew. Chem., Int. Ed. Engl.*, 2006, **45**, 4274.
94. H. Nakamichi and T. Okada, *Proc. Natl. Acad. Sci. U. S. A.*, 2006, **103**, 12729.
95. J. J. Ruprecht, T. Mielke, R. Vogel, C. Villa and G. F. X. Schertler, *EMBO J.*, 2004, **23**, 3609.
96. G. B. Cohen, T. Yang, P. R. Robinson and D. D. Oprian, *Biochemistry*, 1993, **32**, 6111.
97. R. Vogel and F. Siebert, *J. Biol. Chem.*, 2001, **276**, 38487.
98. S. O. Smith, I. Palings, M. E. Miley, J. Courtin, H. de Groot, J. Lugtenburg, R. A. Mathies and R. G. Griffin, *Biochemistry*, 1990, **29**, 8158.
99. M. A. Verhoeven, A. F. L. Creemers, P. H. M. Bovee-Geurts, W. J. De Grip, J. Lugtenburg and H. J. M. de Groot, *Biochemistry*, 2001, **40**, 3282.
100. M. Han, B. S. DeDecker and S. O. Smith, *Biophys. J.*, 1993, **65**, 899.
101. M. Han and S. O. Smith, *Biochemistry*, 1995, **34**, 1425.
102. S. Ahuja, M. Eilers, A. Hirshfeld, E. C. Y. Yan, M. Ziliox, T. P. Sakmar, M. Sheves and S. O. Smith, *J. Am. Chem. Soc.*, 2009, **131**, 15160.
103. T. Okada, Y. Fujiyoshi, M. Silow, J. Navarro, E. M. Landau and Y. Shichida, *Proc. Natl. Acad. Sci. U. S. A.*, 2002, **99**, 5982.
104. S. Lüdeke, R. Beck, E. C. Y. Yan, T. P. Sakmar, F. Siebert and R. Vogel, *J. Mol. Biol.*, 2005, **353**, 345.
105. G. F. J. Salgado, A. V. Struts, K. Tanaka, S. Krane, K. Nakanishi and M. F. Brown, *J. Am. Chem. Soc.*, 2006, **128**, 11067.
106. K. Martinez-Mayorga, M. C. Pitman, A. Grossfield, S. E. Feller and M. F. Brown, *J. Am. Chem. Soc.*, 2006, **128**, 16502.
107. P. R. Robinson, G. B. Cohen, E. A. Zhukovsky and D. D. Oprian, *Neuron*, 1992, **9**, 719.
108. H. J. M. de Groot, S. O. Smith, J. Courtin, E. van den Berg, C. Winkel, J. Lugtenburg, R. G. Griffin and J. Herzfeld, *Biochemistry*, 1990, **29**, 6873.
109. G. Steinberg, M. Ottolenghi and M. Sheves, *Biophys. J.*, 1993, **64**, 1499.
110. I. Palings, J. A. Pardoen, E. van den Berg, C. Winkel, J. Lugtenburg and R. A. Mathies, *Biochemistry*, 1987, **26**, 2544.
111. G. G. Kochendoerfer, S. W. Lin, T. P. Sakmar and R. A. Mathies, *Trends Biochem. Sci.*, 1999, **24**, 300.
112. D. H. Pan and R. A. Mathies, *Biochemistry*, 2001, **40**, 7929.
113. P. Kukura, D. W. McCamant, S. Yoon, D. B. Wandschneider and R. A. Mathies, *Science*, 2005, **310**, 1006.

114. F. Siebert, *Isr. J. Chem.*, 1995, **35**, 309.
115. G. G. Kochendoerfer, P. J. E. Verdegem, I. van der Hoef, J. Lugtenburg and R. A. Mathies, *Biochemistry*, 1996, **35**, 16230.
116. P. J. E. Verdegem, P. H. M. Bovee-Geurts, W. J. De Grip, J. Lugtenburg and H. J. M. de Groot, *Biochemistry*, 1999, **38**, 11316.
117. X. Feng, P. J. E. Verdegem, M. Eden, D. Sandström, Y. K. Lee, P. H. M. Bovee-Geurts, W. J. De Grip, J. Lugtenburg, H. J. M. de Groot and M. H. Levitt, *J. Biomol. NMR*, 2000, **16**, 1.
118. F. Jäger, S. Jäger, O. Kräutle, N. Friedman, M. Sheves, K. P. Hofmann and F. Siebert, *Biochemistry*, 1994, **33**, 7389.
119. R. Vogel, F. Siebert, S. Lüdeke, A. Hirshfeld and M. Sheves, *Biochemistry*, 2005, **44**, 11684.
120. J. W. Lewis, I. Szundi, M. A. Kazmi, T. P. Sakmar and D. S. Kliger, *Biochemistry*, 2006, **45**, 5430.
121. L. Shi, G. Liapakis, R. Xu, F. Guarnieri, J. A. Ballesteros and J. A. Javitch, *J. Biol. Chem.*, 2002, **277**, 40989.
122. A. B. Patel, E. Crocker, P. J. Reeves, E. V. Getmanova, M. Eilers, H. G. Khorana and S. O. Smith, *J. Mol. Biol.*, 2005, **347**, 803.
123. C. E. Elling, T. M. Frimurer, L. O. Gerlach, R. Jorgensen, B. Holst and T. W. Schwartz, *J. Biol. Chem.*, 2006, **281**, 17337.
124. J. A. Goncalves, K. South, S. Ahuja, E. Zaitseva, C. A. Opefi, M. Eilers, R. Vogel, P. J. Reeves and S. O. Smith, *Proc. Natl. Acad. Sci. U. S. A.*, 2010, **107**, 19861.
125. F. F. Davidson, P. C. Loewen and H. G. Khorana, *Proc. Natl. Acad. Sci. U. S. A.*, 1994, **91**, 4029.
126. G. B. Cohen, D. D. Oprian and P. R. Robinson, *Biochemistry*, 1992, **31**, 12592.
127. E. A. Zhukovsky, P. R. Robinson and D. D. Oprian, *Science*, 1991, **251**, 558.
128. T. Matsuyama, T. Yamashita, H. Imai and Y. Shichida, *J. Biol. Chem.*, 2010, **285**, 8114.
129. G. F. Jang, V. Kuksa, S. Filipek, F. Bartl, E. Ritter, M. H. Gelb, K. P. Hofmann and K. Palczewski, *J. Biol. Chem.*, 2001, **276**, 26148.
130. R. Vogel, T. P. Sakmar, M. Sheves and F. Siebert, *Photochem. Photobiol.*, 2007, **83**, 286.
131. J. Ballesteros, S. Kitanovic, F. Guarnieri, P. Davies, B. J. Fromme, K. Konvicka, L. Chi, R. P. Millar, J. S. Davidson, H. Weinstein and S. C. Sealfon, *J. Biol. Chem.*, 1998, **273**, 10445.
132. J. A. Ballesteros, A. D. Jensen, G. Liapakis, S. G. F. Rasmussen, L. Shi, U. Gether and J. A. Javitch, *J. Biol. Chem.*, 2001, **276**, 29171.
133. D. M. Rosenbaum, V. Cherezov, M. A. Hanson, S. G. F. Rasmussen, F. S. Thian, T. S. Kobilka, H. J. Choi, X. J. Yao, W. I. Weis, R. C. Stevens and B. K. Kobilka, *Science*, 2007, **318**, 1266.
134. R. O. Dror, D. H. Arlow, D. W. Borhani, M. O. Jensen, S. Piana and D. E. Shaw, *Proc. Natl. Acad. Sci. U. S. A.*, 2009, **106**, 4689.
135. S. Vanni, M. Neri, I. Tavernelli and U. Rothlisberger, *Biochemistry*, 2009, **48**, 4789.

136. K. Baltensperger, V. Karoor, H. Paul, A. Ruoho, M. P. Czech and C. C. Malbon, *J. Biol. Chem.*, 1996, **271**, 1061.
137. O. G. Kisselev, J. Kao, J. W. Ponder, Y. C. Fann, N. Gautam and G. R. Marshall, *Proc. Natl. Acad. Sci. U. S. A.*, 1998, **95**, 4270.
138. B. W. Koenig, D. C. Mitchell, S. Konig, S. Grzesiek, B. J. Litman and A. Bax, *J. Biomol. NMR*, 2000, **16**, 121.
139. B. W. Koenig, G. Kontaxis, D. C. Mitchell, J. M. Louis, B. J. Litman and A. Bax, *J. Mol. Biol.*, 2002, **322**, 441.
140. M. L. Mak-Jurkauskas, V. S. Bajaj, M. K. Hornstein, M. Belenky, R. G. Griffin and J. Herzfeld, *Proc. Natl. Acad. Sci. U. S. A.*, 2008, **105**, 883.
141. J. A. Goncalves, S. Ahuja, S. Erfani, M. Eilers and S. O. Smith, *Prog. Nucl. Magn. Reson. Spectrosc.*, 2010, **57**, 159.
142. C. G. Tate and G. F. X. Schertler, *Curr. Opin. Struct. Biol.*, 2009, **19**, 386.
143. W. Liu, M. Eilers, A. B. Patel and S. O. Smith, *J. Mol. Biol.*, 2004, **337**, 713.
144. J. Li, P. C. Edwards, M. Burghammer, C. Villa and G. F. X. Schertler, *J. Mol. Biol.*, 2004, **343**, 1409.
145. S. O. Smith, *Ann. Rev. Biophys.*, 2010, **39**, 309.

CHAPTER 3

Signal Transfer from Receptor to G Protein: The Rhodopsin–Transducin Model

O.G. KISSELEV,[a] J.H. PARK,[b] H.-W. CHOE[c] AND O.P. ERNST*[d]

[a] Saint Louis University, School of Medicine, Departments of Ophthalmology and Biochemistry and Molecular Biology, 1402 South Grand Boulevard, St Louis, MO 63104, USA; [b] Charité – Universitätsmedizin Berlin, Institut für Medizinische Physik und Biophysik, Charitéplatz 1, D-10117 Berlin, Germany; [c] Chonbuk National University, College of Natural Science, Department of Chemistry, 561-756 Chonju, South Korea; [d] Oliver P. Ernst, University of Toronto, Departments of Biochemistry and Molecular Genetics, 1 King's College Circle, Toronto, Ontario, M5S 1A8, Canada

3.1 Introduction

More than 130 years ago, rhodopsin was described as the visual pigment of the retina.[1] (For reviews on rhodopsin see refs. 2–4). The pigment is composed of the apoprotein opsin and the covalently linked chromophore 11-*cis*-retinal. Rhodopsin is highly expressed in rod cells, where it localizes as transmembrane protein to plasma and internal membranes of the rod outer segment, a specific cellular compartment dedicated for transformation of light energy into biochemical reactions. The absorption of light by the chromophore leads to transient conformational changes of the receptor protein, which in turn trigger the G protein mediated enzymatic cascade of reactions (termed phototransduction)

RSC Drug Discovery Series No. 8
G Protein-Coupled Receptors: From Structure to Function
Edited by Jesús Giraldo and Jean-Philippe Pin

Published by the Royal Society of Chemistry, www.rsc.org

that result in neuronal signalling. Besides its function as a visual pigment in the detection of light, rhodopsin is also one of the best studied members of the large superfamily of G protein-coupled receptors (GPCRs) which comprises *ca.* 800 members in humans.[5] In the 1980s, the functional relationship to GPCRs became evident when the sequences of rhodopsin and other GPCRs were identified and compared, and interaction of GPCRs with cognate G proteins were described (for review see ref. 6). Unique members of the GPCR superfamily are involved in a vast variety of specific cellular signal transduction processes including visual, taste and odour perceptions, and sensing a variety of hormones. These receptors share a common seven-transmembrane α-helical structure, which can adopt multiple conformations.[7,8] The binding energy of extracellular chemical ligands is used for stabilization of different receptor conformations. Transmembrane signalling, *i.e.* transmission of the extracellular signal into the cell, thus relies on specific conformational changes of GPCRs. GPCR conformations can vary such that, besides G protein signalling, GPCRs can also transmit signals to β-arrestins—another type of GPCR-interacting signalling molecule in the cell (for review see ref. 9). Elucidation of the crystal structures of rhodopsin and transducin, as well as characterization of fundamental aspects of the photoactivation mechanism and rhodopsin/transducin interaction, helps considerably to better understand transmembrane signalling. Here, we give a short overview on current understanding of signal transfer from rhodopsin to transducin.

3.2 Structural Basis

3.2.1 Conformations of Rhodopsin

3.2.1.1 Ground State of Rhodopsin

As a milestone in rhodopsin and GPCR research, Palczewski and colleagues solved in 2000 the first crystal structure of a GPCR—the structure of rhodopsin in its inactive dark state.[10] Together with models of higher resolution, detailed insight into the protein is provided.[11,12] The structures confirmed the basic seven transmembrane helix (7TM) architecture with cytoplasmic and extracellular loops connecting consecutive TMs (Figure 3.1). The crystal structure analysis revealed an amphipathic cytoplasmic helix (H8) that follows TM7 and runs parallel to the membrane surface. The inverse agonist 11-*cis*-retinal is bound deep inside the protein by a protonated Schiff base to Lys296 on TM7. The *N*-terminus and extracellular loops form a compact extracellular domain, termed the retinal plug, which together with the covalent retinal attachment secure the inverse agonist inside its binding pocket. As described in detail in the original literature,[10–12] stabilizing factors are multiple hydrogen bonding networks within the 7TM core, which involve structurally bound waters and residues from conserved motifs in GPCRs.[13–15] All these arrangements contribute to very efficient stabilization of the inactive rhodopsin ground state, such that thermal activation of a single rhodopsin molecule occurs after 420 years on average.[16]

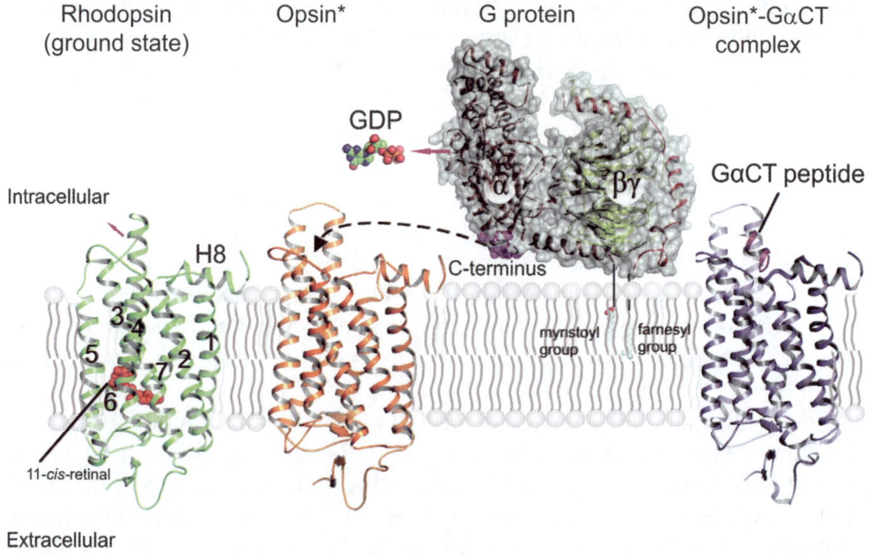

Figure 3.1 Inactive and active conformations of rhodopsin and signal transfer to the G protein. The bound inverse agonist 11-*cis*-retinal constrains rhodopsin in the inactive conformation (left, shown in green, PDB entry 1U19). The structure of the apoprotein opsin represents an active opsin conformation (opsin*, right, shown in blue; PDB entry 3DQB) as identified by a bound synthetic peptide (GαCT peptide, shown in magenta) derived from the membrane-anchored heterotrimeric G protein (Gαβγ). The GαCT peptide represents the key receptor binding site, the *C*-terminus of the Gα subunit. The active opsin* structure was also solved without ligand and GαCT peptide (middle, shown in orange, PDB entry 3CAP) and shows a large cytoplasmic crevice for G-protein binding.

For stabilization of the inactive receptor, residues belonging to the conserved motifs E(D)RY in TM3 and NPxxY$(x)_{5,6}$F in TM7/H8 play key roles (Figure 3.2; reviewed in refs. 14,17). The hydrogen bonding network between Arg135$^{3.50}$ and Glu134$^{3.49}$ of the E(D)RY motif and Glu247$^{6.30}$ on TM6—termed the 'ionic lock' in GPCRs—constrains TM3 and TM6 (superscripts are used for conserved residues in GPCRs and denote the generic numbering of GPCRs according to Ballesteros and Weinstein[18]). Within the NPxxY$(x)_{5,6}$F motif, aromatic side chains of Tyr306$^{7.53}$ and Phe313$^{7.60}$ interact to constrain TM7 and H8.[10,19] As part of a hydrogen-bonding network, Asn302$^{7.49}$ of the NPxxY$(x)_{5,6}$F motif is linked to the most conserved residues in TM1 and TM2, Asn55$^{1.50}$ and Asp83$^{2.50}$, respectively. Additional rhodopsin-specific interactions comprise networks along the retinal binding site and involve electrostatic interactions between the protonated Schiff base and its counterion comprising Glu113 on TM3 and Glu181 in extracellular loop 2 (reviewed in ref. 20).

The crystal structure of opsin, the apoprotein of rhodopsin, represents so far the only structure of a ligand-free GPCR (Figures 3.1 and 3.2).[21] The opsin conformation is characterized by a crevice at the cytoplasmic surface due to a

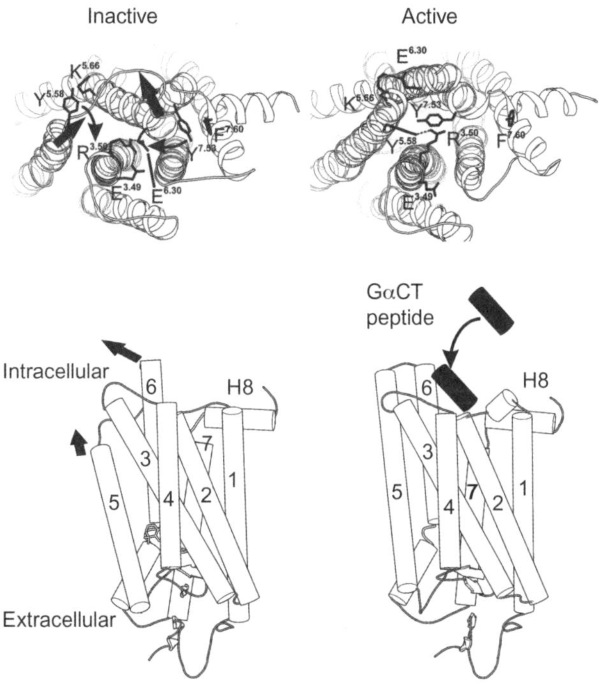

Figure 3.2 Conformational differences between rhodopsin (*left*) and opsin (*right*) viewed from within the membrane (*bottom*) and from the cytoplasmic side (*top*). In rhodopsin, residues of the E(D)RY motif are engaged in a hydrogen-bonding network between TM3 and TM6 (termed 'ionic lock') with the key interaction between Arg135$^{3.50}$ (R$^{3.50}$, superscript denotes Ballesteros Weinstein numbering)[18] and Glu247$^{6.30}$ (E$^{6.30}$). In opsin, a crevice at the cytoplasmic side is observed due to outward motion of TM6 and motion of TM5 towards TM6. Concomitantly, the ionic lock is broken and Arg135$^{3.50}$ and Glu247$^{6.30}$ engage with new interaction partners, the conserved Tyr223$^{5.58}$ (Y$^{5.58}$) and Lys231$^{5.66}$ (K$^{5.66}$), respectively, to form interactions between TM3 and TM5, and TM5 and TM6. The cytoplasmic surface thus opens a crevice into which the GαCT peptide can bind.

rotational tilt of TM6 leading to a 6–7 Å outward movement of TM6 at its cytoplasmic end. The opsin structure further showed an elongation of TM5, which locates 2–3 Å closer to TM6. Due to TM motion, the ionic lock hydrogen-bonding network linking TM3 and TM6 in inactive rhodopsin is broken. In opsin, the released side chains of Arg135$^{3.50}$ and Glu247$^{6.30}$ engage with new interaction partners, the conserved Tyr223$^{5.58}$ and Lys231$^{5.66}$, respectively. As a result new stabilizing interactions between TM3 and TM5, and TM5 and TM6 are characteristic for the opsin structure. In addition, the second conserved motif, the NPxxY(X)$_{5,6}$F motif, shows a rearrangement of the Tyr306$^{7.53}$ and Phe313$^{7.60}$ side chains. The Tyr306$^{7.53}$ side chain rotates away from Phe313$^{7.60}$ towards TM5 and TM6. As a result of all rearrangements, the cytoplasmic surface of opsin shows a more open conformation to which the G protein can bind.[21,22]

From site-directed spin labelling (SDSL) experiments and biochemical crosslinking studies, it was proposed earlier that motion of TM6 relative to TM3 is mandatory for forming the active metarhodopsin II (MII) state capable of catalyzing nucleotide exchange in the G protein.[23,24] The extent of TM6 movement upon formation of MII was recently measured by double electron–electron resonance (DEER) spectroscopy. The measured distance of 5–6 Å is in good agreement with the TM6 position observed in the opsin structure and argues for an active receptor conformation in the opsin crystal.[25] This appraisal was supported by the crystal structure of opsin in complex with a synthetic peptide derived from the Gα *C*-terminus (GαCT peptide; Figure 3.1).[22] The Gα *C*-terminus is known to be the key binding site on transducin.[26,27]

3.2.1.2 Active Conformation of Opsin

Using infrared spectroscopy, an equilibrium between inactive and active conformations of opsin in native membranes was identified.[28] It was found that the active opsin conformation (opsin*) could be stabilized by low pH, in agreement with the protonation dependency of the equilibrium of MII substates (see below). The opsin crystals were grown at low pH, suggesting that the crystal structure represents the active opsin* conformation. An additional line of evidence for the presence of active opsin* in the crystal arises from GαCT peptide binding. The GαCT peptide adopts the same α-helical structure with open reverse turn when bound in the opsin · GαCT peptide crystal and when bound to MII in native membranes as determined by transfer nuclear Overhauser effect NMR spectroscopy.[29,30] From residual dipolar coupling, the orientation of the GαCT peptide was determined for the MII-bound state.[30] The GαCT peptide orientation is in full agreement with the crystal structure of the opsin · GαCT peptide complex.[22] Binding of the GαCT peptide to a receptor conformation with tilted TM6 is also suggested by SDSL data.[31] Another indication for the presence of the opsin* conformation in the opsin and opsin · GαCT peptide crystals arises from the fact that incubation of the crystals with all-*trans*-retinal results in formation of a 380 nm absorbing pigment with retinal Schiff base and high activity towards the G protein.[32] Recent solid-state NMR measurements on rhodopsin and the active MII state now further revealed that the crystal structure of opsin in the region of the ionic lock reflects the active state of the receptor.[33]

3.2.1.3 Receptor States

After light-induced retinal *cis* → *trans* isomerization, rhodopsin can, like other GPCRs, adopt multiple conformations (for reviews see, for example, refs. 8, 14,34). However, the unique light activation process leads to an equilibrium of metarhodopsin states, all containing the covalently bound all-*trans*-retinal agonist. These states develop sequentially on the μs to ms timescale after light activation of rhodopsin, starting with the inactive metarhodopsin I (MI).

$$\mathrm{R} \rightarrow \rightarrow \mathrm{MI} \rightleftarrows \mathrm{MIIa} \rightleftarrows \mathrm{MIIb} \rightleftarrows \mathrm{MIIbH}^{+}$$

Characteristic reaction steps are deprotonation of the retinal Schiff base in the MI→MIIa transition, TM6 motion in the MIIa→MIIb transition, and subsequent proton uptake by Glu134[3.49] to yield $MIIbH^+$.[31,35] It is important to note that the sequential rearrangement of rhodopsin interhelical networks is found both in dodecylmaltoside solution and native membranes.[31,36] The MIIb states with tilted TM6 and open cytoplasmic surface are capable of interacting with the GαCT peptide and are thought to tightly bind G_t.[14,31,37] The crystal structure of a photoactivated rhodopsin state was solved previously.[38] This photoproduct had the 380 nm absorption characteristic for MII and showed only a slight motion of TM6 relative to rhodopsin. According to the above scheme of metarhodopsin equilibria, this structure may represent MIIa. The structure of the photoactivated rhodopsin was also used for modelling of R*·Gt complexes[39] (R* is the active conformation of rhodopsin and ligand-free opsin). Recently a crystal structure of the active photointermediate of rhodospin, MII, was solved for MII alone or in complex with GaCT peptide.[32] The new structures are remarkably similar to the structures of the opsin* and opsin*-GaCT with regard to the overall protein confirmation and movements of helices TM5 and TM6, as well as the opening of the cytoplasmic crevice that collectively represent the active state of the receptor. The structures provide a wealth of new information on the arrangement of all-*trans*-retinal, structural water molecules, and organization of the G-protein binding site.

3.2.2 Conformations of Transducin

3.2.2.1 Conformations of Gα

All members of the G protein superfamily contain a conserved GTP-ase domain responsible for the binding and hydrolysis of GTP (Figure 3.3).[40] This domain also contributes to interfaces of the G protein binding partners such as receptors, βγ-complexes, effectors and regulators of G protein signalling (RGS) proteins. The *N*-terminus of $G_t\alpha$ is post-translationally modified by the heterogeneous mixture of fatty acids ($C<14$)[41] that contribute to the lipid–lipid and lipid–protein interactions.

A distinctive feature of the heterotrimeric G proteins is the helical domain, which occludes the nucleotide-binding site in the interface between the GTP-ase and the helical domains. The stereochemistry of nucleotide binding and hydrolysis for the monomeric and heterotrimeric G proteins has been reviewed.[42,43] Comparison of the GDP- and GTP-bound forms of G_t revealed that only about 14% of the main chain α-carbons change their location. The γ-phosphate induces concerted conformational rearrangements of the so-called switch regions I, II and III at the interface with Gβγ, which are believed to be at the basis of Gβγ dissociation after GTP binding (Figure 3.3b). In the GTP-bound state, the switch regions are held in place by interactions with the γ-phosphate, while in $G_t\alpha$·GDP, the switches show more conformational flexibility.[44,45] In the heterotrimeric form, the helical *N*-terminal domain of $G_t\alpha$ shows clear signs of Gβγ-induced conformational ordering.[46]

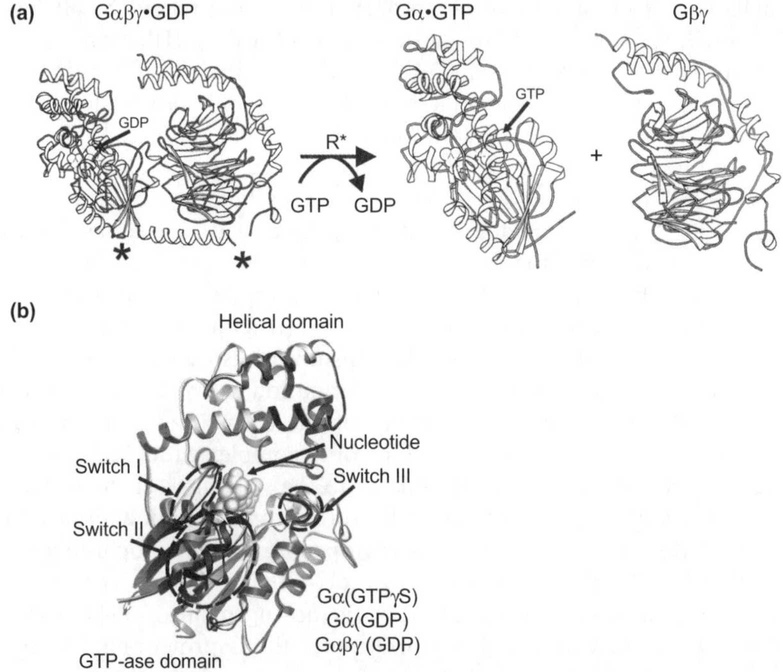

Figure 3.3 Conformations of Gαβγ. (**a**) The inactive Gαβγ · GDP and R*-catalyzed nucleotide exchange leading to the dissociation of the heterotrimer to Gα · GTP and Gβγ-complex. * indicates regions of the G-protein X-ray structure that are unresolved. (**b**) Three conformations of Gα: Gα · GTPγS, Gα · GDP and Gαβγ · GDP are superimposed.

Interestingly, the extreme *C*-terminal region of $G_t\alpha$ is completely disordered in all available high-resolution structures. There is evidence, however, that in the heterotrimeric form the conformation of the *C*-terminus is significantly different compared to $G_t\alpha$ alone. Cys347 preferentially becomes a substrate of pertussis toxin catalyzed ADP ribosylation in the heterotrimeric form.[47] Also, NMR studies showed altered mobility of the $G_t\alpha$ *C*-terminus upon Gtα binding to $G_t\beta\gamma$, which may prepare this region for subsequent interactions with R*.[48]

Perhaps the most important, yet structurally elusive conformation of Gtα is the unique nucleotide-free (Gα-empty) form that is found in the transient R* · $G_t\alpha\beta\gamma$ complex.[49] In the absence of nucleotides this complex remains stable, but currently the high-resolution crystallographic structures of this complex are unavailable and thus it remains an active area of research.[39] Solution NMR studies of a ^{15}N-labeled Gα showed that overall conformation of Gα-empty in the R* · $G_t\alpha\beta\gamma$ complex is similar to the $G_t\alpha$ · GDP-AlF_4/Mg^{2+} state, but shows significantly higher conformational mobility.[50] SDSL studies demonstrated that conformation of Gα-empty is substantially different from the heterotrimeric GDP-bound form of Gα, particularly in changes at helix 4, which is far removed from the Gα/β interface, and in helix 5, as well as in switch I.[45,51]

3.2.2.2 Role of Gα

The G protein α subunits are nucleotide-dependent switches, which control signal relay from the activated heptahelical receptors to the corresponding effector proteins. Four main classes of Gα have been identified: $G\alpha_s$, $G\alpha_i$, $G\alpha_q$ and $G\alpha_{12}$, based on the primary sequence conservation and specificity of receptor and effector coupling. Visual signalling in vertebrate rods is mediated exclusively by $G_t\alpha$, which belongs to the G_i class of the heterotrimeric G proteins. Genetic deletion of $G_t\alpha$ completely eliminates rod-mediated visual responses by uncoupling R* from its effector enzyme cGMP phosphodiesterase.[52]

3.2.2.3 Conformations of Gβγ

The overall molecular architecture of the G protein βγ-complexes is remarkably similar among different isoforms as determined for the visual transducin[53] and for the $G\alpha i_1\beta_1\gamma_2$ complex.[54] The β-subunit forms a toroidal seven-bladed propeller structure with attached *N*-terminal helix. With some overlap, each blade of the propeller corresponds roughly to the repeating WD motifs of about 40 residues in the Gβ primary amino acid sequence. One blade is made of four antiparallel β strands connected by the loops into a symmetrical β-superbarrel. The Gγ subunit, which is dissociable from Gβ only under denaturing conditions, is mostly helical and is found on the side of the β-superbarrel structure opposite to Gα, forming extensive hydrophobic and polar contacts (Figures 3.3 and 3.4). The *N*-terminal helixes of Gβ and Gγ form prominent coiled coil domains present in all Gβγ dimers. Gγ subunits contain the conserved *C*-terminal CAAX sequence, which is the target of post-translational isoprenylation and carboxymethylation.[55] Overall conformation of Gβγ is identical in the free form and in the $G_t\alpha\beta\gamma$ heterotrimer. The Gβγ structure is not entirely static, however, as binding to phosducin[56] and peptides[57] or R*[58,59] introduce distinct conformational changes.

3.2.2.4 Role of $G_t\beta\gamma$

All aspects of $G_t\alpha$ functions, especially its primary interactions with rhodopsin and rhodopsin-catalyzed nucleotide exchange, require the $G_t\beta\gamma$ complex. More than 20 years of research into the molecular role of $G_t\beta\gamma$ in phototransduction have re-asserted, in general, the original reports by Shinozawa *et al.*[60] and Kühn,[61] who identified a G (for GTPase, Gtα) and an H (for Helper, $G_t\beta\gamma$) component of transducin. In fact, the $G_t\beta\gamma$ role was defined to 'provide an allosteric conformational advantage for the PDEase reaction, perhaps by more efficient coupling of rhodopsin, G, and PDEase'.[60]

It is now established that some of the major functions of $G_t\beta\gamma$ are: (i) facilitation of Gtα · GDP interactions with R* and formation of the transient nucleotide-empty R* · G_t complex;[49] (ii) dissociation from $G_t\alpha$ · GTP, allowing $G_t\alpha$ · GTP to activate PDE;[62] (iii) re-association with $G_t\alpha$ · GDP to form an inactive form of transducin, ready for the next activation cycle—$G_t\beta\gamma$ acts as a guanine nucleotide dissociation inhibitor (GDI);[63] and (iv) interactions with phosducin, which regulates the concentration of the free form of $G_t\beta\gamma$, allowing fine modulation of the phototransduction cascade.[64–66] Finally, but not less important: (v) $G_t\beta\gamma$ facilitates membrane attachment of G_t *via* its hydrophobic farnesylated and carboxymethylated *C*-terminus (reviewed in ref. 55). The weaker membrane attachment of the farnesylated $G_t\beta\gamma$, compared to many other $G\beta\gamma$ subtypes that are geranylgeranylated, has been shown to be crucial for proper subcellular translocation of $G_t\beta\gamma$ and the mechanism of light adaptation.[67] Non-retinal $G\beta\gamma$ complexes can also signal through direct interaction with effectors, such as K^+-selective ion channels,[68] but similar functions of $G_t\beta\gamma$ remain poorly understood.

Despite this knowledge, more than a decade of *in vitro* research and recent *in vivo* data have created a growing controversy about the biological function of the G protein $\beta\gamma$-complex. While early 1980s studies implicated $G_t\beta\gamma$ in transducin activation, a number of biochemical experiments explicitly argue that $G_t\alpha$ can be activated by R* without $G_t\beta\gamma$, although with lower efficiency.[69–72] Furthermore, results from a recent physiological study suggested that the $G_t\beta\gamma$-complex does not have any specific role in transducin activation.[73] Thus, a central question of the physiological role of transducin $\beta\gamma$-complex in phototransduction remains undetermined. More studies will be required to determine the role of $G_t\beta\gamma$ in intact retinal rods.

3.2.3 Interaction Sites

Extensive biochemical and biophysical studies over the last decade have contributed to the current map of the receptor-interacting surface of the heterotrimeric G proteins.[27] The potential points of receptor contact have been found an all three subunits of G_t (Figure 3.4). They contribute to a continuous surface that spans approximately 55 Å. At one end of this surface, there are individual domains on $G_t\alpha$, such as the extreme *C*-terminal region and neighbouring loops αN-β1, α2-β4, α3-β5 and α4-β6. On the other distal end of the surface is the extreme *C*-terminal region of the $G_t\gamma$, which is modified post-translationally by farnesylation and carboxymethylation. Due to close spatial proximity, and following the results of the mutational studies, it has been suggested that the $G_t\gamma$ *C*-terminal farnesyl group and $G_t\alpha$ *N*-terminal myristoyl groups form an amphiphilic lipid microdomain that enhances interactions of G_t with lipid membranes and may help to establish initial docking to R*.[74] In the middle of the R*-interacting surface is the *N*-terminal helix of $G_t\alpha$ (αN), which has been shown to interact with R*,[26,75,76] and is also an important site of $G_t\alpha$ interactions with $G_t\beta\gamma$. Furthermore, a peptide corresponding to the third intracellular loop of the α_2-adrenoreceptor crosslinks with the *C*-terminal region of $G\beta$ and the adjacent αN helix of $G\alpha$.[77,78] Taken together, these domains comprise a well-defined surface of the G protein heterotrimer that orient it towards the membrane and the activated receptor.

The *C*-terminal region of $G_t\alpha$ is perhaps the most extensively studied domain involved in signal transfer from R* to G_t. It determines the selectivity of receptor–G protein interactions.[79] It also contains Cys347, which is the site of pertussis toxin catalyzed ADP ribosylation. This covalent modification completely blocks receptor-dependent G protein activation, leading to abnormal

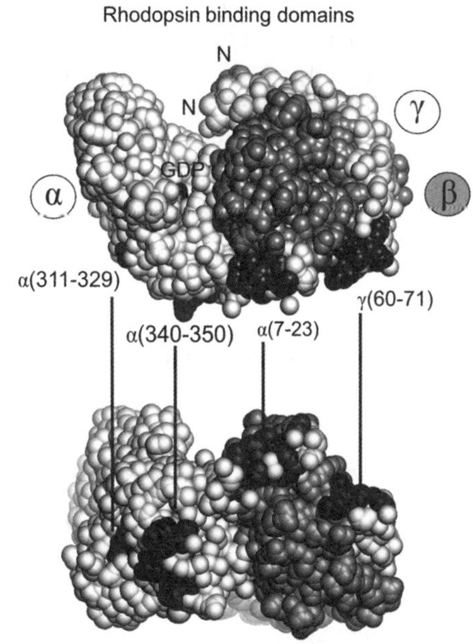

Figure 3.4 The heterotrimeric G protein transducin is composed of the $G_t\alpha$ subunit (white), $G_t\beta$ (grey), and $G_t\gamma$ (white), PDB entry 1AQG. Rhodopsin-binding domains are shown in dark grey.

intracellular concentration of second messengers.[80] A relatively short peptide representing the amino acids 340–350 of $G_t\alpha$ at the *C*-terminal end of the α5 helix, $G_t\alpha$CT, has been used extensively to determine various biological and structural aspects of R*-G_t signal transfer, due to its ability to stabilize the active conformation of MII, and to inhibit R*-Gt interactions.[22,26,29,81–83] Upon binding to R*, the *C*-terminal end of $G_t\alpha$ undergoes dramatic structuring by an induced fit mechanism from a disordered conformation to the α-helix terminated by a reverse turn.[29,30] This R*-induced rearrangement is believed to be transmitted *via* the long α5-helix to the back of the nucleotide binding site situated 30 Å away from the R*-G_t interface.[27]

The extreme farnesylated *C*-terminal region of $G_t\gamma$, amino acids 60–71, binds to R* as well, but the mechanism of its involvement into R*-G_t complex formation or the nucleotide exchange of $G_t\alpha$ is not as well understood. A conformational switch in $G_t\beta\gamma$ was proposed to establish effective R*-G_t coupling.[58,59] Mutations in the *C*-terminal region of $G_t\gamma$ prevent efficient R*-G_t coupling.[84] Furthermore, prenylated synthetic peptides have been shown to stabilize the meta II state of rhodopsin.[85–87] They also interfere with R* catalyzed activation of G_t,[70] as well as in the muscarinic receptor controlled pathway.[88] Photoaffinity labelling experiment identified R* as one of the major interacting partners of $G_t\gamma$ *C*-terminus.[89] More specifically, a recent study determined that the C2 intracellular loop and a part of TM4 of rhodopsin is a specific site of $G_t\gamma$(60–71)farnesyl interaction.[90]

Even brief examination of the R*-interacting surface of G_t identifies a major distance paradox: how can a molecule of R*, which is only about 45 Å in diameter, leave a 55 Å long footprint on G_t? And how does R* catalyze the nucleotide exchange at a site that is 30 Å away? The solution of this major allosteric puzzle may bring us closer to understanding the universal mechanism of interactions and signal transfer from activated GPCRs to heterotrimeric G proteins. Several mechanistic models have been proposed to explain the described experimental results (Figure 3.5).

3.3 Models of Receptor–G Protein Interaction

3.3.1 Lever-arm Model

In the last decade, several hypothetical models have been proposed to explain receptor-induced G protein activation. The lever-arm model (Figure 3.5a) was originally proposed by Bourne's group, who based their analysis on the observation that $G_t\beta\gamma$ occupies the space of a nucleotide exchange factor for the $G_t\alpha$-subunit, similar to the position of a nucleotide exchange factor EF-Ts, crystallized in complex with the nucleotide-empty elongation factor Tu (EF-Tu).[91] In a crystal structure of inactive G_t heterotrimer, $G_t\beta\gamma$ contacts with flexible $G_t\alpha$ switch I and switch II loops, which occlude the potential exit route for GDP, but change conformation depending on the type of the nucleotide bound to Gtα (Figure 3.3b). This provides a tangible opportunity for R* to control the nucleotide binding site *via* $G_t\beta\gamma$. Taken together this analysis lead to the proposal that activating signal from R* propagates *via* well-defined sites on $G_t\alpha$ such as the *C*-terminal domain, the α5-helix, as well as *via* the *C*-terminal domain of $G_t\gamma$, by using the $G_t\alpha$ *N*-terminal helix as a lever arm (Figure 3.5a and arrows 1 and 2 in Figure 3.5e). The end result is the pulling motion by $G_t\beta\gamma$ on the switch I and II regions of $G_t\alpha$ and subsequent release of GDP (Figure 3.5a).

3.3.2 Gear-shift Model

The gear-shift model proposed by Chabre and Cherfils shares common features with the lever-arm model, but differs substantially in explaining the mechanism of the nucleotide release on $G_t\alpha$ (Figure 3.5b).[92] Three putative gear-shift steps are a hallmark of this model:

(1) G_t docks with R* using the *C*-terminal domains of $G_t\gamma$ and the *C*- and *N*-terminal domains of $G_t\alpha$.
(2) A hinge motion of the $G_t\alpha$ *N*-terminus pushes $G_t\beta\gamma$ into close contact with the $G_t\alpha$ switch II region to help stabilize the transient nucleotide-empty state.
(3) The small hinge movement at the $G_t\alpha$ *N*-terminus/$G_t\beta\gamma$ interface translates through the rigid-body motion of $G_t\beta\gamma$ into the large pushing movement of $G_t\gamma$ *N*-terminus to displace the helical domain of $G_t\alpha$.

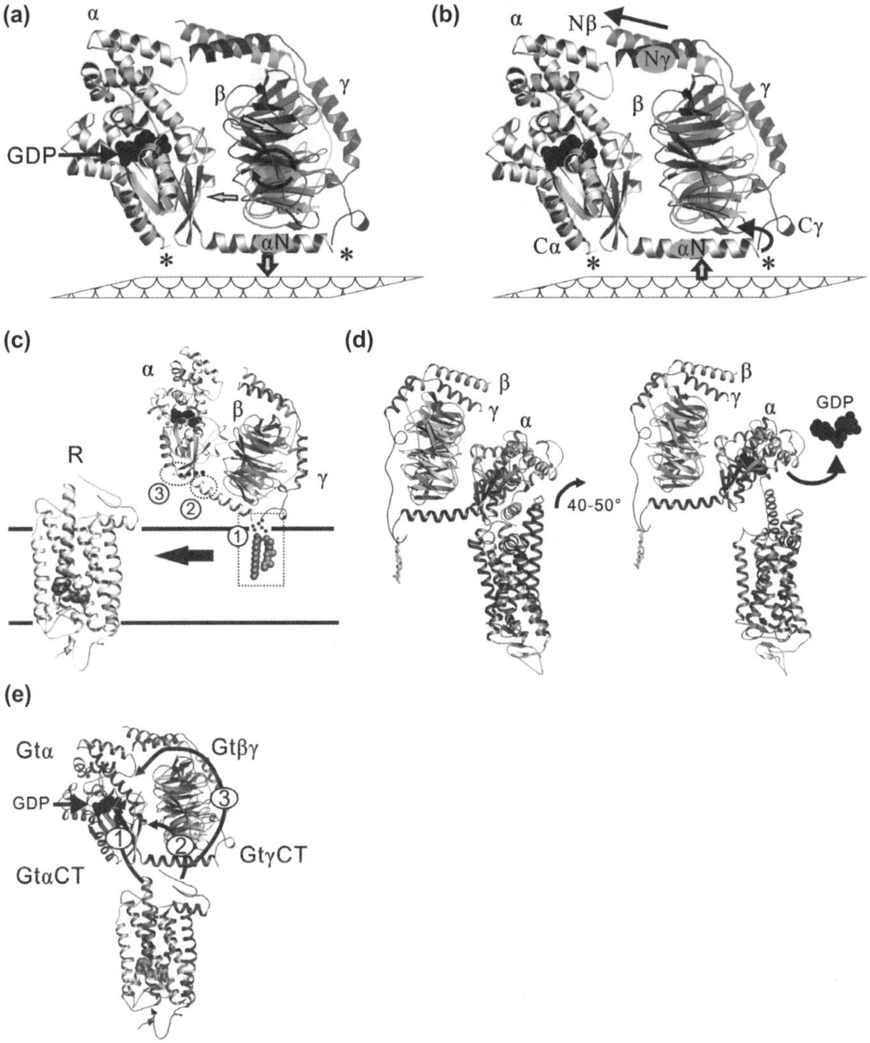

Figure 3.5 Models of receptor–G-protein activation. (**a**) Lever-arm model.[120] (**b**) Gear shift model.[92] In (**a**) and (**b**) * indicate R*-interacting regions that are unresolved in available X-ray structures. (**c**) Sequential fit model.[74,96,97] (**d**) Helix switch model.[37] (**e**) Proposed signal propagation from R* to the nucleotide-binding site of $G_t\alpha$. See text for details.

The resulting opening of the cleft between the GTPase and the helical domains is proposed to trigger GDP release. In this model $G_t\gamma$ is viewed as the main gear-shift for coupling R* to $G_t\alpha$ (arrow 3 on Figure 3.5e).

In major support for the notion that $G_t\beta\gamma$ may regulate nucleotide release *via* interactions with switch II, mutations of the $G_t\beta$ residues at the $G_t\alpha$–$G_t\beta$ interface have been shown to compromise G protein activation.[93] Recent SDSL studies are also consistent with signal propagation *via* $G_t\beta\gamma$ and switch II of $G_t\alpha$.[45,51] A small peptide identified by phage display has been shown to increase spontaneous GDP release *via* direct displacement of switch II.[94] Conformation of another switch region, switch I, also appears to be directly connected to the nucleotide release, at least in the case of Gα_q. Impairment of switch I flexibility by binding of a small-molecule inhibitor has been shown to impair GDP/GTP exchange.[95]

While both the lever-arm and the gear-shift models imply a rigid body rotation of G$_t\beta\gamma$ along the axis parallel to the lipid membrane during interactions with R*, an intriguing hypothesis[27] based on analysis of the SDSL results[45,46] suggests that the axis of rotation may in fact be perpendicular to the plane of the membrane.

3.3.3 Sequential Fit Model

An alternative mechanism that explains an oversized footprint on G$_t$ from R* is proposed in the sequential fit model.[74,96,97] This model postulates temporarily distinct steps involving interactions of individual domains comprising R*-binding surface of G$_t$ during R*-G$_t$ complex formation, resulting in nucleotide exchange on G$_t\alpha$ (Figure 3.5c). The initial docking of G$_t$ to R* is proposed to proceed *via* an amphiphilic microdomain composed of the farnesylated *C*-terminus of G$_t\gamma$, acylated *N*-terminus of G$_t\alpha$ and adjacent terminal residues. This interaction ensures close proximity of G$_t$ to the intracellular loops of R* and is likely to be a key rate limiting step in the G$_t$ activation reaction.[98] Proper docking allows R* to bind the αN-β1 loop of G$_t\alpha$ and release of the G$_t\alpha$ *C*-terminal domain for binding within a specific cavity on R*.[22] Induced-fit rearrangement of G$_t\alpha$(340-350) into a helical conformation stabilized by hydrophobic interactions of residues within the R* binding cavity is translated *via* the α5-helix and the α5-β6 loop to the back of the nucleotide binding site. This signal propagation may cause enough destabilization of the nucleotide-binding site to trigger nucleotide release.

In support of this model, site-directed mutations in the α5 helix of Gαi$_1$ have been shown to dramatically increase the receptor-independent basal rate of GDP release, such as for the mutant Gαi$_1$-T329A.[99] Interestingly, the crystal structure of the Gαi$_1$-T329A · GDP mutant does not reveal any changes in the conformation of the α5 helix, but shows a meaningful conformational shift of switch I and multiple changes in the catalytic site involved in the coordination of Mg^{2+}. These changes may be the major contributors to the GDP release *via* the α5 route. For the R*-catalyzed nucleotide exchange it may, however, still require additional action of G$_t\beta\gamma$ on switch I and switch II to cooperatively induce the release of GDP.

3.3.4 Helix Shift Model

Kinetic analysis of transducin (G$_t\alpha\beta\gamma$) activation showed that an intermediary R* · G$_t\alpha\beta\gamma$ · GDP complex is formed which precedes GDP release and formation of the nucleotide-free R* · G$_t$ complex (Scheme 3.1).[37,98] From the Ops* · GαCT peptide structure a model of the nucleotide-free R* · G$_t$ complex was proposed (Figure 3.5d, *right*).[22] To orient G$_t$ relative to R* and the membrane and fulfil the Ops* · GαCT peptide binding constraints revealed by

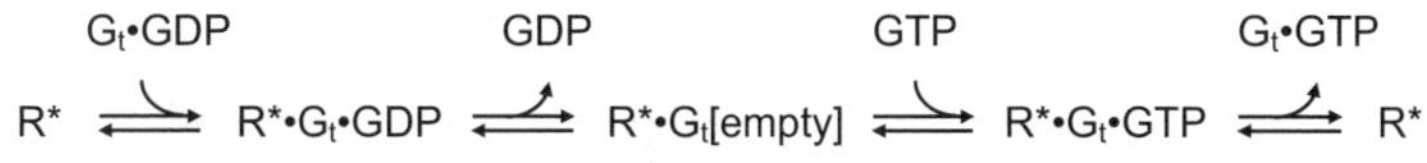

Scheme 3.1 Proposed reaction steps of R*/G_t-interaction.[37,98]

the crystal structure, it was assumed that the α5 helix moves relative to the remainder of G_t. Based on site-directed mutagenesis studies, movement of α5 upon R* binding was proposed to induce GDP release,[99] and SDSL experiments provided direct structural evidence for a rigid body rotational/translational motion of the α5 helix upon R* binding.[51,122]

To obtain insight into the R* · $G_t\alpha\beta\gamma$ · GDP complex, molecular modelling was performed which showed how α5 can be docked to the open binding site of R* represented by the Ops* · GαCT peptide structure. Two modes of interaction were found. One of them closely matches the position of the GαCT peptide in the crystal structure and reproduces the hydrogen bonding network at the C cap of the peptide and conserved E(D)RY and NPxxY(x)$_{5,6}$F regions of the receptor. In the alternative fit, the α5 helix is tilted and rotated to run parallel to the membrane and to bind to the inner surface of the TM5–TM6 helix pair. This was interpreted as a 'helix switch' from an initial transient interaction as in R* · $G_t\alpha\beta\gamma$ · GDP to the final stable interaction within the open crevice of the active receptor in nucleotide-free R* · $G_t\alpha\beta\gamma$. Additional conformational changes in the G protein accompany the motion of α5 and additional interactions between the receptor and G protein are required to provide a firm basis for the α5 switch.[27,37] The helix switch in Figure 3.5d illustrates how α5 might act as a transmission rod to propagate the conformational change from the receptor-G protein interface to the nucleotide binding site.

3.3.5 Receptor Oligomers

The gear-shift model and the sequential fit model assume interactions of G_t with R* monomer. A competing proposal, interaction of a single molecule of G_t with a rhodopsin dimer (Figure 3.6) has been in the centre of considerable debate from the moment of rhodopsin oligomer imaging by atomic force microscopy (AFM).[100] Additional data in support of this model have been recently reviewed.[39,101,102] Perhaps the most important feature of this model is the almost perfect match in size between R*R* or R*R and G_t, which explains the paradox between the large footprint on all three subunits of G_t from a relatively compact cytoplasmic surface of R*.

Recently, the debate has shifted to demonstrating the physiological significance of rhodopsin dimerization. Detergent solubilization conditions that favour rhodopsin oligomers were interpreted to favour faster rates of G_t activation.[103] This observation is consistent with earlier as well as more recent reports from direct binding experiments showing that R*-G_t interactions display significant positive cooperativity, with binding curves modelled with Hill coefficient 2.[104,105] Although the membrane binding step was argued to contribute to this effect,[98] both rhodopsin and transducin oligomerization,[106–108]

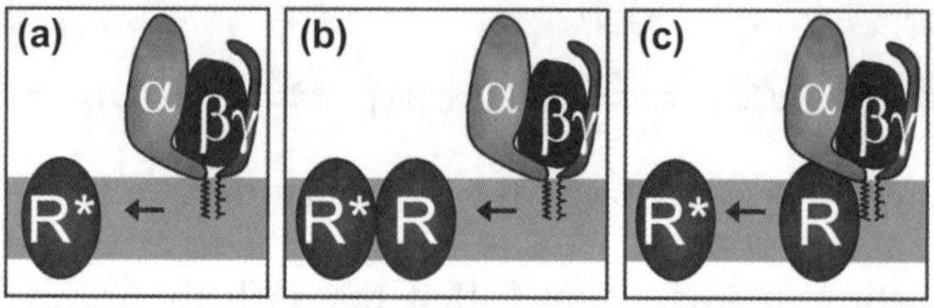

Figure 3.6 Receptor arrangements for signal transfer to the G protein. Potential candidates for the docking step: R* monomer and Gαβγ, preformed R*/R dimer and Gαβγ, and R* and preformed R/Gαβγ. The relative orientation between Gαβγ and membrane/GPCR is suggested by the electrostatic potentials of the corresponding G protein surface.[121] Figure adapted from ref. 74.

induced by the initial binding, can contribute as well, which may ultimately play a role in the shape and speed of the visual response. The general nature of the above conclusion remains controversial, however, because other GPCRs, such as the leukotriene BLT2 receptor has been shown to activate G_{i2} more efficiently as a monomer, compared to its parallel dimer form,[109] confirming similar result obtained earlier with neurotensin receptor.[110]

Computer modelling studies, which integrate the majority of the available biochemical and biophysical data on R*-G_t interactions also favour the R*R*-G_t model,[111,112] while other modelling attempts argue that monomeric rhodopsin contains all necessary determinants for transducin recognition.[113] The latter argument is supported strongly by the recent biochemical data obtained under controlled *in vitro* reconstitution conditions showing that a monomeric form of rhodopsin is fully capable of activating G_t.[114–116] Experiments with monomeric rhodopsin in dodecylmaltoside solution indeed showed that monomeric R* forms a 1 : 1 complex with G_t and activates G_t at the diffusion limit.[117] It is quite possible that both sides of the debate are correct and that under certain conditions rhodopsin can exist in the dimeric or higher oligomeric form,[118] or may form dimers temporarily as suggested for the muscarinic receptor.[119] In any case, it is the single photoactivated receptor molecule that is necessary and sufficient for signal transfer to G_t and effective nucleotide exchange.

In summary, in all current models of R*-G_t complex, $G_t\beta\gamma$ occupies a central position as a potential conduit of the activating conformational changes from metarhodopsin II to $G_t\alpha$, but the mystery of the unified mechanism of the heterotrimeric G protein activation remains to be resolved.

3.4 Open Questions

The universal mechanism of receptor-catalyzed nucleotide exchange by the heterotrimeric G proteins remains one of the most intriguing allosteric puzzles of sensory signalling. Several open questions, as summarized below, need to be resolved in order to understand signal transfer from R* to the G protein:

1. The high-resolution structure of the receptor–G protein complex is undoubtedly among the most awaited pieces of this puzzle which should

resolve crucial mechanistic details of interacting protein domains and conformations. This is a formidable task, however, which may be tempered by the inherent conformational mobility within the complex. More manageable tasks may include high-resolution structural studies of incomplete complexes, or structures of important intermediates of R* and heterotrimeric G proteins stabilized artificially by site-directed mutations or other means.

2. Several competing models of signal transfer from R* to the G protein require careful analysis by mutational, biochemical and biophysical approaches. Signal propagation routes and accompanying conformations changes need to be elucidated.
3. One of the central unresolved issues of the G protein biology is the mechanism of nucleotide release and understanding of the differences between nucleotide release in monomeric *vs.* heterotrimeric G proteins. The biological role of the G protein βγ-complex in this context is especially intriguing. Whether it plays an active role in GDP release is still vigorously debated.

Abbreviations

GPCR = G protein-coupled receptor
$G\alpha\beta\gamma$, G, = heterotrimeric G protein
G_t, $G_t\alpha\beta\gamma$ = transducin
$G_t\alpha$ = α-subunit of G_t
$G_t\beta\gamma$ = βγ-subunit of G_t
$G_t\alpha$(340–350), $G_t\alpha$-CT, GαCT = *C*-terminal peptide of the $G_t\alpha$-subunit, amino acids 340–350
R* = active conformation of rhodopsin and ligand-free opsin
TM = transmembrane helix

Acknowledgements

The work has been supported by the Deutsche Forschungsgemeinschaft (Sfb449 and Sfb740) and the Canada Excellence Research Chair program (O.P.E.), the National Research Foundation of Korea (Basic Science Research Programme funded by the Ministry of Education, Science and Technology; 2010-0002738) and CBNU funds for overseas research 2009 (H.-W.C.), as well as by NIH grants GM063203, EY018107 and by Career Development Awards from Research to Prevent Blindness (RPB) (O.G.K.). We are grateful to Wayne Hubbell and Ned Van Eps (Jules Stein Eye Institute, UCLA, CA) for helpful comments.

References

1. W. Kühne, *Vision Res.*, 1977, **17**, 1269.
2. K. Palczewski, *Annu. Rev. Biochem.*, 2006, **75**, 743.

3. O. P. Ernst, K. P. Hofmann and K. Palczewski, in *Photoreceptors and Light Signalling*, ed. A. Batschauer, Royal Society of Chemistry, Cambridge, 2003, vol. 3, p. 77–123.
4. S. O. Smith, *Annu. Rev. Biophys.*, 2010, **39**, 309.
5. M. C. Lagerstrom and H. B. Schioth, *Nat. Rev. Drug Discov.*, 2008, **7**, 339.
6. R. J. Lefkowitz, *Trends Pharmacol. Sci.*, 2004, **25**, 413.
7. B. K. Kobilka and X. Deupi, *Trends Pharmacol. Sci.*, 2007, **28**, 397.
8. X. Deupi and B. K. Kobilka, *Physiology (Bethesda)*, 2010, **25**, 293.
9. J. D. Violin and R. J. Lefkowitz, *Trends Pharmacol. Sci.*, 2007, **28**, 416.
10. K. Palczewski, T. Kumasaka, T. Hori, C. A. Behnke, H. Motoshima, B. A. Fox, I. Le Trong, D. C. Teller, T. Okada, R. E. Stenkamp, M. Yamamoto and M. Miyano, *Science*, 2000, **289**, 739.
11. T. Okada, M. Sugihara, A. N. Bondar, M. Elstner, P. Entel and V. Buss, *J. Mol. Biol.*, 2004, **342**, 571.
12. J. Li, P. C. Edwards, M. Burghammer, C. Villa and G. F. Schertler, *J. Mol. Biol.*, 2004, **343**, 1409.
13. T. E. Angel, M. R. Chance and K. Palczewski, *Proc. Natl. Acad. Sci. U. S. A.*, 2009, **106**, 8555.
14. K. P. Hofmann, P. Scheerer, P. W. Hildebrand, H.-W. Choe, J. H. Park, M. Heck and O. P. Ernst, *Trends Biochem. Sci.*, 2009, **34**, 540.
15. R. Nygaard, L. Valentin-Hansen, J. Mokrosinski, T. M. Frimurer and T. W. Schwartz, *J. Biol. Chem.*, 2010, **285**, 19625.
16. D. Baylor, *Proc. Natl. Acad. Sci. U. S. A.*, 1996, **93**, 560.
17. R. Nygaard, T. M. Frimurer, B. Holst, M. M. Rosenkilde and T. W. Schwartz, *Trends Pharmacol. Sci.*, 2009, **30**, 249.
18. J. A. Ballesteros and H. Weinstein, *Methods Neurosci.*, 1995, **25**, 366.
19. O. Fritze, S. Filipek, V. Kuksa, K. Palczewski, K. P. Hofmann and O. P. Ernst, *Proc. Natl. Acad. Sci. U. S. A.*, 2003, **100**, 2290.
20. R. Vogel, T. P. Sakmar, M. Sheves and F. Siebert, *Photochem. Photobiol.*, 2007, **83**, 286.
21. J. H. Park, P. Scheerer, K. P. Hofmann, H.-W. Choe and O. P. Ernst, *Nature*, 2008, **454**, 183.
22. P. Scheerer, J. H. Park, P. W. Hildebrand, Y. J. Kim, N. Krauss, H.-W. Choe, K. P. Hofmann and O. P. Ernst, *Nature*, 2008, **455**, 497.
23. D. L. Farrens, C. Altenbach, K. Yang, W. L. Hubbell and H. G. Khorana, *Science*, 1996, **274**, 768.
24. S. P. Sheikh, T. A. Zvyaga, O. Lichtarge, T. P. Sakmar and H. R. Bourne, *Nature*, 1996, **383**, 347.
25. C. Altenbach, A. K. Kusnetzow, O. P. Ernst, K. P. Hofmann and W. L. Hubbell, *Proc. Natl. Acad. Sci. U. S. A.*, 2008, **105**, 7439.
26. H. E. Hamm, D. Deretic, A. Arendt, P. A. Hargrave, B. König and K. P. Hofmann, *Science*, 1988, **241**, 832.
27. W. M. Oldham and H. E. Hamm, *Nat. Rev. Mol. Cell Biol.*, 2008, **9**, 60.
28. R. Vogel and F. Siebert, *J. Biol. Chem.*, 2001, **276**, 38487.
29. O. G. Kisselev, J. Kao, J. W. Ponder, Y. C. Fann, N. Gautam and G. R. Marshall, *Proc. Natl. Acad. Sci. U. S. A.*, 1998, **95**, 4270.

30. B. W. Koenig, G. Kontaxis, D. C. Mitchell, J. M. Louis, B. J. Litman and A. Bax, *J. Mol. Biol.*, 2002, **322**, 441.
31. B. Knierim, K. P. Hofmann, O. P. Ernst and W. L. Hubbell, *Proc. Natl. Acad. Sci. U. S. A.*, 2007, **104**, 20290.
32. H.-W. Choe, Y. J. Kim, J. H. Park, T. Morizumi, E. F. Pai, N. Krauss, K. P. Hofmann, P. Scheerer and O. P. Ernst, *Nature*, 2011, **471**, 651.
33. J. A. Goncalves, K. South, S. Ahuja, E. Zaitseva, C. A. Opefi, M. Eilers, R. Vogel, P. J. Reeves and S. O. Smith, *Proc. Natl. Acad. Sci. U. S. A.*, 2010, **107**, 19861.
34. T. Okada, O. P. Ernst, K. Palczewski and K. P. Hofmann, *Trends Biochem. Sci.*, 2001, **26**, 318.
35. M. Mahalingam, K. Martinez-Mayorga, M. F. Brown and R. Vogel, *Proc. Natl. Acad. Sci. U. S. A.*, 2008, **105**, 17795.
36. E. Zaitseva, M. F. Brown and R. Vogel, *J. Am. Chem. Soc.*, 2010, **132**, 4815.
37. P. Scheerer, M. Heck, A. Goede, J. H. Park, H.-W. Choe, O. P. Ernst, K. P. Hofmann and P. W. Hildebrand, *Proc. Natl. Acad. Sci. U. S. A.*, 2009, **106**, 10660.
38. D. Salom, D. T. Lodowski, R. E. Stenkamp, I. Le Trong, M. Golczak, B. Jastrzebska, T. Harris, J. A. Ballesteros and K. Palczewski, *Proc. Natl. Acad. Sci. U. S. A.*, 2006, **103**, 16123.
39. B. Jastrzebska, Y. Tsybovsky and K. Palczewski, *Biochem. J.*, 2010, **428**, 1.
40. S. R. Sprang, *Annu. Rev. Biochem.*, 1997, **66**, 639.
41. K. Kokame, Y. Fukada, T. Yoshizawa, T. Takao and Y. Shimonishi, *Nature*, 1992, **359**, 749.
42. C. Kleuss, A. S. Raw, E. Lee, S. R. Sprang and A. G. Gilman, *Proc. Natl. Acad. Sci. U. S. A.*, 1994, **91**, 9828.
43. M. Kjeldgaard, J. Nyborg and B. F. Clark, *FASEB J.*, 1996, **10**, 1347.
44. A. Bohm, R. Gaudet and P. B. Sigler, *Curr. Opin. Biotechnol.*, 1997, **8**, 480.
45. N. Van Eps, W. M. Oldham, H. E. Hamm and W. L. Hubbell, *Proc. Natl. Acad. Sci. U. S. A.*, 2006, **103**, 16194.
46. M. Medkova, A. M. Preininger, N. J. Yu, W. L. Hubbell and H. E. Hamm, *Biochemistry*, 2002, **41**, 9962.
47. P. A. Watkins, D. L. Burns, Y. Kanaho, T. Y. Liu, E. L. Hewlett and J. Moss, *J. Biol. Chem.*, 1985, **260**, 13478.
48. N. G. Abdulaev, T. Ngo, C. Zhang, A. Dinh, D. M. Brabazon, K. D. Ridge and J. P. Marino, *J. Biol. Chem.*, 2005, **280**, 38071.
49. F. Bornancin, C. Pfister and M. Chabre, *Eur J. Biochem.*, 1989, **184**, 687.
50. K. D. Ridge, N. G. Abdulaev, C. Zhang, T. Ngo, D. M. Brabazon and J. P. Marino, *J. Biol. Chem.*, 2006, **281**, 7635.
51. W. M. Oldham, N. Van Eps, A. M. Preininger, W. L. Hubbell and H. E. Hamm, *Proc. Natl. Acad. Sci. U. S. A.*, 2007, **104**, 7927.
52. P. D. Calvert, N. V. Krasnoperova, A. L. Lyubarsky, T. Isayama, M. Nicolo, B. Kosaras, G. Wong, K. S. Gannon, R. F. Margolskee, R. L. Sidman, E. N. Pugh, Jr., C. L. Makino and J. Lem, *Proc. Natl. Acad. Sci. U. S. A.*, 2000, **97**, 13913.

53. J. Sondek, A. Bohm, D. G. Lambright, H. E. Hamm and P. B. Sigler, *Nature*, 1996, **379**, 369.
54. M. A. Wall, D. E. Coleman, E. Lee, J. A. Iniguez-Lluhi, B. A. Posner, A. G. Gilman and S. R. Sprang, *Cell*, 1995, **83**, 1047.
55. P. B. Wedegaertner, P. T. Wilson and H. R. Bourne, *J. Biol. Chem.*, 1995, **270**, 503.
56. A. Loew, Y. K. Ho, T. Blundell and B. Bax, *Structure*, 1998, **6**, 1007.
57. A. V. Smrcka, N. Kichik, T. Tarrago, M. Burroughs, M. S. Park, N. K. Itoga, H. A. Stern, B. M. Willardson and E. Giralt, *Proc. Natl. Acad. Sci. U. S. A.*, 2010, **107**, 639.
58. O. Kisselev, A. Pronin, M. Ermolaeva and N. Gautam, *Proc. Natl. Acad. Sci. U. S. A.*, 1995, **92**, 9102.
59. O. G. Kisselev and M. A. Downs, *Structure*, 2003, **11**, 367.
60. T. Shinozawa, S. Uchida, E. Martin, D. Cafiso, W. Hubbell and M. Bitensky, *Proc. Natl. Acad. Sci. U. S. A.*, 1980, **77**, 1408.
61. H. Kuhn, *Nature*, 1980, **283**, 587.
62. T. G. Wensel and L. Stryer, *Proteins*, 1986, **1**, 90.
63. J. Bockaert, P. Deterre, C. Pfister, G. Guillon and M. Chabre, *EMBO J.*, 1985, **4**, 1413.
64. R. Gaudet, J. R. Savage, J. N. McLaughlin, B. M. Willardson and P. B. Sigler, *Mol. Cell*, 1999, **3**, 649.
65. C. D. Thulin, J. R. Savage, J. N. McLaughlin, S. M. Truscott, W. M. Old, N. G. Ahn, K. A. Resing, H. E. Hamm, M. W. Bitensky and B. M. Willardson, *J. Biol. Chem.*, 2001, **276**, 23805.
66. Y. K. Ho, T. D. Ting and R. H. Lee, *Methods Enzymol.*, 2002, **344**, 126.
67. H. Kassai, A. Aiba, K. Nakao, K. Nakamura, M. Katsuki, W. H. Xiong, K. W. Yau, H. Imai, Y. Shichida, Y. Satomi, T. Takao, T. Okano and Y. Fukada, *Neuron*, 2005, **47**, 529.
68. D. E. Clapham and E. J. Neer, *Annu. Rev. Pharmacol. Toxicol.*, 1997, **37**, 167.
69. R. Herrmann, M. Heck, P. Henklein, K. P. Hofmann and O. P. Ernst, *J. Biol. Chem.*, 2006, **281**, 30234.
70. O. G. Kisselev, C. K. Meyer, M. Heck, O. P. Ernst and K. P. Hofmann, *Proc. Natl. Acad. Sci. U. S. A.*, 1999, **96**, 4898.
71. W. J. Phillips, S. C. Wong and R. A. Cerione, *J. Biol. Chem.*, 1992, **267**, 17040.
72. B. K. Fung, *J. Biol. Chem.*, 1983, **258**, 10495.
73. E. S. Lobanova, S. Finkelstein, R. Herrmann, Y. M. Chen, C. Kessler, N. A. Michaud, L. H. Trieu, K. J. Strissel, M. E. Burns and V. Y. Arshavsky, *J. Neurosci.*, 2008, **28**, 3510.
74. R. Herrmann, M. Heck, P. Henklein, K. P. Hofmann and O. P. Ernst, *J. Biol. Chem.*, 2006, **281**, 30234.
75. R. Onrust, P. Herzmark, P. Chi, P. D. Garcia, O. Lichtarge, C. Kingsley and H. R. Bourne, *Science*, 1997, **275**, 381.
76. Y. Itoh, K. Cai and H. G. Khorana, *Proc. Natl. Acad. Sci. U. S. A.*, 2001, **98**, 4883.
77. J. M. Taylor, G. G. Jacob-Mosier, R. G. Lawton, A. E. Remmers and R. R. Neubig, *J. Biol. Chem.*, 1994, **269**, 27618.

78. J. M. Taylor, G. G. Jacob-Mosier, R. G. Lawton, M. VanDort and R. R. Neubig, *J. Biol. Chem.*, 1996, **271**, 3336.
79. B. R. Conklin and H. R. Bourne, *Cell*, 1993, **73**, 631.
80. R. E. West, Jr., J. Moss, M. Vaughan, T. Liu and T. Y. Liu, *J. Biol. Chem.*, 1985, **260**, 14428.
81. F. Bartl, E. Ritter and K. P. Hofmann, *FEBS Lett.*, 2000, **473**, 259.
82. R. Vogel, S. Martell, M. Mahalingam, M. Engelhard and F. Siebert, *J. Mol. Biol.*, 2007, **366**, 1580.
83. N. Van Eps, L. L. Anderson, O. G. Kisselev, T. J. Baranski, W. L. Hubbell and G. R. Marshall, *Biochemistry*, 2010, **49**, 6877.
84. O. G. Kisselev and M. A. Downs, *Biochemistry*, 2006, **45**, 9386.
85. O. G. Kisselev, M. V. Ermolaeva and N. Gautam, *J. Biol. Chem.*, 1994, **269**, 21399.
86. O. G. Kisselev, M. V. Ermolaeva and N. Gautam, *J. Biol. Chem.*, 1995, **270**, 25356.
87. Y. Fukada, T. Matsuda, K. Kokame, T. Takao, Y. Shimonishi, T. Akino and T. Yoshizawa, *J. Biol. Chem.*, 1994, **269**, 5163.
88. I. Azpiazu, H. Cruzblanca, P. Li, M. Linder, M. Zhuo and N. Gautam, *J. Biol. Chem.*, 1999, **274**, 35305.
89. M. Katadae, K. Hagiwara, A. Wada, M. Ito, M. Umeda, P. J. Casey and Y. Fukada, *Biochemistry*, 2008, **47**, 8424.
90. Y. Chen, Y. Wu, P. Henklein, X. Li, K. P. Hofmann, K. Nakanishi and O. P. Ernst, *Chem. Eur. J.*, 2010, **16**, 7389.
91. Y. Wang, Y. Jiang, M. Meyering-Voss, M. Sprinzl and P. B. Sigler, *Nat. Struct. Biol.*, 1997, **4**, 650.
92. J. Cherfils and M. Chabre, *Trends Biochem. Sci.*, 2003, **28**, 13.
93. C. E. Ford, N. P. Skiba, H. Bae, Y. Daaka, E. Reuveny, L. R. Shekter, R. Rosal, G. Weng, C. S. Yang, R. Iyengar, R. J. Miller, L. Y. Jan, R. J. Lefkowitz and H. E. Hamm, *Science*, 1998, **280**, 1271.
94. C. A. Johnston, F. S. Willard, M. R. Jezyk, Z. Fredericks, E. T. Bodor, M. B. Jones, R. Blaesius, V. J. Watts, T. K. Harden, J. Sondek, J. K. Ramer and D. P. Siderovski, *Structure*, 2005, **13**, 1069.
95. A. Nishimura, K. Kitano, J. Takasaki, M. Taniguchi, N. Mizuno, K. Tago, T. Hakoshima and H. Itoh, *Proc. Natl. Acad. Sci. U. S. A.*, 2010, **107**, 13666.
96. O. G. Kisselev, C. K. Meyer, M. Heck, O. P. Ernst and K. P. Hofmann, *Proc. Natl. Acad. Sci. U. S. A.*, 1999, **96**, 4898.
97. R. Herrmann, M. Heck, P. Henklein, P. Henklein, C. Kleuss, K. P. Hofmann and O. P. Ernst, *J. Biol. Chem.*, 2004, **279**, 24283.
98. M. Heck and K. P. Hofmann, *J. Biol. Chem.*, 2001, **276**, 10000.
99. N. Kapoor, S. T. Menon, R. Chauhan, P. Sachdev and T. P. Sakmar, *J. Mol. Biol.*, 2009, **393**, 882.
100. D. Fotiadis, Y. Liang, S. Filipek, D. A. Saperstein, A. Engel and K. Palczewski, *Nature*, 2003, **421**, 127.
101. D. Fotiadis, B. Jastrzebska, A. Philippsen, D. J. Muller, K. Palczewski and A. Engel, *Curr. Opin. Struct. Biol.*, 2006, **16**, 252.

102. K. Palczewski, *Trends Biochem. Sci.*, 2010, **35**, 595.
103. B. Jastrzebska, D. Fotiadis, G. F. Jang, R. E. Stenkamp, A. Engel and K. Palczewski, *J. Biol. Chem.*, 2006, **281**, 11917.
104. B. M. Willardson, B. Pou, T. Yoshida and M. W. Bitensky, *J. Biol. Chem.*, 1993, **268**, 6371.
105. M. A. Downs, R. Arimoto, G. R. Marshall and O. G. Kisselev, *Vision Res.*, 2006, **46**, 4442.
106. V. N. Hingorani, D. T. Tobias, J. T. Henderson and Y. K. Ho, *J. Biol. Chem.*, 1988, **263**, 6916.
107. M. Wessling-Resnick and G. L. Johnson, *Biochem. Biophys. Res. Commun.*, 1989, **159**, 651.
108. R. R. Vaillancourt, N. Dhanasekaran, G. L. Johnson and A. E. Ruoho, *Proc. Natl. Acad. Sci. U. S. A.*, 1990, **87**, 3645.
109. L. Arcemisbehere, T. Sen, L. Boudier, M. N. Balestre, G. Gaibelet, E. Detouillon, H. Orcel, C. Mendre, R. Rahmeh, S. Granier, C. Vives, F. Fieschi, M. Damian, T. Durroux, J. L. Baneres and B. Mouillac, *J. Biol. Chem.*, 2010, **285**, 6337.
110. J. F. White, J. Grodnitzky, J. M. Louis, L. B. Trinh, J. Shiloach, J. Gutierrez, J. K. Northup and R. Grisshammer, *Proc. Natl. Acad. Sci. U. S. A.*, 2007, **104**, 12199.
111. G. V. Nikiforovich, C. M. Taylor and G. R. Marshall, *Biochemistry*, 2007, **46**, 4734.
112. S. Filipek, K. A. Krzysko, D. Fotiadis, Y. Liang, D. A. Saperstein, A. Engel and K. Palczewski, *Photochem. Photobiol. Sci.*, 2004, **3**, 628.
113. D. Dell'Orco, M. Seeber and F. Fanelli, *FEBS Lett.*, 2007, **581**, 944.
114. T. H. Bayburt, A. J. Leitz, G. Xie, D. D. Oprian and S. G. Sligar, *J. Biol. Chem.*, 2007, **282**, 14875.
115. S. Banerjee, T. Huber and T. P. Sakmar, *J. Mol. Biol.*, 2008, **377**, 1067.
116. M. R. Whorton, B. Jastrzebska, P. S. Park, D. Fotiadis, A. Engel, K. Palczewski and R. K. Sunahara, *J. Biol. Chem.*, 2008, **283**, 4387.
117. O. P. Ernst, V. Gramse, M. Kolbe, K. P. Hofmann and M. Heck, *Proc. Natl. Acad. Sci. U. S. A.*, 2007, **104**, 10859.
118. V. I. Govardovskii, D. A. Korenyak, S. A. Shukolyukov and L. V. Zueva, *Mol. Vis*, 2009, **15**, 1717.
119. J. A. Hern, A. H. Baig, G. I. Mashanov, B. Birdsall, J. E. Corrie, S. Lazareno, J. E. Molloy and N. J. Birdsall, *Proc. Natl. Acad. Sci. U. S. A.*, 2010, **107**, 2693.
120. P. Rondard, T. Iiri, S. Srinivasan, E. Meng, T. Fujita and H. R. Bourne, *Proc. Natl. Acad. Sci. U. S. A.*, 2001, **98**, 6150.
121. Z. Zhang, T. J. Melia, F. He, C. Yuan, A. McGough, M. F. Schmid and T. G. Wensel, *J. Biol. Chem.*, 2004, **279**, 33937.
122. W. M. Oldham, N. Van Eps, A. M. Preininger, W. L. Hubbel and H. E. Hamm, *Nat. Struct. Mol. Biol.*, 2006, **13**, 772.

CHAPTER 4

Peptide Hormone Recognition in Class B GPCRs: Role of the Extracellular Domain in Receptor Activation

CHRISTOPH PARTHIER*[a] AND MILTON T. STUBBS*[ab]

[a] Institut für Biochemie und Biotechnologie, Martin-Luther-Universität Halle-Wittenberg, Kurt-Mothes-Straße 3, D-06120 Halle, Germany; [b] ZIK HALOmem, Martin-Luther-Universität Halle-Wittenberg, Kurt-Mothes-Straße 3, D-06120 Halle, Germany

4.1 Introduction: Class B GPCRs

Class B G-protein-coupled receptors (GPCRs) make up a relatively small family (~15 members) of receptors for endogenous peptide hormones—also termed the 'secretin receptor family' after the first identified receptor of this class.[1,2] The hormones act *via* binding and activation of their cognate receptors, thereby mediating transduction of extracellular stimuli across the cell membrane for a variety of physiological processes, including the regulation of blood glucose, growth, bone turnover, calcium homeostasis, neurotransmission, vasodilation and stress. The receptors share the typical structural features of GPCRs from other classes: seven membrane-spanning α-helices (7TM) connected by intracellular and extracellular loops and a *C*-terminal intracellular domain that interacts with the G-protein. The most prominent feature of class

RSC Drug Discovery Series No. 8
G Protein-Coupled Receptors: From Structure to Function
Edited by Jesús Giraldo and Jean-Philippe Pin

Published by the Royal Society of Chemistry, www.rsc.org

Table 4.1 Published structures of class B GPCR ECDs in chronological order (October 2010).

Receptor ECD	*Ligand*	*Ligand type*	*Ref.*
mCRFR2β[a]	–	–	8
	Astressin	Antagonist	9
PAC1R$_S$	PACAP$_{6–38}$	Antagonist[b]	10
GIPR	GIP	Agonist	11
GLP-1R	Exendin$_{9–39}$	Antagonist[b]	12
	GLP–1	Agonist	13
PTH1R	PTH$_{15–34}$	Antagonist[b]	14
	PTHrP$_{12–34}$	Antagonist[b]	15
	– (dimer)	– (cryptic ligand)[c]	20
CRFR1	–	–	16
	CRF$_{22–41}$ CRF$_{27–41}$	Antagonists[b]	16
	αhcCRF	Agonist	17
CGRPR[d]	–	–	18
	Olcegepant Telcagepant	Small molecule antagonists	18
CRFR2α	Urocortin1$_{26–41}$ Urocortin2$_{26–41}$ Urocortin3$_{26–41}$	Antagonists[b]	19
VIPR2	–	–	2X57[e]
GRFR	–	–	2XDG[e]

[a]Murine CRF2β receptor, all other ECDs of human origin.
[b]*N*-terminally truncated agonist fragment, acting as antagonist.
[c]Dimerisation caused by *C*-terminal linker extension to ECD acting as a cryptic ligand,
[d]Complex of calcitonin-like receptor (CLR) and receptor activity-modifying protein 1 (RAMP1).
[e]Unpublished, PDB code given for reference.

B GPCRs is the presence of a characteristic disulphide-linked extracellular domain (ECD, ~100–160 residues) that is implicated in both hormone recognition and binding.[3]

The structural elucidation of class A GPCRs has shown remarkable progress in recent years, greatly widening our perception of the molecular events involved in GPCR activation (reviewed in refs. 4–6). Although the structure of a full-length class B GPCR still waits to be elucidated, considerable advances have been made recently in the structural determination of their ECDs in complex with cognate ligands (Table 4.1), opening a new chapter in class B GPCR research and enhancing our understanding of receptor structure and function to atomic resolution.[7]

4.2 Class B GPCR Ligands: A Family of Peptide Hormones with Helical Propensities

The natural ligands of class B GPCR are peptide hormones of about 30–40 amino acids in length, including glucagons and the glucagon-like peptide 1 (GLP-1), glucose-dependent insulinotropic polypeptide (GIP), parathyroid hormone (PTH) and the parathyroid hormone-related peptide (PTHrP), calcitonin and the calcitonin gene-related peptide (CGRP), growth-hormone releasing factor (GRF), secretin, corticotropin-releasing factor (CRF), pituitary adenylate cyclase activating polypeptide (PACAP) and vasoactive intestinal peptide (VIP).

From this list, it is clear why many class B GPCRs are either validated or potential drug targets. The ligands have attracted considerable pharmaceutical interest for their therapeutic potential in the treatment of, for example, type 2 diabetes (glucagon, GLP-1 and GIP[21,22]), osteoporosis (PTH and calcitonin[23]), migraine (CGRP[24]), chronic stress, anxiety and depression (CRF[25]), neurodegeneration (PACAP[26]) and inflammation (PACAP and VIP[27]). Therapies may require the use of either agonistic or antagonistic ligands, depending on the desired therapeutic strategy for each receptor. In general, however, administration of the natural agonists is hampered by the short half-life of the peptide hormones *in vivo* due to rapid deactivation by endogenous proteases (for example, GLP-1, GIP and PACAP are rapidly cleaved by the enzyme dipeptidyl-peptidase-IV[28–31]), so that pharmaceutical research has focused on the development of stable peptide analogues[32–35] and novel non-peptide ligands[36–38] for class B GPCRs.

As a result, sequence–activity relationships of class B GPCR ligands have been studied in considerable detail. Biochemical analyses involving modified peptide ligands have shown that the *N*-terminal region (the first *ca.* ten residues) is essential for receptor activation, as *N*-terminal ligand truncations result in competitive antagonists.[39–43] By contrast, the *C*-terminal region of the ligand plays an important role in receptor binding, as *C*-terminally truncated ligands remain capable of receptor activation but show a significantly lower affinity for the GPCR.[44–46]

Early structural studies of the peptide hormone glucagon revealed a helical conformation of the hormone.[47] On the other hand, later NMR solution studies showed glucagon to be disordered in solution.[48] Corresponding analyses with other class B GPCR ligands, including PTH,[49,50] GLP-1 and the related antagonist peptide exendin-4 from the Gila monster,[51,52] GIP,[53] PACAP,[10] VIP[54] and CRF, the related urocortins as well as the synthetic CRF agonist stressin and antagonist astressin[55] revealed in each case that the ligands possess only limited secondary structure in aqueous solutions, while α-helical structures are readily adopted through minor changes in the molecular environment, such as in the presence of lipids or organic solvents (*e.g.* trifluorethanol) or in crystals. Interestingly, calcitonin represents an exception, where the isolated hormone exhibits an α-helical conformation in solution, fixed by an intramolecular disulfide bridge between the *N*-terminal cysteine and downstream residues.[56,57] Although helices rarely span the whole peptide hormone (ligands have been observed to adopt continuous as well as kinked α-helices), it is now apparent that class B GPCR ligands share a common propensity for α-helix formation, allowing the induction or extension of helical structures.[7,58,59]

4.3 Extracellular Domains of Class B GPCRs: Dedicated to Ligand Binding

Early extensive biochemical studies involving chimeric receptor constructs and corresponding ligands established the crucial role of the ECD in ligand binding.[60–66] Photochemical crosslinking data revealed that the *C*-terminal ligand residues are recognized by the ECD, whereas the *N*-terminal residues interact with the 7TM region and interconnecting loops on the extracellular side of the receptors.[67–70] Biophysical studies conducted with isolated class B GPCR ECDs have corroborated these findings, confirming the substantial contribution of the ECDs to ligand binding[8,71–76] and leading to a general model for ligand binding and activation of class B GPCRs. This description, termed the 'two domain' model,[58] proposes: (i) binding of a *C*-terminal 'affinity domain' of the endogenous ligand to the ECD; and (ii) interaction of an *N*-terminal 'receptor activation domain' with the remainder of the receptor to trigger activation and subsequent signalling.

4.3.1 Structure of Class B GPCR Extracellular Domains

The NMR structure of the murine CRF receptor 2β-ECD revealed a compact domain structure with two antiparallel β-sheets stabilized by three intramolecular disulphide bonds.[8,9] This architecture resembled the 'short consensus repeat' or SCR fold, a structural motif commonly found in proteins of the complement system.[77] However, subsequent publications concerning the NMR structure of the PACAP receptor PAC1R$_s$-ECD[10] and the X-ray crystal structure of the GIP receptor GIPR-ECD[11] showed that the full ECD fold includes an additional α-helix at the *N*-terminus, linked to the β-sheet core by disulphide bonds (Figure 4.1a). Five loop regions (L1 to L5) interconnect the α-helix and β-strands of the ECD. In addition to the three intramolecular disulphide bonds, the tertiary structure is stabilized by intramolecular interactions between core residues that are conserved among class B GPCRs (Figure 4.1b). A central arginine residue (Arg$_{101}$, residue numbers correspond to the crystal structure of GIPR-ECD) is sandwiched between the side chains of two tryptophan residues (Trp$_{71}$ and Trp$_{109}$), forming a central cluster stabilized by π–π interactions. The strictly conserved aspartic acidic residue Asp$_{66}$ plays a central role in keeping the discontinuous ECD segments in spatial proximity, hydrogen bonding with several ECD residues to fix loop region L2 which in turn forms the base of the ligand binding pocket.

Since then, NMR and crystal structures of seven other class B GPCR ECDs have been reported: GLP-1R-ECD,[12,13] PTH1R-ECD,[14,15] CRFR1-ECD,[16,17] CRFR2α-ECD,[19] CGRPR-ECD,[18] VIPR2-ECD (unpublished, PDB code: 2X57) and GRFR-ECD (unpublished, PDB code 2XDG). Although all structures exhibit the same overall architecture of the ECD, now termed the 'secretin family recognition fold',[7] noticeable variations are observed (Figure 4.1c). The ECDs of the CRF receptors 1 and 2β each possess a significantly shorter *N*-terminal α-helix, as well as an additional glycine residue in loop L2; both regions are implicated in ligand binding (discussed in the following section). In the PTH1 receptor ECD, loop region L1 (which connects the *N*-terminal α-helix with the β-sheet core of the domain) contains a unique insertion of some 50 residues. This region, which exhibits low conservation among class B GPCRs, is disordered in all crystal structures of PTH1R-ECD solved to date.[14,15,20] The role of this

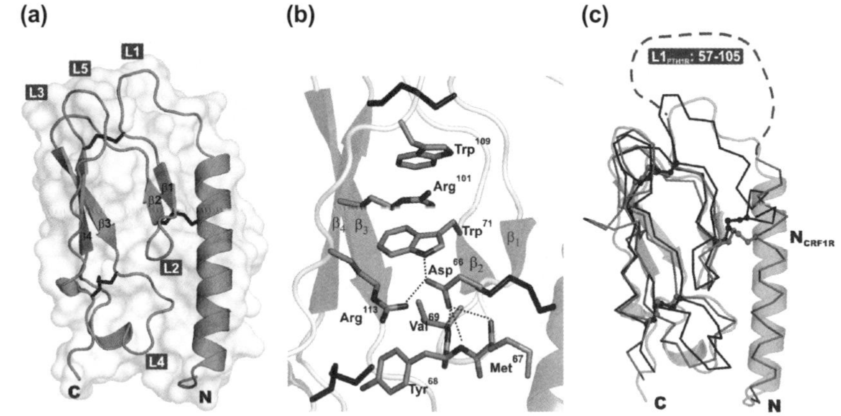

Figure 4.1 Crystal structure of GIPR-ECD—a paradigm for the secretin family recognition fold. (**a**) The secondary structure elements—an *N*-terminal α-helix and two central β-sheets—are connected by five loop regions (L1–L5), stabilized by three conserved disulphide bonds (black sticks). (**b**) Conserved residues in the core of the domain contribute to formation of the characteristic ECD fold (important residues shown as grey sticks, disulphide bonds as black sticks, hydrogen bonds displayed as dotted lines). (**c**) Superposition of the polypeptide backbones of CRFR1-ECD[16] (black; PDB code: 3EHU) and PTH1R-ECD[14] (dark grey, PDB code: 3C4M) with GIPR-ECD[11] (light grey, cartoon representation, PDB code: 2QKII). Disulphide bonds are shown for the ECDs of CRFR1 and PTH1R as balls and sticks in the respective colour. Note the vestigial *N*-terminal helix in CRFR1-ECD and the 48-residue long insertion in loop L1 of PTH1R-ECD (sketched as a dashed line).

extended loop, which is absent in the PTH2R-ECD, is at present unclear; it is possible that this might interact with the extended *C*-terminus of PTH.

4.3.2 Structural Basis of Ligand Recognition and Binding by the ECD

Of the 21 structures solved to date, 13 have been elucidated in the presence of a peptide ligand or analogue (including both agonists and antagonists) (Table 4.1). With the exception of the PAC1R$_s$-ECD : PACAP complex (the results of which are controversial for reasons discussed elsewhere[7]), all ECDs bind their cognate ligands in an equivalent orientation, epitomized by the GIPR-ECD : GIP complex.[11,78] In each case, the ECD binds to *C*-terminal residues of the ligand, whereas *N*-terminal residues (if present) are free of the ECD (Figures 4.2a and 4.2b). Most remarkably, each bound ligand adopts an α-helical conformation, sandwiched between the two β-sheets of the ECD. The helical ligands exhibit an amphipathic character, with hydrophobic residues buried in a complementary binding groove of the ECD formed mainly by residues of loops L2 and L4—in agreement with alanine scanning data for various class B GPCR ligands which highlight the importance of hydrophobic interactions in ligand binding.[9–11,14,16,19]

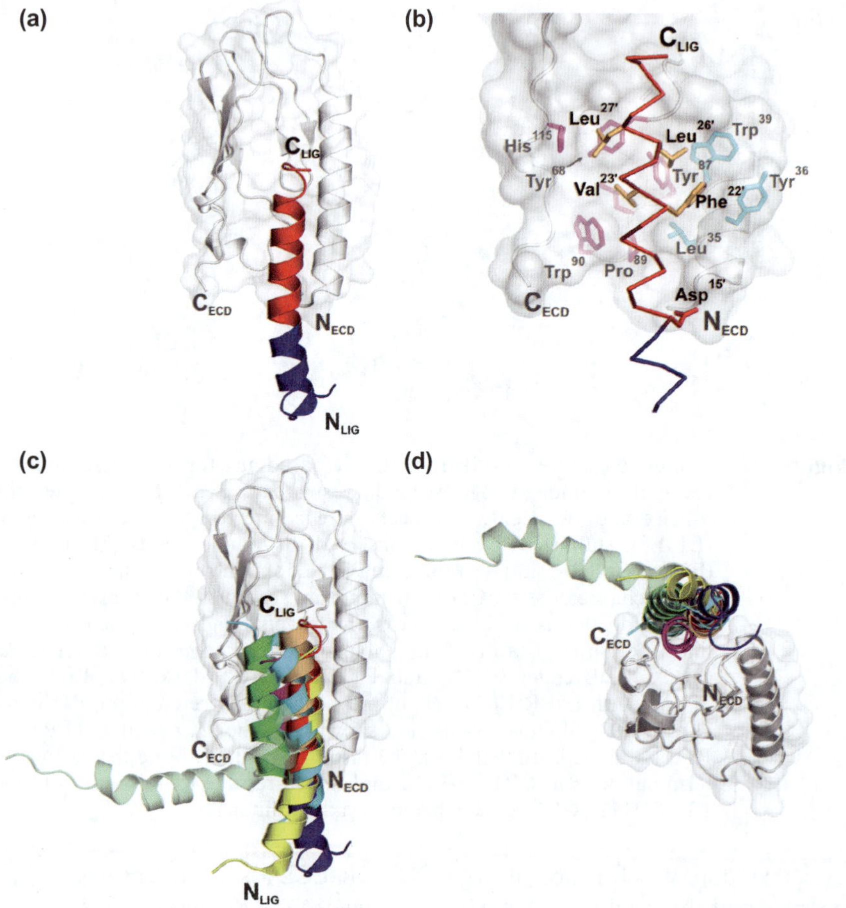

Figure 4.2 Class B GPCR ligands bind in an α-helical conformation. (**a**) GIPR-ECD: GIP complex as paradigm for ligand binding.[11] Only the *C*-terminal ligand residues (red) interact with the ECD (grey), leaving the *N*-terminal residues (blue) free for interaction with other parts of the receptor. (**b**) Hydrophobic residues in the GIPR-ECD core from loop 2, loop 4 and the *C*-terminus (magenta) and from the *N*-terminal helix (cyan) interact with complementary residues (orange sticks) from the *C*-terminal region of the GIP ligand. (**c**) Structural alignment of ECDs reveals similar ligand binding modes (GIPR-ECD shown as grey cartoon with overlaid surface). Ligand colours are as follows: GIP (blue/red; PDB code: 2QKH[11]), astressin (magenta; PDB code: 2JND[9]), GLP-1 (yellow; PDB code: 3IOL[13]), PTH$_{15-34}$ (pink; PDB code: 3C4M[14]), PTHrP$_{12-34}$ (cyan; PDB code: 3H3G[15]), CRF$_{22-41}$ (green; PDB code: 3EHU[16]), αhcCRF (pale green; PDB code: 2L27[17]). *N*- and *C*-termini of ligands and the ECD are labelled. (**d**) View rotated 90° about a horizontal axis.

In most of the complexes, additional contributions to ligand binding are made from residues of the *N*-terminal α-helix of the ECD. An exception to this consensus is seen in the binding mode of CRF and related analogues to their cognate receptors, the ECDs of which possess only a vestigial *N*-terminal helix

(Figure 4.1c). As a result, the positions of the bound CRF-derived ligands show a considerable shift away from the *N*-terminal α-helix (Figures 4.2c and 4.2d). These and other differences observed in the CRF receptor complex structures indicate a further sub-classification of class B GPCRs according to 'glucagon-like' or 'CRF-like' binding modes (discussed in ref. 7).

4.3.3 Specificity of Ligand Binding: Contributions by the ECD

GPCRs of the secretin receptor family share between 21% and 67% sequence identity, with much of the variation located in the ECD.[3] Although this suggests that the basis of receptor specificity resides in the ECD, the intriguing similarity in binding interactions revealed by the recent ECD-ligand structures raises the question: how can a common and seemingly simple principle of ligand binding accommodate specific ligand recognition? Besides the conserved hydrophobic interactions between ECD and ligands, each complex structure exhibits an individual pattern of additional polar and ionic interactions that may be employed to fine-tune affinity and specificity of each ECD towards its cognate ligand. This picture must of necessity be an oversimplification, however, considering that several class B GPCRs may recognize a variety of different hormone ligands (see Table 4.1 for examples).

The recent structure determinations of the CRF receptors 1 and 2α ECDs in complex with derivatives of the natural ligands CRF and the urocortins 1, 2 and 3[16,19] shed some light on this situation. While CRF and urocortin 1 can activate both CRF receptors (CRFR1 and CRFR2), the urocortins 2 and 3 are specific for CRF2R. In this case, the selectivity determinant is a glutamate residue (Glu104) in the CRFR1-ECD (Pro100 in CRFR2) that makes an electrostatic interaction with an arginine (Arg35) of the bound CRF or urocortin 1; this is not possible for the urocortins 2 and 3, as the corresponding residue is an alanine (Ala35). More subtle differences in binding modes are observed in the crystal structures of PTH1R-ECD in complex with derivatives of PTH[14] and the PTH-related peptide (PTHrP).[15] While the majority of interactions are very similar, differences are observed in the intermolecular hydrogen bond network at the *C*-termini of the ligands. This is associated with a *C*-terminal unwinding of the PTHrP α-helix and a slight curvature in the remaining helical region of PTHrP (Figures 4.2c and 4.2d).

As mentioned above, the induction or extension of α-helical structure is a consequence of peptide ligand binding to the ECD. In the GIPR-ECD : GIP complex, the GIP hormone helix extends throughout the peptide, although only the *C*-terminal GIP residues are in contact with the ECD[11] (Figure 4.2a). Similarly, in complex with GLP-1R-ECD, the *N*-terminal residues of exendin-4 and GLP-1 each project away from the ECD, forming a continuous or slightly curved α-helix.[12,13] A similar situation is observed for the complex of CRFR1-ECD with the synthetic agonist αhcCRF,[17] although the ligand exhibits a pronounced kink (Figures 4.2c and 4.2d). It is conceivable that the intrinsic ability of a hormone sequence to form an α-helix—either continuous or kinked, from an unfolded state or by extension of a helical nucleus—may represent an additional layer of 'hidden specificity' of class B GPCR ligands towards their receptors that is 'decoded' upon binding to the ECD.[7]

4.4 A Model for Class B GPCR Activation

The observed helix formation of class B GPCR ligands upon ECD binding suggests a mechanical model for class B receptors[7] (Figure 4.3a). The combined effects of hydrophobic burial, specific polar contacts (hydrogen bonding and

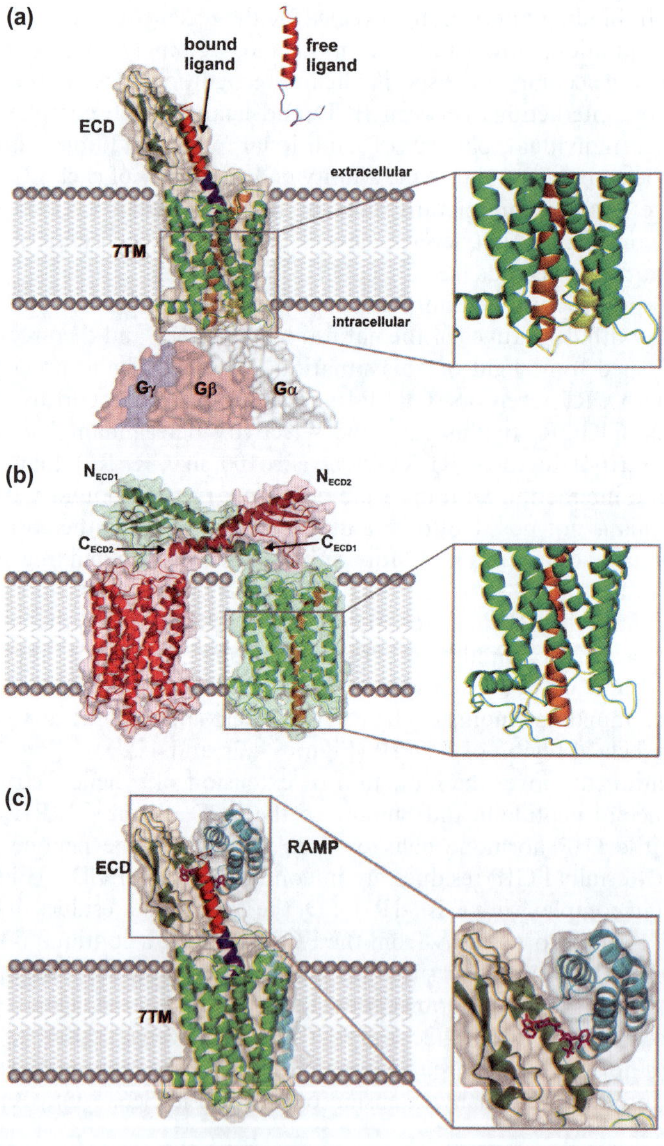

electrostatic interactions) and disposition to helix formation allow not only for specificity and affinity of peptide hormone binding to cognate ECDs, but also for a structuring of disordered regions of the ligand—in particular *N*-terminal residues not in contact with the ECD. The solution structure of $PACAP_{1-21}$ in the PAC1R-bound conformation, which reveals an α-helical conformation spanning residues in the *N*-terminal part of the ligand[79] provides strong support for this scenario. The transition of *N* terminal residues to a helical conformation in the ECD-bound form enables the presentation of this region, required for receptor activation, to the 7TM domain and the bordering extracellular loops.

A combination of ligand stiffening and contraction as a result of helix formation would allow the accurate and specific positioning of these crucial ligand residues to the 'activation site' of the receptor, probably accompanied by ligand-induced conformational changes in ECD loops adjacent to the 7TM domain of the receptor and/or extracellular loops of the 7TM domain.[80] Precise positioning of the *N*-terminal residues may vary between different class B GPCRs, as suggested by comparison of the ligand orientations of GIP in GIPR-ECD[11] and αhcCRF in CRFR1-ECD[17] (Figures 4.2c and 4.2d). Assuming that general structural mechanisms of receptor activation are conserved among GPCRs, ligand recognition and binding could be followed by structural rearrangements of GPCR helices 5 and 6 within the 7TM domain similar to those observed for activated opsin,[81] allowing transduction of the extracellular signal across the membrane and ensuing activation of the intracellular G proteins (Figure 4.3).

Figure 4.3 Structural models depicting the activation steps of full-length class B GPCRs based on *in silico* assembly of the independently solved structures described (see ref. 7). (**a**) Binding of the *C*-terminal region of the hormone (red) to the ECD induces α-helix formation of or extension to the *N*-terminal ligand region (blue). The *N*-terminus can thereby be inserted into the transmembrane domain of the GPCR, triggering conformational changes in the 7TM domain, depicted as a movement of transmembrane helix 6 (coloured in orange) and helical transition within transmembrane helix 5 (yellow, inset) compared to the inactive state (inset, Figure 4.3b) as observed in bovine opsin.[77] This structural transition is in turn sensed by the associated intracellular heterotrimeric G protein, leading to activation and initiation of subsequent steps in signal transduction (PDB codes: GIPR-ECD:GIP, 2QKH[11]; free GIP, 2B4N[51]; opsin, 3DQB[77]; G-protein, 1GOT[78]). (**b**) Putative homodimerization of two ECDs as observed for the unliganded PTH1R-EC[20] could be mediated by a 'domain swap' like arrangement: the *C*-terminal linker sequence between the ECD and the 7TM region of one ECD occupies the ligand binding groove of the other ECD and *vice versa*, so that the linker acts as a 'cryptic ligand' in the inactive GPCR state (monomers coloured red and green, transmembrane helix 6 orange). Hormone binding would lead to dissociation of the dimer and subsequent receptor activation (PDB codes: PTH1R-ECD, 3C4M[20]; rhodopsin, 1U19[75]). (**c**) Heterodimerization of the class B GPCR CGRPR (green) with RAMP1 (cyan) results in a bipartite binding site for a modelled cognate agonist (blue/red), allowing modulation of receptor ligand specificity. The small molecule antagonist telcagepant (magenta, inset) competes for the hormone binding site (PDB codes: CGRPR-ECD-RAMP1 complex, 3N7R[18]; rhodopsin, 1U19[75]; agonist GIP, 2QKH[11]; small molecule antagonist telcagepant, 3N7R[18]).

4.5 Activity Modulation *via* Oligomerization

The oligomerization state of GPCRs is believed to play a role in the regulation of receptor activity, and homo- as well as hetero-oligomers have been reported for representatives of each GPCR class, including the secretin receptor family.[83–86] Although the exact role of receptor oligomerization is still a matter of debate—and may vary among different GPCRs—there is growing evidence that the homodimeric state of class B GPCRs is indeed relevant in the cell.[87,88] Interestingly, a homodimeric arrangement of two ECDs was discovered in the recent crystal structure determination of PTH1R-ECD in the absence of any ligand.[20] The ligand binding groove of each ECD is occupied by an α-helical *C*-terminal extension of the opposite ECD. This *C*-terminal extension—which in the full length receptor represents the linker sequence between the ECD and the first membrane-spanning helix of the 7TM—bears some of the hallmarks of ECD hormone recognition, with suitably placed hydrophobic residues that underpin binding of the α-helix to the ECD. The bound 'cryptic ligand', which results in a 'domain swap' like arrangement of the two ECDs that would lead to dimerization of the full-length receptor, blocks access to the hormone binding site. Thus the dimeric state might represent an inactive state of PTH1R in the absence of the hormone (Figure 4.3b), substantiated by the observation that ligand binding induces dissociation of the receptor dimer.[20]

Class B GPCR receptor activity may also be modulated by the binding of accessory proteins (so-called receptor activity-modifying proteins or RAMPs, reviewed in refs. 89–92). Heterodimerization with RAMPs (which consist of an *N*-terminal extracellular domain, a single transmembrane helix and an intracellular *C*-terminal domain) has been observed *inter alia* for the calcitonin (CTR) and calcitonin-like (CLR) receptors, and modulates both the affinity and specificity of different ligands to the receptors. As an example, CTR forms three distinct heterodimers with RAMP1, 2 and 3, respectively, thereby generating three different receptor subtypes for the ligand amylin. By contrast, the CLR-RAMP1 complex shows high affinity for the ligand CGRP, while heterodimers of CLR with RAMP2 and RAMP3 represent two receptor subtypes for adrenomedullin. The newly solved crystal structure of the CLR-ECD : RAMP1 extracellular domain complex, which represents the ligand binding region of the CGRP receptor heterodimer, provides a first glimpse into the class B GPCR : RAMP interaction.[18,93] Solved in the absence of the natural ligand CGRP, the structure reveals that the RAMP1-derived disulphide-linked three-helix bundle binds to the CLR-ECD *N*-terminal α-helix in a perpendicular orientation. The position of RAMP1 provides room for binding the natural ligand in the helical manner discussed above for other class B GPCR ligands, with the important proviso that the bound hormone would also contact the RAMP1, generating a bi-partite and distinct binding site for the ligand (Figure 4.3c). Of particular interest is the fact that the authors were able to solve structures of CLR-ECD : RAMP1 in complex with the small molecule antagonists, telcagepant and olcegepant (Figure 4.3c, inset). Both ligands bridge the two partners of the complex, CLR-ECD and RAMP1, through distinct interactions with each component, rationalizing the high affinity of these compounds and providing a basis for the development of new treatments for migraine.

4.6 Conclusions

The availability of class B GPCR ECD structures in complex with their cognate ligands has shed light on the principles underlying ligand binding, recognition and selectivity. In contrast to other receptor ligand systems, ECD binding involves a folding or restructuring of the hormone, adding an additional layer of complexity (and thereby regulation) to receptor activation. While these studies have been fruitful in expanding our knowledge of this process, a comprehensive understanding of the key mechanisms governing receptor activity, regulation and modulation will require structural determinations of full-length class B GPCRs in the presence of diverse ligands. Nevertheless, the foundation has been laid for the implementation of structural data in the development of novel synthetic agonist and antagonist drugs acting on this therapeutically important receptor family.[94]

Acknowledgements

Work in our laboratory is supported by the Landesexzellenzinitiative Sachsen-Anhalt 'Strukturen und Mechanismen der biologischen Informationsverarbeitung', DFG Sonderforschungsbereich 610 'Protein-Zustände mit zellbiologischer und medizinischer Relevanz' and the Federal Initiative 'ZIK HALOmem—membrane protein structure and dynamics' of the Bundesministerium für Bildung und Forschung (BMBF). We are indebted to the contributions made to this field by our late friend and colleague Professor Rainer Rudolph.

References

1. T. Ishihara, S. Nakamura, Y. Kaziro, T. Takahashi, K. Takahashi and S. Nagata, *EMBO J.*, 1991, **10**, 1635.
2. R. Fredriksson, M. C. Lagerstrom, L. G. Lundin and H. B. Schioth, *Mol. Pharmacol.*, 2003, **63**, 1256.
3. M. C. Lagerstrom and H. B. Schioth, *Nat. Rev. Drug Discov.*, 2008, **7**, 339.
4. D. M. Rosenbaum, S. G. Rasmussen and B. K. Kobilka, *Nature*, 2009, **459**, 356.
5. W. I. Weis and B. K. Kobilka, *Curr. Opin. Struct. Biol.*, 2008, **18**, 734.
6. M. A. Hanson and R. C. Stevens, *Structure*, 2009, **17**, 8.
7. C. Parthier, S. Reedtz-Runge, R. Rudolph and M. T. Stubbs, *Trends Biochem. Sci.*, 2009, **34**, 303.

8. C. R. Grace, M. H. Perrin, M. R. DiGruccio, C. L. Miller, J. E. Rivier, W. W. Vale and R. Riek, *Proc. Natl. Acad. Sci. U. S. A.*, 2004, **101**, 12836.
9. C. R. Grace, M. H. Perrin, J. Gulyas, M. R. DiGruccio, J. P. Cantle, J. E. Rivier, W. W. Vale and R. Riek, *Proc. Natl. Acad. Sci. U. S. A.*, 2007, **104**, 4858.
10. C. Sun, D. Song, R. A. Davis-Taber, L. W. Barrett, V. E. Scott, P. L. Richardson, A. Pereda-Lopez, M. E. Uchic, L. R. Solomon, M. R. Lake, K. A. Walter, P. J. Hajduk and E. T. Olejniczak, *Proc. Natl. Acad. Sci. U. S. A.*, 2007, **104**, 7875.
11. C. Parthier, M. Kleinschmidt, P. Neumann, R. Rudolph, S. Manhart, D. Schlenzig, J. Fanghanel, J. U. Rahfeld, H. U. Demuth and M. T. Stubbs, *Proc. Natl. Acad. Sci. U. S. A.*, 2007, **104**, 13942.
12. S. Runge, H. Thogersen, K. Madsen, J. Lau and R. Rudolph, *J. Biol. Chem.*, 2008, **283**, 11340.
13. C. R. Underwood, P. Garibay, L. B. Knudsen, S. Hastrup, G. H. Peters, R. Rudolph and S. Reedtz-Runge, *J. Biol. Chem.*, 2010, **285**, 723.
14. A. A. Pioszak and H. E. Xu, *Proc. Natl. Acad. Sci. U. S. A.*, 2008, **105**, 5034.
15. A. A. Pioszak, N. R. Parker, T. J. Gardella and H. E. Xu, *J. Biol. Chem.*, 2009, **284**, 28382.
16. A. A. Pioszak, N. R. Parker, K. Suino-Powell and H. E. Xu, *J. Biol. Chem.*, 2008, **283**, 32900.
17. C. R. Grace, M. H. Perrin, J. Gulyas, J. E. Rivier, W. W. Vale and R. Riek, *J. Biol. Chem.*, 2010.
18. E. ter Haar, C. M. Koth, N. Abdul-Manan, L. Swenson, J. T. Coll, J. A. Lippke, C. A. Lepre, M. Garcia-Guzman and J. M. Moore, *Structure.*, 2010, **18**, 1083.
19. K. Pal, K. Swaminathan, H. E. Xu and A. A. Pioszak, *J. Biol. Chem.*, 2010.
20. A. A. Pioszak, K. G. Harikumar, N. R. Parker, L. J. Miller and H. E. Xu, *J. Biol. Chem.*, 2010, **285**, 12435.
21. D. J. Drucker, *Cell Metab.*, 2006, **3**, 153.
22. S. E. Inzucchi and D. K. McGuire, *Circulation*, 2008, **117**, 574.
23. J. E. Mulder, N. S. Kolatkar and M. S. LeBoff, *Nat. Clin. Pract. Endocrinol. Metab*, 2006, **2**, 670.
24. H. Doods, K. Arndt, K. Rudolf and S. Just, *Trends Pharmacol. Sci.*, 2007, **28**, 580.
25. P. J. Gilligan and Y. W. Li, *Curr. Opin. Drug Discov. Devel.*, 2004, **7**, 487.
26. D. E. Brenneman, *Peptides*, 2007, **28**, 1720.
27. C. Abad, R. P. Gomariz and J. A. Waschek, *Curr. Top. Med. Chem.*, 2006, **6**, 151.
28. R. Mentlein, B. Gallwitz and W. E. Schmidt, *Eur. J. Biochem.*, 1993, **214**, 829.
29. T. J. Kieffer, C. H. McIntosh and R. A. Pederson, *Endocrinology*, 1995, **136**, 3585.
30. S. Bourgault, D. Vaudry, B. Botia, A. Couvineau, M. Laburthe, H. Vaudry and A. Fournier, *Peptides*, 2008, **29**, 919.

31. L. Zhu, C. Tamvakopoulos, D. Xie, J. Dragovic, X. L. Shen, J. E. Fenyk-Melody, K. Schmidt, A. Bagchi, P. R. Griffin, N. A. Thornberry and R. S. Roy, *J. Biol. Chem.*, 2003, **278**, 22418.
32. J. Rivier, J. Gulyas, D. Kirby, W. Low, M. H. Perrin, K. Kunitake, M. DiGruccio, J. Vaughan, J. C. Reubi, B. Waser, S. C. Koerber, V. Martinez, L. Wang, Y. Tache and W. Vale, *J. Med. Chem.*, 2002, **45**, 4737.
33. D. J. Drucker and M. A. Nauck, *Lancet*, 2006, **368**, 1696.
34. C. F. Deacon, L. B. Knudsen, K. Madsen, F. C. Wiberg, O. Jacobsen and J. J. Holst, *Diabetologia*, 1998, **41**, 271.
35. G. A. Spyroulias, S. Papazacharias, G. Pairas and P. Cordopatis, *Eur. J. Biochem.*, 2002, **269**, 6009.
36. J. Kehne and S. De Lombaert, *Curr. Drug Targets CNS. Neurol. Disord.*, 2002, **1**, 467.
37. W. Soudijn, I. van Wijngaarden and A. P. Ijzerman, *Curr. Opin. Drug Discov. Devel.*, 2002, **5**, 749.
38. E. L. Moore, C. S. Burgey, D. V. Paone, A. W. Shaw, Y. S. Tang, S. A. Kane and C. A. Salvatore, *Eur. J. Pharmacol.*, 2009, **602**, 250.
39. J. Rivier, C. Rivier and W. Vale, *Science*, 1984, **224**, 889.
40. B. Thorens, A. Porret, L. Buhler, S. P. Deng, P. Morel and C. Widmann, *Diabetes*, 1993, **42**, 1678.
41. R. W. Gelling, D. H. Coy, R. A. Pederson, M. B. Wheeler, S. Hinke, T. Kwan and C. H. McIntosh, *Regul. Pept.*, 1997, **69**, 151.
42. T. J. Gardella and H. Juppner, *Trends Endocrinol. Metab*, 2001, **12**, 210.
43. J. P. Fortin, Y. Zhu, C. Choi, M. Beinborn, M. N. Nitabach and A. S. Kopin, *Proc. Natl. Acad. Sci. U. S. A.*, 2009, **106**, 8049.
44. S. R. Hoare, S. K. Sullivan, D. A. Schwarz, N. Ling, W. W. Vale, P. D. Crowe and D. E. Grigoriadis, *Biochemistry*, 2004, **43**, 3996.
45. M. H. Perrin, M. R. DiGruccio, S. C. Koerber, J. E. Rivier, K. S. Kunitake, D. L. Bain, W. H. Fischerv and W. W. Vale, *J. Biol. Chem.*, 2003, **278**, 15595.
46. S. Al Sabah and D. Donnelly, *Br. J. Pharmacol.*, 2003, **140**, 339.
47. K. Sasaki, S. Dockerill, D. A. Adamiak, I. J. Tickle and T. Blundell, *Nature*, 1975, **257**, 751.
48. W. Braun, G. Wider, K. H. Lee and K. Wuthrich, *J. Mol. Biol.*, 1983, **169**, 921.
49. U. C. Marx, K. Adermann, P. Bayer, W. G. Forssmann and P. Rosch, *Biochem. Biophys. Res. Commun.*, 2000, **267**, 213.
50. L. Jin, S. L. Briggs, S. Chandrasekhar, N. Y. Chirgadze, D. K. Clawson, R. W. Schevitz, D. L. Smiley, A. H. Tashjian and F. Zhang, *J. Biol. Chem.*, 2000, **275**, 27238.
51. K. Thornton and D. G. Gorenstein, *Biochemistry*, 1994, **33**, 3532.
52. J. W. Neidigh, R. M. Fesinmeyer, K. S. Prickettv and N. H. Andersen, *Biochemistry*, 2001, **40**, 13188.
53. I. Alana, J. C. Parker, V. A. Gault, P. R. Flatt, F. P. O'Harte, J. P. Malthouse and C. M. Hewage, *J. Biol. Chem.*, 2006, **281**, 16370.

54. Y. V. Tan, A. Couvineau, S. Murail, E. Ceraudo, J. M. Neumann, J. J. Lacapere and M. Laburthe, *J. Biol. Chem.*, 2006, **281**, 12792.
55. C. R. Grace, M. H. Perrin, J. P. Cantle, W. W. Vale, J. E. Rivier and R. Riek, *J. Am. Chem. Soc.*, 2007, **129**, 16102.
56. A. Motta, G. Andreotti, P. Amodeo, G. Strazzullo and M. A. Castiglione Morelli, *Proteins*, 1998, **32**, 314.
57. G. Andreotti, B. L. Mendez, P. Amodeo, M. A. Morelli, H. Nakamuta and A. Motta, *J. Biol. Chem.*, 2006, **281**, 24193.
58. S. R. Hoare, *Drug Discov. Today*, 2005, **10**, 417.
59. J. M. Neumann, A. Couvineau, S. Murail, J. J. Lacapere, N. Jamin and M. Laburthe, *Trends Biochem. Sci.*, 2008, **33**, 314.
60. C. Bergwitz, T. J. Gardella, M. R. Flannery, J. T. Potts, Jr., H. M. Kronenberg, S. R. Goldring and H. Juppner, *J. Biol. Chem.*, 1996, **271**, 26469.
61. M. H. Holtmann, E. M. Hadac and L. J. Miller, *J. Biol. Chem.*, 1995, **270**, 14394.
62. M. Laburthe, A. Couvineau and J. C. Marie, *Recept. Channels*, 2002, **8**, 137.
63. S. D. Stroop, R. E. Kuestner, T. F. Serwold, L. Chen and E. E. Moore, *Biochemistry*, 1995, **34**, 1050.
64. S. Runge, B. S. Wulff, K. Madsen, H. Brauner-Osborne and L. B. Knudsen, *Br. J. Pharmacol.*, 2003, **138**, 787.
65. R. W. Gelling, M. B. Wheeler, J. Xue, S. Gyomorey, C. Nian, R. A. Pederson and C. H. McIntosh, *Endocrinology*, 1997, **138**, 2640.
66. C. G. Unson, C. R. Wu, Y. Jiang, B. Yoo, C. Cheung, T. P. Sakmar and R. B. Merrifield, *Biochemistry*, 2002, **41**, 11795.
67. I. Assil-Kishawi and A. B. Abou-Samra, *J. Biol. Chem.*, 2002, **277**, 32558.
68. M. Dong, Z. Li, D. I. Pinon, T. P. Lybrand and L. J. Miller, *J. Biol. Chem.*, 2004, **279**, 2894.
69. M. Dong, M. Zang, D. I. Pinon, Z. Li, T. P. Lybrand and L. J. Miller, *Mol. Endocrinol.*, 2002, **16**, 2490.
70. R. C. Gensure, T. J. Gardella and H. Juppner, *J. Biol. Chem.*, 2001, **276**, 28650.
71. A. Wilmen, B. Goke and R. Goke, *FEBS Lett.*, 1996, **398**, 43.
72. A. Bazarsuren, U. Grauschopf, M. Wozny, D. Reusch, E. Hoffmann, W. Schaefer and S. Panznerv R. Rudolph, *Biophys. Chem.*, 2002, **96**, 305.
73. S. Runge, S. Schimmer, J. Oschmann, C. B. Schiodt, S. M. Knudsen, C. B. Jeppesen, K. Madsen, J. Lau, H. Thogersenv and R. Rudolph, *Biochemistry*, 2007, **46**, 5830.
74. U. Grauschopf, H. Lilie, K. Honold, M. Wozny, D. Reusch, A. Esswein, W. Schafer, K. P. Rucknagel and R. Rudolph, *Biochemistry*, 2000, **39**, 8878.
75. J. Klose, K. Fechner, M. Beyermann, E. Krause, N. Wendt, M. Bienert, R. Rudolph and S. Rothemund, *Biochemistry*, 2005, **44**, 1614.
76. C. M. Koth, N. Abdul-Manan, C. A. Lepre, P. J. Connolly, S. Yoo, A. K. Mohanty, J. A. Lippke, J. Zwahlen, J. T. Coll, J. D. Doran, M. Garcia-Guzman and J. M. Moore, *Biochemistry*, 2010, **49**, 1862.

77. D. G. Norman, P. N. Barlow, M. Baron, A. J. Day, R. B. Sim and I. D. Campbell, *J. Mol. Biol.*, 1991, **219**, 717.
78. T. Okada, M. Sugihara, A. N. Bondar, M. Elstner, P. Entel and V. Buss, *J. Mol. Biol.*, 2004, **342**, 571.
79. H. Inooka, T. Ohtaki, O. Kitahara, T. Ikegami, S. Endo, C. Kitada, K. Ogi, H. Onda, M. Fujino and M. Shirakawa, *Nat. Struct. Biol.*, 2001, **8**, 161.
80. T. Yaqub, I. G. Tikhonova, J. Lattig, R. Magnan, M. Laval, C. Escrieut, C. Boulegue, C. Hewage and D. Fourmy, *Mol. Pharmacol.*, 2010, **77**, 547.
81. P. Scheerer, J. H. Park, P. W. Hildebrand, Y. J. Kim, N. Krauss, H. W. Choe, K. P. Hofmann and O. P. Ernst, *Nature*, 2008, **455**, 497.
82. D. G. Lambright, J. Sondek, A. Bohm, N. P. Skiba, H. E. Hamm and P. B. Sigler, *Nature*, 1996, **379**, 311.
83. K. G. Harikumar, R. M. Happs and L. J. Miller, *Biochim. Biophys. Acta*, 2008, **1778**, 2555.
84. K. G. Harikumar, M. M. Morfis, P. M. Sexton and L. J. Miller, *J. Mol. Neurosci.*, 2008, **36**, 279.
85. T. Seck, R. Baron and W. C. Horne, *J. Biol. Chem.*, 2003, **278**, 23085.
86. O. Kraetke, B. Wiesner, J. Eichhorst, J. Furkert, M. Bienert and M. Beyermann, *J. Recept. Signal Transduct. Res.*, 2005, **25**, 251.
87. V. V. Gurevich and E. V. Gurevich, *Trends Neurosci.*, 2008, **31**, 74.
88. I. Gomes, B. A. Jordan, A. Gupta, C. Rios, N. Trapaidze and L. A. Devi, *J. Mol. Med.*, 2001, **79**, 226.
89. S. L. Ritter and R. A. Hall, *Nat. Rev. Mol. Cell Biol.*, 2009, **10**, 819.
90. A. C. Conner, J. Simms, D. L. Hay, K. Mahmoud, S. G. Howitt, M. Wheatley and D. R. Poyner, *Biochem. Soc. Trans.*, 2004, **32**, 843.
91. P. M. Sexton, M. Morfis, N. Tilakaratne, D. L. Hay, M. Udawela, G. Christopoulos and A. Christopoulos, *Ann. N. Y. Acad. Sci.*, 2006, **1070**, 90.
92. C. S. Walker, A. C. Conner, D. R. Poyner and D. L. Hay, *Trends Pharmacol. Sci.*, 2010, **31**, 476.
93. F. Y. Siu and R. C. Stevens, *Structure*, 2010, **18**, 1067.
94. L. J. Miller, *Ann. N. Y. Acad. Sci.*, 2008, **1144**, 1.

CHAPTER 5

Oligomerization of G Protein-coupled Receptors: Insights from Fluorescent and Luminescent-based Methods

FRANCISCO CIRUELA* AND VÍCTOR FERNÁNDEZ-DUEÑAS

Unitat de Farmacologia, Departament de Patologia i Terapèutica Experimental, Facultat de Medicina, IDIBELL-Universitat de Barcelona, 08907 Barcelona, Spain

5.1 Introduction

Receptor–receptor interactions govern a large array of biological processes. The characterization and visualization of these interactions thus constitute an important step in the understanding of a large number of cellular mechanisms related to the decoding of extracellular signals. Classically by means of a genetic approach like the yeast two-hybrid system[1] one can initially discover potential interactions, but subsequent confirmation by means of classical biochemical approaches (*e.g.* immobilized protein–protein interaction assays such as co-immunoprecipitation and pull-down experiments) is required.[2] Although the combined use of genetic and biochemical approaches has allowed the identification of a large number of protein–protein interactions, the subcellular localization of most of these is still unknown. It is also important to mention that

RSC Drug Discovery Series No. 8
G Protein-Coupled Receptors: From Structure to Function
Edited by Jesús Giraldo and Jean-Philippe Pin

Published by the Royal Society of Chemistry, www.rsc.org

some of these initial methods used to validate interactions still have the disadvantage of altering the natural state of the cell and therefore they may not represent its real structure. Thus, the invasive nature of these technical approaches (*e.g.* the need for detergents to extract membrane proteins) might hamper their use in the study of protein–protein interactions,[3] although when used carefully, these approaches are still very useful to confirm interactions between proteins.

Throughout the last decade, a new array of technologies has been developed in order to overcome the invasive nature of the immobilized protein–protein interaction assays. These new techniques, which are based on the genetic labelling of receptors with fluorescent proteins (FPs),[4,5] focus on the use of various adaptations of resonance energy transfer techniques and protein fragment complementation methods. Interestingly these approaches, which are devoted to the characterization and visualization of protein–protein interactions, have favoured the possibility of carrying out experiments *in vivo* as well as in real time, thus allowing awareness of where and when protein–protein interactions occur in the cell.

In this chapter we focus on these powerful fluorescence and luminescence-based methods which have evolved as useful tools to understand receptor–receptor interactions in living cells.

5.2 The Resonance Energy Transfer Principle

The basic principle of resonance energy transfer (RET) was firstly described by Theodor Förster in the late 1940s[6] and consists of a non-radiative (dipole–dipole) transfer of energy from a chromophore in an excited state—known as the 'donor'—to an 'acceptor' molecule.[7] This transfer of energy, also called Förster resonance energy transfer (FRET), results in a reduction in the donor chromophore emission that matches well with an increase in the emission of the acceptor molecule. Interestingly, the efficiency of this energy transfer (E_{RET}) is inversely proportional to the sixth power of the distance (R) between donor and acceptor molecules according to eqn (1):

$$E_{RET} = 1/(1 + R^6/R_0{}^6) \tag{1}$$

where R_0 is the distance leading to 50% of energy transfer from the donor to the acceptor. R_0 is typically around 5 nm and the effective range of energy transfer is below 10 nm. Thus, RET-based techniques can serve as a 'molecular ruler' for distance calculation in this range.[8] Interestingly, these distances are within the edge of the regular protein dimensions[9] and they are also similar to those described for multimeric protein complexes observed in biological systems.[10]

Conversely, the RET efficiency also depends on other variables, namely the orientation angle between the donor and the acceptor (like how the position of a radio antenna can influence its reception), the degree of spectral overlap between the donor emission and acceptor excitation, the quantum yield of the donor and the extinction coefficient of the acceptor.[6,8] Therefore, when choosing a chromophore RET-pair, we have to think about selecting those with the highest donor–quantum yield and absorbing acceptor together with significant spectral overlapping.[11] For instance, for intermolecular FRET measurements between two GPCRs tagged with the cyan (CFP) and the yellow (YFP) variants of the green fluorescent protein (GFP), respectively, FRET can be monitored as a YFP/CFP emission intensity ratio upon excitation at 433 nm according to eqn (2):

$$\text{Ratio}\left(\frac{F_{\text{YFP}}}{F_{\text{CFP}}}\right) = \frac{F_{\text{YFP}}^{ex433/em527} - \text{a x}\, F_{\text{CFP}}^{ex433/em475} - \text{b x}\, F_{\text{YFP}}^{ex500/em527}}{F_{\text{CFP}}^{ex433/em475}} \quad (2)$$

where:

$F_{\text{YFP}}{}^{\text{ex433/em527}}$ and $F_{\text{CFP}}{}^{\text{ex433/em475}}$ represent the respective emission intensities of YFP (at 527 nm) and CFP (at 475 nm) upon excitation at 433 nm; a and b represent correction factors for the bleed-through of CFP into the 527 nm channel (a), and the crosstalk due to the direct YFP excitation by light at 433 nm (b).

$F_{\text{YFP}}{}^{\text{ex500/em527}}$ represents the emission intensity of YFP (recorded at 527 nm) upon direct excitation at 500 nm, and it is recorded at the beginning of each experiment. Overall, if two RET-tagged proteins do interact, either directly or as part of a multimeric protein complex, these proteins can bring the donor and acceptor into close proximity (within 10 nm) and energy transfer might take place (Figure 5.1), thus increasing the ratio $F_{\text{YFP}}/F_{\text{CFP}}$.

5.3 Use of RET-based Techniques in the Study of GPCR Oligomerization

The classical RET techniques (*i.e.* fluorescence-RET and bioluminescence-RET), which are based on a non-radiative transfer of energy between 'donor' and 'acceptor' fluorescent molecules as a readout of their proximity (see above), have been extensively used in the study of protein–protein interactions in general and in the GPCR oligomerization phenomenon in particular.

5.3.1 Fluorescence-RET

In the fluorescence-RET (FRET) methodology, a fluorophore donor molecule (*i.e.* CFP) is excited with the matching monochromatic light and if an acceptor fluorophore molecule (*i.e.* YFP) is in close proximity and follows the RET principle (see above), an energy transfer may occur between both molecules (Figure 5.1). For instance, the presence of specific receptor homo- or hetero-complexes in transfected cells can be detected by FRET measurements between two receptors (R1 and R2)—usually *C*-terminally tagged with CFP and YFP,

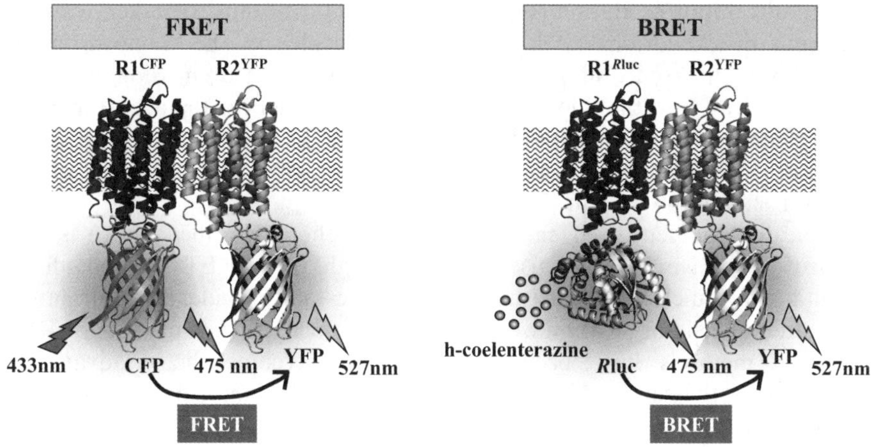

Figure 5.1 Schematic representation of the RET-based methods applied to the study of GPCR oligomerization. Two putative receptors, R1 and R2, bearing a 'donor' CFP (FRET method) or *R*luc (BRET method) and an 'acceptor' YFP chromophore are depicted as R1^{CFP} or R1^{Rluc} and R2^{YFP}, respectively. If R1 and R2 interact, then the donor and acceptor chromophores might be in close proximity ($\leq$10 nm) and energy transfer between the two chromophores can occur after donor excitation at 433 nm (FRET method) or donor-mediated bioluminescence generation (BRET method) after luciferase (*R*luc) oxidation of h-coelenterazine substrate. The schematic GPCR, FP and LP diagrams were prepared using PyMOL (PyMOL Molecular Graphics System, DeLano Scientific, San Carlos, CA, USA) with the crystal structure of the sensory rhodopsin II (PDB 1JGJ), GFP from *Aequorea victoria* (PDB 1EMA) and the luciferase from *Renilla reniformis* (PDB 2PSD) as models. Adapted with permission from ref. 17.

respectively (R1^{CFP} and R2^{YFP})—since dimer formation between R1^{CFP} and R2^{YFP} allows RET between the FPs to occur (Figure 5.1). Typically, the donor and acceptor molecules are fused to the *C*-terminal tail of the GPCRs under study (Figure 5.1), though the attachment of these chromophores to the *N*-terminus or internal attachments has also been described.[12–14] Eventually, the addition of such a large chromophore protein within the receptor's amino acid sequence (*i.e.* GFPs are 4.2 nm long and 2.4 nm diameter[11,15]) might prevent proper receptor functioning, thus making interpretation of results difficult. To overcome this bulk problem, the FP can be substituted by a membrane-permeable yellow-emitting fluorescein analogue termed FlAsH (for fluorescein arsenical hairpin binder)[16] which specifically binds to short cysteine-containing sequences (*e.g.* 'CCPGCC') that are genetically introduced within the sequence of the receptor under study.

As described above, FRET measurements can be expressed as an acceptor/donor emission ratio (*i.e.* YFP/CFP ratio) and can then be compared to samples that express either the donor or the acceptor alone. However, in this FRET approach, because the donor and acceptor emission are measured under continuous donor excitation (Figure 5.1), special care must be taken with the concomitant excitation of the acceptor molecule by the monochromatic light chosen to excite the donor. Similarly, fluorescence emission from the donor can leak into the fluorescence detection channel for the acceptor molecule. Hence, these photophysic crosstalks, together with the sample autofluorescence, result in a 'noise' signal that can give false increases of the acceptor emission—see eqn (2).

Typically, the sensitized emission FRET method described above is performed when live cells are used, but although this FRET method allows simultaneous quick FRET measurements, it does not provide any information about the subcellular location where the FRET is occurring. Thus, other FRET detection methods have been developed to overcome the disadvantages presented by the sensitized emission method. For instance, one approach to improve basic FRET measurements is the temporal disruption of the energy transfer by photobleaching. In donor fluorescence recovery after photobleaching (DFRAP) experiments, which are performed at a single-cell level and using confocal microscopy-based detection systems, the acceptor is directly overexcited in a specific subcellular region and thus irreversibly photodestroyed. Consequently, if there is a real receptor–receptor interaction (*e.g.* FRET between the fluorophores attached to the receptors), after acceptor photodestruction there will be less energy transferred and an increase in the donor emission will be observed (for a review see ref. 17).

Overall, all FRET methods have some advantages and disadvantages which should be taken into account when calculating the FRET efficiency. Thus it is important that, before carrying out any FRET experiment, to determine a clear notion of the desired outcome in order to decide the most appropriate FRET method.[11,17]

5.3.2 Bioluminescence-RET

Bioluminescence-RET (BRET) is a natural RET process that occurs in marine animals (*i.e. Aequorea victoria*, *Renilla reniformis, etc.*) and which involves a bioluminescent enzymatic molecule (*i.e. Renilla* luciferase, *R*luc) as a donor chromophore[18] for the RET process. Thus, the luciferase-mediated substrate oxidation generates bioluminescence activity that can be engaged in a RET process if the proper acceptor molecule is in close proximity and follows the RET principle (Figure 5.1). Interestingly, depending on the substrate that is oxidized, the resultant luciferase-mediated luminescence will better excite one or another acceptor fluorescent protein, thus establishing the BRET method to be used.[19] It is important to mention here that all the substrates currently in use in BRET experiments (*i.e.* coelenterazine derivatives like h-coelenterazine) are small chemical hydrophobic compounds that easily traverse the cell membranes (Figure 5.1), thus allowing experimentation in living cells.[20] The *R*luc-mediated oxidation of h-coelenterazine results in a $\sim$475 nm emission peak (Figure 5.1), with an emission spectrum that does not completely overlap with the YFP excitation spectrum. In addition, the overlap between the emission spectrum generated by h-coelenterazine oxidation and the emission spectrum of YFP is considerable, resulting in a fairly high background signal that has to be taken into account when results are analysed.

To overcome this problem, another coelenterazine derivative, namely DeepBlueC, has been broadly used within the RET approach called $BRET^2$—to distinguish it from the former one, now called $BRET^1$.[21] The use of DeepBlueC as a substrate of *R*luc results in a blue-shifted (~395 nm) emission peak with an emission spectrum that is well-matched for energy transfer to the fluorescent protein GFP.[2] In addition, the emission spectrum of the GFP^2 is well resolved from the *R*luc/DeepBlueC oxidation spectrum because it has a separation of ~115 nm instead the ~52 nm obtained in $BRET^1$. This results in an improvement of the signal-to-noise ratio when the $BRET^2$ approach is employed. However, $BRET^2$ has a major disadvantage, namely the low quantum yield obtained with the oxidation of the DeepBlueC (~100-fold lower compared with h-coelenterazine), thus resulting in much lower absolute signals that also decay rapidly. Consequently, higher expression levels are required for $BRET^2$ to achieve significant luminescence levels. Finally, as BRET assays are based on a substrate-mediated acceptor-excitation, some of the technical problems associated with FRET-based methods (*i.e.* autofluorescence, photobleaching and simultaneous excitation of both donor and acceptor fluorophores[19,22]) are no longer a problem.[17,21,22] But while promising single-cell BRET imaging has been recently performed,[23,24] the subcellular BRET experiments are difficult to achieve since the luciferase substrate cannot be dispensed to a specific subcellular domain. Furthermore, this issue seems to be particularly significant when RET-tagged proteins are overexpressed (*i.e.* in transient transfection experiments) as it could lead to a non-specific RET because random collisions of these proteins occur within the intracellular organelles (*i.e.* endoplasmic reticulum, Golgi apparatus and trafficking vesicles) where they accumulate.[25–27]

Accordingly, the oligomerization between two named GPCRs which are tagged with *R*luc and YFP, respectively ($R1^{Rluc}$ and $R2^{YFP}$), can be detected by means of BRET experiments in a similar way as described for the FRET approach (Figure 5.1). Interestingly, saturation BRET experiments have been extensively used to characterize specific receptor–receptor interactions.[28] Typically, these experiments consist of expressing a constant amount of the donor-labelled protein (*i.e.* $R1^{Rluc}$) with increasing amounts of the acceptor-labelled protein (*i.e.* $R2^{YFP}$) and then detecting the BRET signal after the incubation with the luciferase substrate (for review see ref. 18). The concentration of acceptor that generates half-maximal BRET signal (referred as $BRET_{50}$) is often used as a parameter to express the relative binding affinities between two receptors, although it is difficult to interpret if the association between receptors is irreversible. Overall, with the use of the BRET approach in the study of GPCR oligomerization, it is possible to overcome some of the technical disadvantages presented with the FRET methods, although it also possess some drawbacks (*i.e.* subcellular visualization of protein–protein interactions) that have to be considered.

5.3.3 Fluorescence Lifetime Imaging Microscopy/FRET

One of the most powerful and quantitative approaches to provide direct evidence for the physical interactions between two or more proteins with very high spatial and temporal resolution consists of the combination of the FRET technology with fluorescence lifetime imaging microscopy (FLIM).[29–31] Briefly, FLIM measures the fluorescence lifetime (τ_f), a parameter that defines the average amount of time that a molecule spends in the excited state upon absorption of a photon of light; in the excited state, the molecule is vulnerable to irreversible photodestruction at a rate which, over time, reduces the quantity of chromophores, thus producing an exponential decay in the fluorescence intensity.[29] With FLIM, FRET between a donor and an acceptor can be identified by a shortened fluorescence decay of the donor, since if the acceptor is in close proximity and both chromophores are engaged in a RET process, the donor lifetime is reduced.[32] Interestingly, the reduction of the fluorescence lifetime is a kinetic parameter that can be determined independently of probe concentration, microscope optical path and moderate levels of photobleaching. Furthermore, because only the donor fluorophore lifetime is measured, spectral bleed-through is not a problem in FRET–FLIM. Thus this approach is an extremely robust and quantitative estimate of FRET efficiency and therefore a valuable tool for the study of protein–protein interactions.[29,30,32,33]

Interestingly, the FRET–FLIM approach has been recently used to estimate the size of oligomers formed by the M_2 muscarinic acetylcholine receptor (M_2R).[34] In brief, after expression of the M_2R^{YFP} in Chinese hamster ovary cells, the emission spectra were analysed by spectral deconvolution and the apparent efficiencies estimated by donor-dequenching and acceptor-sensitized emission at different M_2R^{YFP} ratios. Thus, under these experimental conditions, a close analysis of the FRET efficiency determined by FLIM showed that the M_2R^{YFP} is a tetramer.[34]

5.4 Protein–Fragment Complementation Assays to Visualize GPCR Oligomers

Protein–fragment complementation assays (PCAs) consist of the structural and functional reconstitution of an active protein, usually an enzyme or a FP, from two inactive halves that are genetically fused to the interacting protein partners under study.[35,36] If FPs fragments are used for complementation, the assay is known as bimolecular fluorescence complementation (BiFC); alternatively if luminescent proteins (LPs) fragments are used, the assay is called bimolecular luminescence complementation (BiLC) (Figure 5.2). Both assays rely on the generation of a fluorescent/luminescent signal from two non-fluorescent/luminescent fragments of a FP or LP when brought in close proximity by fusion partners (for review see ref. 17). As a fluorophore-based method, the BiFC approach allows direct subcellular visualization of protein–protein interactions in living cells.

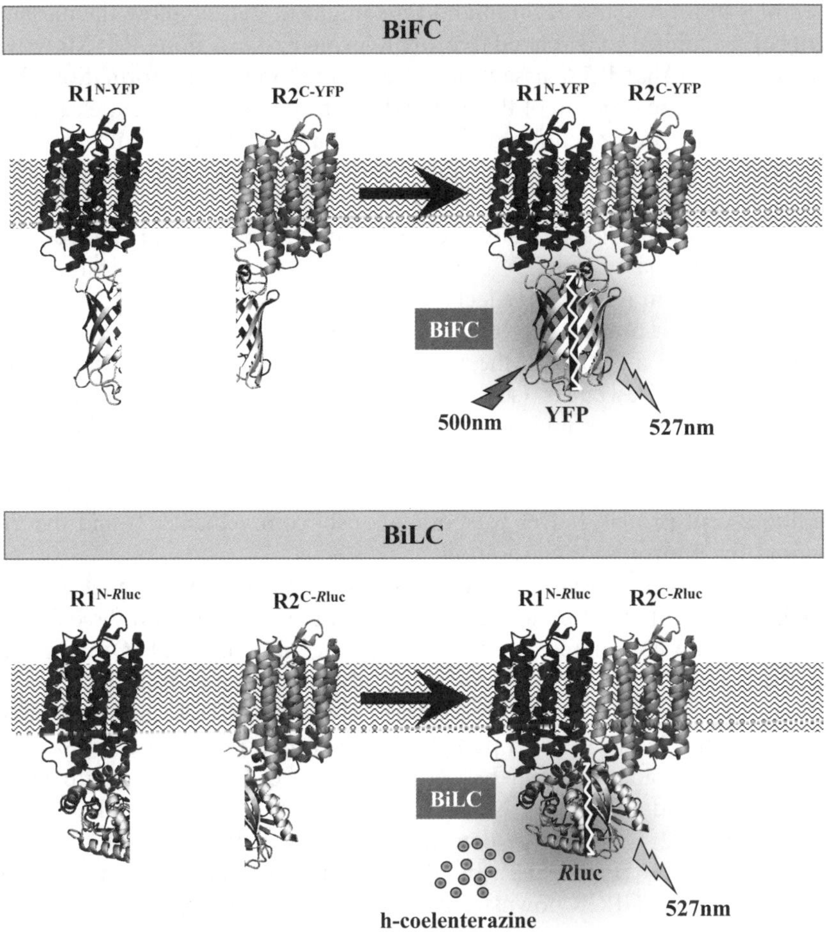

Figure 5.2 Schematic representation of the PCA-based methods applied to the study of GPCR oligomerization. Two putative receptors, R1 and R2, carrying an *N*-terminal YFP (N-YFP) and *C*-terminal YFP (C-YFP) or an *N*-terminal *R*luc (N-*R*luc) and *C*-terminal *R*luc (C-*R*luc) halves are depicted as $R1^{N\text{-}YFP}$ and $R2^{C\text{-}YFP}$ in the BiCF method or as $R1^{N\text{-}Rluc}$ and $R2^{C\text{-}Rluc}$ in the BiLF method, respectively. The interaction of R1 and R2 generates a fluorescent or luminescent complex formed by the two fragments of YFP and *R*luc (N and C), respectively; YFP fluorescence can therefore be directly visualized after excitation at 500 nm and *R*luc luminescence after incubation with the substrate h-coelenterazine. The schematic GPCR, FP and LP diagrams were prepared as described in the legend to Figure 5.1. Adapted with permission from ref. 17.

Interestingly, this approach has been successfully applied to visualize more than 200 protein–protein interactions (see http://sitemaker.umich.edu/kerppola.bifc) in different models of study.[37–40] However, the BiLC approach generates a reconstituted luminescent protein (*i.e.* *R*luc) that commonly assembles in a reversible fashion and thus might in theory allow the measurements of dynamic events related to protein–protein interactions.[41,42] Moreover, despite the fact that BiLC assays do not provide too much information about the subcellular localization of the interaction (the luciferase substrate cannot be dispensed to a specific cellular domain), they have been successfully employed in the detection of protein–protein interactions in living animals because of the high signal-to-noise ratio.[43,44]

Finally, both approaches have some advantages and disadvantages compared with the classical RET techniques described above. For instance, the BiFC assay is a more sensitive technique than the FRET method because the complementation process produces a new fluorescence signal while the FRET method produces a change in the existing fluorescence signal; thus it generates some detection drawbacks (see above). But although the BiFC assays are powerful tools in the study of protein–protein interactions, they have some restrictions that might prevent their wide use, namely the intrinsic irreversible complementation of the fluorescent protein fragments,[39] the inherent ability of the fluorescent protein halves to spontaneously complement[40,45] and the time required for fluorophore maturation.

Consequently, as the FRET methods allow the reversible analysis of the protein complex formation (*i.e.* protein–protein association and dissociation), they should be considered when kinetic studies are required. Overall, RET and BiFC methods have complementary advantages and disadvantages, and consequently we have to decide between dynamics (RET methods) *versus* sensitivity (BiFC assays) when performing our protein–protein studies.

Hence, the oligomerization between two named GPCRs that are tagged with N-YFP and C-YFP or N-*R*luc and C-*R*luc ($R1^{\text{N-YFP}}$ and $R2^{\text{C-YFP}}$ or $R1^{\text{N-}R\text{luc}}$ and $R2^{\text{N-}R\text{luc}}$) can be detected by means of BiFC and BiLC experiments, respectively (Figure 5.2). In addition, these approaches have been used in the study of the GPCR downstream machinery (*i.e.* G proteins or scaffolding proteins).[46,47] For instance, the BiFC assay was applied to visualize specific dimers between the β and γ subunits of G proteins and to demonstrate the functionality of the $G_{\beta\gamma}$ complex in the plasma membrane, thus revealing the role of the different subunits in subcellular targeting.[12,48]

The multicolour BiFC has been developed to visualize simultaneously several interactions in the same cell.[49] Briefly, this assay is based on the generation of complemented FPs by the association of FP fragments coming from different FPs. The 'new' complemented FPs possess distinct spectral properties compared with the native FPs.[49,50] For instance, complementation between Venus *N*-terminal and Cerulean *C*-terminal fragments results in a green-shifted FP compared with Venus or Cerulean native FPs.[49,50] This BiFC approach has been used to visualize the adenosine A_{2A} and the dopamine D_2 receptors homo- and heteromers in living cells and to examine drug-mediated changes in receptor oligomers.[51] Overall, this assay can in theory allow the stoichiometry estimation of a named multiprotein interaction and assist in the screening of drugs that impinge on the stoichiometry of a named GPCR oligomer (*i.e.* $A_{2A}R/D_2R$ heterodimer) which would have high clinical impact.[52]

5.5 GPCR Oligomers at the Surface of Living Cells

As mentioned before, the overexpression of RET-tagged proteins in general and GPCRs in particular might lead to a non-specific RET because of the random collisions of these proteins in the intracellular organelles where they accumulate.[25–27] In addition, as GPCRs are cell surface receptors that detect and transmit a large array of extracellular chemical and sensory signals into cells, the study of the oligomerization of these receptors at the cell surface turns out to be essential. Thus, in order to study GPCR oligomerization at the cell surface of living cells, some further fluorescent-based approaches have been assessed—for instance the time resolved-FRET and the SNAP-taq technology. Interestingly, these approaches can somehow circumvent some of the difficulties encountered with the classical RET techniques (*i.e.* FRET and BRET).

5.5.1 Time-resolved FRET

Time-resolved FRET (TR-FRET) is an outstanding tool for the analysis of membrane protein–protein interactions at the surface of living cells. This technology is based on the use of europium (Eu^{3+}) or terbium (Tb^{3+}) cryptate as a donor molecule with a compatible acceptor chromophore (*i.e.* Alexa Fluor 647, XL665 or allophycocyanin) in a RET process (Figure 5.3).[53–55] These rare earth elements (*i.e.* Eu^{3+}) are characterized by having a long-lived (300–1000 ms) emission fluorescence which facilitates prolonged excitation of the acceptor molecule in the absence of an external continuous excitation source.[53–55] Therefore, as the processes of excitation and detection are separated temporally (time-resolved), the signal-to-noise ratio can be largely increased. This is not only because of the negligible bleed-through emission (from the luminescent lanthanide ion Eu^{3+}) to the acceptor emission wavelength (*e.g.* 665 nm for APC), but also because the long lifetime of the Eu^{3+} cryptate allows a high energy transfer efficiency. Hence, TR-FRET permits circumvention of some of the major limitations encountered with experiments using CFP–YFP as a FRET pair—crosstalk (direct acceptor excitation by light used to excite the donor), bleed-through (partial overlap of donor and acceptor emission wavelengths) and photobleaching.[19]

Typically, in TR-FRET experiments the 'acceptor' and the 'donor' fluorophore molecules (*i.e.* allophycocyanin and Eu^{3+}, respectively) are conjugated to antibodies against epitope tag sequences, allowing the detection of proteins that contain these epitopes within their structure (Figure 5.3).[21,28] Interestingly, this approach has been successfully used in the study of GPCR oligomerization in the cell surface of living cells.[56–59] Briefly, when *N*-terminally epitope-tagged forms of GPCRs are expressed (*i.e.* $R1^{Myc}$ and $R2^{Flag}$) and delivered successfully into the plasma membrane of cells, it is possible to detect these receptor forms in the cell surface of intact cells by means of fluorophore labelled anti-tag antibodies (Figure 5.3). Thus, as the anti-tag antibodies are membrane-impermeable, this approach can be used to exclusively assess cell surface GPCR

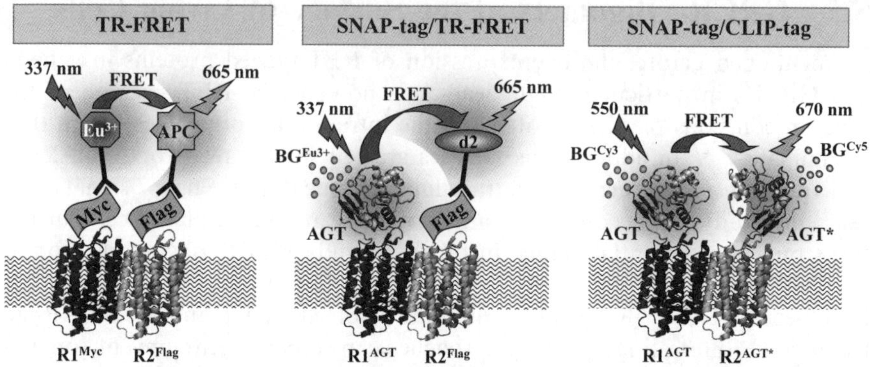

Figure 5.3 Schematic representation of the PCA-based methods applied to the study of GPCR oligomerization. Two putative receptors, R1 and R2, carrying an *N*-terminal YFP (*N*-YFP) and *C*-terminal YFP (*C*-YFP) or an *N*-terminal *R*luc (*N*-*R*luc) and *C*-terminal *R*luc (*C*-*R*luc) halves are depicted as $R1^{N\text{-}YFP}$ and $R2^{C\text{-}YFP}$ in the BiCF method or as $R1^{N\text{-}Rluc}$ and $R2^{C\text{-}Rluc}$ in the BiLF method, respectively. The interaction of R1 and R2 generates a fluorescent or luminescent complex formed by the two fragments of YFP and *R*luc (N and C), respectively; YFP fluorescence can therefore be directly visualized after excitation at 500 nm and *R*luc luminescence after incubation with the substrate h-coelenterazine. The schematic GPCR, FP and LP diagrams were prepared as described in the legend to Figure 5.1. Adapted with permission from ref. 17.

oligomerization by means of FRET measurements between the donor–acceptor pair when the antibodies are in close proximity (Figure 5.3).

One potential caveat of this technology is precisely the use of antibodies, since antibody size (150 kDa) can either increase the FRET signal due to random collisions or hamper oligomer assembly. In addition, it is difficult to ascertain if GPCR oligomerization is not promoted by the bivalent nature of the antibodies,[28] although it is unlikely that the anticipated antibody-mediated receptor clustering would be sufficient to provide enough proximity between the differentially tagged receptors to be responsible for their oligomerization.[56] An additional consideration related to this last point is the fact that, when using TR-FRET techniques, the intrinsically chemical properties of the lanthanide used (*i.e.* Eu^{3+}) should be taken into account. Thus, these tracers have a much larger R_0 values [see eqn (1)], as great as 90 A°, thus increasing the effective range of energy transfer.[54] In addition, the presence of an antibody between the fluorophore molecules and the proteins of interest increases the operative distance of the protein–protein interaction, allowing the detection of larger protein complexes. Overall, association of TR-FRET with antibodies is a powerful tool with which to provide evidence of the existence of GPCR oligomers at the surface of living cells—as has been shown in different studies.[56–59]

Interestingly, a different TR-FRET strategy has been described recently which is based on receptor labelling with selective fluorescent ligands.[60] As such, agonists and antagonists for some GPCRs (*i.e.* oxytocin and dopamine

D2 receptors) were labelled with Eu^{3+} or Lumi4-Tb^{3+} as energy donors or with Alexa-647 or d2 as energy acceptors, thus allowing the detection of GPCR dimerization in the plasma membrane of living cells by means of TR-FRET. In this study the authors first validated the method with different sets of fluorescent ligands in transfected cells expressing GPCRs known to dimerize in a heterologous system; thereafter they applied the approach to native tissues and succeeded in demonstrating the presence of oxytocin receptor dimers and/or oligomers in mammary glands.[60]

5.5.2 SNAP-tag Technology

SNAP-tag technology was firstly described by the group of Kai Johnsson[61] as a general method for the study of protein functions *in vivo*. The technology is based on the enzymatic activity of the human O6-alkylguanine-DNA alkyltransferase (AGT), a 24-kDa protein that acts as a suicide enzyme to covalently transfer modifications from DNA bases onto it.[62]

In essence, the SNAP-tag approach consists firstly of the genetically fusion of AGT to the protein under study and secondly on the staining of the fusion protein with a fluorescent modified AGT substrate. These synthetic AGT substrates are fluorescent derivatives with an O6-benzylguanine (BG) moiety. The BG binds covalently to the SNAP-tag through the transfer of its benzyl group to a cysteine residue within the AGT aminoacidic sequence, thus resulting in a stable thioether bond.[61,63]

A large array of BG derivatives selectively labelled with different chemical fluorophores—both permeable and non-permeable to cell membranes—now exists in the market. In addition, once the AGT fusion proteins are labelled with the fluorescent probes these are extremely stable over periods of hours, thus allowing dynamic studies of cell surface protein complexes in living cells if stained with non-permeable BG probes.[61,63] This technology may also help to overcome some of the limitations encountered with the classical RET methods (*i.e.* FRET and BRET) in which the proteins under study are fused to FPs/LPs (*i.e.* YFP and *R*luc, respectively). Mainly, and as mentioned above, this approach allows the specific labelling of proteins located at the plasma membrane level if desired, and thus differs from the expression of FP-fused proteins where both plasma membrane and intracellular compartments are detected.

This last issue is extremely important as the transient transfection of FP constructs in cultured cells may lead to the possibility of RET occurring within the intracellular compartments (*i.e.* endoplasmic reticulum, Golgi apparatus and trafficking vesicles) where these proteins tend to accumulate if overexpressed, thus making it difficult to demonstrate that RET results from a direct interaction of the proteins at the cell surface rather than the result of random collisions.[56,58,59] Since there is a mutant of the AGT (AGT*)—also called CLIP-tag—that can react with the substrate O6-benzylcytosine (BC), it is possible to perform a double *in vivo* labelling. Thus BC, which is similar to BG but not catabolized by the original AGT, can be labelled with compatible fluorophores to allow specific orthogonal staining of proteins within a named oligomer[64] (Figure 5.3).

Jean-Philippe Pin and colleagues have used this technology in combination with TR-FRET to carry out a rapid, easy and quantitative assessment of cell surface receptor–receptor interactions.[58] In brief, distinct GPCRs (*i.e.* $GABA_B$ receptors) were tagged at their *N*-termini either with AGT ($R1^{AGT}$) or a Flag epitope ($R2^{Flag}$). Next, the cells were stained simultaneously with BG derivatives carrying Eu^{3+} cryptate and with anti-Flag antibodies coupled to the d2 fluorophore (the d2 is an organic motif which emits at ~665 nm). The BG^{Eu3+} can engage in a RET process with the anti-$Flag^{d2}$ antibody if the RET principle is achieved (see Section 5.2). Thus, under GPCR oligomerization conditions, a TR-FRET process occurred when exciting the AGT tagged receptor ($R1^{AGT}$), which is labelled with BG^{Eu3+}, and the d2 tagged receptor ($R2^{d2}$) (Figure 5.3). The SNAP-tag method can be also combined with the CLIP-tag approach to allow the detection of specific receptor homo- or heteromers in the cell surface of transfected cells. Thus, the receptors under study contain AGT or AGT* ($R1^{AGT}$ and $R2^{AGT*}$, respectively) fused at their *N*-terminus and specifically labelled with BG and BC derivatives carrying distinct FRET-compatible fluorophores (*i.e.* Cy3 and Cy5) (Figure 5.3).[64]

In conclusion, the SNAP-tag tool is a powerful strategy for the study of GPCR oligomeric assemblies of distinct GPCRs at the cell surface. Furthermore, the fact that the same fusion construct can be labelled with a variety of chemical compounds gives it a valuable advantage in comparison with the use of fluorescent proteins, since once generated, the same fusion protein can be used—by changing the compound bound to the tag—for multiple types of experiments such as fluorescence microscopy applications, protein purification, or protein–protein interaction analysis.

5.6 Detection of Higher-order GPCR Oligomers in Living Cells

Most fundamental cellular processes involve the formation of multiprotein complexes and so one aspect of post-genomic biomedical research is to systematically record all molecules and their interactions in living cells. Protein–protein interactions constitute an important group of biomolecular interaction networks[65] such as, for example, the neuronal horizontal molecular networks (HMN).[66–68] These HMN take place at the neuronal plasma membrane level, where specific GPCRs interact and form higher-order oligomers or receptor mosaics (RMs) which simultaneously integrate the messages provided by a variety of neurotransmitters. Although the RET techniques described above are well suited to detecting interactions between two proteins when they form homo- or heterodimer complexes, a given protein may also be part of multiprotein complexes involving multiple interactions with several receptor partners and with different stoichiometries. The existence of such higher-order complexes has been demonstrated for some GPCRs[69,70–72] and supported through atomic force microscopy observations of native rhodopsin.[73–75] However, the existence of higher-order complexes has not been demonstrated for some other GPCRs,[69] establishing some controversy.[76] Overall, the formation of highly organized structures could be extremely specific as they might shape the final functional output of a named GPCR oligomer. Indeed, the existence of such specific functional RMs has been historically proposed and recently reviewed.[77–79]

5.6.1 Sequential-RET

RET-based methods were pioneering approaches in demonstrating the occurrence of higher-order GPCRs oligomers in living cells. One of the first fluorescence-based approaches described to analyse interactions between several proteins was the method called three-chromophore RET (3-FRET).[80] This method allowed the detection of trivalent α_{1b} adrenoreceptor complexes in living cells.[70–72] Later, sequential-RET (SRET) was established.[81] In this approach, as described for 3-FRET, the acceptor from an initial RET pair serves as the donor for a second RET process and therefore two consecutive RET processes are engaged. The only difference between 3-FRET and SRET is that in the former the initial donor is a FP and in the latter it is a LP.

Briefly, in the SRET experimental approach shown in Figure 5.4, the bioluminescence emission produced by catabolism of the *R*luc substrate ($R1^{Rluc}$) allows the first energy transfer ($BRET^1$) to a proximal FP acceptor ($R2^{GFP2}$)—assuming the RET principle is achieved (see Section 5.2). The initial FP acceptor may then engage in a second RET process, thus becoming a donor FP in the transfer of energy (FRET) to a downstream acceptor FP ($R3^{YFP}$). This smart RET approach has revealed the existence of distinct higher-order GPCR oligomers in living cells, for instance, the $A_{2A}R/D_2R/CB_1R$ complex and the $A_{2A}R/D_2R/mGlu_5R$ oligomers.[81,82]

Two SRET variants have been described, $SRET^1$ and $SRET^2$; these depend on the initial substrate catabolized by *R*luc (h-coelenterazine or DeepBlueC, respectively). During $SRET$,[1] the luciferase-mediated emission (475 nm) allows energy transfer to a nearby YFP ($BRET^1$ process) and this FP emission (527 nm) can result in a second energy transfer to the FP named DsRed. In $SRET^2$ the *R*luc-mediated emission (395 nm) is able to excite GFP^2 ($BRET^2$ process) and the resulting FP emission (510 nm) can engage in a second energy transfer to the YFP (Figure 5.4).[81] Overall, the SRET method represents a valuable technique for the study of trimeric GPCRs interactions, allowing not only understanding of how these receptors assemble but also how these assemblies are governed by specific allosteric modulators acting on their interfaces.

5.6.2 Integrating PCA Assays and RET Techniques

Recently, some techniques combining PCA and RET assays has been developed for the study of ternary and quaternary protein complexes, making the

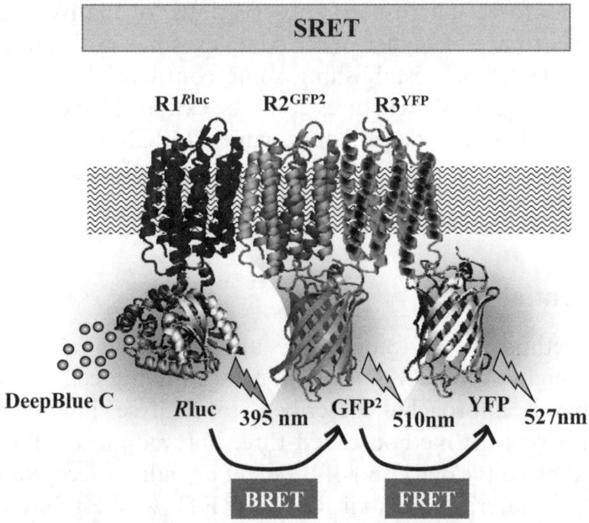

Figure 5.4 Schematic representation of the basic SRET principle applied to the study of GPCR oligomerization. Three putative receptors, R1, R2 and R3 bearing a donor (*R*luc), an acceptor and donor (GFP^2) and acceptor (YFP) chromophores, are depicted as $R1^{Rluc}$, $R2^{GFP2}$ and $R3^{YFP}$, respectively. When the three receptors are in close proximity, a sequential double energy transfer (BRET-FRET) might occur; thus the *R*luc emission at 395 nm excites the GFP^2 which emits at 510 nm and excites the YFP. The schematic GPCR, FP and LP diagrams were prepared as described in the legend to Figure 5.1. Adapted with permission from ref. 17.

detection of higher-order GPCR oligomers or RMs in living cells possible. In brief, PCA/RET approaches (Figure 5.5) centre on the use of a reconstituted FP (*i.e.* YFP) by means of a BiFC technique that serves as an acceptor FP in a RET process. Consequently, BiFC/BRET and BiFC/FRET assays have been readily developed to corroborate the existence of trivalent GPCR complexes.[82–85] These combined PCA/RET approaches have allowed the spatial and temporal resolution of the RM to be detected.[82–85] As the existence of higher-order oligomers is not restricted to the GPCR field (*i.e.* RMs), these technical approaches can be also extended to the study of other protein–protein interacting systems (*i.e.* the heterotrimeric G proteins and other ternary complexes comprising downstream effectors).[86–88]

By simultaneously combining BiLC, BiFC and RET techniques it might be possible in theory to demonstrate the existence of quaternary structures. Indeed, the existence of RMs containing at least four GPCRs has been already described.[89] Briefly, GPCRs genetically fused to *N*- or *C*-terminal fragments of an LP or FP were coexpressed and the formation of a tetravalent protein complex detected by means of a RET process (Figure 5.5). For instance, this approach has allowed the detection of β_2 adrenoreceptors tetramers[90] and the demonstration that at least four D_2Rs are located in close molecular proximity when expressed in living cells, thus giving consistency to the idea that GPCRs

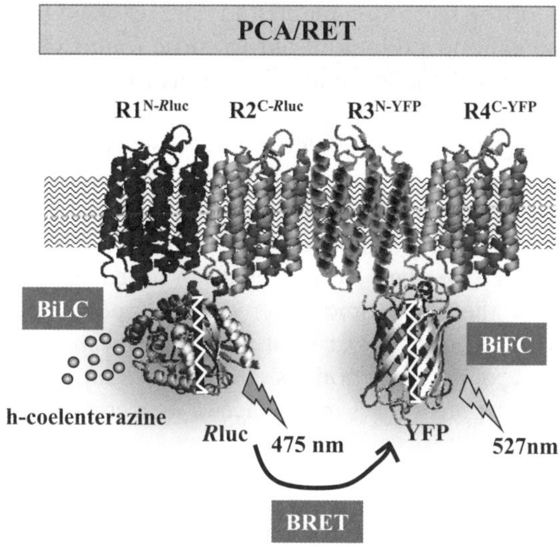

Figure 5.5 Schematic representation of the combination of PCA and RET techniques. By combining the BiLC, BiFC and BRET techniques, the detection of tetrameric GPCRs complexes (R1-R2-R3-R4) is shown. Thus, BiLC generates a complementary luminescent protein (*R*luc) that acts as a donor in a BRET process, with a complementary fluorescent acceptor protein (YFP) generated by BiFC. If these receptors are in close proximity, then donor–acceptor energy transfer can occur after *R*luc substrate (h-coelenterazine) oxidation. Excitation (475 nm) and emission (527 nm) wavelengths are indicated. The schematic GPCR, FP and LP diagrams were prepared as described in the legend to Figure 5.1. Adapted with permission from ref. 17.

are organized as RMs at the plasma membrane.[89] Overall, the combination of PCA and RET approaches has evolved as valuable techniques for the visualization and characterization of ternary and quaternary GPCRs complexes in living cells.

5.7 Conclusions

The existence of GPCR oligomers has been extensively demonstrated in heterologous expression systems. Fluorescence- and luminescence-based assays have played a pivotal role in the molecular characterization of these oligomers. Indeed the research surrounding the study of the GPCR oligomerization phenomenon has catalysed the development of some of these techniques. However, these approaches have some major drawbacks which can preclude proper interpretation of the outcome results: for example, the ectopic overexpression of the respective fusion proteins might prompt the generation of artefacts; and the uselessness of these techniques in the detection of GPCR oligomers in native tissue. The former issue can be easily handled if the expression levels of the receptors under study are assessed (*i.e.* by radioligand binding experiments) and tidily adjusted to those found physiologically. The latest concern is the real challenge in this field as the *in vivo* determination of GPCR oligomers will become essential to fully understand the physiological relevance of the oligomerization phenomenon. The existence of GPCR oligomers in native tissue (*i.e.* in mammary gland) has been demonstrated very recently by means of a TR-FRET approach.[60] Indeed, the *in vivo* occurrence of GPCR oligomers will depend on the spatio-temporal synchronization of the receptors expression (*i.e.* they should be transcribed and translated in the same place and at the same time); therefore a genomic predisposition for GPCR oligomerization might exist for two named receptors. Overall, we can understand GPCR oligomerization as a cellular adaptive mechanism that increases the cell's responsiveness in the face of new environmental challenges.

Acknowledgements

This work was supported by grants SAF2008-01462 and Consolider-Ingenio CSD2008-00005 from Ministerio de Ciencia e Innovación. The authors belong to the 'Neuropharmacology and Pain' accredited research group (Generalitat de Catalunya, 2009 SGR 232).

References

1. J. Miller and I. Stagljar, *Methods Mol. Biol.*, 2004, **261**, 247–262.
2. M. Selbach and M. Mann, *Nat. Methods*, 2006, **3**, 981–983.
3. T. N. Nguyen and J. A. Goodrich, *Nat. Methods*, 2006, **3**, 135–139.
4. P. Wu and L. Brand, *Methods Enzymol.*, 1997, **278**, 321–330.
5. C. F. Becker, R. Seidel, M. Jahnz, K. Bacia, T. Niederhausen, K. Alexandrov, P. Schwille, R. S. Goody and M. Engelhard, *Chembiochem*, 2006, **7**, 891–895.
6. T. Förster, *Ann. Phys. (Weinheim. Ger.)*, 1948, **2**, 55–75.
7. R. A. Cardullo, *Methods Cell Biol.*, 2007, **81**, 479–494.
8. L. Stryer and R. P. Haugland, *Proc. Natl. Acad. Sci. U. S. A.*, 1967, **58**, 719–726.
9. L. Stryer, *Annu. Rev. Biochem.*, 1978, **47**, 819–846.
10. M. Sheng and C. C. Hoogenraad, *Annu. Rev. Biochem.*, 2007, **76**, 823–847.
11. D. W. Piston and G. J. Kremers, *Trends Biochem. Sci.*, 2007, **32**, 407–414.
12. T. R. Hynes, T. E. Hughes and C. H. Berlot, *Methods Mol. Biol.*, 2004, **237**, 233–246.
13. J. P. Vilardaga, M. Bunemann, C. Krasel, M. Castro and M. J. Lohse, *Nat. Biotechnol.*, 2003, **21**, 807–812.
14. D. L. Sheridan, A. Robert, C. H. Cho, J. R. Howe and T. E. Hughes, *Neuroscience*, 2006, **141**, 837–849.
15. N. C. Shaner, G. H. Patterson and M. W. Davidson, *J. Cell. Sci.*, 2007, **120**, 4247–4260.

16. C. Hoffmann, G. Gaietta, M. Bunemann, S. R. Adams, S. Oberdorff-Maass, B. Behr, J. P. Vilardaga, R. Y. Tsien, M. H. Ellisman and M. J. Lohse, *Nat. Methods*, 2005, **2**, 171–176.
17. F. Ciruela, J. P. Vilardaga and V. Fernandez-Duenas, *Trends Biotechnol.*, 2010, **28**, 407–415.
18. K. D. Pfleger and K. A. Eidne, *Nat. Methods*, 2006, **3**, 165–174.
19. J. Gandia, C. Lluis, S. Ferre, R. Franco and F. Ciruela, *Bioessays*, 2008, **30**, 82–89.
20. Y. Xu, D. W. Piston and C. H. Johnson, *Proc. Natl. Acad. Sci. U. S. A.*, 1999, **96**, 151–156.
21. J. Gandia, C. Lluis, S. Ferre, R. Franco and F. Ciruela, *Bioessays*, 2008, **30**, 82–89.
22. F. Ciruela, *Curr. Opin. Biotechnol.*, 2008, **19**, 338–343.
23. X. Xu, M. Soutto, Q. Xie, S. Servick, C. Subramanian, A. G. von Arnim and C. H. Johnson, *Proc. Natl. Acad. Sci. U. S. A.*, 2007, **104**, 10264–10269.
24. H. Hoshino, Y. Nakajima and Y. Ohmiya, *Nat. Methods*, 2007, **4**, 637–639.
25. J. F. Mercier, A. Salahpour, S. Angers, A. Breit and M. Bouvier, *J. Biol. Chem.*, 2002, **277**, 44925–44931.
26. J. R. James, M. I. Oliveira, A. M. Carmo, A. Iaboni and S. J. Davis, *Nat. Methods*, 2006, **3**, 1001–1006.
27. M. Bouvier, N. Heveker, R. Jockers, S. Marullo and G. Milligan, *Nat. Methods*, 2007, **4**, 3–4.
28. G. Milligan and M. Bouvier, *FEBS J.*, 2005, **272**, 2914–2925.
29. P. I. Bastiaens and A. Squire, *Trends Cell Biol.*, 1999, **9**, 48–52.
30. M. Elangovan, R. N. Day and A. Periasamy, *J. Microsc.*, 2002, **205**, 3–14.
31. D. Shcherbo, E. A. Souslova, J. Goedhart, T. V. Chepurnykh, A. Gaintzeva, I. I. Shemiakina, T. W. Gadella, S. Lukyanov and D. M. Chudakov, *BMC Biotechnol.*, 2009, **9**, 24.
32. K. Suhling, P. M. French and D. Phillips, *Photochem. Photobiol. Sci.*, 2005, **4**, 13–22.
33. F. Festy, S. M. Ameer-Beg, T. Ng and K. Suhling, *Mol. Biosyst*, 2007, **3**, 381–391.
34. L. F. Pisterzi, D. B. Jansma, J. Georgiou, M. J. Woodside, J. T. Chou, S. Angers, V. Raicu and J. W. Wells, *J. Biol. Chem.*, 2010, **285**, 16723–16738.
35. T. K. Kerppola, *Nat. Methods*, 2006, **3**, 969–971.
36. S. W. Michnick, P. H. Ear, E. N. Manderson, I. Remy and E. Stefan, *Nat. Rev. Drug Discov.*, 2007, **6**, 569–582.
37. I. Ghosh, A. D. Hamilton and L. Regan, *J. Am. Chem. Soc.*, 2000, **122**, 5658–5659.
38. T. Nagai, A. Sawano, E. S. Park and A. Miyawaki, *Proc. Natl. Acad. Sci. U. S. A.*, 2001, **98**, 3197–3202.
39. C. D. Hu, Y. Chinenov and T. K. Kerppola, *Mol. Cell*, 2002, **9**, 789–798.
40. S. Zhang, C. Ma and M. Chalfie, *Cell*, 2004, **119**, 137–144.
41. I. Remy and S. W. Michnick, *Nat. Methods*, 2006, **3**, 977–979.
42. E. Stefan, S. Aquin, N. Berger, C. R. Landry, B. Nyfeler, M. Bouvier and S. W. Michnick, *Proc. Natl. Acad. Sci. U. S. A.*, 2007, **104**, 16916–16921.

43. T. F. Massoud, R. Paulmurugan, A. De, P. Ray and S. S. Gambhir, *Curr. Opin. Biotechnol.*, 2007, **18**, 31–37.
44. R. Paulmurugan and S. S. Gambhir, *Anal. Chem.*, 2007, **79**, 2346–2353.
45. S. Cabantous, T. C. Terwilliger and G. S. Waldo, *Nat. Biotechnol.*, 2005, **23**, 102–107.
46. R. Sadana, N. Dascal and C. W. Dessauer, *Mol. Pharmacol.*, 2009, **76**, 1256–1264.
47. D. J. Dupre, M. Robitaille, R. V. Rebois and T. E. Hebert, *Annu. Rev. Pharmacol. Toxicol.*, 2009, **49**, 31–56.
48. T. R. Hynes, L. Tang, S. M. Mervine, J. L. Sabo, E. A. Yost, P. N. Devreotes and C. H. Berlot, *J. Biol. Chem.*, 2004, **279**, 30279–30286.
49. C. D. Hu and T. K. Kerppola, *Nat. Biotechnol.*, 2003, **21**, 539–545.
50. Y. J. Shyu, H. Liu, X. Deng and C. D. Hu, *BioTechniques*, 2006, **40**, 61–66.
51. P. A. Vidi, B. R. Chemel, C. D. Hu and V. J. Watts, *Mol. Pharmacol.*, 2008, **74**, 544–551.
52. K. Fuxe, D. Marcellino, D. O. Borroto-Escuela, M. Guescini, V. Fernandez-Duenas, S. Tanganelli, A. Rivera, F. Ciruela and L. F. Agnati, *CNS Neurosci. Ther.*, 2010, **16**, e18–42.
53. G. Mathis, *Clin. Chem.*, 1993, **39**, 1953–1959.
54. G. Mathis, *Clin. Chem.*, 1995, **41**, 1391–1397.
55. H. Bazin, M. Preaudat, E. Trinquet and G. Mathis, *Spectrochim. Acta A: Mol. Biomol. Spectrosc.*, 2001, **57**, 2197–2211.
56. M. McVey, D. Ramsay, E. Kellett, S. Rees, S. Wilson, A. J. Pope and G. Milligan, *J. Biol. Chem.*, 2001, **276**, 14092–14099.
57. F. Ciruela, V. Casado, R. J. Rodrigues, R. Lujan, J. Burgueno, M. Canals, J. Borycz, N. Rebola, S. R. Goldberg, J. Mallol, A. Cortes, E. I. Canela, J. F. Lopez-Gimenez, G. Milligan, C. Lluis, R. A. Cunha, S. Ferre and R. Franco, *J. Neurosci.*, 2006, **26**, 2080–2087.
58. D. Maurel, L. Comps-Agrar, C. Brock, M. L. Rives, E. Bourrier, M. A. Ayoub, H. Bazin, N. Tinel, T. Durroux, L. Prezeau, E. Trinquet and J. P. Pin, *Nat. Methods*, 2008, **5**, 561–567.
59. D. Maurel, J. Kniazeff, G. Mathis, E. Trinquet, J. P. Pin and H. Ansanay, *Anal. Biochem.*, 2004, **329**, 253–262.
60. L. Albizu, M. Cottet, M. Kralikova, S. Stoev, R. Seyer, I. Brabet, T. Roux, H. Bazin, E. Bourrier, L. Lamarque, C. Breton, M. L. Rives, A. Newman, J. Javitch, E. Trinquet, M. Manning, J. P. Pin, B. Mouillac and T. Durroux, *Nat. Chem. Biol.*, 2010, **6**, 587–594.
61. A. Keppler, S. Gendreizig, T. Gronemeyer, H. Pick, H. Vogel and K. Johnsson, *Nat. Biotechnol.*, 2003, **21**, 86–89.
62. A. E. Pegg, *Mutat. Res.*, 2000, **462**, 83–100.
63. A. Keppler, M. Kindermann, S. Gendreizig, H. Pick, H. Vogel and K. Johnsson, *Methods*, 2004, **32**, 437–444.
64. A. Gautier, A. Juillerat, C. Heinis, I. R. Correa Jr, M. Kindermann, F. Beaufils and K. Johnsson, *Chem. Biol.*, 2008, **15**, 128–136.
65. Y. Xia, H. Yu, R. Jansen, M. Seringhaus, S. Baxter, D. Greenbaum, H. Zhao and M. Gerstein, *Annu. Rev. Biochem.*, 2004, **73**, 1051–1087.

66. L. F. Agnati, S. Ferre, C. Lluis, R. Franco and K. Fuxe, *Pharmacol. Rev.*, 2003, **55**, 509–550.
67. J. Bockaert, P. Marin, A. Dumuis and L. Fagni, *FEBS Lett.*, 2003, **546**, 65–72.
68. R. Franco, M. Canals, D. Marcellino, S. Ferre, L. Agnati, J. Mallol, V. Casado, F. Ciruela, K. Fuxe, C. Lluis and E. I. Canela, *Trends Biochem. Sci.*, 2003, **28**, 238–243.
69. C. Brock, N. Oueslati, S. Soler, L. Boudier, P. Rondard and J. P. Pin, *J. Biol. Chem.*, 2007, **282**, 33000–33008.
70. P. S. Park and J. W. Wells, *J. Neurochem.*, 2004, **90**, 537–548.
71. J. F. Lopez-Gimenez, M. Canals, J. D. Pediani and G. Milligan, *Mol. Pharmacol.*, 2007, **71**, 1015–1029.
72. F. Philip, P. Sengupta and S. Scarlata, *J. Biol. Chem.*, 2007, **282**, 19203–19216.
73. Y. Liang, D. Fotiadis, S. Filipek, D. A. Saperstein, K. Palczewski and A. Engel, *J. Biol. Chem.*, 2003, **278**, 21655–21662.
74. D. Fotiadis, B. Jastrzebska, A. Philippsen, D. J. Muller, K. Palczewski and A. Engel, *Curr. Opin. Struct. Biol.*, 2006, **16**, 252–259.
75. B. Jastrzebska, D. Fotiadis, G. F. Jang, R. E. Stenkamp, A. Engel and K. Palczewski, *J. Biol. Chem.*, 2006, **281**, 11917–11922.
76. M. Chabre, R. Cone and H. Saibil, *Nature*, 2003, **426**, 30–31; discussion 31.
77. K. Fuxe, D. Marcellino, D. Guidolin, A. S. Woods and L. F. Agnati, *Physiology (Bethesda)*, 2008, **23**, 322–332.
78. K. Fuxe, M. Canals, M. Torvinen, D. Marcellino, A. Terasmaa, S. Genedani, G. Leo, D. Guidolin, Z. Diaz-Cabiale, A. Rivera, L. Lundstrom, U. Langel, J. Narvaez, S. Tanganelli, C. Lluis, S. Ferre, A. Woods, R. Franco and L. F. Agnati, *J. Neural Transm.*, 2007, **114**, 49–75.
79. L. F. Agnati, D. Guidolin, G. Leo, C. Carone, S. Genedani and K. Fuxe, *Prog. Neurobiol.*, 2009, **90**, 157–175.
80. E. Galperin, V. V. Verkhusha and A. Sorkin, *Nat. Methods*, 2004, **1**, 209–217.
81. P. Carriba, G. Navarro, F. Ciruela, S. Ferre, V. Casado, L. Agnati, A. Cortes, J. Mallol, K. Fuxe, E. I. Canela, C. Lluis and R. Franco, *Nat. Methods*, 2008, **5**, 727–733.
82. N. Cabello, J. Gandia, D. C. Bertarelli, M. Watanabe, C. Lluis, R. Franco, S. Ferre, R. Lujan and F. Ciruela, *J. Neurochem.*, 2009, **109**, 1497–1507.
83. J. Gandia, J. Galino, O. B. Amaral, A. Soriano, C. Lluis, R. Franco and F. Ciruela, *FEBS Lett.*, 2008, **582**, 2979–2984.
84. G. Navarro, P. Carriba, J. Gandia, F. Ciruela, V. Casado, A. Cortes, J. Mallol, E. I. Canela, C. Lluis and R. Franco, *ScientificWorldJournal*, 2008, **8**, 1088–1097.
85. P. A. Vidi, J. Chen, J. M. Irudayaraj and V. J. Watts, *FEBS Lett.*, 2008, **582**, 3985–3990.
86. M. Heroux, M. Hogue, S. Lemieux and M. Bouvier, *J. Biol. Chem.*, 2007, **282**, 31610–31620.

87. R. V. Rebois, M. Robitaille, C. Gales, D. J. Dupre, A. Baragli, P. Trieu, N. Ethier, M. Bouvier and T. E. Hebert, *J. Cell. Sci.*, 2006, **119**, 2807–2818.
88. Y. J. Shyu, C. D. Suarez and C. D. Hu, *Proc. Natl. Acad. Sci. U. S. A.*, 2008, **105**, 151–156.
89. W. Guo, E. Urizar, M. Kralikova, J. C. Mobarec, L. Shi, M. Filizola and J. A. Javitch, *EMBO J.*, 2008, **27**, 2293–2304.
90. R. V. Rebois, M. Robitaille, D. Petrin, P. Zylbergold, P. Trieu and T. E. Hebert, *Methods*, 2008, **45**, 214–218.

CHAPTER 6

Ligand Regulation of GPCR Quaternary Structure

L. SAENZ DEL BURGO AND G. MILLIGAN*

Molecular Pharmacology Group, Institute of Neuroscience and Psychology, College of Medical, Veterinary and Life Sciences, University of Glasgow, Glasgow, G12 8QQ, Scotland, UK

6.1 Introduction

The G protein-coupled receptor (GPCR) family represents one of the largest encoded by the human genome with more than eight hundred genes specifying such polypeptides. At the most basic level, all GPCRs are characterized by the presence of seven transmembrane spanning α-helical segments that are separated by alternating intracellular and extracellular loop regions. Topographically they are arranged with an extracellular *N*-terminus and a cytoplasmic *C*-terminus.

These transmembrane proteins represent the most commonly used signal-transduction system in animals, and various family members are activated by a wide range of endogenously produced ligands including, among others, biogenic amines, peptides, hormones, amino acids and lipids. They are able to transduce different signals through a substantial range of effector mechanisms. Accordingly, GPCRs play a key role in transcellular communication. Furthermore, GPCRs play key roles in perception of the environment as specific family members are activated by light, odorants, pheromones and taste compounds. As a consequence of their importance in the control of so many diverse physiological processes, they are key targets for the development of new drug candidates.

RSC Drug Discovery Series No. 8
G Protein-Coupled Receptors: From Structure to Function
Edited by Jesús Giraldo and Jean-Philippe Pin

Published by the Royal Society of Chemistry, www.rsc.org

Classically, GPCRs have been described as monomeric entities that, upon ligand activation, interact with heterotrimeric G proteins, which in turn interact with and modulate downstream effectors, thus leading to alterations in the concentration of specific second messengers. In recent times, however, a series of functions of GPCRs have been shown to occur in a manner apparently independent of G-protein activation and this has resulted in suggestions that 'GPCR' is too limiting a name for the family. Accordingly, a number of commentators have suggested the terminology of '7TM receptors' cognisant of the fact that this molecular architecture is, perhaps, the defining feature of all family members. Furthermore, it has recently become apparent that many GPCRs can modulate the activity of multiple signalling systems, and different ligands that are able to bind to and activate the same GPCR may display different potencies and relative efficacies at different pathways. This is believed to reflect the fact that there are multiple active states of a GPCR and that different ligands selectively stabilize particular sets of conformations that regulate GPCR-interacting proteins.[1] Although all of the above can be accommodated within a framework of monomeric GPCR polypeptides, it is noticeable that other receptor classes, such as those with intrinsic tyrosine-kinase activity, have been shown to form constitutive or ligand-induced oligomers which are essential for signalling, and are thought to exist in the plasma membrane in monomer–oligomer equilibria.[2–8]

Numerous reports have provided evidence consistent with the oligomerization of GPCRs. However, despite such studies, which have involved observations based on widely different approaches, including migration of GPCRs in SDS-polyacrylamide gels as complexes with the molecular mass predicted for dimers or higher-order oligomers, analysis of ligand binding studies, the capacity to co-immunoprecipitate coexpressed but differentially tagged forms of the same GPCR, the reconstitution of function following coexpression of distinct pairs of mutants that are both individually non-functional and a wide raft of approaches based on the distance constraints of a number of resonance energy transfer techniques, the concept that GPCRs exist as dimers or oligomers remains to be fully validated and accepted.[9–18] Furthermore, whether oligomerization is an integral aspect of the functional biology of GPCRs is still to be completely proven.

Despite this, a large body of evidence indicates that GPCRs might more accurately be described as subunits of large multi-protein signalling complexes because a wide range of GPCR-interacting proteins have been identified.[19–22] Although a great deal of information has being collected regarding many different aspects of GPCR oligomerization, the International Union of Basic and Clinical Pharmacology has suggested types of evidence that should be accumulated prior to acceptance of novel GPCR pairings as authentic heteromers.[23,24] A database of information on GPCR dimerization/oligomerization is maintained at http://data.gpcr-okb.org/gpcr-okb/ and this provides a valuable resource for those with either interest in specific heteromers or those requiring a more general overview of the topic.[25,26]

A number of potential roles for oligomerization in GPCR function have been identified. For example, it has been shown to be important for intracellular trafficking and folding of the receptor protomers and to play a role in cell quality control and export from the endoplasmic reticulum (ER). In this way, incorrectly folded and oligomerized GPCRs can be retained inside the cell and be targeted for degradation.[27–30] There are also indications that oligomerization may play a role in agonist-induced internalization from the cell surface. This reflects a series of studies indicating the co-internalization of pairs of coexpressed GPCRs by agonist ligands with measurable affinity for only one of the receptors.[31] Furthermore, the pathway of agonist-induced internalization can be different for GPCR heteromers compared with that used by the corresponding single receptor.[32–35] Two further roles suggestive of the importance of oligomerization reflect that there have been many observations consistent with the occurrence of cooperativity between GPCR ligand binding sites.[36] Indeed, negative cooperativity has been suggested to have a role in receptor desensitization, as binding of the ligand to such a complex will result in the accelerated dissociation of ligand from interacting binding sites.[37] Oligomerization could also provide a means of signal amplification through the activation by a single ligand of one GPCR protomer that, afterwards, might activate neighbouring GPCRs within an oligomeric complex. Furthermore, since the earliest reports of GPCR hetero-oligomerization, studies have shown ligand-binding and signalling properties of these complexes that are distinct from their constituent protomers.

All this indicates possible mechanisms for the generation of a wide and fascinating diversity of functions and pharmacological properties among coexpressed GPCRs that are able to form heteromers, not least because changes in receptor pharmacology do not seem to follow a common pattern and each receptor pairing may possess unique characteristics. Clear examples where distinct pharmacological properties have been described arising from heteromerization are the δ-κ and δ-μ opioid receptor heteromers.[32,33] However, it is currently unclear whether the panoply of pharmacologically distinct characteristics reported for these heteromers reflects only the wide diversity of ligands available to explore opioid receptor pharmacology or if these are particularly susceptible to allosteric effects within the heteromer.[36] Chemokine receptor pairings have also proved fertile ground for the exploration of differential pharmacology and allosteric effects within GPCR heteromers. In part, this reflects the availability of radiolabelled agonist ligands that have been used at tracer concentrations to allow kinetic effects on ligand cooperativity to be examined in detail.[38–42] As a consequence of such studies on receptor–receptor interactions, an important re-evaluation of the traditionally established model of GPCR organization and their function has been promoted. There is also some evidence that altered levels of GPCR heteromers might represent the molecular basis of certain physiological disorders. These include a significant increase in the amount of angiotensin II AT_1-bradykinin B_2 heteromer together with an increased density of B_2 receptors in pre-eclampsia.[43] However, a number of reports in this area are now rather dated and require independent validation to re-invigorate the potentially important conclusions.

In addition to necessarily changing our view on the quaternary organization and mode of regulation of GPCRs, the existence of homo- and heteromers may have a substantial impact on drug development. As well as the examples noted earlier, the identification of a serotonin 5-HT$_{2A}$ receptor-glutamate mGlu$_2$ receptor heteromer in brain and the effects of mGlu$_2$ receptor ligands on the potency of hallucinogenic 5-HT$_{2A}$ receptor agonists has resulted both in the suggestion that this heteromer might be targeted for the treatment of psychosis and that it might be a molecular target of mGlu$_2$ receptor agonists that are being developed as anti-psychotic medicines.[44–49]

Even though GPCRs are the targets of numerous therapeutic drugs, the link between the techniques routinely used for lead compound identification and drug discovery, and the new possibilities added by GPCR oligomerization are just beginning to be made. New approaches suitable to explore the distinctive pharmacology of GPCR heteromers within ligand screening paradigms have been developed and there is increasing interest in whether incorporation of these might result in the identification of improved or distinct leads that may result eventually in novel therapeutic agents.[50]

Although a literature has developed to suggest that GPCR oligomers and the larger protein complexes they contribute to may be relatively constant, a substantial number of reports have indicated this to be either a transient and/or regulated process.[51–53] Apart from the examples of obligate oligomers that are restricted to the mGluR-related group of GPCRs, where covalent interactions between the extended *N*-terminal 'Venus flytrap' domain of the protomers clearly contribute to oligomer stability, relatively little is known about: the dynamics and regulation of GPCR oligomer formation and/or stability; whether receptor ligands affect the association or dissociation of oligomers; or whether they bind to preformed oligomers early after biosynthesis and alter the oligomeric receptor conformation.

Studies showing that mutant GPCRs can inhibit the cell surface delivery of coexpressed wild-type GPCRs indicate that oligomer assembly often occurs before receptor transport.[27,54] However, it is unclear if dissociation and formation of oligomers can occur at the cell surface. Novel approaches have begun to suggest this may be a dynamic process. For example, by employing total internal reflectance fluorescence microscopy (TIRFM), Hern and co-workers have shown that oligomers of the muscarinic M$_1$ receptor associate and dissociate on the timescale of seconds. Therefore, as the authors have noted, the proportion of transient dimers at the surface of a given cell will depend on the local concentration of protomers, and thus on rates of protomer delivery to and removal from the plasma membrane.[55]

Although a number of reviews have integrated the effects of ligands on GPCR oligomerization into the wider context of the existence and function of GPCR oligomers, we are unaware of efforts to make this topic the central focus. This review thus provides a synopsis of the most recent biochemical and biophysical evidence surrounding the effect of different ligands on the structure of GPCR oligomers. Because most techniques used cannot easily distinguish between dimers and larger oligomers, and many studies on GPCR oligomerization do not make a clear distinction between them, the term oligomer is used in this review as this covers all possible cases.

6.2 Techniques: How are Interactions between Protomers Studied?

Along with observations of cooperativity in ligand binding, co-immunoprecipitation was the earliest approach used to study GPCR oligomerization and the first to have a biochemical basis. In the vast majority of studies and, until recently, unavoidably in those concentrating on receptor homomers, differential epitope-tagging of GPCR cDNA has been used to allow detection of distinct forms of the same GPCR immunoprecipitated by a highly characterized antibody selective for one of the epitope tags from cells transfected to produce their coexpression. Because such studies require solubilization of the GPCRs from cells or cell membranes prior to the immunoprecipitation step, potential concerns over artefactual aggregation have to be overcome as well as concerns that incomplete solubilization might result in bystander pull-down of proteins simply present within the same membrane fragments as the target GPCR. These issues are often addressed by mixing cell populations, each expressing only one of the target GPCRs, prior to solubilization and ensuring no small membrane fragments are present in the solubilized pool by passage through a 0.22 μm filter or extended, high velocity centrifugation.[56] Since the first published work using this strategy to study intermolecular interactions between β_2-adrenoceptor protomers, many other studies have documented the presence of homomers, as well as the existence of heteromers between closely, and also less closely, related GPCR subtypes.[11,12,31–34,57–63] In co-immunoprecipitation studies that explore the effects of the addition of ligands, it is also important to consider that ligand-induced conformational changes in the GPCR partners may alter the accessibility of the specific epitopes used to detect the complex. Therefore, more or less immunoreactive signal can be misinterpreted as changes in the amount of oligomer.

In recent times, approaches based on various resonance energy transfer (RET) techniques, such as bioluminescence resonance energy transfer (BRET) and fluorescence resonance energy transfer (FRET), have become widely used.[56] In part, this reflects that such signals can be measured in populations of intact cells and can even be imaged in single cells.[64,65] The energy transfer donor and acceptor species linked to the respective GPCR protomers need to be in proximity to obtain a signal because the efficiency of energy transfer is extremely dependent on the distance (it decreases with the sixth power of the distance) between the partners.

In practice, although this relationship does not inherently prove that the GPCRs are physically associated because energy transfer is unlikely to be detected if the donor and acceptor are more than 8 nm apart, it is often taken as a direct measure of the presence of a GPCR oligomer. However, in that lack of RET may simply reflect poor orientation of the partner proteins, it has also been argued that detection of RET may simply reflect crowding of monomeric, non-interacting GPCRs particularly when expression levels are high or when using approaches in which the RET donor and acceptor are linked to the intracellular *C*-terminal tail of the potential partner GPCRs because this can result in signals being generated from improperly processed, intracellularly retained receptors.[56,66,67] Also, any change in structural conformation of the receptors may result in movements of the donor and/or acceptor proteins that in turn results in alterations in the signal observed. Indeed, this is the basis of the detection of GPCR activation by ligands when employing intramolecular GPCR FRET reporter constructs and reflects that alterations in distance between the third intracellular loop and the *C*-terminal tail of a GPCR is anticipated due to agonist-induced alteration in the organization of transmembrane domain VI.[68–71]

A distinct FRET approach that takes advantage of the prolonged fluorescence characteristics after excitation of lanthanide compounds that act as energy donor is time-resolved FRET. In a number of forms this approach has measured FRET between appropriately labelled antibodies raised against *N*-terminally tagged recombinant receptors. This generates greater signal to background ratios than conventional FRET and can be used to monitor receptor oligomerization only of receptors that are located at the surface of cells. However, it must be appreciated that the mandatory use of antibodies might provoke receptor clustering as a consequence of their bivalent nature.[56,72,73]

With the caveats above, detection of BRET or FRET signals in intact living cells in the absence of added ligands is generally taken to show unambiguously that GPCRs can form constitutive homomers or heteromers.[15,17,18,74] The potential relevance of such constitutive oligomerization often gains further support if it is observed across a range of expression levels of the GPCR(s) being studied. In the vast majority of studies, however, little attempt is made to assess if a significant proportion of the GPCR is within an oligomeric complex, perhaps because this is a challenging demand. Effects of ligands to increase/decrease detected RET have frequently been recorded and, in certain examples, the detection of such signals have been reported to be dependent entirely upon the presence of an agonist ligand.[14,35,75] However, as noted earlier, the intrinsic sensitivity of RET signals to small changes in distance or orientation of the RET partners places inherent limitations on the interpretation of such data unless clear controls are established.

In order to overcome some limitations of the current techniques, advances in microscopy or the further development of RET-competent ligands will be required to allow wide-ranging studies on cells that express endogenously GPCRs of interest.[76] A wide-ranging discussion of the specific characteristics and latest advances in such technologies is beyond the scope of this chapter. However, we direct the reader to a recently published, detailed review of this topic.[72] Readers are also directed to more detailed and comprehensive overviews of the approaches used to identify populations of GPCR homomers and heteromers.[56,73,77,78]

6.3 Role of GPCR-interacting Proteins in the Regulation and Stability of Oligomers

Before discussing the effect of ligands on the stability of GPCR oligomers, we discuss the potentially important contribution of other receptor-interacting proteins in the formation and regulation of GPCR homo- and heteromers.

Given the key role of heterotrimeric G proteins in signalling from GPCRs, it is likely that interactions with G proteins might regulate the stability of receptor oligomers. The recombinant leukotriene B4 (LTB_4) receptor BLT_1 has been shown to undergo a large increase in its dimeric state upon binding of the natural agonist LTB_4; an effect that requires the heterotrimeric $G\alpha_{i2}\beta_1\gamma_2$ complex.[79,80] Indeed, the pentameric protein complex so formed displays ten times higher receptor affinity for agonists, which might be a consequence of conformational changes induced at the level of the receptor and triggers GDP–GTP exchange at the level of the $G\alpha$ subunit leading to G protein activation.[79] Interestingly, dimer formation and G protein activation is achieved even if only one of the protomers is occupied by the agonist.[81] It has also been shown that the G protein-associated BLT_1 homomer has an asymmetric conformation of the constituent receptor subunits so that only one of them reaches the fully active state. By contrast in the absence of G protein the receptor complex remains in a symmetric configuration.[81] Consequently, interactions between G protein and both subunits of the BLT_1 receptor homomer exert control over the conformational changes of the receptor complex.[81] δ-μ Opioid receptor heteromers have been suggested to play important roles in the regulation of morphine-mediated physiology and development of tolerance to this drug and variation in ligand affinities between opioid receptor heteromers, homomers and monomers are likely to reflect modifications in the binding domain within the receptors.[32,82–84]

Despite the earlier comments on the importance of GPCR oligomerization for effective cellular quality control and of clearance for cell surface delivery, certain studies have suggested that there may be alterations in the molecular make-up of oligomers at the cell surface following synthesis and delivery, observations at odds with the concept that preformed receptor oligomers may exist as such throughout their lifetime.[85,86] Using mutants of the μ-opioid receptor that were retained in the ER it has been suggested that the δ-μ opioid heteromers are not preformed prior to surface delivery. In these studies coexpression of the δ-opioid receptor did not result in its intracellular retention by trafficking-deficient μ-opioid receptor mutants nor did expression of the δ-opioid receptor result in cell surface rescue of the μ-opioid receptor mutants.[83] Similarly, in these studies the endocytic processes of these two receptors appeared to be independent because constitutive internalization of the δ-opioid receptor did not provoke endocytosis of the coexpressed μ-opioid receptor.[83]

Therefore, it was suggested that the specific heteromer properties noted in agonist binding assays must result from transient interactions between the opioid receptors subtypes at the plasma membrane. Furthermore, as uncoupling of the receptors from cellular G proteins by treatment with Pertussis toxin resulted in loss of the heteromer phenotype, these studies also suggested that interaction with G protein might be a requirement for the formation or/and stabilization of such heteromers.[83] Although distinct in conclusions to the studies above, the reported role of receptor transport protein 4 (RTP4)—a Golgi chaperone protein—in interacting with the δ-μ heteromer to increase plasma membrane levels of the heterocomplex[87,88] provides another example of the potential contribution of GPCR-interacting proteins to the generation, stability and potential function of GPCR oligomers. Herein, RTP4 limited ubiquitinylation of the heteromer and, hence, limited its proteolytic destruction, therefore increasing cell surface delivery. Furthermore, in a similar manner, selective opioid ligands were also able to diminish the extent of ubiquitinylation and increase the expression of the heteromer at the plasma membrane. This may be a general mode of action of ligands that act as pharmacological chaperones.[87]

6.4 Effect of Ligands on the Regulation of GPCR Oligomerization

In the last few years, potential roles for oligomerization have been described, including regulation of receptor transport, function and pharmacological specificity.[85,86] However, the extent to which GPCR oligomerization is a regulated process remains unclear and controversial. This reflects the fact that reported effects of ligands on both homomers and heteromers are highly variable. For example, agonist ligands have been reported to increase, decrease or to have no effect on the extent of oligomerization.

Such variability could be attributed to the poor quantitative power of some techniques employed (*e.g.* co-immunoprecipitation), but such variation has also been reported in studies employing a wide range of approaches. The literature in this area is both extensive and, in part, contradictory. As such we have selected examples of various GPCR classes and considered in detail the range of effects reported, rather than attempting to provide a comprehensive review of the field (Table 6.1). In particular, we have not tried to provide comprehensive information on homomers and heteromers involving either opioid receptors or adenosine receptors as these have been covered extensively by other recent reviews.[87,89–104]

6.4.1 Aminergic and Related Receptors

6.4.1.1 Adrenoceptors

The β_2-adrenoceptor was one of the very first GPCRs shown to form constitutive oligomers. Moreover, the extent of oligomerization appeared to be increased upon agonist treatment in a time-dependent manner.[11] Subsequently, it was shown in BRET experiments that this held true even under conditions designed to inhibit receptor internalization.[17] Clearly, a challenge in such experiments is to assess whether clustering of agonist-occupied receptor monomers, first at clathrin-coated pits at the cell surface and, subsequently, within endosomes pinched off from such regions is not the basis of increased BRET signals following agonist treatment. This is particularly relevant because of the recognition that so-called 'bystander' BRET effects can result purely from concentration-induced proximity, and it is also possible that the acidic pH of such endosomes might modulate BRET signals.[56,105,106] In the case of

Table 6.1 Examples of studies using different available approaches that have shed some light on GPCR oligomerization and its regulation by ligands.

Receptor	*Reference*
Aminergic and related receptors	
Adrenoceptors	
α_{1B} homomer	121–124
α_{1B}-α_{1D} heteromer	16, 116, 120
α_{1D}-β_2 heteromer	16, 116, 120
β_1 homomer	109
β_1-β_2 heteromer	108
β_2 homomer	11, 15, 17, 107, 108, 115
β_2-δ heteromer	15
β_2-κ heteromer	31
Dopamine receptors	
D_1-A_1 heteromer	57, 148, 149
D_1-D_2 heteromer	133, 134
D_1-D_3 heteromer	140
D_2 homomer	125–127, 129, 130, 231
D_2-A_{2A} heteromer	144–147
D_2-CB_1 heteromer	132, 141–143
D_2-D_5 heteromer	139
Muscarinic receptors	
M_1 homomer	55, 150–152
M_1-M_2 heteromer	150
M_1-M_3 heteromer	150
M_2 homomer	150, 153–159
M_2-M_3 heteromer	150
M_3 homomer	59, 150, 160
Melatonin receptors	
MT_1 homomer	161
MT_1-MT_2 heteromer	161
MT_2 homomer	161
Chemokine receptors	
CCR2 homomer	162, 167
CCR2-CCR5 heteromer	38
CCR2-CXCR4 heteromer	162
CCR5 homomer	164, 168, 169
CXCR2 homomer	170
CXCR2-δ heteromer	171–174
CXCR4 homomer	162–166
CXCR4-δ heteromer	93, 171–173
Glycoprotein hormone and related receptors	
LH homomer	176–183
GnRH homomer	18, 74, 184, 185
TRH	18, 186
TSH	187–189
Peptide hormone receptors	
Somatostatin receptors	
sst_1 monomer	75
sst_1-sst_5 heteromer	35, 75
sst_2 homomer	63, 192–194
sst_2-sst_3 heteromer	63

Table 6.1 (*Continued*)

Receptor	*Reference*
sst_2-sst_5 heteromer	193, 194
sst_3 homomer	63
sst_5 homomer	35, 190, 192–194
sst_5-D_2 heteromer	14
Bradykinin receptors	
B_1 homomer	198
B_2 homomer	197
B_2-AT_1 heteromer	34, 43, 199–201
Cholecystokinin receptors	
CCK_1 homomer	202–205
Neuropeptide Y receptors	
Y_1 homomer	208–211
Y_2 homomer	208–211
Y_4 homomer	208–212
Vasoactive intestinal peptide and secretin receptors	
Secretin R homomer	213, 214, 216
Secretin R-CLR heteromer	215
Secretin R-GHRH heteromer	215
Secretin R-GLP-1 heteromer	215
Secretin R-GLP-2 heteromer	215
Secretin R-PTH1 heteromer	215
Secretin R-PTH2 heteromer	215
Secretin R-VPAC1 heteromer	213, 214
Secretin R-VPAC2 heteromer	213, 214
VPAC1 homomer	213
VPAC1-VPAC2 heteromer	213
VPAC2 homomer	213
Parathyroid hormone receptors	
PTH1 homomer	217
Neurotensin receptors	
NTS_1 homomer	218
Oxytocin and vasopresin receptors	
OT homomer	76, 219, 220
OT-V_{1A} heteromer	220
OT-V_2 heteromer	220
V_{1A} homomer	220
V_{1A}-V_2 heteromer	220
V_{1B} homomer	220
V_2 homomer	220
Other GPCRs	
P2Y receptors	
$P2Y_{12}$ homomer	222
Thromboxane receptors	
TPα homomer	226, 227
TPα-TPβ heteromer	226, 227
TPβ homomer	226, 227
Nicotinic acid receptors	
NIACR1 homomer	228
NIACR1-NIACR2 heteromer	228
NIACR2 homomer	228

the β_2-adrenoceptor, the agonist effect was observed at the earliest time measured (20 sec), which is too short to be consistent with receptor internalization, and the effect was abolished by the addition of the β_2-adrenoceptor antagonist propanolol.[17]

More recently, FRET saturation experiments have also been used to provide new structural insights into the oligomerization of this receptor following reconstitution into lipid vesicles.[107] These studies analysed not only the specificity of the recorded signals, but also the stoichiometry of the complex by comparing the results with a mathematical model that asserts that high-order oligomers will saturate the FRET signal at lower acceptor/donor ratio.[67] These studies concluded that the β_2-adrenoceptor exists predominantly as a tetramer, although a dynamic equilibrium between monomers and other oligomers might co-exist in this system. Interestingly, neither agonists nor antagonists were reported to modulate the tetramer significantly, although inverse agonists (ICI 118,551, carvedilol and carazolol) were reported to limit conformational movement and the relative orientation of protomers within the complex.[107] These results are consistent with other papers using BRET and fluorescence recovery after photobleaching (FRAP) techniques, which measure the diffusion of non-bleached fluorescent proteins into an irreversibly bleached region.[108,109]

The FRAP studies are of particular interest because, although indicating that the β_2-adrenoceptor forms a stable oligomeric complex, parallel studies on the β_1-adrenoceptor suggested it to form a more transient complex.[109] This would be consistent with interactions between β_1-adrenoceptor protomers being of lower affinity than interactions between β_2-adrenoceptor protomers. However, this is contrary to results derived from saturation BRET studies that have suggested interactions between β_1-adrenoceptor and β_2-adrenoceptor protomers to form homomers are of similar affinity, because $BRET_{50}$ values were similar and, indeed that heteromers between β_1-adrenoceptor and β_2-adrenoceptor protomers might also be of similar affinity.[108] Table 6.2 summarizes the effects of ligands on β_2 adrenoceptor oligomerization.

Although β_1- and β_2-adrenoceptors can clearly interact when coexpressed in heterologous cells, observations on the cell surface location of these receptors in cardiac myocytes and others cells that express endogenously both receptors indicate that the β_2-, but not the β_1-adrenoceptor, is located in 'lipid rafts'. This appears to be required for correct signalling through sequential interactions with G_s and G_i proteins and, therefore, for the proper biphasic control of the contraction rate of myocytes.[110–113] This, however, also indicates that cells must possess the means to deliver individual GPCRs to specific compartments and, presumably, to limit inappropriate heteromeric interactions. In fact, the specific location of these two adrenoceptors within the same cell might explain their differences in functionality as well as in their signalling pathways. Since a myriad of evidence shows that receptors can exist within larger signalling complexes at the plasma membrane, subtype-specific localization of β-adrenoceptors in distinct membrane microdomains may involve their association with different downstream signalling molecules assisted by multidomain

Table 6.2 Effects of ligands on β_2-adrenoceptor oligomerization.

	Ligand	*Effect*	*Technique*	*Reference*
β_2-adrenoceptor as an oligomer	Agonist: Isoproterenol	Increased oligomerization (time-dependent)	WB, Co-IP	11
	Inverse agonist: Timolol	Favoured monomeric state		
	Agonist: Isoproterenol	Increased oligomerization	BRET	17
	Antagonist: Propanolol	Effect of agonist prevented		
β_2-adrenoceptor as a tetramer in equilibrium with other oligomers	Agonist: Isoproterenol Antagonist: Alprenolol	No effect: change in protomer conformation but not in oligomerization	FRET	107
	Inverse agonist: ICI 118,551, Carazolol, Carvedilol	Stabilization of interactions between protomers Formation of higher-order oligomers		
β_2-adrenoceptor as a stable oligomer	Agonist: Isoproterenol Antagonist: Propanolol	No effect	FRAP	109

scaffolding proteins; undoubtedly, this segregation, together with the different interacting proteins colocalizing with the receptors might regulate the formation of heteromeric complexes.[110,111,114]

Like the β_1-adrenoceptor, the β_2-adrenoceptor has been shown to be able to form heteromers with the δ-opioid receptor after transient co-transfection in HEK293 cells as assessed by co-immunoprecipitation and BRET approaches.[15] Nevertheless, and although the BRET signal was smaller than the one obtained when monitoring homo-oligomerization of either receptor, upon agonist treatment the BRET signal recorded was substantially increased. However, the cell surface expression of this heteromer does not seem to be high enough to be recognized in time-resolved FRET experiments.[15] Simultaneously, other authors showed that the β_2-adrenoceptor is able to interact not only with the δ- but also with the κ-opioid receptor, forming heteromers at the cell surface that maintain the same ligand binding, signalling properties and functional coupling as the constituent receptors.[31]

Interestingly, the formation of these heteromers resulted in dramatic alteration of the trafficking properties of these receptors. In cells coexpressing the β_2-adrenoceptor and the δ-opioid receptor, the adrenoceptor underwent endocytosis in response to the opioid receptor agonist etorphine, with 40% reduction of cell surface levels within 60 minutes. In a reciprocal manner, the δ-opioid receptor internalized following addition of a β_2-adrenoceptor agonist (isoproterenol or epinephrine).[31] These studies are consistent with maintenance of the heteromer in the presence of ligand and agonist occupancy of only one protomer being sufficient to promote internalization of the heteromer. Similarly, coexpression of wild-type β_2-adrenoceptor with a mutant unable to respond to conventional agonists but instead to a synthetic ligand that does not have substantial affinity at the wild type receptor resulted in both forms of the receptor being internalized in response to either ligand.[115] By contrast, the β_2-adrenoceptor failed to internalize after agonist treatment when coexpressed with the κ-opioid receptor. Moreover, interaction of the β_2-adrenoceptor with the κ-opioid receptor reduced the ability of the β_2-adrenoceptor to phosphorylate the ERK1/2 MAP kinases and activate this pathway.[31] Therefore, formation of such heteromers might represent a means to regulate β_2-adrenoceptor function in cell coexpressing opioid receptors.

Over the years, the α-adrenoceptors have been widely studied in respect to their signalling pathways, pharmacological properties and tissue distribution. Importantly, heteromerization of these receptors appears to be a key regulatory mechanism for their cell surface expression since, for instance, hetero-interactions involving either α_{1B}- or β_2-adrenoceptors promotes surface localization of the α_{1D} subtype, and these interactions affect significantly not only the trafficking properties but also the pharmacology and signalling of the receptors.[16,116–120] On the other hand, studies on the α_{1B}-adrenoceptor have shown this receptor to form oligomeric complexes, with transmembrane domain IV, together with helix I, being of high importance for the maintenance of this structure and correct maturation and trafficking of the receptor.[121–124] However, and despite α-adrenoceptors playing a key role in the regulation of physiological responses, the effect of ligands on the regulation of oligomerization between pairs of this subfamily has not yet been widely explored.

6.4.1.2 Dopamine Receptors

Because the binding of agonist ligands causes conformational changes in GPCRs and this may be sufficient to modulate oligomerization, in a number of examples crosslinking agents have been used in co-immunoprecipitation or SDS-PAGE experiments to stabilize pre-formed oligomers and/or to detect those newly formed upon ligand treatment (Table 6.3). A generally useful approach in this area has been to employ copper phenanthroline, an oxidizing agent that promotes the formation of disulfide bonds between closely apposed cysteine residues, including those introduced by site directed mutation at strategic locations in the GPCR(s) of interest. For example, this has been used to identify residues in transmembrane domain IV of the dopamine D_2 receptor contributing to a symmetrical dimer interface.[125] Subsequently, a second symmetrical interface at the extracellular end of transmembrane domain I was discovered *via* the same approach. Such multiple points of interaction are consistent with higher-order organization of the D_2 receptor at the plasma

Table 6.3 Effects of different ligands on dopamine D_2 receptor homo- and hetero-oligomerization.

	Ligand	*Effect*	*Technique*	*Reference*
D_2 receptor 'short' isoform oligomerization	Agonist: Propyl-norapomorphine, NPA, Dopamine	Increased oligomerization (concentration-dependent)	FRET	130
	Propyl-norapomorphine, Dopamine	No effect	Co-IP, tr-FRET	146
	Antagonist: Haloperidol	Prevent effect of agonist	FRET	130
	Raclopride	No effect	Co-IP, tr-FRET	146
D_2 receptor 'long' isoform oligomerization	Agonist: Propyl-norapomorphine, NPA, Dopamine	No effect	FRET	130
	Propyl-norapomorphine, Dopamine		Co-IP, tr-FRET	146
D_{2L}-D_{2s} oligomerization	Agonist: Propyl-norapomorphine, Dopamine Antagonist: Raclopride	No effect	Co-IP, tr-FRET	146
CB_1-D_{2L} heteromerization	D_2 Agonist: Quinpirole or/and CB_1 Agonist: HU-210	No effect	FRET	141
	D_2 Agonist: Quinpirole + CB_1 Agonist: CP55,940	Increased heteromerization	Co-IP	142
	D_2 Agonist: Quinpirole or CB_1 Agonist: CP55,940	No effect		
	Persistent D_2 Agonist: Quinpirole or CB_1 Agonist: CP55,940 treatment	Increased D_{2L}-CB_1 heteromerization	FRET, BiFC	143
	D_2 Antagonist: Sulpiride or CB_1 Antagonist: AM281	Prevention of agonist effect: Favoured D_{2L} homomer formation		
A_{2A}-D_{2L} heteromerization	A_{2A} Agonist: CGS21680 and/or D_2 Agonist: Quinpirole	No effect	FRET, BRET	145
	D_2 Agonist. Quinpirole	Reduced D_{2L}-A_{2A} heteromerization and increased D_2 and A_{2A} homomerization	BiFC	144
	D_2 Antagonist: Sulpiride A_{2A} Agonist: MECA	Increased D_{2L}-A_{2A} heteromerization		
	A_{2A} Antagonist: CGS15943	No effect		

membrane of intact cells and at physiological levels of expression,[126,127] as suggested previously for other GPCRs including the complement C5a receptor[128] and the α_{1B}-adrenoceptor.[121] Addition of ligands of varying efficacy to cells expressing D_2 receptors and fusions of these with G protein α subunits noted re-orientation of the receptor complex, as judged by copper phenanthroline crosslinking of distinct residues within transmembrane domain IV, but did not suggest significant variation in oligomer composition upon ligand binding.[127,129]

Earlier, the dopamine D_2 receptor isoforms (S, short, and L, long) were shown to form constitutive dimers in Cos-7 cells as measured in FRET studies.[130] Agonist stimulation of these cells promoted, in a concentration-dependent manner, a significant enhancement in the FRET between D_{2S} protomers that was abolished by adding a dopaminergic antagonist. On the contrary, the D_{2L} isoform required a 27-fold excess in D_{2L} : eCFP *versus* D_{2L} : eYFP plasmid to generate significant FRET signal, and this interaction was completely independent of agonist stimulation.[130] This lack of similarity in the regulation of oligomerization of these two isoforms might be relevant to reported differences in biological function and their distribution in brain has been also shown to be distinct.[131] It is also interesting that the third intracellular loop of the D_2 receptor, where the isoform variation resides, has been suggested to be a element of interaction in heteromers between this receptor and the cannabinoid CB_1 receptor.[132] Indeed the D_{2S} isoform displayed reduced resonance energy transfer signals when coexpressed with the CB_1 receptor than D_{2L}.[132]

The D_1-dopamine receptor is generally considered to generate signals *via* activation of G_s and production of cAMP. However, a series of studies had indicated the existence of a D_1-like pharmacology but with apparent G_q-coupling in brain. The D_1 and D_2 dopamine receptor subtypes can interact when coexpressed,[133] and when they do, activate the G_q-coupled signalling pathway.[134] As a result, dopamine activation of this heteromeric receptor in striatum, or when expressed in heterologous cell lines, induces a rapid increase in intracellular calcium levels that leads to the subsequent initiation of signalling cascades of relevance for synaptic plasticity and long-term effects, such as accelerated morphological maturation and differentiation of striatal neurons as a consequence of an increase in BDNF production.[134–138] Other dopaminergic receptor heteromers may also generate distinct functions. The D_2-D_5 heteromer has recently been shown to have a role in the fine intracellular calcium regulation by dopamine.[139] Apart from the above, D_1 and D_3 dopamine receptors have also been demonstrated to interact in striatal membrane preparations and when coexpressed in HEK293 cells, giving rise to a new complex with different functional and distinct trafficking properties.[140] These provide clear examples of the pivotal importance that GPCR hetero-oligomerization may have in the correct control of physiologically relevant systems through alteration of the signalling properties of the constituent protomers. Undoubtedly, a better understanding of these interactions and their consequences will help to understand and possibly target neurological disorders that involve the dopaminergic system.

Interactions of dopamine receptors with less related GPCRs have also been studied. For example, physical interactions between the D_{2L} receptor and the

CB_1 cannabinoid receptor have been demonstrated by co-immunoprecipitation and FRET experiments.[141,142] In relation to ligand regulation of this heteromer, some studies have reported that exposure to either CB_1 and D_{2L} receptor agonists did not induce significant changes in FRET efficiency between the receptor protomers.[141] By contrast, simultaneous stimulation of both receptors by agonists increased the amount of heteromerization as detected in co-immunoprecipitation assays.[142] Subsequently, using multicolour bimolecular fluorescence complementation (BiFC) in a neuronal cell model, it was shown that formation of the CB_1-D_{2L} hetero-complex is favoured by persistent agonist treatment (up to 30 hours), either by dopaminergic or cannabinergic ligands, relative to the formation of D_{2L}-D_{2L} homomers at the cell surface and/or intracellularly. In contrast, the dopaminergic antagonist sulpiride induced an increase in the formation of D_{2L}-D_{2L} homomers over heteromers.[143]

The significance of this for drug treatments remains uncertain. BiFC has also been employed to investigate the formation and regulation of homo- and hetero-oligomers formed between the dopamine D_{2L} and the adenosine A_{2A} receptors because of the potential physiological relevance of the interactions between these two systems.[144] When coexpressed in the Cath.a differentiated (CAD) neuronal cell model, these two receptors co-exist as homo- and heteromers.[144] Studies on the effect of different acute ligand-treatments had previously shown these not to affect the constitutively formed homo- or heteromers.[145–147] However, prolonged stimulation with quinpirole, a selective dopaminergic D_2 agonist, for 18 hours led to a decrease in the levels of A_{2A}-D_{2L} heteromers relative to A_{2A}-A_{2A} and to D_{2L}-D_{2L} homomers. By contrast, antagonists of the D_2 receptor as well as the A_{2A} receptor agonist 5′-*N*-methylcarboxamidoadenosine had an opposite effect, increasing the proportion of A_{2A}-D_{2L} heteromers.[144] Therefore, it seems that the oligomerization of at least these two interacting receptors might be modulated by their ligands, depending on the duration of the treatment that is applied, without any overall alteration in receptor density. This would appear to favour a model in which heteromers are not of fixed composition throughout their lifetime.

On the other hand, interactions between other subtypes of these receptor families, the adenosine A_1 and the dopamine D_1 receptor, have also been suggested to underlie and clarify the reported antagonism between these two systems *in vivo*, since ligands for the adenosine A_1 receptor appear to regulate the effects of dopaminergic agonists in behavioural studies in animal models.[148,149] These two GPCRs are able to associate and form heteromers in a constitutive manner, but this has also been shown to be dynamically regulated by ligands. The D_1 receptor agonist SFK-38393 resulted in a decrease in the amount of detected heteromers. In contrast, treatment with A_1 and D_1 receptor agonists caused no changes in this heteromeric association in co-transfected Ltk^- fibroblast cells. However, in primary cultures of neurons, SFK-38393, as well as the A_1 receptor agonist *R*-(2)*N*^{6}-(2-phenylisopropyl)adenosine), provoked the clustering of both receptors.[57] The basis for this discrepancy is unclear but may reflect the different relative amounts of each receptor expressed in the two systems. Undoubtedly, further understanding of the regulation of such interactions, not only among closely related receptors of the same family but also between receptors of different systems, by a wide range of ligands will provide insight into the treatment of highly prevalent diseases, such as Parkinson's disease, in the future.

6.4.1.3 Muscarinic Acetylcholine Receptors

Five individual genes encode muscarinic acetylcholine receptors and these have been studied extensively in terms of their potential for oligomerization. M_1, M_2 and M_3 muscarinic receptors, when coexpressed individually or in pairs, have been shown to form high-affinity homo- as well as heteromers (M_1-M_2, M_2-M_3, M_1-M_3) *via* BRET, with a minor proportion of monomers.[150] However, in these studies, treatment with the non-selective acetylcholine mimetic carbachol for up to 30 minutes did not promote any change in the recorded BRET signal for any of the receptor pairs tested.[150] Similarly, the addition of carbachol to Cos-7 cells expressing a M_3 receptor mutant, neither destabilized nor further promoted receptor dimer formation.[59]

More recently, however, an interesting study on the oligomerization state of the human M_1 muscarinic receptor and the influence of different ligands has been published.[151] Using various cell lines and a variety of approaches, it was suggested that the muscarinic toxin 7, which has a high affinity and specificity for the M_1 receptor subtype, stabilizes a constitutive dimeric form of the receptor by inducing conformational rearrangements.[151]

A challenging question is whether, and if so in what proportions, might oligomeric receptors also co-exist with monomeric forms.[150,152] The presence of the antagonist *N*-methylscopolamine was recently reported to exert a similar effect to muscarinic toxin 7, as was binding of the selective M_1 receptor antagonist pirenzepine.[151,152] However, the agonist acetylcholine did not appear to modify the oligomerization status of the M_1 receptor in these studies.[151,152] A related recent study by Hern *et al.* used TIRFM in living cells to visualize individual M_1 receptor molecules expressed in CHO cells and to study their mobility and interaction kinetics with high spatial and temporal resolution.[55] Although a fluorescently labelled form of the antagonist telenzepine was used to visualize the receptor and moving fluorescent spots with intensities anticipated for single receptor molecules were the principal species observed, the authors also observed some spots consistent with the intensity of two fluorophores, *i.e.* a receptor dimer. These observations were interpreted to indicate that only some 15% of the M_1 receptor population was present as dimers and that changes in intensity that were observed during the study were compatible with dimers forming and then dissociating with rapid kinetics.[55] This possibility was further studied using two-colour TIRFM. This suggested the M_1 receptor may exist in a dynamic equilibrium between monomeric and dimeric states that undergo inter-conversion on the timescale of seconds. Interestingly these studies failed to observe higher order oligomeric combinations of the receptor.[55]

This evidence for transient dimerization could provide a molecular explanation for the conflicting data often found in the literature. By contrast, recent studies on the M_2 receptor aimed to elucidate the oligomeric state of this receptor on the basis of fluorescence properties and mathematical analysis.[153,154] The results indicated that the M_2 receptor has a tetrameric structure when expressed at the cell surface of CHO cells, consistent with previous results using a range of approaches.[153,155–158] However, the oligomeric structure of M_2 receptor expressed in Sf9 cells has also been reported to be regulated. The antagonist quinuclidinylbenzilate has been suggested to lead to either an increase in the number of functional oligomers or a stabilization of previously existing complexes.[159]

Because of the wide range of assays available nowadays to study the interactions between GPCRs and the regulation of their structure by ligand treatment, it is important to employ a range of approaches before attempting to draw conclusions, since different approaches may bias the observations in different ways. An interesting example of this is recently published work on the M_3 muscarinic receptor where strong FRET signals were recorded between a M_3 wild-type receptor and a mutated M_3 receptor activated solely by synthetic ligand variant that responds only to a synthetic ligand which has virtually no potency at the wild-type receptor. With this combination and in single cell FRET imaging studies, signals were substantially reduced in response to carbachol.[160] Such results can be interpreted to reflect either a modification in the structure of the constitutively formed oligomer or the disruption of such complexes. However, when cell surface homogeneous time-resolved FRET was applied to study the effect of ligands on M_3 receptor oligomerization, a significant increase in the recorded signal was observed that may mirror either changes in the organization of the oligomeric complex or an increase in the extent of oligomerization of the receptor[160] (see Figure 6.1).

6.4.1.4 Melatonin Receptors

The MT_1 and MT_2 melatonin receptors have also been used to address questions about whether their homo- and hetero-oligomeric state depends on their activation. Using BRET assays, Ayoub *et al.* coexpressed appropriate variants of the melatonin MT_1 and MT_2 receptors and noted that significant BRET could be measured.[161] Herein, antagonist and inverse agonist ligands, as well as agonists, promoted concentration-dependent increases in the measured BRET signal corresponding to pre-existing MT_2 homomers with no change in the constitutive basal BRET signal being observed for MT_1 receptor homomers. As such, ligand-promoted BRET changes did not seem to be related to the activation state of the MT_1 receptor.[161] Interestingly, ligands also increased energy transfer between coexpressed MT_1 and MT_2 receptors but only when one of the two possible energy transfer orientations was employed, *i.e.* either MT_1 or MT_2 acting as the energy acceptor. Such observations highlight the benefit of performing such studies in multiple formats, although this is frequently lacking from many reports. These results are, consequently, more compatible with the idea that ligands induce conformational changes of pre-assembled oligomers and that these can be more easily detected in specific orientations, than with the suggestion of ligands promoting the extent of oligomer formation.[161]

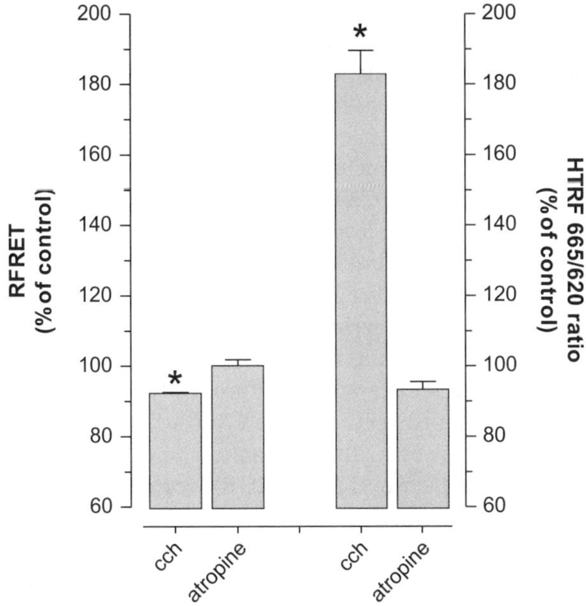

Figure 6.1 Ligand effects on the quaternary organization of the M_3 muscarinic acetylcholine receptor. This graph shows ratiometric FRET at the cell surface (on the left, $p<0.0001$) and homogeneous tr-FRET (on the right, $p<0.0001$) measurements after treatment with the ligands, carbachol (cch) 10^{-3}M and atropine 10^{-5}M, expressed as percentage of the interaction between M_3 receptor protomers in absence of drugs. While carbachol causes a decrease in FRET that would be consistent with reduction in interactions between protomers of the pre-formed oligomeric complexes, it produced a significant increase in the measured homogeneous tr-FRET signal after 20-minute treatment that is at least consistent with enhanced interactions. On the contrary, no effect of atropine was detected. This is a clear example of how the application of different approaches can result in contradictory conclusions on the regulation of oligomers by ligand treatment. Adapted from ref. 160.

6.4.2 Chemokine Receptors

RET techniques have been at the heart of many efforts to study the effect of ligands on the oligomeric structure of GPCRs. Frequently, an increase in the signal obtained after ligand treatment has been interpreted as an increase in the formation of oligomers although, as pointed out earlier, technical limitations must be taken into consideration. The chemokine CXCR4 and CCR2 receptors, for instance, have been shown to form constitutive homo- and heteromers in BRET assays. Furthermore, treatment with ligands of different intrinsic efficacy induced significant modifications in the maximal BRET signals. However, as the direction of change in signal did not correlate with ligand efficacy, such alterations have been interpreted as stabilization of preformed dimers/oligomers in a variety of conformational configurations.[162] In addition, as ligands had little effect on the energy acceptor/energy donor ratio ($BRET_{50}$) required to produce half-maximal signal and, because $BRET_{50}$ is considered to provide a measure of the affinity of the protomers to interact, ligands were concluded not to enhance the association or dissociation of oligomers of these receptors.[162]

In agreement with these conclusions, combinations of BRET and sedimentation studies indicated the CXCR4 receptor to be a constitutive homomer over a wide range of expression levels and completely unaffected by binding of either SDF-1α (stromal cell-derived factor) or HIV-1 gp120.[163,164] Nevertheless, this contrasts with earlier work indicating that SDF-1α was able to induce CXCR4 receptor oligomerization that was otherwise almost undetectable in co-immunoprecipitation assays.[165] Interestingly, a recent study by Wu *et al.* reported on five different crystal structures of the CXCR4 receptor bound with either of two antagonists (IT1t and CVX15).[166] In all cases the structure was of a homodimer. However, the interaction interface(s) depended on the ligand bound: when occupied by ITt1, CXCR4 receptor monomers interacted through the extracellular element of helices V and VI. By contrast with CVX15 bound interactions involved the intracellular extensions of helices II and IV together with the second intracellular loop. Table 6.4 provides a summary of the effects of ligands on chemokine receptor oligomerization.

In the same way, the quaternary organization of the CCR2 receptor was shown to be affected by agonist stimulation, since the natural ligand monocyte chemoattractant protein-1 (MCP-1) triggered receptor homo-oligomerization. In fact, this CCR2 complex formation was suggested to be a necessary step to initiate signal–transduction pathways.[167] It has also been suggested that CCR5 homo-oligomerization, in this case triggered by the natural agonist RANTES (regulated upon activation, normal T cell expressed and secreted), has an important function in limiting HIV-1 infection, potentially giving rise to alternative approaches for the development of anti-HIV-1 drugs that would improve the prevention and treatment of AIDS infection.[168] Interestingly, CCR2-CCR5 heteromers seem also to be formed when these two GPCRs are coexpressed in HEK293 cells but only when agonists for both receptors are present.[38] Indeed, these authors suggested that such heteromers were efficient at inducing biological responses, as the concentration of chemokines required to induce them was ten to a hundred fold lower that required for function at the individual receptors.[38] However, Issafras *et al.*, using another experimental approach, were unable to corroborate these results for the CCR5 receptor.[164] They observed strong, constitutive BRET signals when the CCR5 receptor was expressed at 'physiological' levels in HEK293 cells, at both the plasma membrane and also in the endoplasmic reticulum indicating, as alluded to earlier, that constitutive oligomers of the CCR5 receptor are formed early after biosynthesis. In fact, these observations are in agreement with a early report in which co-immunoprecipitation experiments were consistent with CCR5 oligomerization in absence of ligand stimulation.[169] Moreover, in the studies of Issafras *et al.*, addition of ligands did not modify the extent of oligomerization.[164]

Table 6.4 Effects of ligands on chemokine receptor homo- and hetero-oligomerization.

	Ligand	*Effect*	*Technique*	*Reference*
CCR2 oligomer	Agonist: MCP-1	Stabilization of preformed oligomers Conformational changes	BRET	162
		Induction of oligomerization	Co-IP	38, 167
CCR2-CCR5 heteromer	CCR2 Agonist: MCP-1 + CCR5 Agonist: RANTES	Induction of CCR2 and CCR5 homo-oligomerization and CCR2-CCR5 hetero-oligomerization	Co-IP	38
CCR2-CXCR4 heteromer	CCR2 Agonist: MCP-1 CXCR4 Agonist: SDF-1 CXCR4 Inverse agonist: TC14012	Stabilization of preformed oligomers Conformational changes	BRET	162
CCR5 oligomer	Agonist: RANTES	Induction of oligomerization	Co-IP	38
	Anti CCR5-mAb			168
	Agonist: RANTES, MIP-1β	Constitutive oligomerization No effect of agonists	BRET	164
	Not studied	Constitutive oligomerization	CO-IP	169
CXCR2 oligomer	Agonist: CXCL8	Constitutive oligomerization No effect of agonist	CO-IP	170
CXCR4 oligomer	Full agonist: SDF-1 Partial agonist: AMD3100 Inverse agonist: TC14012	Stabilization of preformed heteromers Conformational changes	BRET	162
	Agonist: SDF-1α HIV-1 gp120	Constitutive oligomer No effect of ligands	BRET, CO-IP, sedimentation studies	163
	Agonist: SDF-1α	Induction of oligomerization	CO-IP	165
CXCR2-δ opioid receptor heteromer	Not studied	Constitutive oligomerization	BRET, FRET, tr-FRET, CO-IP, functional complementation	174
CXCR4-δ opioid receptor heteromer	CXCR4 Agonist: CXCL12 or δ Agonist: DPDPE	Disruption of heteromers and induction of CXCR4 homomer formation	FRET, CO-IP	93
	CXCR4 Agonist: CXCL12 + δ Agonist: DPDPE	Stabilization of heteromers		

Another example of agonist-independent chemokine receptor oligomerization is provided by the CXCR2 receptor, which was been found to form predominantly dimers when expressed in HEK293 cells as well as in cerebellar neuron cells.[170] In fact, this dimeric structure was argued to be the active form of the CXCR2 receptor because various mutants of this receptor, when coexpressed with the wild type, had a dominant negative effect in impairing wild-type receptor function.[170]

More recently interactions between both the CXCR4 and CXCR2 receptors and the δ-opioid receptor have been studied. In part, this reflects the fact that opioid agonists display a broad range of effects on the immune system, including the regulation of inflammatory processes, and chemokines are also regulators of pain and inflammation.[171–173] CXCR4 and δ receptors form stable heteromers in the absence of ligands when analysed in transiently transfected cells and also in monocytes from healthy donors.[93] Interestingly, addition of either the CXCR4 agonist CXCL12 or the δ-opioid selective peptide [D-Pen2, D-Pen5]enkephalin was able to disrupt the heteromer and triggered the formation of new functional CXCR4 and δ-opioid receptor homomers. By contrast, simultaneous addition of both ligands stabilized the CXCR4 and δ-opioid heteromers, and blocked individual receptor responses. This was suggested, rather speculatively, to represent a means to alert the organism of inflammatory processes by increasing pain sensation through the reduction of the antinociceptive activity of the opioid receptors.[93] The CXCR2 receptor has also been identified as a partner for the δ-opioid receptor. Herein, BRET saturation assays showed that the tendency to form heteromeric complexes was higher than for either receptor to form homomers.[174] In this study the application of fusion proteins, in which a G protein α subunit was linked in-frame to the *C*-terminal tail of a GPCR to generate a bi-functional polypeptide, was employed as a distinct tool to explore the interaction between these two receptors. When coexpressing such a pair of receptor fusion proteins, a CXCR2 blocker was able to increase the maximal activity of different opioid receptor agonists. This allosteric effect of the CXCR2 antagonist was produced in a concentration-dependent manner.[174] Therefore, both pharmacological and biophysical and biochemical data support the idea that these two receptors are able to interact physically and that specific ligands for each receptor are capable of regulating not only the receptor they interact with, but also the heteromeric partner of that receptor. This concept may open new opportunities for heterodimer-specific allosteric ligands to be exploited in therapeutic settings.[36,175]

6.4.3 Glycoprotein Hormone and Related Receptors

6.4.3.1 Lutropin Receptor

Various studies employing a range of different approaches, including studies on lateral diffusion by FRAP, have shed light on interactions between lutropin (luteinizing hormone) receptor protomers.[176–182] This receptor has been shown to self-associate at the plasma membrane and to form higher-order oligomers and complexes, with probably other non-receptor proteins, following agonist-dependent desensitization.[177,178] Indeed, it has been suggested that dissociation of self-associated receptors must take place after desensitization, before the receptor can respond again to the hormone.[181] Interestingly, the interaction between monomers of the lutropin receptor and their maturation state appears to depend on where the receptor is expressed. In transiently transfected cells, most of the lutropin receptor complexes are formed between immature intracellular receptors and are not affected by human choriogonadotrophin, presumably because the hormone is unable to access them. In contrast, in cells stably expressing the receptor, it is mostly located at the plasma membrane as a mature protein and, in this case, agonist promotes an increase in receptor dimerization at the cell surface.[176]

Recently, the first clear demonstration of the requirement for intramolecular interactions between LH receptor polypeptides for function *in vivo* was demonstrated by Rivero-Muller *et al.*[183] These workers employed two previously characterized mutant LH receptors, a binding-deficient one and a signalling-deficient one, that had been shown to be capable of specific interaction and restoration of function in transiently transfected cells. Each mutant was used to generate 'knock in' mice in a lutropin 'knock out' background. When these lines were crossed, mice expressing both forms displayed almost full functional complementation of receptor function. Indeed, the male mice displayed normal sexual development, gametogenesis, reproductive behaviour and fertility levels.[179,180,183] Consequently, this work implies that at least this glycoprotein receptor is able to form dimers or oligomers *in vivo* since complementation of these two lutropin mutant receptors is required to re-establish normal luteinizing hormone functionality.

6.4.3.2 Gonadotropin-releasing Hormone Receptor

Analysis of the lateral diffusion of the gonadotropin-releasing hormone (GnRH) receptor *via* FRAP showed that it moves more slowly after addition of agonist or antagonist ligands than in the absence of ligands.[184] In fact, agonist treatment led not only to a concentration-dependent slowing of receptor lateral diffusion but also to a reduction in the percentage of laterally mobile receptors.[184] Using resonance energy transfer approaches, it was seen that increasing concentrations of agonist induced a concentration-dependent increase in energy transfer efficiencies between GnRH receptor protomers, consistent with self-association of the receptors and consistent with the observed reduction in receptor mobility.[18,74,184,185]

6.4.3.3 Thyrotropin-releasing Hormone Receptor

Thyrotropin-releasing hormone (TRH) receptor has been shown to form constitutive homomers in the absence of ligand.[18] However, agonist stimulation resulted in a time- and concentration-dependent increase in recorded energy transfer signals. This could not be explained simply by clustering of the receptors within clathrin-coated pits because disruption of the internalization mechanisms did not alter such effects.[18] In a similar way TRH addition to HEK293 cells expressing low levels of the TRH receptor, and in a pituitary cell model system, has supported the idea that the abundance of TRH receptor homomers can be increased, or maybe stabilized, by agonist binding in concert with an increase in receptor phosphorylation.[186]

6.4.3.4 Thyroid-stimulating Hormone Receptor

Contrary to the previous examples, another glycoprotein hormone receptor, human thyroid-stimulating hormone (TSH) receptor displayed ligand-independent homomer formation in CHO cells, giving rise to specific FRET and co-immunoprecipitation signals.[187,188] In this case, agonist binding produced a concentration-dependent disruption of the oligomeric complexes and, as a consequence, an increase in the monomeric form of the receptor.[187,188] These results were suggested to imply that oligomeric complexes inhibited the constitutive activity of the TSH receptor and that, on agonist binding, disruption of such complexes, generated the receptor active state.[188] Later, TSH receptor dimers were also described at the plasma membrane of living cells through the application of homogeneous time-resolved FRET.[189] However, the authors were not able to confirm the disruption of constitutively formed dimers after stimulation with either TSH or with monoclonal antibodies. In fact, they suggested that the dimer was capable of working as a single functional unit, since recovery of functionality was achieved after coexpression of two different, non-functional mutants of the receptor.[189] Moreover, this receptor appears to be maintained in a dimeric configuration in the presence of agonist, based on observations of negative cooperativity of ligand binding.[189]

6.4.4 Peptide Hormone Receptors

6.4.4.1 Somatostatin Receptors

The somatostatin (sst) receptor family consists of five subtypes, sst_1–sst_5, and all couple selectively to Gi/o G proteins. Each receptor is differentially distributed throughout the brain and periphery, and they have been shown to associate to form homomers and heteromers of various combinations with other members of the family.[35,75,190] However, it has been reported that the sst_1 receptor seems to persist as a monomer regardless of somatostatin-14 (SST-14) stimulation.[75] Furthermore, unlike other sst receptors, the sst_1 subtype is resistant to agonist-induced endocytosis. In fact, after maintained agonist treatment, the expression of this receptor is upregulated at the cell surface.[191]

Interestingly, inhibition of sst_2–sst_2 dimer disruption after agonist-treatment by pre-incubation of cells with a crosslinking agent alters receptor internalization. Therefore, the regulation of the oligomeric structure of the sst_2 receptor may be a prerequisite for correct trafficking of this receptor after agonist-exposure.[192] However, it remains to be clarified whether there might be differences among species orthologues in this regard because other studies observed no effect of ligands on the dimeric structure of the rat sst_2 receptor.[63]

Similarly, the sst_3 receptor has been reported to exist as both a monomer as well as dimers in membrane extracts, and SST-14 stimulation did not modulate the proportion of monomers *versus* homomers.[63] In the same set of studies, the sst_2–sst_3 heteromer was reported to destabilize after agonist-treatment, and only sst_2 underwent internalization after stimulation, possibly due to a conformational change to allow G protein-regulated kinase (GRK) mediated phosphorylation of only the sst_2 receptor.

The low FRET efficiency detected for the human sst_5 receptor in the absence of ligand has been interpreted as evidence that it exists predominantly as a monomer when expressed at levels similar to those found endogenously.[35,190,193] In this case the endogenous agonist SST-14 increased the homomer proportion in a concentration-dependent manner when measured by western blot, and also increased FRET efficiency, suggesting that agonist binding induces oligomerization of the sst_5 receptor.[35,190,193]

Because of the distinct pharmacology and characteristics of internalization of the sst_5 receptor when compared to the sst_1 subtype, hetero-oligomerization between these two receptors has also been investigated. As observed in studies on the sst_5 receptor, a low FRET efficiency in basal conditions may reflect low levels of preformed sst_1–sst_5 hetero-complexes. However, this signal increased in response to agonist.[35,75] Similarly, interactions between sst_2 and sst_5 receptors in cells coexpressing both somatostatin receptor subtypes only occurred after sst_2- but not sst_5-selective agonist treatment or co-stimulation of both receptors.[194] Heteromeric interactions between these receptors also interfered with trafficking of the sst_2 subtype since, as a consequence of being within the hetero-complex, this receptor exhibited a less stable interaction with β-arrestins and a faster recycling rate.[194]

Another factor that is reported to modulate the oligomeric structure of these receptors is the activation state of the G proteins to which they are coupled.[193] While inactivation of receptor-G protein interactions *via* Pertussis toxin treatment had no effect on ligand-induced homo-interactions of the sst_5 receptor, it induced the dissociation of constitutively formed sst_2 homomers in a similar manner as agonist stimulation.[192–194] In contrast, in cells stably expressing both sst_2 and sst_5 receptors, Pertussis toxin treatment provoked an increase in heteromerization of this pair.[193] Therefore, not only interactions with specific ligands, but also communication of the receptor with G proteins in reported to play a key role in regulation of sst receptor oligomerization.

As described above, members of the somatostatin receptor family are regulated in distinct ways. While the human sst_5 receptor forms both homo- and heteromers with sst_2 and sst_1 receptors as a consequence of agonist binding, the latter stays in a monomeric form and, as pointed out earlier, some pieces of information support evidence for agonist-induced dissociation of constitutively self-associated sst_2 receptor oligomers.[192,193] That ligand stimulation appears to modulate the oligomeric state of receptors within the same family in different ways suggests that ligands might generate specific signals depending on the coexpression of specific receptor subtypes in particular cells or tissues. Table 6.5 summarizes the effects of ligands on somatostatin receptor oligomerization.

Apart from the formation of a variety of heteromers within this receptor family, the sst_5 receptor also appears to be able to form agonist-induced heteromers in low density expression systems with more distantly related GPCRs, including the dopamine D_2 receptor. Upon treatment with either SST-14 or dopamine (or simultaneous treatment with both agonists), heteromers with a higher affinity for binding both agonists were formed.[14] Consequently, the behavioural and clinical evidence indicating interactions between the somatostatin and dopaminergic systems could be explained, at least in part, by such molecular interactions.

6.4.4.2 *Bradykinin Receptors*

Bradykinin receptors bind the endogenous peptide agonist within the amino terminus of the receptor and the extracellular loops connecting transmembrane domains IV and V and domains VI and VII.[195,196] The bradykinin B_2 receptor was shown to form agonist-induced dimers/oligomers in PC12 cells that endogenously express this receptor, as well as in transiently transfected HEK293 cells.[197] The amino terminal domain of this receptor has been suggested to play a key role in the generation of these complexes upon agonist activation, since no significant amount of oligomer was detected when the B_2 receptor had a truncated amino terminus.[197] However, since this early publication on the bradykinin B_2 receptor and a later study on residues critical for constitutive homomeric interactions of the B_1 subtype, no other work has explored regulation by ligands of the homomer as well as, if any, hetero-oligomerization characteristics of these bradykinin receptors,[197,198] although interactions between the bradykinin B_2 receptor and the angiotensin AT_1 receptor have been reported and suggested to be modulated in extent in disease.[34,43,199] However, other independent groups have failed to replicate those findings and neither physical interaction between these receptors, assessed by several approaches, nor functional interactions between the two systems could be reproduced.[200]

One explanation to reconcile all the available data is to suggest that this interaction only takes place, and is relevant at a functional level, in specific cellular systems where other yet unidentified proteins need to be involved. At this respect, one of the protein suggested to be important for AT_1-B_2 receptor heterodimerization is the chaperone calreticulin.[201] Due to the possible involvement of the interaction between these two systems in hypertension and preeclampia,[43,199] future studies are needed to clarify what truly happens at a physiological level.

6.4.4.3 *Cholecystokinin Receptors*

Cholecystokinin, along with gastrin, plays key roles in gastrointestinal physiology and the maintenance of nutritional homeostasis. Constitutive homo-oligomerization has been described for the cholecystokinin CCK_1 receptor in the

Table 6.5 Effects of ligands on somatostatin receptor homo- and hetero-oligomerization.

	Ligand	*Effect*	*Technique*	*Reference*
sst_1 as a monomer	Agonist: SST-14	No effect	pbFRET, FCS	75
low levels of sst_1-sst_5 heteromer	Agonist: SST-14	Increased heteromerization (concentration-dependent)	pbFRET, FRET, FCS	35, 75
sst_2 oligomer	Agonist: SST-14	Disruption of homomers (concentration-dependent)	pbFRET, WB	192, 193
		No effect	CO-IP, WB	63
	Pertussis toxin	Disruption of homomers	CO-IP, pbFRET	193, 194
sst_2-sst_3 heteromer	Agonist: SST-14	Destabilization of heteromers	CO-IP, WB	63
sst_2-sst_5 low levels/no preformed heteromer	sst_2 Agonist: L-779,976	Induced heteromerization (concentration-dependent)	CO-IP, pbFRET	193, 194
	sst_5 Agonist: L-817,818	No effect		194
	No selective Agonist: SST-14	No effect		
	Pertussis toxin	Induced heteromerization	CO-IP, pbFRET	193
sst_3 as monomers and oligomers in equilibrium	Agonist: SST-14	No effect	CO-IP, WB	63
sst_5 predominantly as a monomer	Agonist: SST-14	Increased oligomerization (concentration-dependent)	FCS, FRET. pbFRET, Co-IP, WB	75, 35, 190, 194
	Pertussis toxin	No effect	pbFRET	193
D_2-sst_5 heteromer	sst_5 Agonist: SST-14	Increased heteromerization	pbFRET	14
	D_2 Agonist: Dopamine			
	sst_5 Agonist: SST-14 + D_2 Agonist: Dopamine			
	D_2 Antagonist: sulpiride or eticlopride	Modest or no receptor heteromerization		

absence of ligands in membranes of living cells using BRET and co-immunoprecipitation approaches.[202] In this case, a rapid reduction in BRET signal was recorded in response to addition of different agonists in a concentration-dependent manner, with reductions in BRET signal correlating with the affinity and potency of the agonists.[202,203] As such effects were observed when placing the energy donor and acceptor reporters in different positions within the CCK_1 receptor, these results are consistent with agonist occupancy of this receptor inducing dissociation of the homomer, and this effect seemed to be independent of the phosphorylation state of the receptor.[202] In contrast, partial agonists were less effective, while antagonist treatment had no effect on interaction between the protomers.[202] Subsequently, using peptides representing different transmembrane domains of the receptor as potential competitors, it was observed that those corresponding to transmembrane domains VI and VII were able to disrupt receptor BRET signals, suggesting these regions as important interfaces for receptor oligomerization.[204] Equally, studies employing analysis of the decay of time-resolved fluorescence anisotropy (a measure of rotational motion) complemented the previous work in supporting the view that the CCK receptor is able to form constitutive oligomers that are disrupted upon agonist occupancy, whilst antagonists do not show any effects on receptor organization.[205]

6.4.4.4 Neuropeptide Y Receptors

Neuropeptide Y receptors have been shown to form homomers that can be isolated in association with G proteins.[206,207] Furthermore, Y_2 receptor oligomeric complexes are reported to be disrupted to monomers after agonist addition.[208] Pertussis toxin treatment of cells expressing Y_1, Y_2 or Y_4 receptor homomers is also reported to disrupt such dimers, suggesting, as for other receptors considered previously, a role for G protein interaction in stabilization of the complex [208–211] Therefore, the maintenance of functional neuropeptide Y receptors in a dimeric state depends on the presence of $G\alpha_i$ subunits that the receptors can interact with. Ligand-independent formation of neuropeptide Y Y_4 receptor homomers is also reported to be disrupted by pancreatic polypeptide binding in a concentration-dependent manner.[212]

Therefore, at least for the different subtypes of the neuropeptide Y receptor family, it seems that a common pattern can be observed in which constitutively formed dimers are maintained in a complex together with heterotrimeric G_i proteins, while disruption of such dimers, by either agonist stimulation or inactivation of G_i proteins, makes the constituent protomers able to couple to G_q proteins thus changing the signalling pathway they activate.

6.4.4.5 Receptors for Vasoactive Intestinal Peptide and the Secretin Receptor

The $VPAC_1$ and $VPAC_2$ receptors for the vasoactive intestinal peptide (VIP) form constitutive homo- as well as heteromers of living cells early in the biosynthetic pathway.[213] These oligomeric complexes have also been shown to be modulated by agonist, but not antagonist, occupancy as upon VIP treatment a reduction in BRET signal for $VPAC_1$ and $VPAC_2$ homo- as well as heteromers was observed in a concentration-dependent manner.[213] In contrast, secretin receptor homomers are not regulated by secretin treatment whilst the secretin-$VPAC_1$ and secretin-$VPAC_2$ receptor heteromers are unaffected either by either VIP or secretin binding alone or in combination.[213,214]

In the same way, glucagon-like peptide 1 receptor (GLP-1)-, glucagon-like peptide 2 receptor (GLP-2)-, growth hormone-releasing hormone receptor (GHRH)- and calcitonin receptor-like receptor (CLR)-secretin heteromers do not seem to be regulated by the stimulation of the complexes with any ligand tested.[215] By contrast, it has been reported that parathyroid hormone (PTH) significantly reduced BRET signal corresponding to PTH1-secretin and PTH2-secretin heteromers, whilst secretin reduced energy transfer signal only for the PTH2-secretin receptor complex.[215] The implications of these observations remain unclear.

The secretin receptor has been one of the most carefully studied in terms of understanding the basis of homomeric interactions. Using BRET to explore the potential disruption of such interactions upon addition of peptides representing different transmembrane domains of the receptor, it was observed that homomeric interactions require key residues of transmembrane domain IV.[216] This results are similar to studies on the α_{1B}-adrenoceptor that pointed to specific residues within transmembrane domain IV as critical determinants for oligomerization.[121] Importantly, disruption of the oligomeric structure of the secretin receptor did not interfere with agonist binding although its potency was reduced and activation of the cAMP signalling pathway was less effective.[216]

6.4.4.6 Parathyroid Hormone Receptors

The crystal structure of the PTH1 receptor extracellular domain (ECD) in the absence of ligand shows a dimeric configuration, where the ECD C-terminal α2-helix of one protomer occupies the peptide-binding pocket of the opposing protomer, therefore mimicking the binding of the endogenous ligand. As a consequence, addition of the agonist PTH provokes dissociation of the dimer.[217] In order to confirm that this dimeric structure was not an artefact, the authors studied the full-length PTH receptor expressed in COS cells and observed homo-oligomer formation with the ECD α2-helix playing a critical role. They also described a decrease in BRET signal after PTH binding, consistent with the earlier atomic level structural data. Consistent with a model where agonist binding induces the dissociation of the dimers, the monomeric PTH receptor was shown to be capable of coupling to G_s proteins to activate intracellular signaling.[217]

6.4.4.7 Neurotensin Receptors

Most information about GPCR oligomerization has been obtained from studies carried out in transfected cell lines. However, *in vitro* systems that allow the control of receptor density or lipid composition, for example, are beginning to facilitate interpretation of results obtained from cellular systems. In this regard, FRET experiments on the neurotensin 1 (NTS_1) receptor have shown that the receptor is able to form oligomers when expressed at low receptor densities in brain polar lipid liposomes in an agonist-independent manner.[218] The results obtained from these experiments, when compared with theoretical calculations and modelled FRET curves, suggested that the NTS_1 receptor is more likely to exists as homodimers than as higher-order oligomers when expressed in this lipid bilayer model.[218] Undoubtedly, the strength of this work derives from the various controls, not only competition experiments with untagged receptor, but also of the effect of variation in the receptor density to maintain them at low levels. However, inclusion of other signalling components into the system would also allow possible effects of intracellular signalling pathways or receptor-interacting proteins on the oligomerization of the NTS_1 receptor to be explored.

6.4.4.8 Oxytocin and Vasopressin Receptors

Oxytocin (OT) and vasopressin, V_{1A}, V_{1B} and V_2, receptors have all been studied in relation to their capacity to form specific homo- and heteromers.[219,220] Apparently, these receptors have the same propensity to form homo- and heterocomplexes with no differences in their pharmacological or signalling properties.[220] Therefore, the relative expression levels of each of the receptor subtypes is an important factor for determining the amount of homomers *versus* heteromers at the cell surface. Moreover, since molecular interactions were found for both the glycosylated mature and immature forms of the receptors, this process seems to take place before the receptors reach the cell surface, during early steps of their synthesis.[220] Terrillon *et al.* also observed that neither agonist nor antagonist treatment of cells coexpressing different pairs of these receptors had effects on the oligomerization state.[220] However, Devost and Zingg observed that agonist treatment led to a rapid decrease in the amount of detectable surface OT receptor homo-oligomers when a co-immunoprecipitation assay was used, but not when the effect of agonist-stimulation was studied by BRET.[219] Once again, different results were derived when using different techniques and cell lines, and obviously, it is possible that agonist binding induces conformational changes in the receptor structure, rather than physical monomerization, and that this is accompanied by a decrease in the stability of receptor oligomers.

Demonstrating homo- and heteromer formation in native tissues is now the most important challenge in moving forward questions about the role and importance of GPCR dimerization. One means to achieve this is *via* FRET between selective receptor ligands. Recently Albizu *et al.* labelled oxytocin receptors in the mammary glands of lactating rats and measured homogeneous time-resolved FRET signals between the bound ligands.[76] This is consistent with the existence of oxytocin homomers in native tissues and at physiological expression levels. For a more extensive review on oligomerization of these receptors and its modulation, see the recently published study by Cottet *et al.*[221]

6.4.5 Other GPCRs

6.4.5.1 P2Y₁₂ Receptor

The importance of the dynamic regulation of receptor oligomerization and its potential modulation by therapeutic medicines was underlined by a study of the $P2Y_{12}$ receptor.[222] The authors showed that the clinically used antithrombotic agent clopidogrel, and more specifically its active metabolite, disrupted constitutively formed $P2Y_{12}$ receptor oligomers into dimeric and monomeric receptor species at the surface of HEK293 cells as well as in platelets isolated from clopidogrel-treated rats.[222] Interestingly, the study indicated that the basal, oligomeric form of the receptor was located in membrane lipid rafts, whilst following drug-treatment, dimers and monomers of the receptor were relocated outside lipid rafts. This disruption of the quaternary structure of the $P2Y_{12}$ receptor, together with the induced change in cellular localization was suggested to potentially underlie the irreversible anti-platelet activity of this drug.[222]

6.4.5.2 Thromboxane Receptor

The thromboxane (TP) A_2 receptor is a member of the prostanoid receptor subfamily. Two isoforms, TPα and TPβ, derived from alternative splicing of a single gene and differing in their *C*-terminal sequence, are differentially expressed throughout the body and regulate adenylate cyclase activity in opposite fashions.[223–225] The receptor isoforms have been shown to form homo- and heteromers that appear to be the signalling unit of the receptor when they coexpressed in HEK293 cells and also in lysates from human platelets.[226,227] This constitutive oligomerization process is not affected, however, by ligand stimulation or coexpression of GRKs or β-arrestins, but it does affect patterns of endocytosis. The TPα isoform is unable to do so when expressed alone but does when coexpressed with the TPβ isoform.[226]

6.4.5.3 Nicotinic Acid Receptors

Recently, the human nicotinic acid receptor subtypes NIACR1 (previously known as GPR109A) and NIACR2 (GPR109B) have also been shown to form constitutive homo- and heteromers during early steps in biosynthesis. However, once again, these complexes seem to be stable, as they do not seem to change their oligomeric structure after addition of selective agonists, even at high concentrations and with extended incubation periods.[228]

6.5 Conclusions

Oligomerization has emerged as a general trademark of GPCRs that is supported by multiple lines of evidence following expression in cell lines. Evidence indicates that all the steps of the life of a GPCR can be affected by oligomerization and this underlines its importance in understanding the function and regulation of these receptors. If hetero-oligomerization can be more extensively validated—perhaps using recently described approaches included ligand-based FRET[76] and characterization of heteromer-selective antibodies[99]—to be widespread among GPCRs occurring in native tissues, it will offer an exciting, but undoubtedly challenging avenue, for the design of new drugs that may offer tissue selectivity and improved safety.[50,229,230]

As many of the studies described in this review would seem to indicate, GPCR oligomers are unlikely to be unchanging species and may be modulated in disease states and by ligand interaction, although both of these issues require significant further analysis. Whether effects of ligands are predominantly at the level of limited conformational changes associated with receptor activation and signalling or more major alterations in quaternary organization also remains to be further clarified. Indeed, this may be quite different between distinct GPCR homomer and heteromer pairs. In this regard several recent studies have shed light on the idea that GPCR oligomers might be short-lived entities at the cell surface.[55,109,231] However, no general rule concerning the effects of different ligands on the formation and/or stability of GPCR oligomers can so far be defined and therefore effects need be studied on a case by case basis. Undoubtedly, ligand regulation of GPCR quaternary structure will be both physiologically and pharmacologically of great importance if the oligomers behave differently from the corresponding monomers.

Acknowledgements

Laura Saenz del Burgo is a postdoctoral researcher at Professor Milligan's lab and recipient of a scholarship from the Spanish Ministry of Innovation and Science. Work in this area in the Milligan laboratory is supported by the UK Medical Research Council (grant G0900050) and the Biotechnology and Biosciences Research Council (grant BB/E006302/1).

References

1. S. Galandrin, G. Oligny-Longpre and M. Bouvier, *Trends Pharmacol. Sci.*, 2007, **28**(8), 423.
2. C. H. Heldin, *Cell*, 1995, **80**(2), 213.
3. S. R. Hubbard and W. T. Miller, *Curr. Opin. Cell. Biol.*, 2007, **19**(2), 117.
4. R. A. Seifert, C. E. Hart, P. E. Phillips, J. W. Forstrom, R. Ross, M. J. Murray and D. F. Bowen-Pope, *J. Biol. Chem.*, 1989, **264**(15), 8771.
5. C. H. Heldin, A. Ernlund, C. Rorsman and L. Ronnstrand, *J. Biol. Chem.*, 1989, **264**(15), 8905.
6. S. Bishayee, S. Majumdar, J. Khire and M. Das, *J. Biol. Chem.*, 1989, **264**(20), 11699.
7. K. M. Ferguson, *Annu. Rev. Biophys.*, 2008, **37**, 353.

8. A. C. Greenlund, R. D. Schreiber, D. V. Goeddel and D. Pennica, *J. Biol. Chem.*, 1993, **268**(24), 18103.
9. D. Armstrong and P. G. Strange, *J. Biol. Chem.*, 2001, **276**(25), 22621.
10. E. A. Nimchinsky, P. R. Hof, W. G. Janssen, J. H. Morrison and C. Schmauss, *J. Biol. Chem.*, 1997, **272**(46), 29229.
11. T. E. Hebert, S. Moffett, J. P. Morello, T. P. Loisel, D. G. Bichet, C. Barret and M. Bouvier, *J. Biol. Chem.*, 1996, **271**(27), 16384.
12. S. Cvejic and L. A. Devi, *J. Biol. Chem.*, 1997, **272**(43), 26959.
13. J. J. Carrillo, J. Pediani and G. Milligan, *J. Biol. Chem.*, 2003, **278**(43), 42578.
14. M. Rocheville, D. C. Lange, U. Kumar, S. C. Patel, R. C. Patel and Y. C. Patel, *Science*, 2000, **288**(5463), 154.
15. M. McVey, D. Ramsay, E. Kellett, S. Rees, S. Wilson, A. J. Pope and G. Milligan, *J. Biol. Chem.*, 2001, **276**(17), 14092.
16. L. Stanasila, J. B. Perez, H. Vogel and S. Cotecchia, *J. Biol. Chem.*, 2003, **278**(41), 40239.
17. S. Angers, A. Salahpour, E. Joly, S. Hilairet, D. Chelsky, M. Dennis and M. Bouvier, *Proc. Natl. Acad. Sci. U. S. A.*, 2000, **97**(7), 3684.
18. K. M. Kroeger, A. C. Hanyaloglu, R. M. Seeber, L. E. Miles and K. A. Eidne, *J. Biol. Chem.*, 2001, **276**(16), 12736.
19. J. Bockaert, J. Perroy, C. Becamel, P. Marin and L. Fagni, *Annu. Rev. Pharmacol. Toxicol.*, 2010, **50**, 89.
20. J. Bockaert, A. Dumuis, L. Fagni and P. Marin, *Curr. Opin. Drug Discov. Devel.*, 2004, **7**(5), 649.
21. J. Bockaert, G. Roussignol, C. Becamel, S. Gavarini, L. Joubert, A. Dumuis, L. Fagni and P. Marin, *Biochem. Soc. Trans.*, 2004, **32**(5), 851.
22. A. E. Brady and L. E. Limbird, *Cell. Signal.*, 2002, **14**(4), 297.
23. J. P. Pin, R. Neubig, M. Bouvier, L. Devi, M. Filizola, J. A. Javitch, M. J. Lohse, G. Milligan, K. Palczewski, M. Parmentier and M. Spedding, *Pharmacol. Rev.*, 2007, **59**(1), 5.
24. S. Ferre and R. Franco, *Curr. Opin. Pharmacol.*, 2010, **10**(1), 1.
25. L. Skrabanek, M. Murcia, M. Bouvier, L. Devi, S. R. George, M. J. Lohse, G. Milligan, R. Neubig, K. Palczewski, M. Parmentier, J. P. Pin, G. Vriend, J. A. Javitch, F. Campagne and M. Filizola, *BMC Bioinf.*, 2007, **8**, 177.
26. G. Khelashvili, K. Dorff, J. Shan, M. Camacho-Artacho, L. Skrabanek, B. Vroling, M. Bouvier, L. A. Devi, S. R. George, J. A. Javitch, M. J. Lohse, G. Milligan, R. R. Neubig, K. Palczewski, M. Parmentier, J. P. Pin, G. Vriend, F. Campagne and M. Filizola, *Bioinformatics*, 2010, **26**(14), 1804.
27. S. P. Lee, B. F. O'Dowd, G. Y. Ng, G. Varghese, H. Akil, A. Mansour, T. Nguyen and S. R. George, *Mol. Pharmacol.*, 2000, **58**(1), 120.
28. D. Brown and S. Breton, *Kidney Int.*, 2000, **57**(3), 816.
29. F. Coge, S. P. Guenin, A. Renouard-Try, H. Rique, C. Ouvry, N. Fabry, P. Beauverger, J. P. Nicolas, J. P. Galizzi, J. A. Boutin and E. Canet, *Biochem. J.*, 1999, **343 Pt** 1, 231.

30. K. D. Karpa, R. Lin, N. Kabbani and R. Levenson, *Mol. Pharmacol.*, 2000, **58**(4), 677.
31. B. A. Jordan, N. Trapaidze, I. Gomes, R. Nivarthi and L. A. Devi, *Proc. Natl. Acad. Sci. U. S. A.*, 2001, **98**(1), 343.
32. S. R. George, T. Fan, Z. Xie, R. Tse, V. Tam, G. Varghese and B. F. O'Dowd, *J. Biol. Chem.*, 2000, **275**(34), 26128.
33. B. A. Jordan and L. A. Devi, *Nature*, 1999, **399**(6737), 697.
34. S. AbdAlla, H. Lother and U. Quitterer, *Nature.*, 2000, **407**(6800), 94.
35. M. Rocheville, D. C. Lange, U. Kumar, R. Sasi, R. C. Patel and Y. C. Patel, *J. Biol. Chem.*, 2000, **275**(11), 7862.
36. N. J. Smith and G. Milligan, *Pharmacol. Rev.*, 2010, **62**(4), 701.
37. A. Pizard, J. Marchetti, J. Allegrini, F. Alhenc-Gelas and R. M. Rajerison, *J. Biol. Chem.*, 1998, **273**(3), 1309.
38. M. Mellado, J. M. Rodriguez-Frade, A. J. Vila-Coro, S. Fernandez, A. Martin de Ana, D. R. Jones, J. L. Toran and A. C. Martinez, *EMBO J.*, 2001, **20**(10), 2497.
39. J. Y. Springael, E. Urizar and M. Parmentier, *Cytokine Growth Factor Rev.*, 2005, **16**(6), 611.
40. D. Sohy, M. Parmentier and J. Y. Springael, *J. Biol. Chem.*, 2007, **282**(41), 30062.
41. J. Y. Springael, P. N. Le Minh, E. Urizar, S. Costagliola, G. Vassart and M. Parmentier, *Mol. Pharmacol.*, 2006, **69**(5), 1652.
42. L. El-Asmar, J. Y. Springael, S. Ballet, E. U. Andrieu, G. Vassart and M. Parmentier, *Mol. Pharmacol.*, 2005, **67**(2), 460.
43. S. AbdAlla, H. Lother, A. el Massiery and U. Quitterer, *Nat. Med.*, 2001, **7**(9), 1003.
44. J. Gonzalez-Maeso, R. L. Ang, T. Yuen, P. Chan, N. V. Weisstaub, J. F. Lopez-Gimenez, M. Zhou, Y. Okawa, L. F. Callado, G. Milligan, J. A. Gingrich, M. Filizola, J. J. Meana and S. C. Sealfon, *Nature*, 2008, **452**(7183), 93.
45. J. M. Uslaner, S. M. Smith, S. L. Huszar, R. Pachmerhiwala, R. M. Hinchliffe, J. D. Vardigan and P. H. Hutson, *PsychoPharmacology*, 2009, **206**(4), 641.
46. S. T. Patil, L. Zhang, F. Martenyi, S. L. Lowe, K. A. Jackson, B. V. Andreev, A. S. Avedisova, L. M. Bardenstein, I. Y. Gurovich, M. A. Morozova, S. N. Mosolov, N. G. Neznanov, A. M. Reznik, A. B. Smulevich, V. A. Tochilov, B. G. Johnson, J. A. Monn and D. D. Schoepp, *Nat. Med.*, 2007, **13**(9), 1102.
47. J. Gonzalez-Maeso and S. C. Sealfon, *Curr. Med. Chem.*, 2009, **16**(8), 1017.
48. J. Gonzalez-Maeso and S. C. Sealfon, *Trends Neurosci.*, 2009, **32**(4), 225.
49. J. L. Moreno, S. C. Sealfon and J. Gonzalez-Maeso, *Cell Mol. Life Sci.*, 2009, **66**(23), 3777.
50. L. Saenz del Burgo and G. Milligan, *Expert Opin. Drug Discov.*, 2010, **5**(5), 461.
51. D. J. Dupre, M. Robitaille, N. Ethier, L. R. Villeneuve, A. M. Mamarbachi and T. E. Hebert, *J. Biol. Chem.*, 2006, **281**(45), 34561.

52. D. J. Dupre and T. E. Hebert, *Cell. Signal.*, 2006, **18**(10), 1549.
53. M. David, M. Richer, A. M. Mamarbachi, L. R. Villeneuve, D. J. Dupre and T. E. Hebert, *Cell. Signal.*, 2006, **18**(12), 2172.
54. X. Zhu and J. Wess, *Biochemistry*, 1998, **37**(45), 15773.
55. J. A. Hern, A. H. Baig, G. I. Mashanov, B. Birdsall, J. E. Corrie, S. Lazareno, J. E. Molloy and N. J. Birdsall, *Proc. Natl. Acad. Sci. U. S. A.*, 2010, **107**(6), 2693.
56. G. Milligan and M. Bouvier, *FEBS J.*, 2005, **272**(12), 2914.
57. S. Gines, J. Hillion, M. Torvinen, S. Le Crom, V. Casado, E. I. Canela, S. Rondin, J. Y. Lew, S. Watson, M. Zoli, L. F. Agnati, P. Verniera, C. Lluis, S. Ferre, K. Fuxe and R. Franco, *Proc. Natl. Acad. Sci. U. S. A.*, 2000, **97**(15), 8606.
58. G. Y. Ng, B. F. O'Dowd, S. P. Lee, H. T. Chung, M. R. Brann, P. Seeman and S. R. George, *Biochem. Biophys. Res. Commun*, 1996, **227**(1), 200.
59. F. Y. Zeng and J. Wess, *J. Biol. Chem.*, 1999, **274**(27), 19487.
60. J. H. White, A. Wise, M. J. Main, A. Green, N. J. Fraser, G. H. Disney, A. A. Barnes, P. Emson, S. M. Foord and F. H. Marshall, *Nature*, 1998, **396**(6712), 679.
61. K. A. Jones, B. Borowsky, J. A. Tamm, D. A. Craig, M. M. Durkin, M. Dai, W. J. Yao, M. Johnson, C. Gunwaldsen, L. Y. Huang, C. Tang, Q. Shen, J. A. Salon, K. Morse, T. Laz, K. E. Smith, D. Nagarathnam, S. A. Noble, T. A. Branchek and C. Gerald, *Nature*, 1998, **396**(6712), 674.
62. K. Kaupmann, B. Malitschek, V. Schuler, J. Heid, W. Froestl, P. Beck, J. Mosbacher, S. Bischoff, A. Kulik, R. Shigemoto, A. Karschin and B. Bettler, *Nature*, 1998, **396**(6712), 683.
63. M. Pfeiffer, T. Koch, H. Schroder, M. Klutzny, S. Kirscht, H. J. Kreienkamp, V. Hollt and S. Schulz, *J. Biol. Chem.*, 2001, **276**(17), 14027.
64. R. Hovius, P. Vallotton, T. Wohland and H. Vogel, *Trends Pharmacol. Sci.*, 2000, **21**(7), 266.
65. K. D. Pfleger and K. A. Eidne, *Nat. Methods*, 2006, **3**(3), 165.
66. M. Bouvier, N. Heveker, R. Jockers, S. Marullo and G. Milligan, *Nat. Methods*, 2007, **4**(1), 3.
67. J. R. James, M. I. Oliveira, A. M. Carmo, A. Iaboni and S. J. Davis, *Nat. Methods*, 2006, **3**(12), 1001.
68. C. Hoffmann, G. Gaietta, M. Bunemann, S. R. Adams, S. Oberdorff-Maass, B. Behr, J. P. Vilardaga, R. Y. Tsien, M. H. Ellisman and M. J. Lohse, *Nat. Methods*, 2005, **2**(3), 171.
69. M. J. Lohse, M. Bunemann, C. Hoffmann, J. P. Vilardaga and V. O. Nikolaev, *Curr. Opin. Pharmacol.*, 2007, **7**(5), 547.
70. R. Nygaard, T. M. Frimurer, B. Holst, M. M. Rosenkilde and T. W. Schwartz, *Trends Pharmacol. Sci.*, 2009, **30**(5), 249.
71. B. Holst, R. Nygaard, L. Valentin-Hansen, A. Bach, M. S. Engelstoft, P. S. Petersen, T. M. Frimurer and T. W. Schwartz, *J. Biol. Chem.*, 2010, **285**(6), 3973.
72. E. Alvarez-Curto, J. D. Pediani and G. Milligan, *Anal. Bioanal. Chem.*, 2010, **398**(1), 167.

73. C. Harrison and P. H. van der Graaf, *J. Pharmacol. Toxicol. Methods*, 2006, **54**(1), 26.
74. A. Cornea, J. A. Janovick, G. Maya-Nunez and P. M. Conn, *J. Biol. Chem.*, 2001, **276**(3), 2153.
75. R. C. Patel, U. Kumar, D. C. Lamb, J. S. Eid, M. Rocheville, M. Grant, A. Rani, T. Hazlett, S. C. Patel, E. Gratton and Y. C. Patel, *Proc. Natl. Acad. Sci. U. S. A.*, 2002, **99**(5), 3294.
76. L. Albizu, M. Cottet, M. Kralikova, S. Stoev, R. Seyer, I. Brabet, T. Roux, H. Bazin, E. Bourrier, L. Lamarque, C. Breton, M. L. Rives, A. Newman, J. Javitch, E. Trinquet, M. Manning, J. P. Pin, B. Mouillac and T. Durroux, *Nat. Chem. Biol.*, 2010, **6**(8), 587.
77. J. Gandia, J. Galino, O. B. Amaral, A. Soriano, C. Lluis, R. Franco and F. Ciruela, *FEBS Lett*, 2008, **582**(20), 2979.
78. J. Gandia, C. Lluis, S. Ferre, R. Franco and F. Ciruela, *Bioessays*, 2008, **30**(1), 82.
79. J. L. Baneres and J. Parello, *J. Mol. Biol.*, 2003, **329**(4), 815.
80. J. L. Baneres, A. Martin, P. Hullot, J. P. Girard, J. C. Rossi and J. Parello, *J. Mol. Biol.*, 2003, **329**(4), 801.
81. M. Damian, A. Martin, D. Mesnier, J. P. Pin and J. L. Baneres, *EMBO J.*, 2006, **25**(24), 5693.
82. I. Gomes, B. A. Jordan, A. Gupta, N. Trapaidze, V. Nagy and L. A. Devi, *J. Neurosci.*, 2000, **20**(22), RC110.
83. P. Y. Law, L. J. Erickson-Herbrandson, Q. Q. Zha, J. Solberg, J. Chu, A. Sarre and H. H. Loh, *J. Biol. Chem.*, 2005, **280**(12), 11152.
84. J. F. Lopez-Gimenez and G. Milligan, *CNS Neurol. Disord. Drug Targets*, 2010, **9**(5).
85. G. Milligan, *Br J. Pharmacol.*, 2008, **153** Suppl 1, S216.
86. G. Milligan, *Curr. Opin. Pharmacol.*, 2010, **10**(1), 23.
87. F. M. Decaillot, R. Rozenfeld, A. Gupta and L. A. Devi, *Proc. Natl. Acad. Sci. U. S. A.*, 2008, **105**(41), 16045.
88. X. Zhang, L. Bao and J. S. Guan, *Trends Pharmacol. Sci.*, 2006, **27**(6), 324.
89. R. M. van Rijn, J. L. Whistler and M. Waldhoer, *Curr. Opin. Pharmacol.*, 2010, **10**(1), 73.
90. K. Alfaras-Melainis, I. Gomes, R. Rozenfeld, V. Zachariou and L. Devi, *Front. Biosci.*, 2009, **14**, 3594.
91. W. Walwyn, S. John, M. Maga, C. J. Evans and T. G. Hales, *Mol. Pharmacol.*, 2009, **76**(1), 134.
92. Y. Zheng, E. Akgun, K. G. Harikumar, J. Hopson, M. D. Powers, M. M. Lunzer, L. J. Miller and P. S. Portoghese, *J. Med. Chem.*, 2009, **52**(2), 247.
93. O. M. Pello, L. Martinez-Munoz, V. Parrillas, A. Serrano, J. M. Rodriguez-Frade, M. J. Toro, P. Lucas, M. Monterrubio, A. C. Martinez and M. Mellado, *Eur. J. Immunol.*, 2008, **38**(2), 537.
94. D. Hereld and T. Jin, *Eur. J. Immunol.*, 2008, **38**(2), 334.
95. G. Pascal and G. Milligan, *Mol. Pharmacol.*, 2005, **68**(3), 905.
96. K. Fuxe, D. Marcellino, G. Leo and L. F. Agnati, *Curr. Opin. Pharmacol.*, 2010, **10**(1), 14.

97. R. Franco, *J. Neural Transm.*, 2009, **116**(8), 983.
98. S. Ferre, F. Ciruela, J. Borycz, M. Solinas, D. Quarta, K. Antoniou, C. Quiroz, Z. Justinova, C. Lluis, R. Franco and S. R. Goldberg, *Front. Biosci.*, 2008, **13**, 2391.
99. A. Gupta, J. Mulder, I. Gomes, R. Rozenfeld, I. Bushlin, E. Ong, M. Lim, E. Maillet, M. Junek, C. M. Cahill, T. Harkany and L. A. Devi, *Sci. Signal.*, 2010, **3**(131), ra54.
100. D. Marcellino, G. Navarro, K. Sahlholm, J. Nilsson, L. F. Agnati, E. I. Canela, C. Lluis, P. Rhem, R. Franco and K. Fuxe, *Biochem. Biophys. Res. Commun.*, 2010, **394**(4), 988.
101. D. O. Borroto-Escuela, D. Marcellino, M. Narvaez, M. Flajolet, N. Heintz, L. Agnati, F. Ciruela and K. Fuxe, *Biochem. Biophys. Res. Commun.*, 2010, **394**(1), 222.
102. S. Ferre, A. S. Woods, G. Navarro, M. Aymerich, C. Lluis and R. Franco, *Curr. Opin. Pharmacol.*, 2010, **10**(1), 67.
103. N. Cabello, J. Gandia, D. C. Bertarelli, M. Watanabe, C. Lluis, R. Franco, S. Ferre, R. Lujan and F. Ciruela, *J. Neurochem.*, 2009, **109**(5), 1497.
104. G. Navarro, P. Carriba, J. Gandia, F. Ciruela, V. Casado, A. Cortes, J. Mallol, E. I. Canela, C. Lluis and R. Franco, *ScientificWorldJournal*, 2008, **8**, 1088.
105. V. Coulon, M. Audet, V. Homburger, J. Bockaert, L. Fagni, M. Bouvier and J. Perroy, *Biophys. J.*, 2008, **94**(3), 1001
106. F. F. Hamdan, Y. Percherancier, B. Breton and M. Bouvier, *Curr. Protoc. Neurosci.*, 2006, Chapter 5, Unit 5 23.
107. J. J. Fung, X. Deupi, L. Pardo, X. J. Yao, G. A. Velez-Ruiz, B. T. Devree, R. K. Sunahara and B. K. Kobilka, *EMBO J.*, 2009, **28**(21), 3315.
108. J. F. Mercier, A. Salahpour, S. Angers, A. Breit and M. Bouvier, *J. Biol. Chem.*, 2002, **277**(47), 44925.
109. S. Dorsch, K. N. Klotz, S. Engelhardt, M. J. Lohse and M. Bunemann, *Nat. Methods*, 2009, **6**(3), 225
110. Y. Xiang, V. O. Rybin, S. F. Steinberg and B. Kobilka, *J. Biol. Chem.*, 2002, **277**(37), 34280.
111. V. O. Rybin, X. Xu, M. P. Lisanti and S. F. Steinberg, *J. Biol. Chem.*, 2000, **275**(52), 41447.
112. R. P. Xiao, W. Zhu, M. Zheng, K. Chakir, R. Bond, E. G. Lakatta and H. Cheng, *Trends Pharmacol. Sci.*, 2004, **25**(7), 358.
113. E. Devic, Y. Xiang, D. Gould and B. Kobilka, *Mol. Pharmacol.*, 2001, **60**(3), 577.
114. O. G. Shcherbakova, C. M. Hurt, Y. Xiang, M. L. Dell'Acqua, Q. Zhang, R. W. Tsien and B. K. Kobilka, *J. Cell Biol.*, 2007, **176**(4), 521.
115. N. Sartania, S. Appelbe, J. D. Pediani and G. Milligan, *Cell. Signal.*, 2007, **19**(9), 1928.
116. M. A. Uberti, R. A. Hall and K. P. Minneman, *Mol. Pharmacol.*, 2003, **64**(6), 1379.
117. C. Hague, M. A. Uberti, Z. Chen, R. A. Hall and K. P. Minneman, *J. Biol. Chem.*, 2004, **279**(15), 15541.

118. M. A. Uberti, C. Hague, H. Oller, K. P. Minneman and R. A. Hall, *J. Pharmacol. Exp. Ther.*, 2005, **313**(1), 16.
119. Z. J. Chen and K. P. Minneman, *Acta Pharmacol. Sin.*, 2005, **26**(11), 1281.
120. C. Hague, S. E. Lee, Z. Chen, S. C. Prinster, R. A. Hall and K. P. Minneman, *Mol. Pharmacol.*, 2006, **69**(1), 45.
121. J. F. Lopez-Gimenez, M. Canals, J. D. Pediani and G. Milligan, *Mol. Pharmacol.*, 2007, **71**(4), 1015.
122. M. Canals, J. F. Lopez-Gimenez and G. Milligan, *Biochem. J.*, 2009, **417**(1), 161.
123. J. J. Carrillo, J. F. Lopez-Gimenez and G. Milligan, *Mol. Pharmacol.*, 2004, **66**(5), 1123.
124. S. Cotecchia, L. Stanasila, D. Diviani, K. Bjorklof, O. Rossier and F. Fanelli, *Biol. Cell*, 2004, **96**(5), 327.
125. W. Guo, L. Shi and J. A. Javitch, *J. Biol. Chem.*, 2003, **278**(7), 4385.
126. W. Guo, L. Shi, M. Filizola, H. Weinstein and J. A. Javitch, *Proc. Natl. Acad. Sci. U. S. A.*, 2005, **102**(48), 17495.
127. W. Guo, E. Urizar, M. Kralikova, J. C. Mobarec, L. Shi, M. Filizola and J. A. Javitch, *EMBO J.*, 2008, **27**(17), 2293.
128. J. M. Klco, T. B. Lassere and T. J. Baranski, *J. Biol. Chem.*, 2003, **278**(37), 35345.
129. Y. Han, I. S. Moreira, E. Urizar, H. Weinstein and J. A. Javitch, *Nat. Chem. Biol.*, 2009, **5**(9), 688.
130. T. Wurch, A. Matsumoto and P. J. Pauwels, *FEBS Lett.*, 2001, **507**(1), 109.
131. Z. U. Khan, L. Mrzljak, A. Gutierrez, A. de la Calle and P. S. Goldman-Rakic, *Proc. Natl. Acad. Sci. U. S. A.*, 1998, **95**(13), 7731.
132. G. Navarro, S. Ferre, A. Cordomi, E. Moreno, J. Mallol, V. Casado, A. Cortes, H. Hoffmann, J. Ortiz, E. I. Canela, C. Lluis, L. Pardo, R. Franco and A. S. Woods, *J. Biol. Chem.*, 2010, **285**(35).
133. C. H. So, G. Varghese, K. J. Curley, M. M. Kong, M. Alijaniaram, X. Ji, T. Nguyen, B. F. O'Dowd and S. R. George, *Mol. Pharmacol.*, 2005, **68**(3), 568.
134. S. P. Lee, C. H. So, A. J. Rashid, G. Varghese, R. Cheng, A. J. Lanca, B. F. O'Dowd and S. R. George, *J. Biol. Chem.*, 2004, **279**(34), 35671.
135. J. Ng, A. J. Rashid, C. H. So, B. F. O'Dowd and S. R. George, *Neuroscience*, 2010, **165**(2), 535.
136. A. Hasbi, T. Fan, M. Alijaniaram, T. Nguyen, M. L. Perreault, B. F. O'Dowd and S. R. George, *Proc. Natl. Acad. Sci. U. S. A.*, 2009, **106**(50), 21377.
137. A. J. Rashid, C. H. So, M. M. Kong, T. Furtak, M. El-Ghundi, R. Cheng, B. F. O'Dowd and S. R. George, *Proc. Natl. Acad. Sci. U. S. A.*, 2007, **104**(2), 654.
138. S. R. George and B. F. O'Dowd, *ScientificWorldJournal*, 2007, **7**, 58.
139. A. Hasbi, B. F. O'Dowd and S. R. George, *Curr. Opin. Pharmacol.*, 2010, **10**(1), 93.
140. C. Fiorentini, C. Busi, E. Gorruso, C. Gotti, P. Spano and C. Missale, *Mol. Pharmacol.*, 2008, **74**(1), 59.

141. D. Marcellino, P. Carriba, M. Filip, A. Borgkvist, M. Frankowska, I. Bellido, S. Tanganelli, C. E. Muller, G. Fisone, C. Lluis, L. F. Agnati, R. Franco and K. Fuxe, *NeuroPharmacology*, 2008, **54**(5), 815.
142. C. S. Kearn, K. Blake-Palmer, E. Daniel, K. Mackie and M. Glass, *Mol. Pharmacol.*, 2005, **67**(5), 1697.
143. J. A. Przybyla and V. J. Watts, *J. Pharmacol. Exp. Ther.*, 2009, **332**(3), 710.
144. P. A. Vidi, B. R. Chemel, C. D. Hu and V. J. Watts, *Mol. Pharmacol.*, 2008, **74**(3), 544.
145. M. Canals, D. Marcellino, F. Fanelli, F. Ciruela, P. de Benedetti, S. R. Goldberg, K. Neve, K. Fuxe, L. F. Agnati, A. S. Woods, S. Ferre, C. Lluis, M. Bouvier and R. Franco, *J. Biol. Chem.*, 2003, **278**(47), 46741.
146. L. Gazi, J. F. Lopez-Gimenez, M. P. Rudiger and P. G. Strange, *Eur. J. Biochem.*, 2003, **270**(19), 3928.
147. M. Canals, J. Burgueno, D. Marcellino, N. Cabello, E. I. Canela, J. Mallol, L. Agnati, S. Ferre, M. Bouvier, K. Fuxe, F. Ciruela, C. Lluis and R. Franco, *J. Neurochem.*, 2004, **88**(3), 726.
148. S. Ferre, B. B. Fredholm, M. Morelli, P. Popoli and K. Fuxe, *Trends Neurosci.*, 1997, **20**(10), 482.
149. T. Kanda, M. J. Jackson, L. A. Smith, R. K. Pearce, J. Nakamura, H. Kase, Y. Kuwana and P. Jenner, *Ann. Neurol.*, 1998, **43**(4), 507.
150. J. C. Goin and N. M. Nathanson, *J. Biol. Chem.*, 2006, **281**(9), 5416.
151. C. Marquer, C. Fruchart-Gaillard, G. Mourier, O. Grandjean, E. Girard, M. le Maire, S. Brown and D. Servent, *Biol. Cell*, 2010, **102**(7), 409.
152. B. Ilien, N. Glasser, J. P. Clamme, P. Didier, E. Piemont, R. Chinnappan, S. B. Daval, J. L. Galzi and Y. Mely, *J. Biol. Chem.*, 2009, **284**(29), 19533.
153. L. F. Pisterzi, D. B. Jansma, J. Georgiou, M. J. Woodside, J. T. Chou, S. Angers, V. Raicu and J. W. Wells, *J. Biol. Chem.*, 2010, **285**(22), 16723.
154. V. Raicu, *J. Biol. Phys*, 2007, **33**(2), 109.
155. A. W. Ma, D. S. Redka, L. F. Pisterzi, S. Angers and J. W. Wells, *Biochemistry*, 2007, **46**(26), 7907.
156. P. S. Park and J. W. Wells, *J. Neurochem.*, 2004, **90**(3), 537.
157. A. T. Colozo, P. S. Park, C. S. Sum, L. F. Pisterzi and J. W. Wells, *Biochem. Pharmacol.*, 2007, **74**(2), 236.
158. P. S. Park, C. S. Sum, A. B. Pawagi and J. W. Wells, *Biochemistry*, 2002, **41**(17), 5588.
159. P. S. Park and J. W. Wells, *Biochemistry*, 2003, **42**(44), 12960.
160. E. Alvarez-Curto, R. J. Ward, J. D. Pediani and G. Milligan, *J. Biol. Chem.*, 2010, **285**(30).
161. M. A. Ayoub, C. Couturier, E. Lucas-Meunier, S. Angers, P. Fossier, M. Bouvier and R. Jockers, *J. Biol. Chem.*, 2002, **277**(24), 21522.
162. Y. Percherancier, Y. A. Berchiche, I. Slight, R. Volkmer-Engert, H. Tamamura, N. Fujii, M. Bouvier and N. Heveker, *J. Biol. Chem.*, 2005, **280**(11), 9895.
163. G. J. Babcock, M. Farzan and J. Sodroski, *J. Biol. Chem.*, 2003, **278**(5), 3378.

164. H. Issafras, S. Angers, S. Bulenger, C. Blanpain, M. Parmentier, C. Labbe-Jullie, M. Bouvier and S. Marullo, *J. Biol. Chem.*, 2002, **277**(38), 34666.
165. A. J. Vila-Coro, J. M. Rodriguez-Frade, A. Martin De Ana, M. C. Moreno-Ortiz, A. C. Martinez and M. Mellado, *FASEB J.*, 1999, **13**(13), 1699.
166. B. Wu, E. Y. Chien, C. D. Mol, G. Fenalti, W. Liu, V. Katritch, R. Abagyan, A. Brooun, P. Wells, F. C. Bi, D. J. Hamel, P. Kuhn, T. M. Handel, V. Cherezov and R. C. Stevens, *Science*, 2010, **330**(6007), 1066.
167. J. M. Rodriguez-Frade, A. J. Vila-Coro, A. M. de Ana, J. P. Albar, A. C. Martinez and M. Mellado, *Proc. Natl. Acad. Sci. U. S. A.*, 1999, **96**(7), 3628.
168. A. J. Vila-Coro, M. Mellado, A. Martin de Ana, P. Lucas, G. del Real, A. C. Martinez and J. M. Rodriguez-Frade, *Proc. Natl. Acad. Sci. U. S. A.*, 2000, **97**(7), 3388.
169. M. Benkirane, D. Y. Jin, R. F. Chun, R. A. Koup and K. T. Jeang, *J. Biol. Chem.*, 1997, **272**(49), 30603.
170. F. Trettel, S. Di Bartolomeo, C. Lauro, M. Catalano, M. T. Ciotti and C. Limatola, *J. Biol. Chem.*, 2003, **278**(42), 40980.
171. T. J. Rogers and P. K. Peterson, *Trends Immunol.*, 2003, **24**(3), 116.
172. C. Abbadie, *Trends Immunol.*, 2005, **26**(10), 529.
173. H. L. Rittner and A. Brack, *Curr. Opin. Investig. Drugs*, 2006, **7**(7), 643.
174. G. Parenty, S. Appelbe and G. Milligan, *Biochem. J.*, 2008, **412**(2), 245.
175. G. Milligan and N. J. Smith, *Trends Pharmacol. Sci.*, 2007, **28**(12), 615.
176. Y. X. Tao, N. B. Johnson and D. L. Segaloff, *J. Biol. Chem.*, 2004, **279**(7), 5904.
177. D. A. Roess, R. D. Horvat, H. Munnelly and B. G. Barisas, *Endocrinology*, 2000, **141**(12), 4518.
178. M. Hunzicker-Dunn, G. Barisas, J. Song and D. A. Roess, *J. Biol. Chem.*, 2003, **278**(44), 42744.
179. C. Lee, I. Ji, K. Ryu, Y. Song, P. M. Conn and T. H. Ji, *J. Biol. Chem.*, 2002, **277**(18), 15795.
180. I. Ji, C. Lee, Y. Song, P. M. Conn and T. H. Ji, *Mol. Endocrinol.*, 2002, **16**(6), 1299.
181. R. D. Horvat, B. G. Barisas and D. A. Roess, *Mol. Endocrinol.*, 2001, **15**(4), 534.
182. M. Zhang, X. Feng, R. Guan, T. E. Hebert and D. L. Segaloff, *Cell. Signal.*, 2009, **21**(11), 1663.
183. A. Rivero-Muller, Y. Y. Chou, I. Ji, S. Lajic, A. C. Hanyaloglu, K. Jonas, N. Rahman, T. H. Ji and I. Huhtaniemi, *Proc. Natl. Acad. Sci. U. S. A.*, 2010, **107**(5), 2319.
184. R. D. Horvat, D. A. Roess, S. E. Nelson, B. G. Barisas and C. M. Clay, *Mol. Endocrinol.*, 2001, **15**(5), 695.
185. J. A. Janovick and P. M. Conn, *Endocrinology*, 1996, **137**(8), 3602.
186. C. C. Zhu, L. B. Cook and P. M. Hinkle, *J. Biol. Chem.*, 2002, **277**(31), 28228.

187. R. Latif, P. Graves and T. F. Davies, *J. Biol. Chem.*, 2001, **276**(48), 45217.
188. R. Latif, P. Graves and T. F. Davies, *J. Biol. Chem.*, 2002, **277**(47), 45059.
189. E. Urizar, L. Montanelli, T. Loy, M. Bonomi, S. Swillens, C. Gales, M. Bouvier, G. Smits, G. Vassart and S. Costagliola, *EMBO J.*, 2005, **24**(11), 1954.
190. R. C. Patel, D. C. Lange and Y. C. Patel, *Methods*, 2002, **27**(4), 340.
191. N. Ilukovic, M. Rocheville, U. Kumar, R. Sasi, S. Khare and Y. C. Patel, *J. Biol. Chem.*, 1999, **274**(35), 24550.
192. M. Grant, B. Collier and U. Kumar, *J. Biol. Chem.*, 2004, **279**(35), 36179.
193. M. Grant and U. Kumar, *Regul. Pept.*, 2010, **159**(1–3), 3.
194. M. Grant, H. Alturaihi, P. Jaquet, B. Collier and U. Kumar, *Mol. Endocrinol.*, 2008, **22**(10), 2278.
195. S. abu Alla, U. Quitterer, S. Grigoriev, A. Maidhof, M. Haasemann, K. Jarnagin and W. Muller-Esterl, *J. Biol. Chem.*, 1996, **271**(3), 1748.
196. E. A. Novotny, D. L. Bednar, M. A. Connolly, J. R. Connor and T. M. Stormann, *Biochem. Biophys. Res. Commun.*, 1994, **201**(2), 523.
197. S. AbdAlla, E. Zaki, H. Lother and U. Quitterer, *J. Biol. Chem.*, 1999, **274**(37), 26079.
198. D. S. Kang, C. Gustafsson, M. Morgelin and L. M. Leeb-Lundberg, *Mol. Pharmacol.*, 2005, **67**(1), 309.
199. S. AbdAlla, A. Abdel-Baset, H. Lother, A. el Massiery and U. Quitterer, *J. Mol. Neurosci.*, 2005, **26**(2–3), 185.
200. J. L. Hansen, J. T. Hansen, T. Speerschneider, C. Lyngso, N. Erikstrup, E. S. Burstein, D. M. Weiner, T. Walther, N. Makita, T. Iiri, N. Merten, E. Kostenis and S. P. Sheikh, *J. Biol. Chem.*, 2009, **284**(3), 1831.
201. J. Abd Alla, K. Reeck, A. Langer, T. Streichert and U. Quitterer, *Biochem. Biophys. Res. Commun.*, 2009, **387**(1), 186.
202. Z. J. Cheng and L. J. Miller, *J. Biol. Chem.*, 2001, **276**(51), 48040.
203. S. V. Wu, K. G. Harikumar, R. J. Burgess, J. R. Reeve, Jr. and L. J. Miller, *Am. J. Physiol Gastrointest. Liver Physiol.*, 2008, **295**(3), G641.
204. K. G. Harikumar, M. Dong, Z. Cheng, D. I. Pinon, T. P. Lybrand and L. J. Miller, *Biochemistry*, 2006, **45**(49), 14706.
205. K. G. Harikumar and L. J. Miller, *Ann. N. Y. Acad. Sci.*, 2008, **1144**, 21.
206. M. C. Dinger, J. E. Bader, A. D. Kobor, A. K. Kretzschmar and A. G. Beck-Sickinger, *J. Biol. Chem.*, 2003, **278**(12), 10562.
207. S. L. Parker and M. S. Parker, *Regul. Pept.*, 2000, **87**(1–3), 59.
208. S. L. Parker, M. S. Parker, F. R. Sallee and A. Balasubramaniam, *Regul. Pept.*, 2007, **144**(1–3), 72.
209. S. L. Parker, M. S. Parker, R. Sah, F. R. Sallee and A. Balasubramaniam, *Regul. Pept.*, 2007, **139**(1–3), 128.
210. S. L. Parker, M. S. Parker, R. Sah, A. Balasubramaniam and F. R. Sallee, *Eur. J. Pharmacol.*, 2008, **579**(1–3), 13.
211. A. M. Estes, K. McAllen, M. S. Parker, R. Sah, T. Sweatman, E. A. Park, A. Balasubramaniam, F. R. Sallee, M. W. Walker and S. L. Parker, *Amino Acids*, 2010, [Epub ahead of print]PMID: 20577889.

212. M. M. Berglund, D. A. Schober, M. A. Esterman and D. R. Gehlert, *J. Pharmacol. Exp. Ther.*, 2003, **307**(3), 1120.
213. K. G. Harikumar, M. M. Morfis, C. S. Lisenbee, P. M. Sexton and L. J. Miller, *Mol. Pharmacol.*, 2006, **69**(1), 363.
214. W. Q. Ding, Z. J. Cheng, J. McElhiney, S. M. Kuntz and L. J. Miller, *Cancer Res.*, 2002, **62**(18), 5223.
215. K. G. Harikumar, M. M. Morfis, P. M. Sexton and L. J. Miller, *J. Mol. Neurosci.*, 2008, **36**(1–3), 279.
216. K. G. Harikumar, D. I. Pinon and L. J. Miller, *J. Biol. Chem.*, 2007, **282**(42), 30363.
217. A. A. Pioszak, K. G. Harikumar, N. R. Parker, L. J. Miller and H. E. Xu, *J. Biol. Chem.*, 2010, **285**(16), 12435.
218. P. J. Harding, H. Attrill, J. Boehringer, S. Ross, G. H. Wadhams, E. Smith, J. P. Armitage and A. Watts, *Biophys. J.*, 2009, **96**(3), 964.
219. D. Devost and H. H. Zingg, *J. Mol. Endocrinol*, 2003, **31**(3), 461.
220. S. Terrillon, T. Durroux, B. Mouillac, A. Breit, M. A. Ayoub, M. Taulan, R. Jockers, C. Barberis and M. Bouvier, *Mol. Endocrinol*, 2003, **17**(4), 677.
221. M. Cottet, L. Albizu, S. Perkovska, F. Jean-Alphonse, R. Rahmeh, H. Orcel, C. Mejean, S. Granier, C. Mendre, B. Mouillac and T. Durroux, *Curr. Opin. Pharmacol.*, 2010, **10**(1), 59.
222. P. Savi, J. L. Zachayus, N. Delesque-Touchard, C. Labouret, C. Herve, M. F. Uzabiaga, J. M. Pereillo, J. M. Culouscou, F. Bono, P. Ferrara and J. M. Herbert, *Proc. Natl. Acad. Sci. U. S. A.*, 2006, **103**(29), 11069.
223. M. K. Raychowdhury, M. Yukawa, L. J. Collins, S. H. McGrail, K. C. Kent and J. A. Ware, *J. Biol. Chem.*, 1994, **269**(30), 19256.
224. T. Hirata, F. Ushikubi, A. Kakizuka, M. Okuma and S. Narumiya, *J. Clin. Invest.*, 1996, **97**(4), 949.
225. M. T. Walsh and B. T. Kinsella, *Br. J. Pharmacol.*, 2000, **131**(3), 601.
226. G. Laroche, M. C. Lepine, C. Theriault, P. Giguere, V. Giguere, M. A. Gallant, A. de Brum-Fernandes and J. L. Parent, *Cell. Signal.*, 2005, **17**(11), 1373.
227. M. Sasaki, K. Miyosawa, S. Ohkubo and N. Nakahata, *J. Pharmacol. Sci.*, 2006, **100**(4), 263.
228. I. Mandrika, R. Petrovska and J. Klovins, *Biochem. Biophys. Res. Commun.*, 2010, **395**(2), 281.
229. G. Milligan, *Drug Discov. Today*, 2006, **11**(11–12), 541.
230. G. Milligan, *Br. J. Pharmacol.*, 2009, **158**(1), 5.
231. J. M. Fonseca and N. A. Lambert, *Mol. Pharmacol.*, 2009, **75**(6), 1296.

CHAPTER 7

Lipid–Protein Interactions in G Protein Signal Transduction

DAVID J. LÓPEZ, RAFAEL ÁLVAREZ AND PABLO V. ESCRIBÁ*

Molecular Cell Biomedicine, Department of Biology-IUNICS, Universitat de les Illes Balears, E-07122, Palma de Mallorca, Spain

7.1 Introduction

According to the human genome sequence, transmembrane proteins may account for 30% of the total pool of proteins in the cell.[1] Among such proteins, G protein-coupled receptors (GPCRs) represent by far the largest family of membrane proteins involved in cell signalling. Indeed, their importance is highlighted by the fact that these proteins are encoded by around 950 genes[2] and that they currently constitute the main target for rational drug design.[3] Heterotrimeric and monomeric G proteins (including proteins of the Ras family) bind to GPCRs and are responsible for the initial amplification of the signals triggered through these receptors by a variety of messengers. In this process, many G protein molecules can be activated by a single receptor molecule, each of which can interact with effector proteins that in turn can produce thousands of second messenger molecules to regulate downstream proteins.

G protein-associated cell signalling, like many other processes within the cell, starts at the cell membrane. Lipid bilayers are essential components of the cell, forming a semi-permeable barrier for the flux and exchange of different substances between the extracellular environment, the cytoplasm and the various

RSC Drug Discovery Series No. 8
G Protein-Coupled Receptors: From Structure to Function
Edited by Jesús Giraldo and Jean-Philippe Pin

Published by the Royal Society of Chemistry, www.rsc.org

organelles in the cell. Since the publication of the fluid mosaic model by Singer and Nicolson in 1972,[4] our knowledge of the lipid bilayer has increased notably. Biological membranes are no longer considered as mere structural scaffolds where proteins are simply held so that they can fulfil their functions. The complex dynamics of the plasma membrane goes far beyond these basic functions, since it also participates in lipid–lipid and lipid–protein interactions. These interactions modify the lipid bilayer itself, influencing protein activity. In turn, the presence of proteins also affects the structure of membranes.

GPCRs and G proteins are regulated by co- and post-translational modifications. Some such modifications have a relevant influence on protein–lipid interactions and on GPCR-associated signalling such as myristoylation, palmitoylation and isoprenylation—all of which have been observed in GPCRs,[5] G proteins,[6] their effector proteins[7] and receptor kinases.[8] These lipid modifications involve the covalent binding of acyl or isoprenyl groups to the amino acid backbone, and they modulate the membrane structure, protein activity and trafficking across the membrane. The structure of the lipid bilayer affects cell signal transduction by modulating the binding of signalling proteins to the membrane, as well as their activity. The heterogeneous biophysical properties of lipids leads to the generation of microdomains with specific compositions, giving rise to membrane regions with singular physical characteristics in terms of the degree of fluidity and curvature. Membrane proteins have a preference for different environments in the membrane, in part since the localization and activity of proteins is influenced by the chemical nature of membrane lipids.[9,10] In turn, both integral and peripheral proteins may change the lipid environment where they are localized by modulating the degree of mobility of the surrounding lipids, thereby modifying the properties and organization of the plasma membrane.[11,12]

In summary, the lipid–protein interactions related to G protein-associated signalling are not fortuitous events, but rather they are defined processes with important implications in signal transduction. This chapter reviews how GPCRs and G proteins modulate, and are modulated by, the lipid environment in which they are found.

7.2 Interactions between Lipid Molecules and GPCRs

GPCRs are also known as seven-transmembrane domain receptors, heptahelical receptors, serpentine receptors and G protein-linked receptors. These proteins constitute a large family of transmembrane receptors that bind extracellular molecules and activate signal transduction pathways, ultimately regulating many cellular responses. GPCRs are only found in eukaryotes and they are mostly involved in biological cell functions. This almost ubiquitous presence in cell signalling events is due to their wide distribution in different cell types, the diversity of agonists that they may bind (*e.g.* light-sensitive compounds, flavours, pheromones, hormones, neurotransmitters, *etc.*), and the variety of downstream effectors they modulate.

GPCRs are integral membrane proteins with seven transmembrane α-helices that are associated with three extracellular and three intracellular loops. The *N*-terminal extracellular region of the protein may contain glycosylated moieties, as well as highly conserved cysteines that form disulfide bonds to stabilize the receptor's structure. The first crystal structure of a mammalian GPCR to be resolved was that of bovine rhodopsin[13] and, in 2007, the β_2-adrenergic receptor was the first human GPCR to be crystallized.[14]

7.2.1 Lipid Modifications of GPCRs

Due to the highly hydrophobic nature of the transmembrane α-helices of GPCRs, no co- or post-translational lipid modifications are required for them to bind to the lipid membrane. However, GPCRs are still subjected to protein lipidation and indeed, the trafficking, folding within the membrane and the activity of the GPCRs are partially controlled by the lipids that are covalently bound to their peptide backbone. This protein lipidation mainly involves thio(S)-acylation, although isoprenylation of some receptors has also been observed. Thio(S)-acylation is performed *via* a thioester bond that links a palmitate molecule (C16:0) to a cysteine residue. Rhodopsin was the first GPCR shown to be palmitoylated,[5] and although palmitoylation is not necessary for the receptor to bind to the membrane, it is an important modification for protein folding because it generates an additional intracellular loop in the *C*-terminal region.[15] Moreover, the correct conformation of the protein possibly controls the interaction of GPCRs with specific signalling proteins.

The finding that the β_2-adrenergic receptor is also palmitoylated at a cysteine residue near the *C*-terminal tail, a position analogous to that of bovine rhodopsin,[16] further supports the involvement of GPCR lipidation in receptor structure and function. In fact, 80% of these receptors possess a palmitoylation site at a similar location and the detailed palmitoylation sites of some such proteins[17–23] are shown in Table 7.1. Although most GPCRs are palmitoylated in their *C*-terminal region, some receptors might have additional palmitoylation sites. Indeed, the μ-opioid receptor[24] and the V1a vasopressin receptor[25] are still palmitoylated after their cysteine residues in the carboxyl tail are mutated, and the incorporation of this palmitate suggests the existence of palmitoylated cysteines in the intracellular loops that connect the transmembrane domains.

It is worth mentioning that prostacyclin receptors can undergo dual lipidation, namely isoprenylation and palmitoylation. A farnesyl (C15) group is thioesterified to cysteine 383, while palmitate molecules are added at cysteines 308 and 311.[26,27] The isoprenylation site is located in the *C*-terminus, which contains a conserved putative isoprenylation CAAX motif (C = cysteine, A = aliphatic, X = any amino acid). This conserved sequence is also found in the Gγ subunits of G proteins[28] (see Section 7.3.2) and in monomeric G proteins of the Ras family[29,30] (see Section 7.3.3). Interestingly, isoprenylation always occurs before palmitoylation in this dual lipid modification,[31] meaning that if isoprenylation does not take place, the membrane-attached enzyme in charge of the palmitoylation of the protein, palmitoyltransferase, cannot act.

Table 7.1 Palmitoylation for G protein coupling (adapted from ref. 98).

	Cysteines	*Selected sites of modification*[a]
GPCRs subjected to palmitoylation-dependent G protein coupling		
Bovine rhodopsin	322, 323	QFRNCMVTTL**CC**GKNPLGDDEA
Human β_2-adrenergic receptor	341	DFRIAFQELL**C**LRRSSLKAYG
M2 muscarinic acetylcholine receptor	457	KKTFKHLLM**C**HYKNIGATR
Human somatostatin receptor, type 5	320	DNFRQSFQKVL**C**RLKGSGAKDADA
Human endothelin receptor, subtype A	383, 385-388	KFKNCFQS**C**L**CCCC**YSKSLMTSV
Human endothelin receptor, subtype B	402, 403, 405	RFKNCFKSCL**CC**W**C**QSFEEKQSLE
	[+400]	RFKNCFKS**C**L**CC**W**C**QSFEEKQSLE[b]
Human prostacyclin receptor, IP	308, 311, 383	AVFQRLKLWV**C**CL**C**LGPAHGDSQT PSAVGTSSKAEASVA**C**SLC[c]
GPCRs not subjected to palmitoylation-dependent G protein coupling		
Porcine α_{2A}-adrenergic receptor	442	HDFRRAFKKIL**C**RGDRKRIV
Rat LH/hCG receptor	621, 622	DFLLLLSRFG**CC**KRRAELYRRK
Human dopamine D_1 receptor	347, 351	RKAFSTLLG**C**YRLCPATNNAIETV
Human adenosine A_1 receptor	309	FLKIWNDHFR**C**QPAPPIDEDL
Human thyrotropin receptor	699	VFILLSKFGI**C**KRQAQAYRGQ

[a]Bold and boxed cysteine residues represent the site for thioester-linked palmitoylation, as identified by mutagenesis.
[b]Palmitoylated cysteine residues were detected by mass spectrometry.
[c]Bold and underlined cysteine residue represents the site for thioether-linked isoprenylation as identified by mutagenesis.

7.2.2 Lipid Modifications Influence GPCR Trafficking

GPCR palmitoylation has been observed in different membrane compartments within the cell and indeed, caveolin-1[32] and the transferrin receptor[33] are thio(S)-palmitoylated at the plasma membrane while palmitoylation of viral membrane glycoproteins takes place in the Golgi.[34,35] In addition, some proteins are acylated in the endoplasmic reticulum, such as the α_7-nicotinic acetycholine receptor.[36] The δ-opioid receptor is an interesting case since it is palmitoylated at various stages during its lifecycle.[37] First, it is palmitoylated during or immediately after its exit from the endoplasmic reticulum and then again at the plasma membrane. The inhibition of the first palmitoylation step blocks the receptors' export to the plasma membrane and similarly other receptors also accumulate inside the cell when their palmitoylation residues are mutated (*e.g.* the bovine opsin receptor,[38] the lutropin/choriogonadotropin receptor,[39] the canine H_2 histamine receptor[40] and the chemokine CCR5 receptor[41]). All these data show that certain palmitoylation processes are vital for the trafficking of GPCRs and its correct sorting to the cell surface.

Besides this constitutive GPCR palmitoylation, a second type of palmitoylation is distinguished in mature receptors once they have reached their specific localization. The acyl–thioester bond that binds the palmitate molecule to the cysteine residue is considered to be labile, favouring the palmitoylation/depalmitoylation of a protein several times during its lifecycle.[42,43] Therefore, like phosphorylation, palmitoylation is a regulatory mechanism that affects proteins by modifying their localization to plasma membrane microdomains and their interactions with other proteins. In fact, the palmitoylation/depalmitoylation cycle of some GPCRs such as the β_2-adrenergic and α_{2A}-adrenergic receptors (among others) depends on their activation.[44–47]

7.2.3 Lipid Modifications Influence GPCR Signalling

GPCR palmitoylation affects their interaction with G proteins, thereby modulating signal transduction in the cell. The mutation of cysteine 342 in the β_2-adrenergic receptor produces its uncoupling from G proteins[16] and, consequently, it impairs signal propagation. By contrast, mutations of the palmitoylation sites in the α_{2A}-adrenergic receptor do not alter receptor-G protein coupling.[48] Interestingly, some GPCRs exert a preference for certain G proteins depending on their degree of palmitoylation. Thus, while the native endothelin receptor subtype B can bind to $G\alpha_q$ and $G\alpha_i$ subunits, its binding to $G\alpha_i$ is abolished when cysteine 403 and cysteine 405 are mutated. If additionally a third cysteine (Cys402) is mutated, then binding to any $G\alpha$ is inhibited.[18] The interruption of cell signalling upon depalmitoylation of a GPCR may be explained by the increase in receptor phosphorylation since G protein-coupled receptor kinases (GRKs) can phosphorylate and inactivate GPCRs.[49,50] Desensitization of GPCRs commences with their phosphorylation and it is followed by their uncoupling from the interacting G protein. Subsequently, the receptor is internalized where it is degraded or recycled back to the membrane. Phosphorylation of GPCRs is involved in the tolerance/resistance to some drugs and long-term opioid treatment, including that of morphine, provokes the internalization of the μ-opioid receptor with the concomitant reduction in their response.[51]

In this context, phosphorylation of a rhodopsin receptor expressed in mice is highly favoured when palmitoylation is deficient, leading to decreased phototransduction upon stimulation.[52] A higher degree of phosphorylation and inactivation has also been observed in 5-HT_{4a} hydroxytryptamine[53] and luteinizing hormone receptor[54] that lack their palmitoylation sites. These data led to the hypothesis that palmitoylation regulates access to the phosphorylation sites in GPCRs and highlights the relevance of receptor palmitoylation in the regulation of GPCR-mediated signal transduction. However, results obtained with mutated cysteine residues must be interpreted with care since the lack of a sulfhydryl group (-SH) may not only hinder receptor palmitoylation but the formation of disulfide bonds as well, which might be essential for the correct structure and activity of the protein.

7.3 Lipid Modification of G Proteins

Both small monomeric G proteins like Ras, and heterotrimeric G proteins (Gαβγ), undergo lipidation co- or post-translationally. Lipid moieties participate in G protein–lipid interactions and they are involved in the translocation of G proteins to membranes and further mobilization to different microdomains. Lipid modifications of heterotrimeric G proteins involve the covalent binding of three different types of hydrophobic molecules to the peptide chain, namely myristic acid (C14:0), palmitic acid (C16:0) and isoprenyl groups (*i.e.* farnesyl or geranylgeranyl) (Table 7.2). In addition, Ras proteins are isoprenylated and some of them are also palmitoylated.

7.3.1 Gα Subunit Lipid Modifications

Myristoylation occurs co-translationally in some members of the $G\alpha_i$ and the $G\alpha_t$ subfamilies and it involves the formation of an amide bond between a glycine residue and the saturated fatty acid myristate (*N*-myristoylation).[55] This lipid modification is irreversible due to the chemically stable structure of the amide bond. Myristoylation takes place on glycine 2 (Gly^2) in the *N*-terminus of the mature Gα subunit and it is catalysed by *N*-myristoyltranferase. Removal of the initial methionine must occur prior to lipidation, although the

Table 7.2 Lipid modifications of heterotrimeric G proteins (adapted from ref. 98).

Gα	*Lipid modification*	*Selected sites of modification*[a]
α_s, α_{olf}	Palmitoylation and *N*-palmitoylation[b]	α_s MGCLGNSKT-
α_q, α_{11}, α_{14}, α_{16}	Palmitoylation (2 or more sites)	α_q MTLESIMACCLSEEA – α_{16}MARSLRWRCCPWCL-[c]
α_{i1}, α_{i2}, α_{i3}, α_o, α_z	Myristoylation and palmitoylation	α_{i1} MGCTLSAED-
α_{t1}, α_{t2}, α_{gust}	Myristoylation	α_{i1} MGAGASAEE-
α_{12}, α_{13}	Palmitoylation	α_{12} MSGVVRTLSRCLLPAE- α_{13} MADFLPSRSVLSVCFPGCV-
Gγ	*Lipid modification*[d]	*Selected sites of modification*[e]
γ_1, γ_9, γ_{11}	Farnesylation	γ_1-KNPFKELKGGCVIS
γ_2, γ_3, γ_4, γ_5, γ_7, γ_8, γ_{10}, γ_{12}, γ_{13}	Geranylgeranylation	γ_2-ENPFREKKFFCAIL

[a]Selected *N*-termini of Gα are shown. Underlined glycine residues represent sites for *N*-myristoylation or *N*-palmitoylation (for α_s) after removal of the initiating methionine. Bold and boxed cysteine residues represent the sites for thioester-linked palmitoylation, as identified by mutagenesis and palmitate labeling studies.

[b]*N*-terminal palmitoylation has not been tested for α_{olf}.

[c]Bold and boxed cysteine residues represent the probable sites for palmitoylation of α_{16}; however, mutagenesis of these cysteines has not yet been reported.

[d]Although all mammalian Gγ are modified by a single isoprenyl group, the Gγ from *S. cerevisiae*, Stel 8, is modified by both farnesylation and palmitoylation at the *C*-terminus.

[e]Selected *C*-termini of Gγ are shown. Bold and boxed cysteine residues represent the sites for isoprenylation. The residues highlighted indicate the three amino acids removed by a CAAX protease.

presence of a free *N*-terminal glycine is not sufficient for myristoylation of the Gα subunit. For example, $G\alpha_s$ is not myristoylated even though it possesses a glycine at position 2, and this is due to the influence of the sixth amino acid in the protein sequence (methionine being the first one). Indeed, myristoylated proteins usually contain a serine or threonine residue at the sixth position[56] and only those members of the G_i subfamily that contain a glycine at position 2 and serine at position 6 ($G\alpha_1$, $G\alpha_2$, $G\alpha_3$, $G\alpha_o$, $G\alpha_z$) are myristoylated. One exception to this rule is $G\alpha_t$, which may be acylated at Gly2 by the addition of myristate (C16:0), laurate (C14:0), (*cis*-delta 5)-tetradecenoate (C14:$1^{\Delta5}$) or (*cis, cis*-delta 5, delta 8)-tetradecadienate (C14:$2^{\Delta5,\ 8}$) bound at the same position.[57] This reaction is performed by *N*-myristoyltransferase and it seems to be specific to retinal rod cells.[58,59]

Protein palmitoylation is a post-translational modification which affects all mammalian Gα subunits except $G\alpha_t$ and $G\alpha_{gust}$.[60] It involves the *S*-acylation (covalent binding) of palmitate to a cysteine *via* a thioester bond. This reaction is catalysed by a palmitoyltransferase, and although a specific enzyme for Gα has yet to be identified, palmitoyltransferases appear to exist for yeast and mammalian Ras.[61] In contrast to *N*-myristoylation, palmitoylation is a more dynamic modification due to its chemically labile bond, which favours the rapid turnover of palmitoylation in the cell.[62,63] This lipid modification occurs at one or several cysteine residues within the first 20 amino acids of the *N*-terminal domain of Gα subunits. Although no consensus signal for this lipidation has been identified, the presence of myristic acid (in the case of myristoylated Gα subunits) or an interaction with a Gβγ subunit seems to be required for G protein palmitoylation. Thus, there are some Gα subunits that can be lipidated by both the attachment of myristate to Gly2 and of palmitate to one or various cysteines within the *N*-terminal region, while others are only palmitoylated. Besides *S*-palmitoylation, a palmitate molecule can also be bound covalently to Gly2 in G_s *via* an amide bond.[64] This dual palmitoylation, involving *N*-palmitoylation on Gly2 and *S*-palmitoylation on Cys3, might be considered analogous to the dual myristoylation and palmitoylation of some members of the $G\alpha_i$ subfamily.

In this context, reversible palmitoylation has been shown to regulate cell signalling. Thus, the activation of Gα by GPCRs induces an increase in G protein subunit palmitoylation/depalmitoylation. For example, depalmitoylated $G\alpha_s$ translocates from the plasma membrane into the cytoplasm upon activation by its GPCR.[65–68] While the exact mechanism behind this process is not fully understood, the relationship between lipid modification of the Gα subunit and its localization and activation status remains patent.

7.3.2 Lipid Modifications of Gγ Subunits

Isoprenylation is a multistep post-translational process that affects all 12 known mammalian Gγ subunits. This lipidation starts with the formation of a covalent thioester bond between a farnesyl (C15) or geranylgeranyl group (C20) and a cysteine in a CAAX motif at the *C*-terminus. This motif consists of a cysteine to be thioesterified (C), two aliphatic residues (AA) and one amino acid (X) that determines the type of isoprenyl group to be bound to C. If the last residue of the conserved motif is serine or methionine, the cysteine is farnesylated. By contrast, a geranyltransferase modifies the protein if the last residue is leucine. Therefore, Gγ_1, Gγ_9 and Gγ_{11} are farnesylated and the remaining nine Gγ are all geranylgeranylated.[28] Isoprenylation of Gγ subunits is an irreversible process that takes place in the cytosol. As a result, the Gβγ translocates to the endoplasmic reticulum (ER) where the three terminal–AAX amino acids are hydrolysed by the Ras-converting enzyme (Rce)[69] and the isoprenylated cysteine is methylated by the isoprenylcysteine carboxyl methyltransferase (Imct).[70] The addition of an extra methyl group to the peptide structure seems to increase the hydrophobicity of the protein, and consequently, its binding to the lipid bilayer.

However, these modifications are not sufficient for translocation of the Gβγ dimer from the ER to the plasma membrane. Indeed, palmitoylation has been determined as a key signal for the trafficking of these proteins to the membrane. In HEK293 cells that do not express the Gα subunit, the G$\beta_1\gamma_2$ remains bound to the ER, while coexpression of the three subunits induces the localization of the Gαβγ heterotrimer to the plasma membrane. Interestingly, the Gβγ dimer alone can localize to the plasma membrane upon palmitoylation.[71] Gβγ dimers need a Gα palmitate signal for their correct trafficking, while Gα also needs to bind Gβγ for its palmitoylation and correct translocation to the plasma membrane. In this context, Gα_s and Gα_q subunits lacking the *N*-terminal region involved in the binding to the Gβγ dimer are neither palmitoylated nor translocated to the plasma membrane.[72]

7.3.3 Lipid Modifications of Ras: An Example of Lipidation of a Small Monomeric G Protein

Small monomeric G proteins are GTPases which propagate signals in a monomeric state without undergoing the transit from trimer to monomer and *vice versa*. This is in part due to the similarity between the *C*-terminal regions of Gγ and Ras proteins that contain a CAAX motif for isoprenylation. On the other hand, Ras proteins have a GTP binding site with GTPase activity that switches the protein between the active and inactive states, like Gα subunits. Therefore, Ras proteins contain features typical of G protein α-monomers and βγ-dimers. The Ras superfamily is formed by more than a hundred proteins[73] that are divided into eight main families based on structure, sequence and function (Ras, Rho, Rab, Rap, Arf, Ran, Rheb, Rad and Rit), each of which can be further divided into subfamilies.

Ras proteins are found in the inner monolayer of the cell plasma membrane, and like heterotrimeric G proteins, they influence signal transduction. These proteins act as molecular transducers that propagate external stimuli to the cytoplasm and nucleus upon activation through their association with growth factor receptors (or adaptor/scaffolding proteins bound to them).[74]

Mammalian Ras proteins have a molecular weight of 21 kDa and three functional domains define their structure:[75]

A. The G domain is a highly conserved sequence formed by the first 166 amino acids where GTP–GDP exchange or hydrolysis takes place, and where binding to downstream effectors occurs.
B. The next 24–25 amino acids form the poorly conserved hypervariable region (HVR) with less than 15% sequence identity, the most divergent domain in the different Ras isoforms. The HVR is involved in the binding of Ras to the inner leaflet of the plasma membrane, as well as in the trafficking of the protein from the cytosolic monolayer of the endoplasmic reticulum to the plasma membrane.
C. The four *C*-terminal amino acids formed by a CAAX motif act as a isoprenylation site and they participate in membrane binding as in Gγ subunits.

Like the CAAX motif of Gγ, the last amino acid defines the type of isoprenyl group attached to the cysteine residue. Thus, while proteins with serine or methionine are farnesylated, those with leucine are geranylgeranylated. Although most Ras isoforms undergo farnesylation, they can also be geranylgeranylated when farnesyltranferase is inhibited.[76] This alternative isoprenylation explains the therapeutic failure of drugs based on farnesyltranferase inhibitors to impair Ras activity in tumours with excessive Ras activity.

Besides isoprenylation, some isoforms can also be single (N-Ras) or double-palmitoylated (H-Ras), while others, like K-Ras4B lack a palmitoylation site. In addition to the CAAX motif, the presence of palmitoylation or a polybasic domain is required to localize the protein to specific microdomains within the plasma membrane.[77]

7.4 Biophysical Properties of Lipid Membranes

7.4.1 Membrane Structure

Biological membranes are vital to organize biochemical processes that require compartmentalization. Not only is the structure of a cell defined by the plasma membrane, but also the different compartments in eukaryotic organisms are bound by lipid bilayers such as nuclei, mitochondria, chloroplasts, endoplasmic reticulum and the Golgi apparatus.[78] Biological membranes are composed of lipids, proteins and carbohydrates, organized in such a manner that they form a selective barrier for a variety of chemical compounds. This selective permeability permits the controlled exchange of matter, energy and information between the intracellular and extracellular milieu, while it maintains the capacity of the cell to function.

Our understanding of the structure and properties of biomembranes has increased steadily since 1895 when Ernest Overton first described the cell membrane as a covering structure made up of lipid molecules. Years later, in 1925, Gorter and Grendel concluded that the erythrocyte membranes were formed by two lipid layers[79] and, in 1972, Singer and Nicolson's fluid mosaic model represented a qualitative leap in our understanding of the membrane structure and functionality.[4] Nowadays, the lipid membrane is no longer defined as just a passive film that blocks the passage of water and solutes, and in which the 'truly' regulatory elements (proteins) are inserted. The development of novel biophysical techniques has made it possible to attribute new properties to membranes not previously defined in Singer and Nicolson's model.[80,81]

The protein density seems to be higher than was originally thought, and lipids and proteins are clearly not homogeneously distributed since lipid–lipid, lipid–protein and protein–protein interactions induce the formation of domains of specific lipid and protein compositions (Figure 7.1). In addition, interactions between different molecules and cytoskeletal proteins in these domains also play an important role in defining membrane microdomains. Moreover, Singer–Nicolson's model did not contemplate the existence of transmembrane proteins with a hydrophobic region thicker than that of the phospholipid bilayer. The currently accepted model postulates that proteins are not the elements that adapt to the thickness of the membrane but rather, lipids have

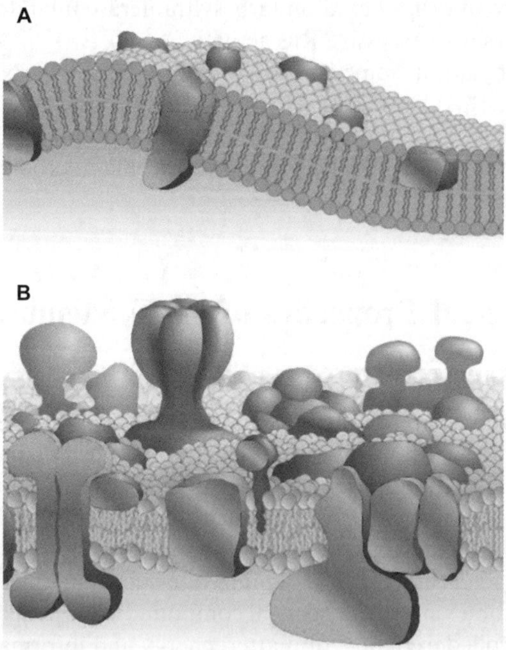

Figure 7.1 Structural model of a fluid mosaic membrane. Panel **A** shows the original Singer-Nicolson's fluid mosaic membrane model, whereas panel **B** shows some modifications of this model in the light of more recent discoveries. Adapted from ref. 81.

certain flexibility to accommodate proteins with different sized hydrophobic transmembrane regions. Thus, the bilayer thickness may vary over the entire surface of the membrane.

7.4.2 Lipid Distribution within the Membrane

As previously mentioned, lipids are not homogeneously distributed in the membrane. The membrane lipid composition not only varies in different types of cells, but also between the membranes surrounding the distinct organelles within a cell.[78] Moreover, the lipid composition of the inner and the outer leaflet of some membranes may also be asymmetric. For example, while phospholipids are distributed equally in both monolayers of the endoplasmic reticulum, where most lipids are synthesized, the lipid composition of the extracellular leaflet of the plasma membrane is completely different from the cytosolic leaflet. In this latter membrane, choline-containing phospholipids sphingomyelin (SM) and phosphatidylcholine (PC) are localized preferentially in the outer monolayer, while the aminophospholipids phosphatidylserine (PS) and phosphatidylethanolamine (PE) are present in the internal face due to the action of various enzymes.[82,83]

Besides this cross-sectional asymmetry, many membranes also show an important lateral asymmetry. Microdomains coexist in a single membrane, maintaining their own special biophysical properties by restricting or impairing the intermixing of their lipid and protein components.[84] Such domains have recently become a focus of increasing interest due to their implication in many cellular processes such as signal transduction, vesicular trafficking and viral infection. Over a decade ago, the lipid raft was first defined as a detergent-resistant transient microdomain composed of cholesterol (Cho), SM, glycosphingolipids and different proteins that attach to the lipid structure *via* a glycosylphophoinositol (GPI) anchor, a fatty acid modification or a hydrophobic amino acid sequence.[85] The carbonyl group of sphingolipids forms a hydrogen bond with the 3β-OH of Cho, forming a rigid structure with the phospholipid acyl chains fully extended, but with certain rotational and bending mobility. This liquid-ordered (Lo) structure is more fluid than the gel lamellar phase (L_β), although it is more rigid than the liquid crystalline phase ($L\alpha$). By contrast, a liquid-disordered phase (L_d) contains a small amount of cholesterol with unsaturated PCs, producing a less compact structure that resembles the $L\alpha$ structure.[86] In L_d regions, the surface packing is looser than in Lo regions; it is especially loose in regions with high PE content, where proteins with bulky membrane anchors (*e.g.* isoprenyl moieties) can find an appropriate docking space for their membrane binding.

7.4.3 Lipid Polymorphism

When dispersed in aqueous solutions, lipids can organize in different ways depending on their molecular structure, water concentration, pH, ionic strength or system pressure.[87,88] Such polymorphism among lipids is important in cell processes such as membrane fusion and fission, vesicular trafficking, macromolecule transport through the membrane and the stabilization of protein complexes in the lipid bilayer.[89,90] Moreover, the way in which lipids are organized affects their interactions with membrane proteins, thereby modulating their activity.[91,92]

A lipid phase is a thermodynamic concept that defines each of the different structural stages of matter, like water in the solid or liquid state. A lipid phase refers to a specific conformation adopted by lipids in an aqueous solution (*i.e.* how lipids organize into supramolecular structures). Lipid phases may be classified according to three criteria: (i) the type of network; (ii) the packing of the acyl chains; and (iii) the curvature of the whole structure.

The most widely used nomenclature for their designation involves the use of a letter and a subscript, as proposed by Luzzati.[93] The type of network may be: unidimensional, as in lamellar (L) or micellar structures; bidimensional, like in hexagonal (H) phases; or cubic (Q) and crystalline three-dimensional (C) structures. The subscript indicates the degree of acyl chain packing: α refers to disordered hydrocarbon chains (fluid); β, ordered (gel); β', rippled-ordered; and c, crystalline. In addition, the lipid structure may adopt a positive curvature with the phospholipid acyl chains facing inward (type I), or a negative curvature where the acyl chains are outwards (type II or inverted). The most relevant lipid structures from a biological point of view are the lamellar (L), micellar, inverted hexagonal (H_{II}) and inverted cubic (Q_{II}) phases (Figure 7.2). Lamellar phases include the gel (L_β), the liquid-crystalline or fluid phase (L_α), the liquid-ordered (Lo) and the liquid-disordered phase (L_d).

The coexistence of large amounts of Cho (over 20 mol%), SM and another glycerophospholipids (*e.g.* PC) leads to the formation of Lo structures, like lipid rafts, where Cho and SM molecules are tightly packed. As mentioned previously, Lo structures are less fluid than the gel phase (L_β) but more rigid than the fluid phase (L_α).[94] In contrast, PC adopts a liquid-disordered (L_d) structure which resembles the L_α phase in this mixture.

The macroscopic lipid organization within a membrane in part depends on the monomeric structure of its lipids[95] (Figure 7.3). Lipids with a cylindrical shape, like SM and PC, form lamellar structures with a global curvature of zero. Other lipids, such as PE, Cho or diacylglycerol, form membranes with negative curvature strain due to their truncated cone shape. These molecules induce the formation of hexagonal (non-lamellar) phases *in vitro*. Finally, molecules whose hydrophilic region occupies a larger area than the hydrophobic moiety (*e.g.* detergents and lysophospholipids) possess an inverted cone shape and they induce a positive curvature in the lipid structure.

On the whole, biological membranes adopt a lamellar structure, though certain transient regions with a high concentration of specific lipids may exist that induce a local curvature other than zero. This modulation in membrane lipid composition is essential to many cell processes including membrane fusion-fission,[96] the formation of proteolipidic pores[97] or the binding of membrane proteins to the lipid bilayer.[92,98]

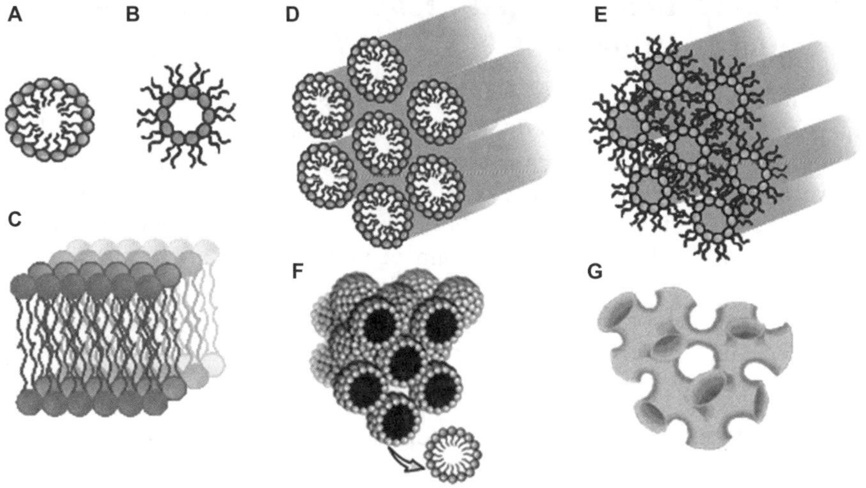

Figure 7.2 Lipid polymorphism. (**A**) Normal micelle (type I). (**B**) Inverted micelle (type II). (**C**) Lamellar structure (Lα). (**D**) Normal hexagonal phase (H_I) (**E**) Inverted hexagonal phase (H_{II}). (**F**) Normal cubic phase (Q_I). (**G**) Bicontinuous inverted cubic phase (Q_{II}).

Lipid	Phase	Shape and curvature
Lysophospholipids Detergents	Micellar 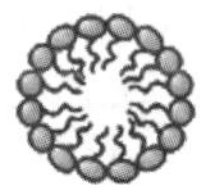	S<1 Inverted cone Positive curvature 
PC SM PS PI PG CL PA	Lamellar 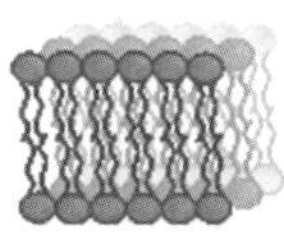	S=1 Cylinder Zero curvature 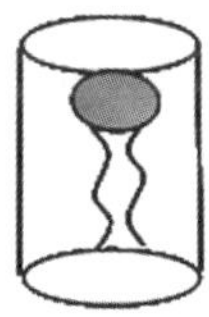
PE (unsaturated) CL –Ca^{2+} PA –Ca^{2+} PS (pH<5.0) Ch DAG	Hexagonal (H_{II}) 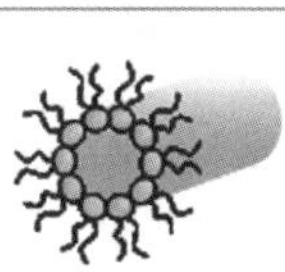	S>1 Cone Negative curvature 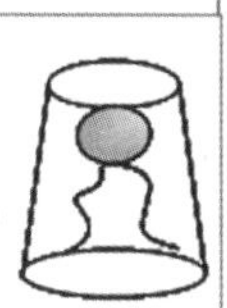

Figure 7.3 Relationship between lipid shape, intrinsic curvature and lipid phase. Membrane lipids with a cylindrical shape tend to organize into lamellar phases. Lipids with other shapes induce a positive or negative curvature strain on membranes and favour the formation of non-lamellar phases *in vitro*. Adapted from ref. 95.

7.5 Interactions of Lipid Membranes with GPCRs and G Proteins

Isoprenyl and/or fatty acid moieties in G proteins and GPCRs are not the only lipid–protein interactions in GPCR-associated signalling. The regions of lipid membranes to which these proteins are bound also constitute a meeting point between lipids and proteins. The regulatory effect of these lipid–protein interactions are bidirectional and as such, both the activities of G proteins and GPCRs can be modulated by the lipid environment in which they are localized, while these proteins are also capable of modulating the lipid structure and the organization of the membrane. This section centres on the reciprocal effects of each of these elements on the other molecular entity in the context of GPCR signalling.

7.5.1 Effects of Membrane Structure on G Protein Signalling

Membrane fluidity was one of the first physical parameters of a membrane used to demonstrate how certain protein activities could be modulated by a lipid bilayer.[99] Membrane fluidity depends on the lipid composition, the nature of the phospholipid hydrocarbon chains, the amount of Cho and other factors. As a rule of thumb, fluidity is directly related to the number of unsaturated bonds in the fatty acids present in the lipids forming a given membrane mixture.[100] It has been seen that metabolic or nutritional disorders alter the lipid composition of membranes, thereby changing its fluidity.[101,102] Moreover, alterations of this membrane parameter have been associated with disorders, such as Alzheimer's disease and hypertension, where G protein-associated signalling plays an important role.[103,104] However, while changes in membrane fluidity are clearly related to certain diseases, this parameter contemplates the membrane as a homogeneous mixture of lipids and proteins, disregarding the presence of domains with distinct compositions and structures, as well as other physico-chemical properties of membranes. Therefore, while the bulk fluidity of membranes may remain constant, other biophysical properties of membrane microdomains might be altered, and *vice versa*.

A more correct way to study the influence of the lipid bilayer on G protein signalling would be to analyse the properties of the domains that interact directly with the GPCRs and G proteins. It is known that GPCRs bind to the membrane at defined sites in the bilayer. When an extracellular signal mediated by an agonist ligand induces a conformational change in the receptor, the heterotrimeric G protein bound to the GPCR becomes activated. Subsequently, the Gα subunit of the heterotrimeric complex exchanges GDP for GTP and it dissociates from the G$\beta\gamma$ dimer, enabling both the monomer and dimer to regulate the activity of different effectors. The overall process amplifies the original signal in a cascade, whereby a single agonist-activated receptor can activate several G protein molecules, and so on. The first amplification step whereby several G proteins are activated by a single receptor molecule is produced through the presence of a large number of inactive (pre-active) heterotrimeric G proteins in the receptor vicinity.

The influence of alterations in different lipid membrane parameters on the functional efficacy, structural stability or oligomeric state of membrane proteins has been already shown.[105] But until the structure of the β2-adrenergic receptor was published in 2007,[14] rhodopsin was the only GPCR known at atomic resolution.[13] As a result, many works have been developed to study the activity of the light-activated receptor mentioned above. Upon activation, rhodopsin initiates rapid conformational changes resulting in the active intermediate state, metarhodopsin II (meta II), which coexists with the inactive intermediate state metarhodpsin I (meta I).[106] Meta I–meta II conformational equilibrium is sensitive to membrane physical properties such as curvature stress,[107] acyl chain packing[108] and hydrophobic thickness.[109] This 'on' and 'off' state is also modulated by chemical parameters of the lipid membrane including the type of headgroup, acyl chain length and unsaturation degree of the phospholipids.[110,111]

The interaction of Cho with rhodopsin has been explored thoroughly. It is known that the presence of Cho favours the stabilization of the inactive state of rhodopsin, thus, reducing the kinetics of the photocycle.[112] This inactivation effect of Cho over GPCRs is in agreement with the increase coupling of the β2-adrenergic receptor with its partnering G-protein in Cho-depleted cardiac myocites.[113] However, docohexaenoic acid (DHA) containing phospholipids stabilizes the active state of rhodopsin, resulting in an enhancement of the signalling efficiency.[114,115] Molecular dynamics studies have allowed investigation of the modulation of rhodopsin stability and kinetics by specific direct interactions with polyunsaturated acyl chains.[116] On the other hand, microsecond molecular dynamics simulations have also been used to describe a detailed mechanism by which cholesterol specifically and directly interacts with rhodopsin.[117] The interaction with cholesterol modulates the behaviour of the transmembrane segments 1, 2, 7 and helix 8 (TM1-TM2-TM7-H8), which constitutes a functional network essential for the GPCR activation.

Lipids with a propensity to form non-lamellar phases (*e.g.* PE, DAG, Cho) can also modulate the physical properties of biomembranes by inducing alterations in the intrinsic curvature of the monolayer, the lateral surface pressure and the hydration of membranes.[118] The inner leaflet of plasma membranes is usually enriched in PE,[82] a non-lamellar prone lipid that induces an inverted hexagonal (H_{II}) structure and influences the binding of G proteins to the lipid bilayer. The physiological amount of PE in natural membranes is sufficiently high to induce a negative curvature *in vivo*, which might be stabilized by the presence of certain membrane proteins.[119] The use of liposomes as membrane models constitutes a useful tool to study the binding preferences of G proteins to lipid bilayers due to the possibility of controlling the lipid composition of the vesicles.[120–122] Using model membranes (liposomes), it has been shown that heterotrimeric G proteins and the Gβγ dimer show a preference to bind to membranes with a higher propensity to form H_{II} phases (*e.g.* with high PE content), while the $G\alpha_i$ subunit uncoupled from the Gαβ dimer preferentially binds to lamellar prone lipids (*e.g.* with high PC content) (Figure 7.4).[120,123] G protein lipid modifications (fatty acid and/or isoprenyl moieties) are probably

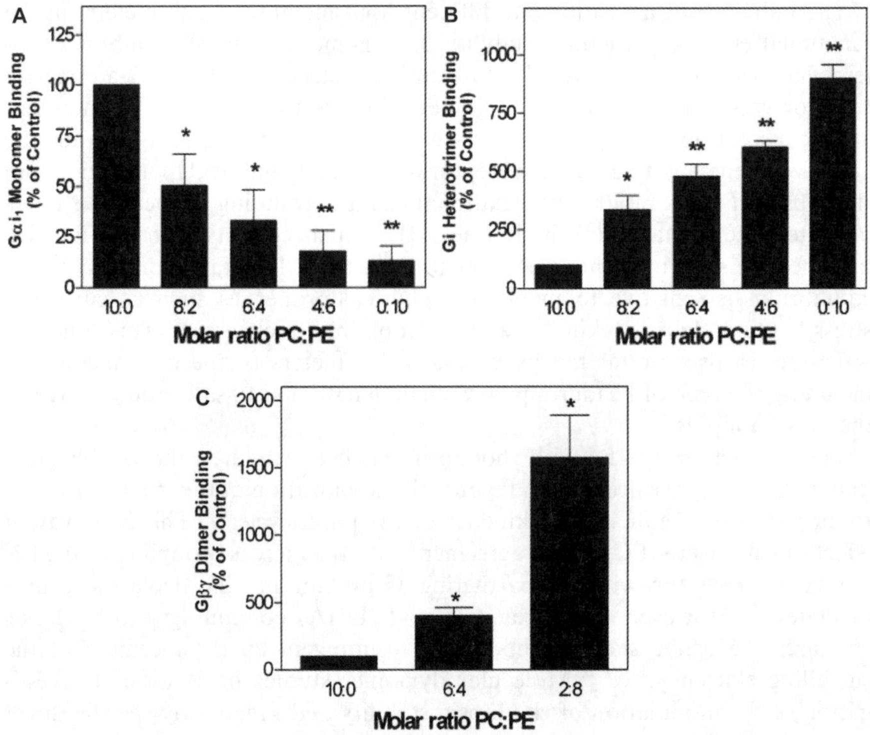

Figure 7.4 Effect of PE on the binding of G proteins to model membranes. Liposomes were prepared from PC in the absence (PC : PE 10 : 0, mol/mol) or in the presence of PE (PC : PE 8 : 2, 6 : 4, 4 : 6, 2 : 8 and 0 : 10; mol/mol) and incubated with heterotrimeric G proteins (**A**), myristoylated $G\alpha_{i1}$ monomers (**B**) and Gβγ dimers (**C**) at 25 °C. The proteins bound to the liposomes were quantified by immunoblotting and the data are the mean ± S.E.M. values of four independent experiments. * $p < 0.05$; ** $p < 0.01$ *versus* pure PC membranes. The Gβγ dimer defines the interaction of heterotrimeric G proteins with the membranes, both showing opposite behaviour to that of the alpha subunit. Adapted from ref. 120.

involved in this effect, through which the different forms of G proteins are sorted into different membrane domains.

There is considerable evidence that the biophysical properties of the lipid bilayer modulate the activity of membrane proteins. Lipids with a negative spontaneous curvature favour the elongation of the GPCR rhodopsin during activation and these lipids facilitate the conformational change of the receptor to its activated state in model membranes.[124] In addition, PE, the non-lamellar prone lipid used in such studies, increased the receptor affinity for transducin ($G\alpha_t$) upon light activation, while its affinity in dark conditions (inactive receptor) remained unchanged.[125]

However, when pure PC membrane bilayers were used, the affinity between light-activated rhodopsin and transducin was considerably lower.[120]

Rhodopsin is found in retinal outer segment membranes, which are enriched in DHA, a highly unsaturated fatty acid with a strong propensity to form non-lamellar structures.[126] Rhodopsin activation was also observed when PE was used in the lipid mixtures instead of DHA, indicating that the physical curvature stress induced by these lipids is the feature that most likely facilitates the conformational change of this GPCR, rather than a direct lipid-specific reaction between DHA and rhodopsin. Once again, the membrane structure is associated with the regulation of activity status of given protein. This mechanism is additional to that of specific interactions between proteins and specific lipids in cells (not necessarily at the membrane).

By contrast, other GPCRs and heterotrimeric G proteins have also been found in membrane domains with a strong lamellar membrane organization, such as lipid rafts and caveolae. The latter are endocytic-projection microdomains in a liquid-ordered (Lo) state (midway between a flat membrane region and an endocytic vesicle) stabilized by a matrix of caveolin molecules that serves as a scaffold for a variety of different proteins in the GPCR signalling cascade.[127] In cardiomyocytes, the β_2-adrenoreceptor (β_2-AR) and $G\alpha_i$ are found in caveolae, while β_1-AR and $G\alpha_s$ are predominantly present at non-caveolae regions of the plasma membrane.[128] Furthermore, the correct localization of β_2-AR in caveolae is essential for efficient signalling of this receptor subtype.[113] In the case of β_3-AR, the two subtypes (β_{3A}-AR and β_{3B}-AR) may even adopt a differential membrane localization since there is evidence that β_{3A}-AR but not β_{3B}-AR localizes to caveolae.[129]

The specific localization of membrane proteins is crucial for cell signalling because, in each membrane microdomain, a given molecular entity (*e.g.* GPCRs, G proteins, *etc.*) may interact productively with different upstream and downstream signalling proteins. By extension, drugs targeted to membrane lipids can be used to regulate membrane structure and to reverse pathological malfunctions whose etiology is associated with alterations in membrane protein signalling. This approach has been called membrane–lipid therapy and it has been successfully applied to the development of drugs for the treatment of cancer, obesity, hypertension, neurodegeneration, inflammation, metabolic, diseases, *etc.*[130,131]

The $G\alpha$ protein monomer dissociates from the $G\beta\gamma$ dimer and the receptor, and it localizes to lamellar structures.[120] Mobilization of the $G\alpha$ protein far from the receptor environment may facilitate its interaction with effector proteins (*e.g.* adenylyl cyclase) localized at other membrane areas with greater lamellar-phase propensity. This hypothesis is supported by the fact that a large number of effector proteins are localized to lipid rafts. In addition, $G\alpha_q$ is enriched in detergent-resistant membranes of human platelets upon receptor activation by thrombin,[132] and transducin ($G\alpha_t$) translocates to lipid rafts areas after rhodopsin photoactivation.[133] In endothelial and epithelial cells from lung tissue, both $G\alpha_i$ and $G\alpha_s$ localize to lipid rafts while $G\alpha_q$ is found in caveolae.[134] Therefore, these liquid-ordered microdomains seem to act as platforms with specific biophysical properties that enhance the effectiveness of the second step in the signalling cascade, concentrating activated $G\alpha$ subunits and their effectors in a small area of the membrane.

Despite the dissociation of Gα from the trimer, Gβγ subunits maintain a high affinity for membrane regions with a non-lamellar propensity. Accordingly, Gβγ dimers are not localized to lipid rafts,[135] in agreement with the fact that most prenylated proteins are excluded from these microdomains.[136] Since Gα prefers lamellar structures and heterotrimeric Gαβγ binds to non-lamellar prone structures, the Gβγ dimer most probably determines the characteristics of the lipid–protein interaction of heterotrimeric G proteins. As such, the Gβγ dimer defines the preference of complete Gαβγ heterotrimers for the hexagonal phase, thereby masking the lamellar membrane affinity of the Gα subunit. As a result, the Gβγ dimer might shuttles the Gα subunit from its preferred lamellar membrane phase to the vicinity of GPCRs. Receptor activation would provoke the dissociation of the Gα subunit and its migration to lamellar regions with tight surface packing (Figure 7.5).

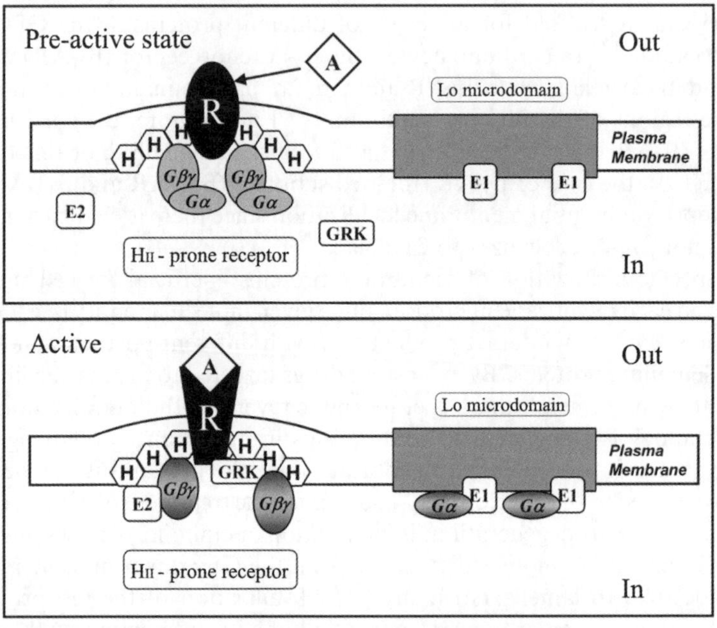

Figure 7.5 Membrane structure and GPCR-associated signalling. (**Upper panel**) GPCRs (R) induce the formation of hexagonal phases (H) in the membrane in their vicinity and these non-lamellar membrane regions attract heterotrimeric (inactive) G proteins, driven by the Gβγ-subunit. (**Lower panel**) Upon agonist (A) binding, several heterotrimeric G protein molecules are activated by one GPCR (R) molecule. The Gα-subunit then dissociates from the Gβγ-dimer and is targeted to specialized regions of the plasma membrane, such as lipid rafts, due to their greater affinity for ordered lamellar structures. It is these structures where they may activate their corresponding effector proteins (E1). Gβγ-dimers remain in the non-lamellar prone regions, where they can interact with their specific effectors (E2) or recruit GRKs directly to the receptors, promoting the GPCR phosphorylation that leads to receptor inactivation. Adapted from ref. 92.

The differential localization of G proteins in the membrane is influenced by the lipids they contain. Isoprenylated G proteins like Gγ and Ras preferentially localize to H_{II}-prone membrane regions where the farnesyl or geranylgeranyl groups may be better accommodated in a membrane with a weaker lateral surface pressure. In turn, palmitic and myristic acids in Gα subunits sit in membrane regions with a higher degree of rigidity with dense surface packing, such as Lo phases. This is in agreement with the fact that isoprenylated proteins (*e.g.* Gγ subunits) are excluded from lipid rafts, whereas fatty acylated proteins (*e.g.* Gα subunits) are often found in these ordered lamellar domains.

Besides the membrane lipid packing, the charge of membrane lipids also influences the localization of G proteins and GPCRs in the membrane. Thus, the negative charge of the phosphatidylserine (PS) seems to drive the binding of the Gα and Gγ subunits through electrostatic interactions with positive amino acids in the *N*- and *C*-terminal regions of these proteins, respectively.

7.5.2 Effects of G Proteins on Membrane Structure

Certain transmembrane peptides with α-helical structures, including the transmembrane domains of GPCRs, may promote the formation of hexagonal phases in a lipid bilayer.[137,138] G proteins are also able to modulate the membrane lipid structure and organization. For example, the farnesyl or geranylgeranyl groups of Gγ subunits (an other membrane proteins) affect the thermotropic behaviour of the non-lamellar prone phospholipid dielaidoyl-phosphatidylethanolamine (DEPE) by altering the gel-to-fluid lamellar and lamellar-to-hexagonal (H_{II}) phase transitions.[139] Not only do isoprenoids regulate membrane lipid structure but also the complete *C*-terminal region of Gγ subunits that contains this isoprenyl moiety.[121] The isoprenyl groups tend to stabilize non-lamellar structures and this moiety, along with the 12 *C*-terminal amino acids of the Gγ subunit, generate H_{II}-prone microdomains enriched in this peptide.[121] Accordingly, it has been concluded that: (i) the Gβγ dimer has a strong influence on the binding of the alpha subunit in the heterotrimer; (ii) the *C*-terminal region of the gamma subunit is critical in this interaction; and (iii) the presence of the latter in membranes stabilizes microdomains with negative curvature strain and induces the formation of domains rich and poor in Gγ-subunit peptides, respectively.

7.6 Conclusions

Lipid–protein interactions are critical for GPCR-associated cell signalling due to their involvement in modulating the ligand-mediated propagation of messages and through the regulatory effects they have on the initial steps of numerous transduction pathways. Lipid–protein interactions not only involve the binding of G protein and GPCR lipid moieties to the lipid bilayer but also other interactions between specific amino acids and membrane lipids. Gα subunits may be *N*-myristoylated and/or *S*-palmitoylated, which influences the distribution of these proteins. However, the presence of a fatty acyl moiety (*e.g.* myristic acid) is insufficient to translocate Gα subunits from the cytosolic to membrane fractions. Myristoylation and/or palmitoylation of Gα subunits not only regulates their targeting to specific cell membrane regions but also their interaction with other signalling proteins. Additionally, all Gγ subunits are *C*-terminally prenylated and this lipid modification is also important for the proper membrane localization of the Gβγ dimer and the Gαβγ trimer. GPCRs can be lipidated by the addition of a palmitate acyl chain and, in some cases, an isoprenyl group has also been detected. In both G proteins and GPCRs, these lipid modifications are actively involved in the trafficking of these proteins and in their activation/deactivation during signalling.

The structure and the physical properties of membranes are also relevant in the localization and activity of GPCRs and G proteins. The presence of non-lamellar prone lipids seems to be crucial for the binding of Gαβγ heterotrimers to the lipid bilayer, and for the localization and tertiary structure of GPCRs, while active Gα monomers prefer lamellar structures. With this in mind, and considering that GPCRs and associated signalling proteins currently constitute the main target for pharmaceutical drugs, it has become apparent that manipulating the lipid composition of the membrane might also be used to regulate important cellular processes and thus reverse certain diseases. Significantly, some lipids present in diets have beneficial effects against the onset or progression of a variety of illnesses by modifying the plasma membrane lipid composition.[140–144]

Indeed, it was recently shown that oleic acid is responsible for the hypotensive effects of olive oil due to its regulatory effects on GPCR-associated signalling mediated by the membrane.[144] Moreover, synthetic lipids that modify the composition and structure of membranes have been successfully used to develop drugs that control blood pressure,[145] body weight[146] and cancer.[122] Thus, the regulatory effects of 2-hydroxyoleic acid (Minerval®) on membranes induce the translocation of PKC to membranes and of Ras from the plasma membrane to the cytosol, which induces differentiation, inhibits cancer cell growth and causes death of cancer cells.[122] In addition, polyunsaturated fatty acids (PUFA) derivatives, which induce different cell membrane structure modifications, have been developed to treat inflammation and Alzheimer's disease. This fact indicates that the regulation of the plasma membrane structure can induce specific effects on the cell's physiology. Moreover, subtle changes in the structure of lipids, such as the replacement of a H atom by a OH moiety or the steric configuration of double bonds (*cis* or *trans*), induce huge changes in the pharmacological efficacy of these molecules.[147,148] This novel approach has been called membrane–lipid therapy, which is based on the regulatory effects of drugs on membrane composition and the subsequent modification of cell membrane structure to revert pathological processes.[130]

Acknowledgements

This work was supported by Grant BFU2007-61071 and BIO2010-21132 from the 'Ministerio de Ciencia e Innovación' (Spain). The authors thank Dr L. R. Montes for her help in the preparation of the manuscript.

References

1. T. J. Stevens and I. T. Arkin, *Proteins*, 2000, **39**, 4.
2. S. Takeda, S. Kadowaki, T. Haga, H. Takaesu and S. Mitaku, *FEBS Lett.*, 2002, **520**, 1.
3. D. Filmore, *Mod. Drug Discov.*, 2004, **7**, 11.
4. S. J. Singer and G. L. Nicolson, *Science*, 1972, **175**, 23.
5. P. J. O'Brien and M. Zatz, *J. Biol. Chem.*, 1984, **259**, 8.
6. J. E. Buss, S. M. Mumby, P. J. Casey, A. G. Gilman and B. M. Sefton, *Proc. Natl. Acad. Sci. U. S. A.*, 1987, **84**, 21.
7. J. M. Sugars, S. Cellek, M. Manifava, J. Coadwell and N. T. Ktistakis, *J. Biol. Chem.*, 1999, **274**, 42.
8. R. H. Stoffel, R. R. Randall, R. T. Premont, R. J. Lefkowitz and J. Inglese, *J. Biol. Chem.*, 1994, **269**, 45.
9. D. A. Brown and E. London, *Annu. Rev. Cell Dev. Biol.*, 1998, **14**, 111.
10. K. Simons and D. Toomre, *Nat. Rev. Mol. Cell. Biol.*, 2000, **1**, 1.
11. A. C. Rowat, J. Brask, T. Sparrman, K. J. Jensen, G. Lindblom and J. H. Ipsen, *Eur. Biophys. J.*, 2004, **33**, 4.
12. S. Janosch, C. Nicolini, B. Ludolph, C. Peters, M. Volkert, T. L. Hazlet, E. Gratton, H. Waldmann and R. Winter, *J. Am. Chem. Soc.*, 2004, **126**, 24.
13. K. Palczewski, T. Kumasaka, T. Hori, C. A. Behnke, H. Motoshima, B. A. Fox, I. Le Trong, D. C. Teller, T. Okada, R. E. Stenkamp, M. Yamamoto and M. Miyano, *Science*, 2000, **289**, 5480.
14. S. G. Rasmussen, H. J. Choi, D. M. Rosenbaum, T. S. Kobilka, F. S. Thian, P. C. Edwards, M. Burghammer, V. R. Ratnala, R. Sanishvili, R. F. Fischetti, G. F. Schertler, W. I. Weis and B. K. Kobilka, *Nature*, 2007, **450**, 7168.
15. S. J. Moench, J. Moreland, D. H. Stewart and T. G. Dewey, *Biochemistry*, 1994, **33**, 19.
16. B. F. O'Dowd, M. Hnatowich, M. G. Caron, R. J. Lefkowitz and M. Bouvier, *J. Biol. Chem.*, 1989, **264**, 13.
17. H. Jin, Z. Xie, S. R. George and B. F. O'Dowd, *Eur. J. Pharmacol.*, 1999, **386**, 2.
18. Y. Okamoto, H. Ninomiya, M. Tanioka, A. Sakamoto, S. Miwa and T. Masaki, *J. Biol. Chem.*, 1997, **272**, 34.
19. C. Stannard, P. Lehenkari and J. Godovac-Zimmermann, *Biochemistry*, 2003, **42**, 47.
20. A. Horstmeyer, H. Cramer, T. Sauer, W. Muller-Esterl and C. Schroeder, *J. Biol. Chem.*, 1996, **271**, 34.
21. M. K. Hayashi and T. Haga, *Arch. Biochem. Biophys.*, 1997, **340**, 2.
22. N. Hukovic, R. Panetta, U. Kumar, M. Rocheville and Y. C. Patel, *J. Biol. Chem.*, 1998, **273**, 33.
23. N. Kawate and K. M. Menon, *J. Biol. Chem.*, 1994, **269**, 48.
24. C. Chen, V. Shahabi, W. Xu and L. Y. Liu-Chen, *FEBS Lett.*, 1998, **441**, 1.

25. S. R. Hawtin, A. B. Tobin, S. Patel and M. Wheatley, *J. Biol. Chem.*, 2001, **276**, 41.
26. S. M. Miggin, O. A. Lawler and B. T. Kinsella, *Eur. J. Biochem.*, 2002, **269**, 6.
27. S. M. Miggin, O. A. Lawler and B. T. Kinsella, *J. Biol. Chem.*, 2003, **278**, 9.
28. J. B. Higgins and P. J. Casey, *Cell. Signal.*, 1996, **8**, 6.
29. J. F. Hancock, *Nat. Rev. Mol. Cell Biol.*, 2003, **4**, 5.
30. J. F. Hancock, A. I. Magee, J. E. Childs and C. J. Marshall, *Cell*, 1989, **57**, 7.
31. J. T. Dunphy and M. E. Linder, *Biochim Biophys. Acta*, 1998, **1436**, 1.
32. M. O. Parat and P. L. Fox, *J. Biol. Chem.*, 2001, **276**, 19.
33. M. B. Omary and I. S. Trowbridge, *J. Biol. Chem.*, 1981, **256**, 10.
34. M. Berger and M. F. Schmidt, *FEBS Lett.*, 1985, **187**, 2.
35. S. Bonatti, G. Migliaccio and K. Simons, *J. Biol. Chem.*, 1989, **264**, 21.
36. R. C. Drisdel, E. Manzana and W. N. Green, *J. Neurosci.*, 2004, **24**, 46.
37. U. E. Petaja-Repo, M. Hogue, T. T. Leskela, P. M. Markkanen, J. T. Tuusa and M. Bouvier, *J. Biol. Chem.*, 2006, **281**, 23.
38. S. S. Karnik, K. D. Ridge, S. Bhattacharya and H. G. Khorana, *Proc. Natl. Acad. Sci. U. S. A.*, 1993, **90**, 1.
39. H. Zhu, H. Wang and M. Ascoli, *Mol. Endocrinol.*, 1995, **9**, 2.
40. Y. Fukushima, T. Saitoh, M. Anai, T. Ogihara, K. Inukai, M. Funaki, H. Sakoda, Y. Onishi, H. Ono, M. Fujishiro, T. Ishikawa, K. Takata, R. Nagai, M. Omata and T. Asano, *Biochim Biophys. Acta*, 2001, **1539**, 3.
41. Y. Percherancier, T. Planchenault, A. Valenzuela-Fernandez, J. L. Virelizier, F. Arenzana-Seisdedos and F. Bachelerie, *J. Biol. Chem.*, 2001, **276**, 34.
42. T. Iwanaga, R. Tsutsumi, J. Noritake, Y. Fukata and M. Fukata, *Prog. Lipid Res.*, 2009, **48**, 3–4.
43. J. H. Skene and I. Virag, *J. Cell Biol.*, 1989, **108**, 2.
44. G. James and E. N. Olson, *J. Biol. Chem.*, 1989, **264**, 35.
45. B. Mouillac, M. Caron, H. Bonin, M. Dennis and M. Bouvier, *J. Biol. Chem.*, 1992, **267**, 30.
46. M. E. Kennedy and L. E. Limbird, *J. Biol. Chem.*, 1994, **269**, 50.
47. G. Y. Ng, B. Mouillac, S. R. George, M. Caron, M. Dennis, M. Bouvier and B. F. O'Dowd, *Eur. J. Pharmacol.*, 1994, **267**, 1.
48. M. E. Kennedy and L. E. Limbird, *J. Biol. Chem.*, 1993, **268**, 11.
49. E. Kelly, C. P. Bailey and G. Henderson, *Br. J. Pharmacol.*, 2008, **153**(Suppl 1), S379.
50. S. L. Ritter and R. A. Hall, *Nat. Rev. Mol. Cell. Biol.*, 2009, **10**, 12.
51. Y. Zhang, W. Xiong, X. Lin, X. Ma and L. C. Yu, *Neurosci. Biobehav. Rev.*, 2009, **33**, 8.
52. Z. Wang, X. H. Wen, Z. Ablonczy, R. K. Crouch, C. L. Makino and J. Lem, *J. Biol. Chem.*, 2005, **280**, 26.
53. E. Ponimaskin, A. Dumuis, F. Gaven, G. Barthet, M. Heine, K. Glebov, D. W. Richter and M. Oppermann, *Mol. Pharmacol.*, 2005, **67**, 5.

54. U. M. Munshi, C. L. Clouser, H. Peegel and K. M. Menon, *Mol. Endocrinol.*, 2005, **19**, 3.
55. T. A. Farazi, G. Waksman and J. I. Gordon, *J. Biol. Chem.*, 2001, **276**, 43.
56. S. Maurer-Stroh, B. Eisenhaber and F. Eisenhaber, *J. Mol. Biol.*, 2002, **317**, 4.
57. R. S. Bhatnagar, O. F. Schall, E. Jackson-Machelski, J. A. Sikorski, B. Devadas, G. W. Gokel and J. I. Gordon, *Biochemistry*, 1997, **36**, 22.
58. K. Kokame, Y. Fukada, T. Yoshizawa, T. Takao and Y. Shimonishi, *Nature*, 1992, **359**, 6397.
59. T. A. Neubert, R. S. Johnson, J. B. Hurley and K. A. Walsh, *J. Biol. Chem.*, 1992, **267**, 26.
60. P. B. Wedegaertner, *Biol. Signals Recept.*, 1998, **7**, 2.
61. D. A. Mitchell, A. Vasudevan, M. E. Linder and R. J. Deschenes, *J. Lipid Res.*, 2006, **47**, 6.
62. M. Y. Degtyarev, A. M. Spiegel and T. L. Jones, *J. Biol. Chem.*, 1993, **268**, 32.
63. T. P. Loisel, L. Adam, T. E. Hebert and M. Bouvier, *Biochemistry*, 1996, **35**, 49.
64. C. Kleuss and E. Krause, *Embo J.*, 2003, **22**, 4.
65. P. B. Wedegaertner, H. R. Bourne and M. von Zastrow, *Mol. Biol. Cell*, 1996, **7**, 8.
66. J. Z. Yu and M. M. Rasenick, *Mol. Pharmacol.*, 2002, **61**, 2.
67. T. R. Hynes, S. M. Mervine, E. A. Yost, J. L. Sabo and C. H. Berlot, *J. Biol. Chem.*, 2004, **279**, 42.
68. J. A. Allen, J. Z. Yu, R. J. Donati and M. M. Rasenick, *Mol. Pharmacol.*, 2005, **67**, 5.
69. V. L. Boyartchuk, M. N. Ashby and J. Rine, *Science*, 1997, **275**, 5307.
70. Q. Dai, E. Choy, V. Chiu, J. Romano, S. R. Slivka, S. A. Steitz, S. Michaelis and M. R. Philips, *J. Biol. Chem.*, 1998, **273**, 24.
71. S. Takida and P. B. Wedegaertner, *J. Biol. Chem.*, 2003, **278**, 19.
72. D. S. Evanko, M. M. Thiyagarajan and P. B. Wedegaertner, *J. Biol. Chem.*, 2000, **275**, 2.
73. K. Wennerberg, K. L. Rossman and C. J. Der, *J. Cell Sci.*, 2005, **118**, Pt 5.
74. A. Ehrhardt, G. R. Ehrhardt, X. Guo and J. W. Schrader, *Exp. Hematol.*, 2002, **30**, 10.
75. J. F. Hancock and R. G. Parton, *Biochem. J.*, 2005, **389**, Pt 1.
76. D. B. Whyte, P. Kirschmeier, T. N. Hockenberry, I. Nunez-Oliva, L. James, J. J. Catino, W. R. Bishop and J. K. Pai, *J. Biol. Chem.*, 1997, **272**, 22.
77. J. F. Hancock, H. Paterson and C. J. Marshall, *Cell*, 1990, **63**, 1.
78. D. Voet and J. G. Voet, *Biochemistry*, John Wiley & Sons, Hoboken, NJ, 3rd edn, 2004.
79. E. Gorter and F. Grendel, *J. Exp. Med.*, 1925, **41**.
80. G. Vereb, J. Szollosi, J. Matko, P. Nagy, T. Farkas, L. Vigh, L. Matyus, T. A. Waldmann and S. Damjanovich, *Proc. Natl. Acad. Sci. U. S. A.*, 2003, **100**, 14.
81. D. M. Engelman, *Nature*, 2005, **438**, 7068.

82. A. J. Verkleij, R. F. Zwaal, B. Roelofsen, P. Comfurius, D. Kastelijn and L. L. van Deenen, *Biochim. Biophys. Acta*, 1973, **323**, 2.
83. G. van Meer, D. R. Voelker and G. W. Feigenson, *Nat. Rev. Mol. Cell. Biol.*, 2008, **9**, 2.
84. K. Simons and W. L. Vaz, *Annu. Rev. Biophys. Biomol. Struct.*, 2004, **33**.
85. K. Simons and E. Ikonen, *Nature*, 1997, **387**, 6633.
86. B. G. Tenchov, R. C. MacDonald and D. P. Siegel, *Biophys. J.*, 2006, **91**, 7.
87. P. R. Cullis and B. de Kruijff, *Biochim. Biophys. Acta*, 1979, **559**, 4.
88. S. M. Gruner, P. R. Cullis, M. J. Hope and C. P. Tilcock, *Annu. Rev. Biophys. Biophys. Chem.*, 1985, **14**, 211.
89. P. R. Cullis, M. J. Hope and C. P. Tilcock, *Chem. Phys. Lipids*, 1986, **40**, 2.
90. G. Basañez, *Cell. Mol. Life Sci.*, 2002, **59**, 9.
91. R. M. Epand, *Biochim. Biophys. Acta*, 1998, **1376**, 3.
92. O. Vögler, J. M. Barceló, C. Ribas and P. V. Escribá, *Biochim. Biophys. Acta*, 2008, **1778**, 7–8.
93. V. Luzzati, in *Biological Membranes*, ed. D. Chapman, Academic Press, London, 1968, pp. 71–123.
94. D. A. Brown and E. London, *J. Membr. Biol.*, 1998, **164**, 2.
95. J. N. Israelachvili, S. Marcelja and R. G. Horn, *Q. Rev. Biophys.*, 1980, **13**, 2.
96. J. L. Nieva, A. Alonso, G. Basañez, F. M. Goñi, A. Gulik, R. Vargas and V. Luzzati, *FEBS Lett.*, 1995, **368**, 1.
97. O. Terrones, B. Antonsson, H. Yamaguchi, H. G. Wang, J. Liu, R. M. Lee, A. Herrmann and G. Basañez, *J. Biol. Chem.*, 2004, **279**, 29.
98. P. V. Escribá, P. B. Wedegaertner, F. M. Goñi and O. Vögler, *Biochim. Biophys. Acta*, 2007, **1768**, 4.
99. D. A. Vessey and D. Zakim, *Horiz. Biochem. Biophys.*, 1974, **1**.
100. D. Chapman and P. J. Quinn, *Proc. Natl. Acad. Sci. U. S. A.*, 1976, **73**, 11.
101. K. M. Goebel and J. Schneider, *Acta Biol. Med, Ger.*, 1981, **40**, 4–5.
102. E. Berlin, E. J. Matusik, Jr. and C. Young, Jr, *Lipids*, 1980, **15**, 8.
103. K. Wells, A. A. Farooqui, L. Liss and L. A. Horrocks, *Neurochem. Res.*, 1995, **20**, 11.
104. J. Zicha, J. Kunes and M. A. Devynck, *Am. J. Hypertens.*, 1999, **12**, 3.
105. A. G. Lee, *Biochim. Biophys. Acta*, 2004, **1666**, 1.
106. R. G. Matthews, R. Hubbard, P. K. Brown and G. Wald, *J. Gen. Physiol.*, 1963, **47**.
107. T. Huber, A. V. Botelho, K. Beyer and M. F. Brown, *Biophys. J.*, 2004, **86**, 4.
108. B. J. Litman and D. C. Mitchell, *Lipids*, 1996, **31**(Suppl), S193.
109. A. V. Botelho, T. Huber, T. P. Sakmar and M. F. Brown, *Biophys. J.*, 2006, **91**, 12.
110. N. J. Gibson and M. F. Brown, *Biochemistry*, 1993, **32**, 9.
111. D. C. Mitchell, S. L. Niu and B. J. Litman, *Lipids*, 2003, **38**, 4.
112. S. L. Niu, D. C. Mitchell and B. J. Litman, *J. Biol. Chem.*, 2002, **277**, 23.
113. Y. Xiang, V. O. Rybin, S. F. Steinberg and B. Kobilka, *J. Biol. Chem.*, 2002, **277**, 37.

114. D. C. Mitchell, S. L. Niu and B. J. Litman, *J. Biol. Chem.*, 2001, **276**, 46.
115. S. L. Niu, D. C. Mitchell and B. J. Litman, *J. Biol. Chem.*, 2001, **276**, 46.
116. A. Grossfield, S. E. Feller and M. C. Pitman, *Proc. Natl. Acad. Sci. U. S. A.*, 2006, **103**, 13.
117. G. Khelashvili, A. Grossfield, S. E. Feller, M. C. Pitman and H. Weinstein, *Proteins*, 2009, **76**, 2.
118. G. Basañez, J. L. Nieva, F. M. Goñi and A. Alonso, *Biochemistry*, 1996, **35**, 48.
119. W. R. Perkins, R. B. Dause, R. A. Parente, S. R. Minchey, K. C. Neuman, S. M. Gruner, T. F. Taraschi and A. S. Janoff, *Science*, 1996, **273**, 5273.
120. O. Vögler, J. Casas, D. Capo, T. Nagy, G. Borchert, G. Martorell and P. V. Escribá, *J. Biol. Chem.*, 2004, **279**, 35.
121. F. Barceló, J. Prades, J. A. Encinar, S. S. Funari, O. Vögler, J. M. González-Ros and P. V. Escribá, *Biophys. J.*, 2007, **93**, 7.
122. J. Martínez, O. Vögler, J. Casas, F. Barceló, R. Alemany, J. Prades, T. Nagy, C. Baamonde, P. G. Kasprzyk, S. Terés, C. Saus and P. V. Escribá, *Mol. Pharmacol.*, 2005, **67**, 2.
123. P. V. Escribá, M. Sastre and J. A. García-Sevilla, *Proc. Natl. Acad. Sci. U. S. A.*, 1995, **92**, 16.
124. A. V. Botelho, N. J. Gibson, R. L. Thurmond, Y. Wang and M. F. Brown, *Biochemistry*, 2002, **41**, 20.
125. I. D. Alves, G. F. Salgado, Z. Salamon, M. F. Brown, G. Tollin and V. J. Hruby, *Biophys. J.*, 2005, **88**, 1.
126. S. E. Feller and K. Gawrisch, *Curr. Opin. Struct. Biol.*, 2005, **15**, 4.
127. E. J. Smart, G. A. Graf, M. A. McNiven, W. C. Sessa, J. A. Engelman, P. E. Scherer, T. Okamoto and M. P. Lisanti, *Mol. Cell. Biol.*, 1999, **19**, 11.
128. V. O. Rybin, X. Xu, M. P. Lisanti and S. F. Steinberg, *J. Biol. Chem.*, 2000, **275**, 52.
129. M. Sato, D. S. Hutchinson, B. A. Evans and R. J. Summers, *Biochem. Soc. Trans.*, 2007, **35**, Pt 5.
130. P. V. Escribá, *Trends Mol. Med.*, 2006, **12**, 1.
131. P. V. Escriba, J. M. Gonzalez-Ros, F. M. Goni, P. K. Kinnunen, L. Vigh, L. Sanchez-Magraner, A. M. Fernandez, X. Busquets, I. Horvath and G. Barcelo-Coblijn, *J. Cell. Mol. Med.*, 2008, **12**, 3.
132. S. Bodin, S. Giuriato, J. Ragab, B. M. Humbel, C. Viala, C. Vieu, H. Chap and B. Payrastre, *Biochemistry*, 2001, **40**, 50.
133. K. S. Nair, N. Balasubramanian and V. Z. Slepak, *Curr. Biol.*, 2002, **12**, 5.
134. P. Oh and J. E. Schnitzer, *Mol. Biol. Cell*, 2001, **12**, 3.
135. S. Moffett, D. A. Brown and M. E. Linder, *J. Biol. Chem.*, 2000, **275**, 3.
136. K. A. Melkonian, A. G. Ostermeyer, J. Z. Chen, M. G. Roth and D. A. Brown, *J. Biol. Chem.*, 1999, **274**, 6.
137. P. C. van der Wel, T. Pott, S. Morein, D. V. Greathouse, R. E. Koeppe, 2nd and J. A. Killian, *Biochemistry*, 2000, **39**, 11.
138. J. Prades, J. A. Encinar, S. S. Funari, J. M. Gonzalez-Ros, P. V. Escriba and F. Barcelo, *Mol. Membr. Biol.*, 2009, **26**, 5.

139. S. S. Funari, J. Prades, P. V. Escribá and F. Barceló, *Mol. Membr. Biol.*, 2005, **22**, 4.
140. M. A. Martínez-González and A. Sánchez-Villegas, *Eur. J. Epidemiol.*, 2004, **19**, 1.
141. R. Bartoli, F. Fernandez-Banares, E. Navarro, E. Castella, J. Mane, M. Alvarez, C. Pastor, E. Cabre and M. A. Gassull, *Gut*, 2000, **46**, 2.
142. V. Ruiz-Gutiérrez, F. J. Muriana, A. Guerrero, A. M. Cert and J. Villar, *J. Hypertens.*, 1996, **14**, 12.
143. M. C. de Wilde, E. Hogyes, A. J. Kiliaan, T. Farkas, P. G. Luiten and E. Farkas, *Brain Res.*, 2003, **988**, 1.
144. S. Terés, G. Barceló-Coblijn, M. Benet, R. Alvárez, R. Bressani, J. E. Halver and P. V. Escribá, *Proc. Natl. Acad. Sci. U. S. A.*, 2008, **105**, 37.
145. R. Alemany, O. Vögler, S. Terés, C. Egea, C. Baamonde, F. Barceló, C. Delgado, K. H. Jakobs and P. V. Escribá, *J. Lipid Res.*, 2006, **47**, 8.
146. O. Vögler, A. López-Bellán, R. Alemany, S. Tofe, M. González, J. Quevedo, V. Pereg, F. Barceló and P. V. Escribá, *Int. J. Obes.*, 2008, **32**, 3.
147. J. Martínez, A. Gutiérrez, J. Casas, V. Lladó, A. López-Bellán, J. Besalduch, A. Dopazo and P. V. Escribá, *J. Pharmacol. Exp. Ther.*, 2005, **315**, 1.
148. V. Lladó, S. Terés, M. Higuera, R. Álvarez, M. A. Noguera-Salva, J. E. Halver, P. V. Escribá and X. Busquets, *Proc. Natl. Acad. Sci. U. S. A.*, 2009, **106**, 33.

CHAPTER 8

Use of Model Membranes to Study GPCR Signalling Units: Insights into Monomers and Oligomers

D. M. CALINSKI, E. EDWALD AND R. K. SUNAHARA*

University of Michigan Medical School, Department of Pharmacology, 1301 MSRB III, 1150 W. Medical Center Drive, Ann Arbor, MI 48109-5632, USA

8.1 Introduction

The field of G protein-coupled receptors (GPCRs) advanced significantly with the use of model membranes. Isolation of GPCRs from the cellular membranes results in a propensity of the protein to aggregate, as well as lose activity, hindering *in vitro* experimentation. In order to perform minimalistic analyses of GPCRs, it is essential to re-establish the natural *milieu* of the protein and thus the use of 'model membranes'. The goal of a model membrane is to mimic the environment of the cellular membrane, while removing any unknown components so that precise analysis of the GPCR can be completed. Advantageously, once purified GPCRs that are reconstituted into a lipid bilayer are more stable and soluble, and recover most of the activity that was lost during the purification process.

RSC Drug Discovery Series No. 8
G Protein-Coupled Receptors: From Structure to Function
Edited by Jesús Giraldo and Jean-Philippe Pin

Published by the Royal Society of Chemistry, www.rsc.org

This chapter describes two different models used to complete *in vitro* studies with integral membrane proteins: reconstituted high density lipoprotein (rHDL) particles—which are also known as nanodics, or nanoscale apolipoprotein-bound bilayers (NABBs)—and lipid vesicles (or lamellae). The advantages and disadvantages, as well as major contributions to the field of GPCR research, are introduced for each of these models.

8.1.1 Why Study GPCRs in Model Lipid Membranes?

It is difficult to isolate and maintain GPCRs in solution because the receptors are integral membrane proteins and therefore contain large hydrophobic areas that are typically buried within the cellular membrane.[1] Detergents are used to solubilize the membrane and extract GPCRs, but there are problems with this process. For instance, GPCR activity is exquisitely sensitive to detergent as exemplified by compromised ligand binding and a significant impairment of G protein coupling.[2–5] These effects are due to structural and charge alterations that occur during the solubilization process.[6] The structural changes include the loss of support and rigidity that the membrane contributed to the receptor, as well as alterations to the ligand binding pocket of the receptor. The charge modifications relate to the loss of the charged phospholipid head groups of the membrane. It is known that these charges are critical for receptor-interacting proteins, and the absence of charge can be noted in the inability of the detergent-solubilized receptor to interact with other proteins.[27,77,78] The importance of phospholipid charges is also demonstrated by the crystallographic successes of the adenosine 2A (A2AR), β_1-adrenergic (β_1AR), β_2-adrenergic (β_2AR) and chemokine (CXCR4) receptors dependent on the generation of stable lattices in lipid environments in the form of bicelles or lipidic cubic phases.[7–11] These points draw attention to the need for model membrane bilayers when studying GPCR activity.

With the incorporation of GPCRs into lipid bilayers, detergent can be removed from the sample and the phospholipid head group charges are restored. An example of detergent and reconstitution effects is shown in Figure 8.1. Here [^{3}H]diprenorphin ([^{3}H]DPN) binds poorly to the μ-opioid receptor in *n*-dodecyl β-D-maltoside (DDM) detergent micelles compared with the receptor reconstituted in a model lipid bilayer.[2] Therefore, *in vitro* activity studies on purified GPCRs should be conducted in lipid bilayers when possible.

8.2 Reconstitution of Monomeric GPCRs into Model Lipid Bilayers

Historically, analysis of GPCR oligomerization relied on evidence from detergent-solubilized preparations. These studies suggested that the monomeric rhodopsin, a prototypic GPCR, is sufficient to activate a G protein (transducin) and that monomeric rhodopsin is likely the major species in native rod outer segments.[12] However, there is a growing body of evidence to the contrary; these

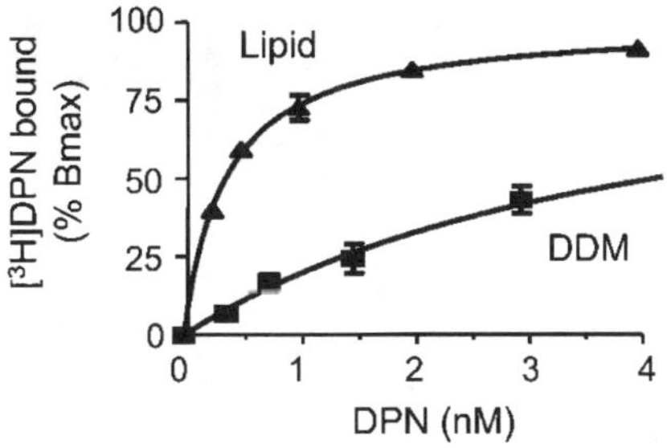

Figure 8.1 Deleterious effect of detergents on GPCR function. Illustrated is the capacity of opioid receptor antagonist [^{3}H]diprenorphin ([^{3}H]DPN) to bind to the μ-opioid receptor (μ-OR) in detergent micelles (*n*-dodecyl-β-D-maltoside, DDM) or reconstituted into rHDL particles (lipid). Opioid receptor bound [^{3}H]DPN with a K_d > 20 nM when solubilized DDM, compared to a K_d ~0.5 nM when reconstituted into phospholipids. Figure derived from data by Kuszak *et al.* (2009).[2]

data are discussed in the next section.[13–16] In addition, there is now an appreciation of the deleterious effects of detergent on GPCRs, as described earlier.[6] The introduction of rHDL technology to study receptor oligomerization has circumvented many of these issues, as it is now possible to isolate and study monomeric and dimeric forms of GPCRs in a phospholipid bilayer. Moreover, rHDL reconstitution creates an environment conducive to assessing receptor-interacting proteins as the phospholipid bilayer provides an optimal *milieu* for studying receptors.

8.2.1 The rHDL

8.2.1.1 High Density Lipoproteins in vivo

High density lipoproteins (HDL) are naturally occurring lipoproteins in the human body that are responsible for the reverse transport of cholesterol from arteries to the liver for excretion or reuse. An HDL particle is composed of a dimer of apolipoprotein A-1 (apoA-1) that surrounds a planar bilayer of roughly 160 phospholipids (see Figure 8.2A).[4,5] The apoA-1 dimer serves as a helical belt surrounding the lipid bilayer and therefore maintains a fixed and uniform discoidal-shaped particle of approximately 100–120 Å in diameter.

Since the early 1980s, scientists have been able to reconstitute HDL particles from purified components for study and observation.[17,18] However, the application of rHDL particles to reconstitute and study the activity of membrane proteins has only recently been established. Since this discovery a number of membrane proteins have been studied within this model lipid bilayer including cytochrome P450, the protein pump bacteriorhodopsin, SecYEG heterotrimer, bacterial chemoreceptors, voltage-dependent anion channel (VDAC-1), as well as a number of GPCRs.[2,4,5,17,19–24]

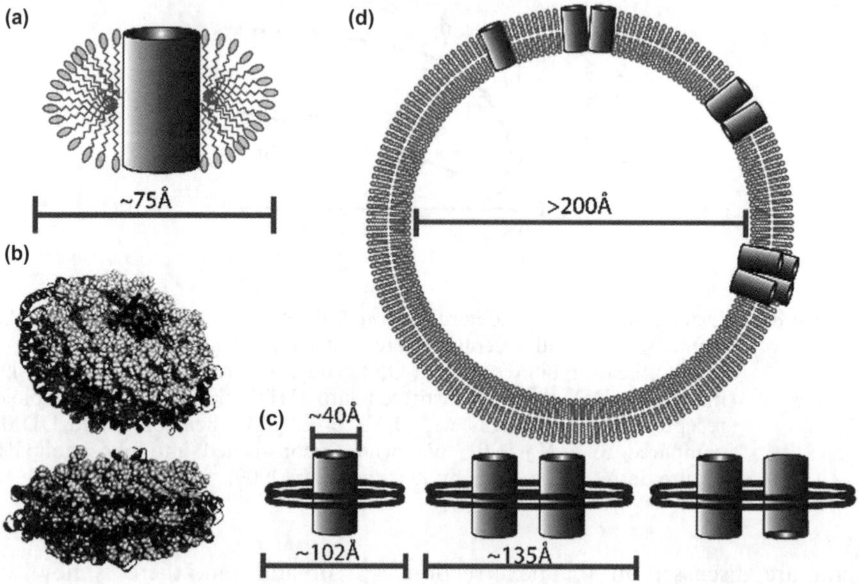

Figure 8.2 Chapter overview. An overview of the detergent micelle and lipid model membranes discussed in this chapter. (**a**) Schematic representation and approximate diameter size of a receptor (grey cylinder) in a DDM detergent micelle. (**b**) A receptor stabilized HDL particle. Apolipoprotein A1 dimer (model, black) forms a belt containing the lipid bilayer (model, light grey) in which the β_2AR crystal structure is imbedded (pdb 3NY8, black.) (**c**) Representation of an HDL particle and expanded HDL-like particles incorporating monomeric, dimeric parallel and dimeric anti-parallel receptors. 102 Å refers to outer diameter of HDL particle; not labeled is the inner diameter which is approximately 80 Å. The larger particles will have outer and inner diameters of 135 Å and 120 Å, respectively. (**d**) Depiction of a phospholipid vesicle with receptors in various oligomeric states and both 'outside–out' and 'inside–out' orientations. The smallest vesicles are 200 Å in diameter, but giant lamellar vesicles can be as large as 300 μm. The coordinates for rHDL were generously provided by Dr Stephen Harvey; coordinates for the HDL model from Segrest *et al.*[28] were obtained from http://rumour.biology.gatech.edu/Publications/).

8.2.1.2 Advantages of the rHDL Approach

The development of the rHDL technique provided control over the number of receptors within the model system. This is a consequence of the size of rHDL as well as the reconstitution conditions. The thickness of the bilayer is 40–50 Å, which is similar to a cell membrane. This dimension depends on the type of lipid used for reconstitution; 16 and 18 carbon acyl chain lengths are most commonly used.[5] The inner diameter of the rHDL particle is approximately 80–90 Å and can therefore accommodate a monomeric GPCR, which has a predicted diameter of 40–45 Å.[5,7,9] Although it is possible to fit two GPCRs into a single rHDL, it requires tight packing of the receptors and leaves little room for annular lipids within the rHDL particle.[25] To further ensure that only single receptors are reconstituted into rHDL, investigators bias the initial reconstitution ratio so that a vast excess of apoA-1 and lipid is present compared to receptor (apoA-1 : receptor ratio of 80–100 : 1, or an rHDL : receptor ratio of 40-50 : 1).[2,4,5] Reconstitution of monomeric receptors has been confirmed by colocalization experiments using total internal reflection fluorescence microscopy (TIRFm) and fluorescence resonance energy transfer (FRET) experiments, both of which are discussed in more detail below.

Electron microscopy (EM) and atomic force microscopy (AFM) show that receptor containing particles are of uniform size.[4,5] This is also shown by size exclusion chromatography (SEC) and sucrose density ultracentrifugation.[21,22] Particle homogeneity is essential in studies using extrinsic membrane proteins, as an uneven particle size would allow for a variable number of interacting proteins per particle, complicating analyses of experiments.

Finally the rHDL system allows for accessibility of both the extracellular and intracellular faces of the GPCR to ligands, G proteins, or other receptor-interacting proteins. Most model membranes, such as phospholipid vesicles, prohibit access to one side of the reconstituted membrane protein (Figure 8.2), making accurate conclusions about interactions with the receptor difficult to complete and interpret. In the rHDL approach, the orientation of a single incorporated protein is irrelevant.

8.2.2 Methodology of Incorporating Monomeric Receptors

A more thorough description of GPCR incorporation into rHDL is given by Whorton *et al.* or Bayburt *et al.*[4,26] Apo A-1 is purified from either recombinant sources or human serum.[5,28] The purified apoA-1 is then incubated with detergent-solubilized phospholipids and the purified protein of interest. The choice of the lipid mixture, derived from reconstitution studies in vesicle preparations, is typically palmitoylphosphatidylcholine (POPC) and palmitoylphosphatidylglycerol (POPG) combinations. Upon detergent removal, either by addition of hydrophobic adsorbent Bio-Beads, dialysis or gel filtration, the rHDL particles form spontaneously. At an rHDL : receptor ratio of approximately 40 : 1, a single receptor will be incorporated into a single rHDL particle greater than 98% of the time.[2,4,5] This procedure must be optimized per protein of interest; variables include phospholipid type, ratio of apo A-1 : receptor, ratio of apo A-1 : lipid, incubation times, temperature, choice of detergent and rate of detergent removal.

8.2.3 Evidence and Functional Studies of Monomeric Receptors in rHDL

8.2.3.1 Reconstituted Receptors are Monomeric

To determine the oligomeric state of receptors in rHDL, a mixture of Cy3-labeled β_2AR (Cy3-β_2AR) and Cy5-labeled β_2AR (Cy5-β_2AR) were reconstituted under conditions that should favour a monomeric incorporation. Analysis by TIRFm indicated that >98% of the receptor containing rHDL had only Cy3-β_2AR or Cy5-β_2AR, but not both. Analysis by FRET of the same sample confirmed that the receptors were not within the Förster distance to transfer energy. These data confirm the existence of monomeric receptors within the particles when reconstituted in the appropriate conditions.[4] Reconstitution of the μ-opioid receptor also resulted in isolation of monomeric receptors, as analysed by the same assays.[2] The photoreceptor from bovine retina, rhodopsin, has also been successfully reconstituted into rHDL particles as a monomer; ensemble absorbance measurements based on the known extinction coefficient at 500 nm of rhodopsin, and apoA-1 with its known extinction coefficient at 280 nm, give the approximation that the majority of the particles contain only one rhodopsin.[5] EM imaging by negative staining and the use of specific antibodies against rhodopsin also confirm that, depending on the reconstitution conditions, only one rhodopsin per particle is obtained.[29]

As discussed later, particles containing two rhodopsins may also be created using a higher receptor : rHDL ratio and mutants of apoA-1. The mutants of apoA-1 contain insertions of tandem repeats found in apoA-1 protein. These tandem repeat insertions create longer apoA-1, and therefore larger rHDL particles (diameters of 130–140 Å), which are large enough to accommodate two seven-transmembrane (7TM) domain proteins.[26]

8.2.3.2 Functional Studies on Monomeric Receptors

Reconstitution of GPCRs into rHDL reverses the deleterious effects of detergents, as evidenced by the reinstated ligand binding properties as well as G protein coupling. Specifically, reconstituted receptors exhibit binding affinities for antagonists and agonists with K_i values that are more consistent for GPCRs in biological membranes than solubilized in detergent.[2,4,5,30]

In the rHDL system, allostery is observed between monomeric GPCRs and G proteins. In the prototypical case of the reconstituted β_2AR with its stimulatory G protein (G_s), efficient coupling and agonist-dependent (isoproterenol) stimulation of guanine nucleotide exchange are observed. Additionally, agonist binding to monomeric GPCRs in rHDL is allosterically modulated by G proteins.[31] Simply stated, G protein coupling stabilizes the high affinity state of the receptor for agonists. When the G protein is uncoupled using GTP or non-hydrolysable GTP analogues such as GTPγS or GppNHp, the receptor is in the low affinity agonist binding state, as predicted by radioactive ligand competition data. In summary, the allosteric role of G proteins on agonist binding, a hallmark of G protein coupling as seen for many GPCRs in membrane assays is also maintained with monomeric GPCRs reconstituted in rHDL.

The rHDL system was recently used to investigate rhodopsin and arrestin interactions.[30,32] Arrestin participates in the internalization of receptors from the cell surface to endosomes and leads to degradation or recycling of GPCRs.[33,34] Arrestin recruitment also serves as a trigger to recruit and activate other important signalling proteins such as Src and ERK, leading to downstream cellular events that can be 'ligand specific'. The identification of 'biased' or 'signalling-specific' ligands suggests that not all 7TM receptors are GPCRs.[33,35] Although structural analyses of arrestin has led to the suggestion that arrestin may bind two closely positioned receptors (*e.g.* a dimer), recent data suggest that it is the monomeric form of the rhodopsin that is the minimal unit and that dimerization appears not to cause arrestin binding cooperativity.[30,32] Analysis of monomeric rhodopsin in rHDL reveals that visual arrestin, or β-arrestin1, facilitates the transition to and stabilization of the active meta-II state (metarhodopsin II) which is the hallmark of rhodopsin bound to arrestin.[30,32] Interestingly, when rhodopsin was reconstituted into liposomes, less meta-II was observed, promoting the idea that arrestin may favour monomeric rhodopsin. Of course this speculation should be taken with caution because receptor orientation cannot be controlled during liposome reconstitution. Nevertheless these studies unequivocally demonstrate that rhodopsin oligomers are *not* necessary for arrestin binding.

8.2.4 Thoughts on Monomeric GPCRs

The rHDL reconstitution approach allowed for the isolation of single receptors, which leads to the resolution of an important question in GPCR research: what is a minimal signalling unit? The studies outlined here undeniably demonstrate that a monomeric GPCR can bind ligand, activate a G protein, and bind arrestin. The question that remains is whether it is the monomeric form or the oligomeric form of GPCRs that is responsible for signalling *in vivo*.

Support for monomeric receptors *in cellulo* has recently been studied using single molecule spectroscopy with fluorescently labelled inverse agonists. Here, the stoichiometry of fluorescent ligand binding to muscarinic M1 receptors in live cells was studied using step photobleaching.[15] When excited, fluorescent molecules will photobleach at a specific rate in a step-wise manner that is proportional to the number of fluorophores excited. When observing fluorescent ligand-bound receptors on the cell surface by TIRFm, discrete puncta are observed. Step photobleaching of the majority (~70%) of these puncta using a fluorescent probe at saturating concentrations reveals that each spot contains only a single fluorophore. The remaining puncta (~30%) have two or more fluorophores. Based on these observations, the authors suggest that the puncta that bleach with a single step (70% of total) represent monomeric receptors. Careful analysis of the remaining 30% that display multistep photobleaching suggests that these species are dynamic, oscillating between a dimeric and monomeric forms with time. One caveat to this study is that a single ligand may be binding to an M1 dimer; in this scenario, a dimer would appear as a monomer and distinction between the species would be impossible. Similarly, a tetramer would appear as a dimer if a single ligand were binding to a dimer.

The physiological relevance of monomeric receptors is questionable as it is difficult to resolve the oligomeric state of the signalling GPCRs in intact cells. A plethora of FRET and bioluminescence resonance energy transfer (BRET) studies have suggested that GPCRs exist largely as oligomers, the minimal oligomer being the dimer.[13–15] Several studies have also suggested that GPCR oligomers are formed during protein processing and that oligomerization is required for proper presentation at the cellular membrane (as summarized in chapter 11 of this book).[16] Yet, these studies are in stark contrast with the study mentioned above where 70% of receptors remain in the monomeric form on the cell surface.[15]

A potential prejudice of the rHDL approach is the extreme *in vitro* nature of the system. The addition of G proteins or arrestins to receptors embedded in rHDL particles may represent a biased approach, in that the membrane proteins may be concentrated onto the small, $\sim$6000 Å^2 (radius $\sim$42 Å) phospholipid bilayer surface. The higher relative concentrations of interacting proteins with respect to receptors may mask a weaker affinity and result in a higher *apparent* affinity, and consequently affect binding of ligands that are allosterically modulated by the interacting proteins. The impact of this potential will not be known until an alternative method to isolate reconstituted monomeric receptors and facilitate G protein or arrestin coupling is established.

The fact that monomeric receptors in rHDL are capable of performing as GPCRs in biological membranes does *not* disprove the existence of receptor oligomers. It only suggests that the monomeric form is the minimal unit required to activate G proteins or recruit arrestin. An obvious next step is to study the contributions of receptor oligomers in an isolated system, beginning with a dimer, and perform the same series of experiments that have been described here. The methodology and significance of these studies are discussed in the following section.

8.3 Reconstitution of Oligomeric GPCRs into Model Lipid Bilayers

Although detergent-based studies and rHDL technology have demonstrated the functionality of monomeric GPCRs, there are microscopic-, biophysical- and pharmacological-based observations that suggest many, if not most, GPCRs exist in an oligomeric arrangement in cells.[14,37,48,59,60] Most clearly, AFM imaging of native rod disks from bovine eyes, in which rhodopsin accounts for >98% of protein content, revealed a uniform array of rhodopsin dimers, or even higher order oligomers.[37] This concept was revisited using lateral diffusion analysis of rhodopsin in rod disks and it was determined that rhodopsin may exist in distinct forms, with distinct lateral diffusion coefficients, the slowest of which may represent rhodopsin paracrystalline arrays.[38] FRET and BRET studies of GPCRs in cells suggest that receptors exist as either transient dimers or permanent dimers on the cellular membrane, providing further evidence for physiological oligomeric GPCRs.[58,61] It is known that oligomerization is critical for efficient targeting of $GABA_B$ and chemokine CXCR4 receptor to the plasma membrane.[79–81] Moreover, recent crystallographic evidence suggests that the CXCR4 is a dimer, as the same dimer was observed under different crystallization conditions and in different crystal forms.[11] Lipid vesicles or lamellae are used to assess the function of GPCR oligomers in a controlled manner.

8.3.1 A Good Model for Isolating GPCR Oligomers

As with monomeric GPCRs, isolation of oligomeric receptors in a model lipid bilayer system must be designed in a highly specific manner in order to provide a quantitative assessment of the oligomeric state and also to study receptor function. To date, the rHDL approach has not been optimized for the isolation of oligomers. Colleagues have reported the formation of larger rHDL particles, which hold two or more receptors; however, as noted by Banerjee *et al.*, the essential problem lies with the orientation of the receptors within the particles.[29] There is roughly a 50% probability that two receptors will be reconstituted in a bilayer in a parallel manner, *i.e.* with both *N*-termini located on the same side of the membrane (see Figures 8.2D and 8.2E). The remaining 50% represent a non-physiological, anti-parallel species. EM and immunogold labelling of rhodopsin dimers reconstituted in rHDL confirmed this notion, where only ~50% display a parallel orientation.[29] Isolation of the receptor dimer proved that the dimer is less efficacious than the monomer at coupling to G proteins and arrestins.[29,30] However, deciphering the contributions of the parallel and anti-parallel reconstituted dimers is imprecise and arduous. For instance, anti-parallel dimers should behave as two monomeric receptors each coupling to a G protein on opposite sides of the bilayer. Yet, the large *N*-termini and extracellular domains of each of the receptors, especially the presence of glycosylation groups, may impose steric effects on G protein and/or arrestin binding that cannot be accounted for in this model. Furthermore, the effects of anti-parallel receptor- receptor interactions on receptor conformation, ligand binding, or G protein/arrestin interactions are unknown. Thus, until a methodology is developed that can separate and isolate parallel from anti-parallel dimers, ascertaining the behavior of GPCR oligomerization in rHDL particles remains challenging.

Reconstituting oligomeric receptors into liposomes provides an alternative, but not necessarily optimal strategy to study GPCR oligomerization. As phospholipid vesicles are larger than rHDL particles, there is potential to accommodate multiple receptors. Unfortunately there is no way to control the orientation of the inserted GPCRs. More recently, biophysical methodologies to detect parallel oligomers using FRET have been developed and used to screen reconstitution methods.[44] The following paragraphs summarize the methodologies used to isolate parallel oligomers in vesicles and discuss the implications of receptor oligomerization in ligand binding as well as G protein coupling.

8.3.2 Methodology and Characterization of GPCR Reconstitution into Vesicles

8.3.2.1 Methodology of GPCR incorporation into Vesicles

The reconstitution of membrane proteins in phospholipid vesicles or in lamellae is an effective method to displace detergents and their deleterious effects by reintroducing lipids (see Figure 8.2E). The methodology has been the mainstay for GPCR biochemistry for decades. Typically, detergent-solubilized lipid mixtures composed of dioleoylphosphatidylcholine (DOPC) or POPC and POPG, together with cholesterol hemisuccinate, are added to purified receptor preparations in detergent micelles. Spontaneous reconstitution of GPCRs is initiated by the removal of detergent by gel filtration.

8.3.2.2 Characterization of GPCRs in Vesicles

Rhodopsin, β_2AR, neurotensin receptor 1 and muscarinic receptors have been reconstituted into vesicles and the receptor orientation analysed.[39–43] The orientation of rhodopsin reconstituted in vesicles was assessed by digestion with Asp-N, taking advantage of protease sites at the C-terminus. Treatment with Asp-N revealed that 90% of rhodopsin reconstituted in vesicles adopted an 'inside–out' orientation.[41] A similar strategy was used to determine the topology of β_2AR in vesicles by taking advantage of a Factor Xa protease site located in third intracellular loop of the receptor. Here, 90% of receptors reconstituted were resistant to treatment with Factor Xa, suggesting that the third intracellular loop was protected in the lumen of the vesicle in an 'outside–out' orientation.[44]

FRET of fluorophore-labelled receptors was used in order to study the oligomeric state of β_2AR in vesicles. Cy3 or Cy5 fluorophores were strategically placed in one of three positions on the intracellular loops or *C*-terminus.[44] Receptors labelled with Cy3 at positions T66 (IL1), A265 (IL2) or R333 (IL3) were co-reconstituted with Cy5-labelled receptors and FRET efficiency determined. Each combination of Cy3 and Cy5-labelled receptor pair displayed FRET with varying efficiencies, depending on the receptor pair. The Förster radius of a Cy3–Cy5 FRET pair was estimated at approximately 37–56 Å, suggesting that the receptors pairs were close enough to be considered oligomers.[45] The observed FRET may be diminished by solubilization with the detergent DDM. In addition, the observed FRET was not altered by increasing the overall lipid concentration, strongly supporting the notion that the receptors spontaneously form oligomers.

To further investigate the oligomeric arrangement of the receptors in vesicles, FRET saturation analysis was performed. By increasing the acceptor : donor ratios while maintaining the total number of receptors within the vesicles, FRET saturation occurred at a lower ratio; based on mathematical modelling of the FRET data, this implies the existence of higher order oligomers.[39,46–49] In fact these data indicate that β_2AR in vesicles follows the theoretical model for tetramers.[44] It should be noted that FRET saturation is taken as an ensemble, meaning that there may be populations of monomers and higher order oligomers, but that the predominant form within this particular sample is tetrameric.

8.3.3 Functional Studies on Oligomeric GPCRs in Vesicles

8.3.3.1 Ligand Binding to Oligomeric GPCRs

The intrinsic effect of ligands to either initiate or eliminate GPCR oligomerization has been observed in several receptor systems.[50–54] This was tested in vesicles with the β_2AR tetramers mentioned above. The influences of three classes of ligands (agonist, antagonist, and inverse agonist) were tested against six different β_2AR FRET pairs, based on the earlier mentioned mutants of the intracellular loops in vesicles. While only small changes in FRET efficiency were observed for the full agonist (isoproterenol) and antagonist (alprenolol), the inverse agonist (ICI-118,551) produced a large change in the FRET efficiency. Furthermore, ICI-118,551 induced a detergent-sensitive FRET saturation curve predictive of a higher order oligomer, suggesting that the inverse agonist may stabilize a higher order oligomer. No effect of either agonist or antagonist on FRET saturation was observed.[44] Certainly, the higher order oligomers of GPCRs in the inactive state (inverse agonist bound in this study) are reminiscent of the AFM images of native rod discs that revealed organized arrays of rhodopsin dimers.[37,44] The ability to isolate oligomeric forms of receptors in vesicles and monitor their transitions to different ordered states upon binding ligands that stabilize specific conformations is striking. The fact that ligands such as inverse agonists, which stabilize inactive conformations, can also stabilize higher order oligomers strongly suggests that oligomerization may be directly related to regulating receptor activity.

8.3.3.2 G Proteins and Oligomeric GPCRs

Recall that G protein coupling to receptors, as demonstrated using the rHDL approach, requires *only* a monomeric receptor. More importantly the rHDL data clearly demonstrated that G proteins themselves are responsible for the positive allosteric influence on agonist binding. Interestingly, co-reconstitution of G proteins and Cy3-β_2AR/Cy5-β_2AR in vesicles, or simply the addition of G proteins to preformed Cy3-β_2AR/Cy5-β_2AR-containing vesicles, appeared to disrupt FRET and hence disrupt the oligomeric structure.[44] Uncoupling the G protein using a non-hydrolysable GTP analogue GTPγS restored FRET, allowing the higher order oligomeric structure to re-establish.

8.3.4 Thoughts on Oligomeric GPCRs

The observations made in lipid vesicles are in good agreement with studies in other systems—specifically, biochemical studies on the dimeric extracellular domains of the metabotropic glutamate receptor (mGluR) and the GABA$_B$ receptor (both class C GPCRs) and on the family B GPCR, the parathyroid receptor.[55–57] In these studies, agonist binding induces dramatic conformational changes in the extracellular domains that are likely to affect the 7TM domain oligomeric structure. Furthermore, Maurel *et al.* observed an enhancement in G protein coupling that accompanies disruption of higher order oligomerization of the GABA$_B$ receptor.[58] These are interpreted such that G protein coupling alters GABA$_B$ receptor oligomerization. Both the ligand binding data and G protein coupling data on class C GPCRs are similar to the observations made in lipid vesicles.

In biological membranes, GPCRs are observed as oligomers.[14,37,48,59–61] Moreover, *in vitro* analyses of GPCRs reconstituted into vesicles reveal that the receptors may spontaneously form oligomers.[39–41,44] The reconstituted receptors in vesicles are fully functional as demonstrated by ligand binding and G protein coupling. However, the dynamics of the oligomeric state are only now being investigated. Together these data support the notion that GPCRs in a cell may be organized as oligomers, with a dynamic degree of order that varies perhaps in an activity-dependent manner.

8.4 Synthesizing Monomer and Oligomer Data from *in vitro* Models and *in vivo* Studies

8.4.1 Summary

Model membranes have given rise to the reductionist approach for studying integral membrane proteins. While removing the complicated *milieu* of the cell, the model membrane provides the necessary environment for integral membrane proteins to remain functional. In many ways, the reconstitution approach has revolutionized the resolution at which GPCR function can be studied. Direct measurements of protein–protein interactions can now be accomplished using receptors in various oligomeric states. The role of ligand binding or receptor conformation on G protein or arrestin binding, and *vice versa*, can be investigated in a refined and quantitative manner. Model membranes have also redirected attention to the old debate: what is the minimal GPCR signalling unit? Now, with the use of model lipid bilayers this question has been satisfactorily addressed, but new questions have been raised with regards to physiological relevance.

The rHDL approach allowed, for the first time, isolation of monomeric forms of GPCRs within a membrane. Prior to the development of the system, studies were conducted in detergents, sacrificing the receptors' function, or in cellular membranes and lipid vesicle preparations where oligomeric state of GPCRs cannot be controlled. The rHDL approach has unequivocally demonstrated that the monomeric form of the GPCR is the minimal signalling unit.[2,4,5,22,26] These studies also illustrate that both G proteins and arrestins directly and allosterically modify receptor conformation resulting in stabilization of the active state. Stabilization of the active conformation will result in enhanced agonist binding to hormone receptors, a notion that serves as the foundation of the 'ternary complex' coined by DeLean and Lefkowitz almost 30 years ago.[63] Importantly, these data do *not* disprove the existence or contributions of GPCR oligomers with regards to signalling. It is likely that a monomer is capable of binding an agonist and G protein but it does not preclude that, in a more *physiological* environment, the receptor may exist in an oligomeric form. GPCRs may exist in some oligomeric forms in cells, as evidenced by the growing number of reports supporting the existence of higher order oligomeric GPCRs *in vivo*.[14,37,48,59–61]

A challenging question that arises relates to the function of oligomerization. Studies from vesicle preparations indicate that inverse agonists, which stabilize the inactive state for G protein-mediated signalling, appear to stabilize higher order oligomerization (larger than tetramer).[44] These analyses predicted that, in vesicles, receptors may exist in an ensemble of various oligomeric states that average out to be predicted tetramers. These data are in good agreement with AFM imaging of rhodopsin, BRET and FRET analysis of various receptor oligomers in cells, and receptor crystal structure information, as discussed earlier.[37] However, a clear role for oligomerization with regards to ligand binding, G protein coupling and signalling events has yet to be addressed.

Although the field has not established a perfect platform to monitor the dynamics of monomer–dimer assembly and stabilization, the development and refinement of *in vitro* reconstitution methodologies have helped to shape general mechanisms of GPCR activation. Unfortunately, isolated systems do not take into account potential cellular factors such as scaffolding, chaperones, lipid rafts and others that may contribute toward oligomer formation or destruction, or determination of the oligomeric unit.[16,64] Classical functional assays using these *in vitro* reconstitutions systems have recapitulated many of the canonical behaviours of GPCRs that are observed in native membranes and in cells. These reconstitution methodologies revealed what receptors *can* and *cannot* do. Now *in vivo* or *in cellulo* studies must identify components that may be missing in the reductionist approaches. As new components, new ligands and refined methodologies are identified, they may be retested in the more controlled *in vitro* systems.

8.4.2 Conclusions and Future Directions

Putative receptor dimers have been captured in several receptor crystals.[7–11] Whether the nature of the receptor–receptor interaction is physiologically relevant or an artefact of crystallization has to be determined. Data suggests that the receptor to G protein ratio in a complex is 2 : 1. The leukotriene B4 (LTB_4) receptor dimer co-purifies with a single G protein heterotrimer.[65,66] Subsequent studies in reconstitution studies in rHDL with rhodopsin, as well as isolation of LTB_4 dimers, suggest that the dimer is less efficient than the monomeric form at activating G proteins. Note that, as stated earlier, assessing the activity of rhodopsin dimers in rHDL-like particles should be interpreted with caution due to the random nature of the parallel *vs.* anti-parallel dimer reconstitution. With this caveat in mind, the data does strongly suggest that only one receptor within the dimer is responsible in activating the G protein.

Current thinking is that the *C*-terminus of the G protein α-subunit, the G protein, induces a conformational change in only one of the dimeric receptors and stabilizes high affinity agonist binding. It is important to remember that a G protein heterotrimer is large, approximately 80 Å along the proposed membrane-binding surface.[67,68] Similarly GPCRs, based on several crystal structures are approximately 35–40 Å, making dimers in the 70–80 Å range.[7–11] Studies suggest that within a dimer, one receptor behaves as a monomer, activating the G protein by promoting GDP release on the α-subunit (see Figure 8.3).[69–71] The same receptor experiences stabilization of the high affinity agonist state by the nucleotide-free G protein. There is no apparent influence of the G protein on the other receptor within the dimer other than preventing, through steric occlusion, another G protein α-subunit from interacting productively. Thus, rather than invoking a model of negative cooperativity of agonist binding, the asymmetric nature of the dimer (*i.e.* simultaneous presence of high affinity and low affinity sites) is afforded by the G protein. This proposal was demonstrated by the Baneres and Pin laboratories with the LTB4 and mGluR dimers, respectively, and more recently using *in vivo* complementation studies with the luteinizing hormone receptor (LHR).[69–71] An asymmetric arrangement was observed in the crystal structure of the ligand-binding domain of the type II mGluR co-crystallized with agonists. Here multiple conformations of the mGluR dimer were found including complexes containing only a single agonist, as well as complexes containing two agonists, despite crystallization conditions with saturating agonist concentrations (1–4 mM).[72] However, Kniazeff *et al.* demonstrated that full activation—in terms of efficacy of activation of phospholipase C (PLC)—requires both Venus Fly Trap Domains (VFTs) to be occupied by agonist.[73–75] Partial activation may be achieved through agonist occupation of only one VFT. The implications of a dimer with symmetrical ligand binding domains and an asymmetrical transmembrane domain are unclear but insinuate the notion that higher order oligomers may be involved. This notion is discussed in more detail in chapter 11.

Arrestins, like G proteins, impose an asymmetric nature on a GPCR homodimer and also allosterically modulate agonist binding and stabilize high affinity agonist binding.[76] Similarly, arrestins stabilize the active state (meta II)

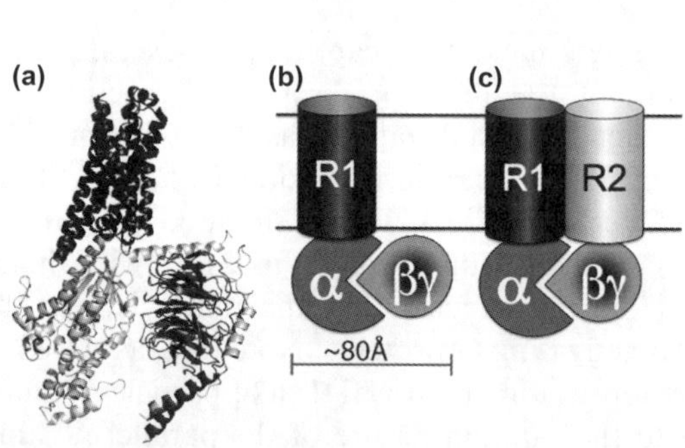

Figure 8.3 GPCR monomer and dimer coupled to G protein. (**a**) Model of the β_2AR coupling to stimulatory G protein heterotrimer, G_s. The β_2AR was modelled onto the structure of opsin bound to the *C*-terminus of the transducin α-subunit.[68] The *C*-terminus of the G_s α-subunit was modeled on to the structure of the transducin heterotrimer.[67] (**b**, **c**) Cartoon of the monomeric (**b**) and asymmetric dimeric GPCR forms (**c**) coupling to a G protein heterotrimer. G proteins apply an allosteric modulatory effect on agonist binding by stabilizing the high affinity state of only one receptor (R1) within the dimer. Steric effects of the large G protein heterotrimer prevent coupling of another G protein to R2, thus creating an asymmetric dimer.

of rhodopsin and appear to prefer the monomeric rather than the dimeric form of rhodopsin, when assessed in rHDL particles.[30,32] The decreased capacity of arrestin to bind and stabilize two rhodopsins within the dimer would be analogous to having only one G protein bound and coupling to a dimer. Note that arrestins, like G proteins, are also quite large with the longest dimensions being ~80 Å and the shortest ~40 Å (see Figure 8.3). It is physically possible that two arrestins may bind to a GPCR dimer in a side-by-side arrangement, as the width of arrestin is close to the diameter of a GPCR. Although the rHDL-like particles used in these studies can accommodate such an arrangement, the binding data suggest otherwise. Conceptually, G protein and arrestins thus appear to stabilize similar active conformations of GPCRs and may, through different mechanisms, prefer to have a G protein (or arrestin): receptor stoichiometry of 1 : 2.

While it is not clear what the general roles of oligomerization are for GPCRs, great progress in the field has been made in the last decade. Evidence from *in vitro* reconstitution studies and *in vivo* (*in cellulo*) biophysical studies have provided a framework for how GPCRs function and the understanding of oligomeric state of GPCRs. Future studies, using more refined approaches, may delineate the intricacies of how receptor oligomerization regulates GPCR function.

Acknowledgements

This research was made possible by support from NIGMS (GM068603) and the Chemical Biology Program at the University of Michigan. Additional funding was provided by a Pilot and Feasibility Grant from the Michigan Diabetes Research and Training Center, NIH/NIDDK Grant P60-DK020572 (R.K.S.).

References

1. B. S. Selinsky, *Membrane Protein Protocols: Expression, Purification, and Characterization*, Humana Press, Totowa, NJ, 2003.
2. A. J. Kuszak, S. Pitchiaya, J. P. Anand, H. I. Mosberg, N. G. Walter and R. K. Sunahara, *J. Biol. Chem.*, 2009, **284**, 26732.
3. B. Strulovici, R. A. Cerione, B. F. Kilpatrick, M. G. Caron and R. J. Lefkowitz, *Science*, 1984, **225**, 837.
4. M. R. Whorton, M. P. Bokoch, S. G. Rasmussen, B. Huang, R. N. Zare, B. Kobilka and R. K. Sunahara, *Proc. Natl. Acad. Sci. U. S. A.*, 2007, **104**, 7682.
5. M. R. Whorton, B. Jastrzebska, P. S. Park, D. Fotiadis, A. Engel, K. Palczewski and R. K. Sunahara, *J. Biol. Chem.*, 2008, **283**, 4387.
6. T. H. Bayburt and S. G. Sligar, *FEBS Lett.*, 2010, **584**, 1721.
7. V. Cherezov, D. M. Rosenbaum, M. A. Hanson, S. G. Rasmussen, F. S. Thian, T. S. Kobilka, H. J. Choi, P. Kuhn, W. I. Weis, B. K. Kobilka and R. C. Stevens, *Science*, 2007, **318**, 1258.
8. V. P. Jaakola, M. T. Griffith, M. A. Hanson, V. Cherezov, E. Y. Chien, J. R. Lane, A. P. Ijzerman and R. C. Stevens, *Science*, 2008, **322**, 1211.

9. D. M. Rosenbaum, V. Cherezov, M. A. Hanson, S. G. Rasmussen, F. S. Thian, T. S. Kobilka, H. J. Choi, X. J. Yao, W. I. Weis, R. C. Stevens and B. K. Kobilka, *Science*, 2007, **318**, 1266.
10. T. Warne, M. J. Serrano-Vega, J. G. Baker, R. Moukhametzianov, P. C. Edwards, R. Henderson, A. G. Leslie, C. G. Tate and G. F. Schertler, *Nature*, 2008, **454**, 486.
11. B. Wu, E. Y. Chien, C. D. Mol, G. Fenalti, W. Liu, V. Katritch, R. Abagyan, A. Brooun, P. Wells, F. C. Bi, D. J. Hamel, P. Kuhn, T. M. Handel, V. Cherezov and R. C. Stevens, *Science*, 2010, **330**, 1066.
12. H. Kuhn, N. Bennett, M. Michel-Villaz and M. Chabre, *Proc. Natl. Acad. Sci. U. S. A.*, 1981, **78**, 6873.
13. L. Albizu, M. N. Balestre, C. Breton, J. P. Pin, M. Manning, B. Mouillac, C. Barberis and T. Durroux, *Mol. Pharmacol.*, 2006, **70**, 1783.
14. S. Angers, A. Salahpour, E. Joly, S. Hilairet, D. Chelsky, M. Dennis and M. Bouvier, *Proc. Natl. Acad. Sci. U. S. A.*, 2000, **97**, 3684.
15. J. A. Hern, A. H. Baig, G. I. Mashanov, B. Birdsall, J. E. Corrie, S. Lazareno, J. E. Molloy and N. J. Birdsall, *Proc. Natl. Acad. Sci. U. S. A.*, 2010, **107**, 2693.
16. A. Salahpour, S. Angers, J. F. Mercier, M. Lagace, S. Marullo and M. Bouvier, *J. Biol. Chem.*, 2004, **279**, 33390.
17. B. J. Baas, I. G. Denisov and S. G. Sligar, *Arch. Biochem. Biophys.*, 2004, **430**, 218.
18. A. Jonas, *Methods Enzymol.*, 1986, **128**, 553.
19. M. Alami, K. Dalal, B. Lelj-Garolla, S. G. Sligar and F. Duong, *EMBO J.*, 2007, **26**, 1995.
20. D. N. Amin and G. L. Hazelbauer, *J. Bacteriol.*, 2010, **192**, 1193.
21. T. H. Bayburt and S. G. Sligar, *Protein Sci.*, 2003, **12**, 2476.
22. A. J. Leitz, T. H. Bayburt, A. N. Barnakov, B. A. Springer and S. G. Sligar, *Biotechniques*, 2006, **40**, 601.
23. L. Z. Mi, M. J. Grey, N. Nishida, T. Walz, C. Lu and T. A. Springer, *Biochemistry*, 2008, **47**, 10314.
24. T. Raschle, S. Hiller, T. Y. Yu, A. J. Rice, T. Walz and G. Wagner, *J. Am. Chem. Soc.*, 2009, **131**, 17777.
25. K. Palczewski, T. Kumasaka, T. Hori, C. A. Behnke, H. Motoshima, B. A. Fox, I. Le Trong, D. C. Teller, T. Okada, R. E. Stenkamp, M. Yamamoto and M. Miyano, *Science*, 2000, **289**, 739.
26. T. H. Bayburt, A. J. Leitz, G. Xie, D. D. Oprian and S. G. Sligar, *J. Biol. Chem.*, 2007, **282**, 14875.
27. S. M. Hanson and V. V. Gurevich, *J. Biol. Chem.*, 2006, **281**, 3458.
28. J. P. Segrest, M. K. Jones, A. E. Klon, C. J. Sheldahl, M. Hellinger, H. De Loof and S. C. Harvey, *J. Biol. Chem.*, 1999, **274**, 31755.
29. S. Banerjee, T. Huber and T. P. Sakmar, *J. Mol. Biol.*, 2008, **377**, 1067.
30. H. Tsukamoto, A. Sinha, M. Dewitt and D. L. Farrens, *J. Mol. Biol.*, 2010.
31. X. J. Yao, G. Velez Ruiz, M. R. Whorton, S. G. Rasmussen, B. T. DeVree, X. Deupi, R. K. Sunahara and B. Kobilka, *Proc. Natl. Acad. Sci. U. S. A.*, 2009, **106**, 9501.

32. T. H. Bayburt, S. A. Vishnivetskiy, M. A. McLean, T. Morizumi, C. C. Huang, J. J. Tesmer, O. P. Ernst, S. G. Sligar and V. V. Gurevich, *J. Biol. Chem.*, 2010.
33. C. A. Moore, S. K. Milano and J. L. Benovic, *Annu. Rev. Physciol.*, 2007, **69**, 451.
34. M. E. Sommer and D. L. Farrens, *Vision Res.*, 2006, **46**, 4532.
35. S. M. DeWire, S. Ahn, R. J. Lefkowitz and S. K. Shenoy, *Annu. Rev. Physciol.*, 2007, **69**, 483.
36. J. A. Hirsch, C. Schubert, V. V. Gurevich and P. B. Sigler, *Cell*, 1999, **97**, 257.
37. D. Fotiadis, Y. Liang, S. Filipek, D. A. Saperstein, A. Engel and K. Palczewski, *Nature*, 2003, **421**, 127.
38. V. I. Govardovskii, D. A. Korenyak, S. A. Shukolyukov and L. V. Zueva, *Mol. Vision*, 2009, **15**, 1717.
39. P. J. Harding, H. Attrill, J. Boehringer, S. Ross, G. H. Wadhams, E. Smith, J. P. Armitage and A. Watts, *Biophys. J.*, 2009, **96**, 964.
40. A. W. Ma, D. S. Redka, L. F. Pisterzi, S. Angers and J. W. Wells, *Biochemistry*, 2007, **46**, 7907.
41. L. Niu, J. M. Kim and H. G. Khorana, *Proc. Natl. Acad. Sci. U. S. A.*, 2002, **99**, 13409.
42. B. K. Fung and W. L. Hubbell, *Biochemistry*, 1978, **17**, 4403.
43. B. K. Fung and W. L. Hubbell, *Biochemistry*, 1978, **17**, 4396.
44. J. J. Fung, X. Deupi, L. Pardo, X. J. Yao, G. A. Velez-Ruiz, B. T. Devree, R. K. Sunahara and B. K. Kobilka, *EMBO J.*, 2009, **28**, 3315.
45. S. E. Mansoor, K. Palczewski and D. L. Farrens, *Proc. Natl. Acad. Sci. U. S. A.*, 2006, **103**, 3060.
46. K. G. Harikumar, R. M. Happs and L. J. Miller, *Biochim. Biophys. Acta*, 2008, **1778**, 2555.
47. J. R. James, M. I. Oliveira, A. M. Carmo, A. Iaboni and S. J. Davis, *Nat. Methods*, 2006, **3**, 1001.
48. J. F. Mercier, A. Salahpour, S. Angers, A. Breit and M. Bouvier, *J. Biol. Chem.*, 2002, **277**, 44925.
49. W. Veatch and L. Stryer, *J. Mol. Biol.*, 1977, **113**, 89.
50. Z. J. Cheng and L. J. Miller, *J. Biol. Chem.*, 2001, **276**, 48040.
51. S. Dorsch, K. N. Klotz, S. Engelhardt, M. J. Lohse and M. Bunemann, *Nat. Methods*, 2009, **6**, 225.
52. R. Latif, P. Graves and T. F. Davies, *J. Biol. Chem.*, 2002, **277**, 45059.
53. D. A. Roess and S. M. Smith, *Biol. Reprod.*, 2003, **69**, 1765.
54. C. C. Zhu, L. B. Cook and P. M. Hinkle, *J. Biol. Chem.*, 2002, **277**, 28228.
55. N. Kunishima, Y. Shimada, Y. Tsuji, T. Sato, M. Yamamoto, T. Kumasaka, S. Nakanishi, H. Jingami and K. Morikawa, *Nature*, 2000, **407**, 971.
56. A. A. Pioszak, K. G. Harikumar, N. R. Parker, L. J. Miller and H. E. Xu, *J. Biol. Chem.*, 2010, **285**, 12435.
57. P. Rondard, S. Huang, C. Monnier, H. Tu, B. Blanchard, N. Oueslati, F. Malhaire, Y. Li, E. Trinquet, G. Labesse, J. P. Pin and J. Liu, *EMBO J.*, 2008, **27**, 1321.

58. D. Maurel, J. Kniazeff, G. Mathis, E. Trinquet, J. P. Pin and H. Ansanay, *Anal. Biochem.*, 2004, **329**, 253.
59. S. Angers and M. Bouvier, *Trends Pharmacol. Sci.*, 2000, **21**, 326.
60. W. Guo, E. Urizar, M. Kralikova, J. C. Mobarec, L. Shi, M. Filizola and J. A. Javitch, *EMBO J.*, 2008, **27**, 2293.
61. G. Milligan and M. Bouvier, *FEBS J.*, 2005, **272**, 2914.
62. H. R. Bourne, B. Beiderman, F. Steinberg and V. M. Brothers, *Mol. Pharmacol.*, 1982, **22**, 204.
63. A. DeLean and R. J. Lefkowitz, *Nature*, 1980, **283**, 109.
64. J. A. Allen, R. A. Halverson-Tamboli and M. M. Rasenick, *Nat. Rev. Neurosci.*, 2007, **8**, 128.
65. J. L. Baneres, A. Martin, P. Hullot, J. P. Girard, J. C. Rossi and J. Parello, *J. Mol. Biol.*, 2003, **329**, 801.
66. J. L. Baneres and J. Parello, *J. Mol. Biol.*, 2003, **329**, 815.
67. D. G. Lambright, J. Sondek, A. Bohm, N. P. Skiba, H. E. Hamm and P. B. Sigler, *Nature*, 1996, **379**, 311.
68. P. Scheerer, J. H. Park, P. W. Hildebrand, Y. J. Kim, N. Krauss, H. W. Choe, K. P. Hofmann and O. P. Ernst, *Nature*, 2008, **455**, 497.
69. M. Damian, A. Martin, D. Mesnier, J. P. Pin and J. L. Baneres, *EMBO J.*, 2006, **25**, 5693.
70. C. Goudet, J. Kniazeff, V. Hlavackova, F. Malhaire, D. Maurel, F. Acher, J. Blahos, L. Prezeau and J. P. Pin, *J. Biol. Chem.*, 2005, **280**, 24380.
71. A. Rivero-Muller, Y. Y. Chou, I. Ji, S. Lajic, A. C. Hanyaloglu, K. Jonas, N. Rahman, T. H. Ji and I. Huhtaniemi, *Proc. Natl. Acad. Sci. U. S. A.*, 2010, **107**, 2319.
72. T. Muto, D. Tsuchiya, K. Morikawa and H. Jingami, *Proc. Natl. Acad. Sci. U. S. A.*, 2007, **104**, 3759.
73. J. Kniazeff, A. S. Bessis, D. Maurel, H. Ansanay, L. Prezeau and J. P. Pin, *Nat. Struct. Mol. Biol.*, 2004, **11**, 706.
74. J. Kniazeff, T. Galvez, G. Labesse and J. P. Pin, *J. Neurosci.*, 2002, **22**, 7352.
75. J. Kniazeff, P. P. Saintot, C. Goudet, J. Liu, A. Charnet, G. Guillon and J. P. Pin, *J. Neurosci.*, 2004, **24**, 370.
76. V. V. Gurevich, R. Pals-Rylaarsdam, J. L. Benovic, M. M. Hosey and J. J. Onorato, *J. Biol. Chem.*, 1997, **272**, 28849.
77. S. K. Milano, Y. M. Kim, F. P. Stefano, J. L. Benovic and C. Brenner, *J. Biol. Chem.*, 2006, **281**, 9812.
78. E. M. Ross, *J. Biol. Chem.*, 1982, **257**, 10751.
79. G. J. Babcock, M. Farzan and J. Sodroski, *J. Biol. Chem.*, 2003, **278**, 3378.
80. K. A. Bones, B. Borowsky, J. A. Tamm, D. A. Craig, M. M. Durkin, M. Dai, W. J. Yao, M. Johnson, C. Gunwaldsen, L. Y. Huang, C. Tang, Q. Shen, J. A. Salon, K. Morse, T. Laz, K. E. Smith, D. Nagarathnam, S. A. Noble, T. A. Brancheck and C. Gerald, *Nature*, 1998, **396**, 674.
81. J. H. White, A. Wise, M. J. Main, A. Green, N. J. Fraser, G. H. Disney, A. A. Barnes, P. Emson, S. M. Foord and F. H. Marshall, *Nature*, 1998, **396**, 679.

Section II
G Protein-coupled Receptors: Multifaceted Functional Machines

CHAPTER 9

Kinetics and Mechanisms of GPCR Activation

MANUELA AMBROSIO AND MARTIN J. LOHSE*

University of Würzburg, Institute of Pharmacology and Toxicology and Rudolf Virchow Center, Versbacher Str. 9, 97078 Würzburg, Germany

9.1 Introduction

G protein-coupled receptors (GPCRs) represent the largest family of cell surface proteins in the human genome. They are key regulators of cellular functions and prime targets of therapeutic drugs and are characterized by a common structure and, presumably, common activation mechanisms.[1] These common characteristics include: (a) a shared structure (*i.e.* a bundle of seven membrane-spanning α-helices with an extracellular amino and a cytosolic carboxy terminus); (b) several conserved sequence motifs, which can also be used to define subclasses of GPCRs: (c) common signalling mechanisms which result from the activation of heterotrimeric guanine nucleotide-binding proteins (G proteins); and (d) various common regulatory and desensitization mechanisms.[2–6] Despite these similarities, individual receptors undergo various forms of dimerization and oligomerization,[7] and they also engage distinct signal-transduction pathways involving different G protein subtypes as well as G protein-independent signalling pathways.[4] How specificity is generated that directs signals from GPCRs to specific downstream pathways is still not well understood.

In spite of the great significance of rapid regulation of biological systems, little was known until recently about the kinetics of GPCR activation and

RSC Drug Discovery Series No. 8
G Protein-Coupled Receptors: From Structure to Function
Edited by Jesús Giraldo and Jean-Philippe Pin

Published by the Royal Society of Chemistry, www.rsc.org

signalling, primarily due to a lack of suitable methods.[8] If we consider physiological responses to agonists (*e.g.* increases in cardiac frequency in response to sympathetic activation) or if we use classical biochemical assays usually performed on membrane preparations, then GPCRs appear to signal with intermediate speed.[9,10] Faster responses to external stimuli are mediated *via* ligand or voltage gated ion channels (for example the nicotinic acetylcholine or the γ-amino butyric acid (GABA) receptors), which respond with millisecond rates and mediate neuronal and neuro-muscular transmission.[11,12] In contrast, slower responses to external stimuli are mediated *via* receptors of the enzymatic tyrosine kinase or guanylyl cyclase types, which give rise to slow and protracted intracellular signals that persist over minutes to hours and are often involved in metabolic and growth responses.[13]

However, it seems that the kinetics of GPCR activation and signalling are generally underestimated. Thus, it has become evident that a complete GPCR-signalling cascade can be activated within less than 500 ms. This has most clearly been shown by electrophysiological recordings of GPCR-regulated channels (*i.e.* opening of GIRK K^+-channels by M_2 muscarinic or by α_2-adrenergic receptors (α_2ARs)).[14,15] Even faster, the activation of rhodopsin, a class A GPCR, occurs within a time span of ~1 ms, and the downstream closure of the cGMP-gated cation channel that follows the intermediate steps of rhodopsin-/transducin(G_t)-activated phosphodiesterase occurs within ~200 ms.[16,17] These data illustrate that the activation of GPCRs must, in general, occur within one to a few hundred milliseconds.

During recent years, the development of several fluorescent approaches has allowed us to follow the activation speed of numerous GPCRs in real-time, both in reconstituted systems and in intact cells. These studies permit a much more detailed temporal and spatial analysis of receptor signalling than non-optical techniques and confirm the potential of GPCRs to become activated faster than previously thought. Such fluorescence-based assays were first established for purified, chemically labelled and reconstituted receptors, and later developed to monitor the activation of GPCRs in intact cells.[18,19]

9.2 Kinetic Analysis of Isolated GPCRs

The first method to directly monitor agonist-induced conformational changes in a GPCR other than rhodopsin was developed by Kobilka and co-workers (reviewed in ref. 20). In order to detect the active conformations of β_2-adrenergic receptors (β_2ARs), they applied fluorescence-quenching techniques. Taking advantage of the observation that the fluorescence emission of many fluorophores is highly affected by changes in their immediate surrounding, it is possible to follow the orientation of fluorophores within a protein, thus gaining information about the conformational movements occurring within activated receptors. This approach was established using purified and detergent-solubilized β_2ARs that were then specifically labelled with the environmentally sensitive fluorophore, fluorescein maleimide, at the endogenous cysteine

Cys265 (or alternatively at artificially introduced cysteines in a cysteine-free background). The use of purified receptors was necessary to achieve specificity in this direct labelling approach.

Cys265 is located at the cytoplasmic end of the sixth transmembrane (TM6) α-helix—an important domain for G protein activation. Furthermore, this site has been shown with many different mutational and biophysical approaches (and later also on the basis of the inactive (rhodopsin) and active (opsin) structures of rhodopsin[21]) to undergo major conformational changes upon activation.[20] Thus, the fluorophore covalently bound at this site is well positioned to detect agonist-induced conformational changes relevant to G protein activation. The use of a second label, which quenches fluorescence, at different positions of such a fluorescein-labelled construct permits the monitoring of the relative movements of the two labels to each other. The results of these studies suggest that the conformational changes associated with β_2-AR activation are similar to those previously observed with the rhodopsin system (Figure 9.1) (thus indicating a common mechanism of GPCR activation) and are compatible with a clockwise rotation of TM6 and a tilting of the cytoplasmic end of TM6 toward TM5[18,22]—data recently confirmed for rhodopsin by X-ray structural analyses.[21]

Following these first studies, the system was further developed to elucidate differences between the effects produced by stimulation with different classes of ligands, *i.e.* antagonists, partial and full agonists. Examining ligand-induced changes in the fluorescence lifetime of Cys265-bound fluorescein, it was possible to detect that β_2-AR appears to exist in multiple conformational states depending on the nature of the ligand. Partial agonists caused only partial changes in the receptor fluorescence compared to those induced by full agonists, and antagonist did not alter the receptor fluorescence and, hence, presumably the conformation.[23] Subsequent kinetic studies showed that, upon catecholamine binding, β_2-ARs undergo transitions into two kinetically distinguishable conformational states.[24] Whereas the response to the full agonists, epinephrine and isoproterenol, was fitted by a two-component exponential function comprising a rapid phase followed by a slower one, the response induced by partial agonists (*i.e.* dopamine) comprised only the rapid component, while the second slower component was only observed for the full agonists. These distinct conformational changes correlated with biological responses in cellular assays: dopamine was more efficacious in activating G_s than in inducing receptor internalization, whereas full agonists showed a similar efficacy in activating both G_s and receptor internalization. These observations supported the idea that partial agonists are able to induce only the first steps of a series of conformational rearrangements of the receptor in contrast to full agonists, which can promote further changes resulting in an active receptor conformation that is able to interact with all downstream proteins.

For reasons that are not entirely clear, the kinetics of these changes are relatively slow. The rates that were observed were several seconds for the fast phase and almost a minute for the slow phase of β_2-AR activation, which is clearly slower than the biological response to GPCR activation. Several

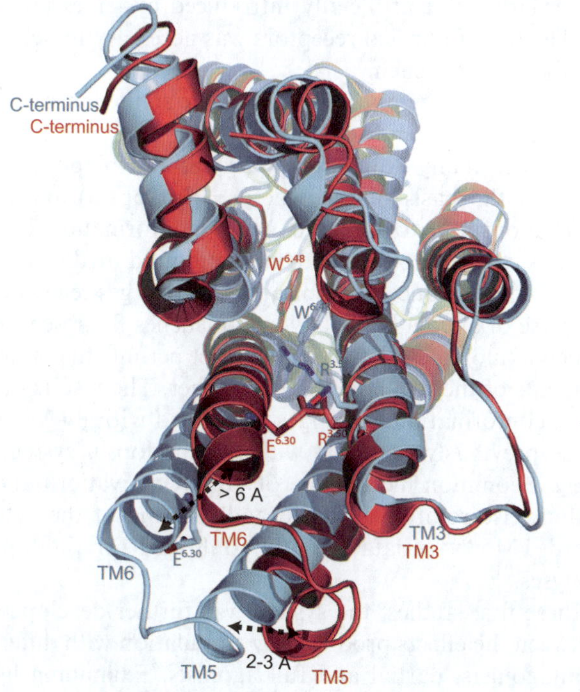

Figure 9.1 Cytosolic view of the superposition of the X-ray structure of inactive rhodopsin (PDB accession 1GZM, red) and of its partially active form opsin (PDB accession 3CAP, light blue). The main structural differences can be observed at the cytosolic ends of TM5 and TM6 (black arrows). The structural rearrangement of the receptor during activation causes the alteration of several key interactions including the disruption of the TM3–TM6 ionic lock (*i.e.* the salt bridge between $R135^{3.50}$ and $E247^{6.30}$) and the displacement of the side chain of $W265^{6.48}$ (of the conserved $CWxP^{6.50}$ motif on TM6) involved in the rotamer toggle switch mechanism. The final consequence of rhodopsin activation is the formation of a cavity between TM3, TM5 and TM6, into which the G protein can bind, and has been suggested to be a common mechanism of GPCR activation.

possible reasons might explain these kinetic differences. Most importantly, purified, reconstituted receptors may lack their natural environment such as essential constituents of the cell membrane but also G proteins and possibly other cellular components. Using carefully reconstituted β_2-ARs, it has been possible to observe faster receptor conformational switches, with changes in intramolecular receptor fluorescence on the timescale of ~30 s.[25] The latter studies also showed two distinct conformational switches occurring during the β_2-AR activation process; taking advantage of a label at the cytoplasmic end of TM6 in position 271, it was possible to detect an ionic lock mechanism that involves an interaction of the cytoplasmic ends of TM6 and TM3 (*i.e.* the salt bridge between $R^{3.50}$ and $E^{6.30}$), whereas labelling of the β_2-AR in position 265 reported a 'rotamer toggle' switch mechanism caused by the displacement of the side chain of $W^{6.48}$ (Figure 9.1). Because the efficacy of the partial agonist dopamine in causing these two conformational movements were comparable to those of full agonists, additional conformational changes were proposed to be required for the complete activation of β_2-AR induced by full agonists.

9.3 Kinetic Studies in Intact Cells by FRET

In order to be able to study GPCR activation kinetics in a more native environment, we have developed approaches based on fluorescence resonance energy transfer (FRET) which allow the study of this process in intact cells. FRET involves the energy transfer from a donor fluorophore to an acceptor fluorophore, which requires that their emission and excitation spectra overlap and that they are in very close proximity (<10 nm). FRET is very sensitive to changes in distance and thus can be used to observe such changes in the positions of the fluorophores and, hence, report conformational changes in a suitably labelled protein. As an alternative, when the two labels are attached to two different proteins, FRET can be used to measure protein–protein interactions.

The most frequently used fluorophore pair for these FRET studies consists of two variants of the green fluorescent protein (GFP), namely the donor CFP (cyan fluorescent protein) and the acceptor YFP (yellow fluorescent protein). The emission spectrum of CFP overlaps sufficiently with the excitation spectrum of YFP, thus enabling FRET. In addition, the emission maxima of the two dyes are separated enough to permit individual detection of the fluorescence intensity of the two proteins. CFP is excited at its excitation maximum (436 nm) and emits with a maximum at 480 nm. When FRET occurs, excitation of CFP induces transfer of energy to YFP, which becomes excited and emits with a maximum wavelength at 535 nm. At the same time, emission of CFP at 480 nm decreases. Using an appropriate microscopic FRET setup, variations in the distance or orientation between CFP and YFP can be detected as changes in FRET in single intact cells with great temporal resolution (<5 ms; depending on the brightness of fluorescence and, thus, the time required to obtain sufficient signal intensities).

Because FRET is very sensitive to the orientation and the distance between the fluorophores (FRET efficiency is proportional to $1/r^6$) and works only over ~ 10 nm or less, these methods are ideally suited to reveal the small conformational changes occurring in a receptor carrying the two labels. The crucial step in this approach is the choice of the sites for insertion of the two fluorophores. As represented in Figure 9.2A, we have generally fused one fluorescent protein to the *C*-terminus and the second one into the third intracellular loop of the receptors. The positions for insertion of the fluorescent probes were determined assuming that the movement of the TM6 produced by the agonist-induced receptor activation would be transmitted to the third intracellular loop, causing a shift of the fluorophores and a subsequent change in the distance between them, which should result in a change of the FRET signal. It is worth noting that the large GFP variants (27 kDa) always carry the risk of altering the characteristics of the labelled receptor (*i.e.* surface expression, ligand binding,

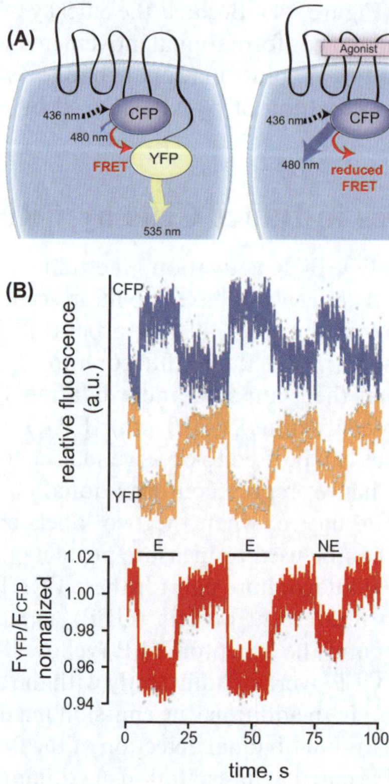

Figure 9.2 (A) Schematic representation of a CFP/YFP FRET-based receptor sensor. FRET requires the presence of two fluorophores: one with a shorter emission wavelength (donor, $\lambda em_{CFP} = 480$ nm) and another with a longer emission wavelength (acceptor, $\lambda em_{YFP} = 535$ nm). Here, CFP is inserted into the third intracellular loop whereas YFP is fused to the *C*-terminus. In basal conditions (left), excitation of the donor ($\lambda ex_{CFP} = 436$ nm) induces transfer of energy (FRET, red arrow) to the acceptor and results in emission at the wavelength typical for the acceptor ($\lambda em_{YFP} = 535$ nm). Stimulation of receptors by agonists causes a conformational change of the receptor that is transmitted to the third intracellular loop and results in a decrease of the FRET signal, presumably due to an increased distance between the two fluorophores (right). (**B**) FRET trace recorded using the β_2-AR FRET CFP/YFP construct expressed in HEK293 cells: stimulation of the cells with two endogenous agonists (E, epinephrine; NE, norepinepherine) causes an increase of the CFP emission coincident with a decrease of the YFP emission, thus resulting in a quantifiable decrease of the FRET signal. Analogous traces have been obtained with other receptor sensors.

signalling). Furthermore, in our experience, while labelling the *C*-terminus of GPCRs rarely influenced their function, the insertion of fluorophores into their third intracellular loop is more difficult, as it can lead to significant loss of cell surface expression and G protein activation.[26] However, knowledge about the structure and functional sites of the receptors can facilitate the selection of the labelling sites, and in several cases, a significant fraction of G protein signalling can be preserved.[27,28] Somewhat surprisingly, these functional effects can vary greatly between very similar receptor constructs.

One variant of FRET strategies are the approaches based on bioluminescence resonance energy transfer (BRET) developed by Bouvier and co-workers.[29] Because in BRET, energy is transferred from a luminescent moiety to a fluorophore, CFP is replaced in this technique by a light-emitting luciferase. BRET has the advantage of lower background signals, but because of the lower intensities of the emitted light compared with illumination by a light source, it generally requires longer periods of light acquisition—which reduces the temporal resolution. This makes it more difficult to use BRET for the detection of fast kinetics.

The CFP/YFP FRET-based system was the first successful attempt for real-time measurements of GPCR activation in living cells.[27] The placement of CFP and YFP into the third intracellular loop and at the *C*-terminus, respectively, led to the generation of receptor constructs with preserved ligand binding and signalling properties. If appropriately constructed, these receptors respond to agonists with an alteration—usually a reduction—in FRET between CFP and YFP (Figure 9.2B). Swapping the location of the two fluorophores leads to constructs with comparable features. Such FRET-based sensors have been developed for a number of GPCRs including the parathyroid hormone (PTH), bradykinin B2, α_{2A}-, β_1- and β_2-adrenergic and several muscarinic receptors.[30–32]

An extension of this approach replaces YFP with the small fluorescein-derived analog FlAsH (fluorescein arsenical hairpin binder).[26,33] This fluorophore (0.7 kDa) is significantly smaller than YFP, and it has the additional advantage that it only becomes fluorescent when bound to two flanking cysteines present in a specific sequence (CCPGCC), which needs to be inserted into the protein of interest (Figure 9.3A). The similar fluorescent properties of FlAsH and YFP make FlAsH a valid substitute of YFP in FRET experiments in association with CFP.

This approach was first tried for intramolecular FRET in GPCRs using the A_{2A}-adenosine receptor. The six amino acid sequence, CCPGCC, was introduced in the third intracellular loop of an A_{2A}-adenosine receptor construct carrying a CFP at the *C*-terminus (Figure 9.3B).[26] A side-by-side comparison of the two approaches (*i.e.* FlAsH/CFP *vs.* YFP/CFP) revealed that, in the specific case of the A_{2A}-adenosine receptor construct, the FlAsH label increased the amplitude of the FRET signal (by about five-fold) and much improved the ability of the modified receptor to couple to and activate its G proteins. However, somewhat surprisingly, the apparent kinetics of the agonist-induced receptor activation were the same for the two labelling approaches, suggesting that either the YFP-label does not impair receptor activation more than the much smaller FlAsH label, or that other factors are rate limiting in these assays.

More recently, we successfully used this approach for the development of other FlAsH/CFP-based receptor sensors (*i.e.* α_{2A}-adrenergic, β_1-adrenergic and M_1- M_2-, M_3- and M_5-muscarinic receptors).[30,34–37] Again, the activation kinetics of these receptors were no different than those of other ones based on CFP and YFP, *i.e.* with activation half times on the order of $\sim$60 ms.

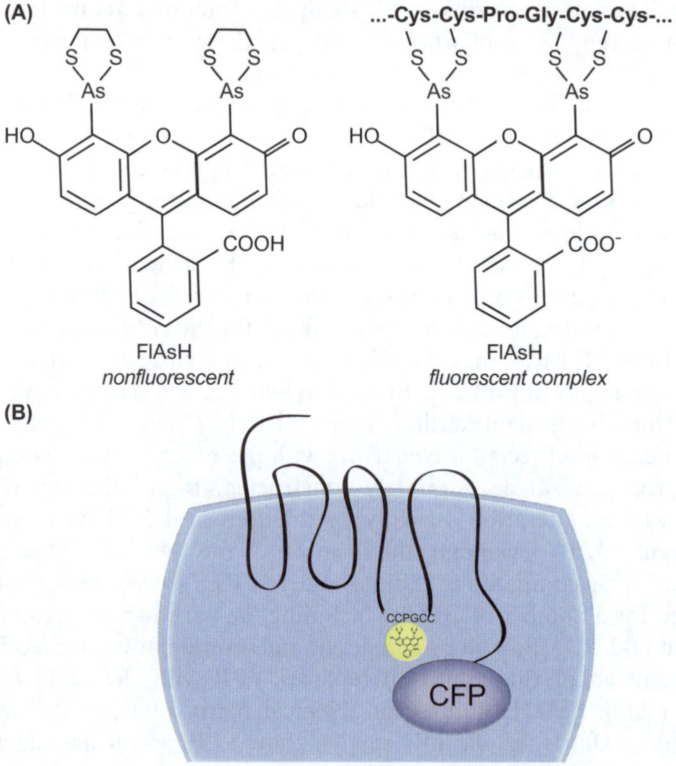

Figure 9.3 (**A**) Molecular structure of the small fluorescein-derived analogue FlAsH. This small ligand (0.7 kDa) is membrane-permeant and remains nonfluorescent until it binds with high affinity and specificity two flanking cysteines on both side of the specific sequence, Cys-Cys-Pro-Gly-Cys-Cys, previously inserted in the protein of interest. (**B**) Overall membrane topology of a FlAsH-labelled FRET receptor sensor characterized by the label site CCPGCC in the third intracellular loop.[26] A detailed protocol for such labelling has been recently described.[67]

In principle, FlAsH can even be combined with a second similar but red dye, abbreviated as ReAsH, to measure FRET between two specifically labelled distinct tetracysteine motifs[38]—but this approach has not yet been employed for intramolecular FRET studies.

9.4 Trans-conformational Switching Within GPCR Heterodimers

Even though GPCRs have been traditionally considered as monomeric membrane proteins, several lines of evidence obtained during the years suggest that they can exist and function as homo- and heterodimers[7] (and possibly also higher order oligomers[39]). An intriguing aspect of receptor heterodimerization is the possibility that the dimer may display functional properties that differ from those characterizing the individual parental receptors. The ability of such heterodimers to modulate receptor function has led to the idea that the single protomers constituting a heterodimer may communicate by trans-conformational changes, thus regulating the functional properties of the receptor complex.

An example of a functional and presumably physiologically occurring heterodimer is the one formed by the α_{2A}-AR and the μ-opioid receptor (MOR), which has been reported to exist at the plasma membrane of both transfected cells and native neurons.[40] Despite increasing morphine-dependent MOR signalling, the formation of such heterodimers appears to result in a decrease of the opioid response when morphine is applied in combination with α_{2A}-AR agonists. A direct functional interaction between α_{2A}-AR and MOR may contribute to the complex interactions of the effects of α_{2A}-AR agonists with those of morphine.[41,42]

Recently, taking advantage of the FlAsH/CFP-based α_{2A}-AR sensor described above, we succeeded in monitoring trans-conformational switches induced by MOR on α_{2A}-AR.[43] Our study revealed that, in cells coexpressing the α_{2A}-AR FRET-sensor and the wild-type MOR, morphine did not affect the FRET signal in the absence of norepinephrine, nor did it have an effect on the α_{2A}-AR FRET-sensor when expressed alone. However, when both receptors were coexpressed, morphine caused a reduction of the norepinephrine-induced FRET signal by ~25%. This effect appears to be due to a direct interaction between the two receptors: morphine binding to the MOR appears to induce a MOR conformational change which is directly propagated to the neighbouring norepinephrine-activated α_{2A}-AR FRET sensor and alters its conformation (Figure 9.4). This hypothesis is supported by the fact that the MOR-induced trans-conformational change of the α_{2A}-AR occurred with a time constant of

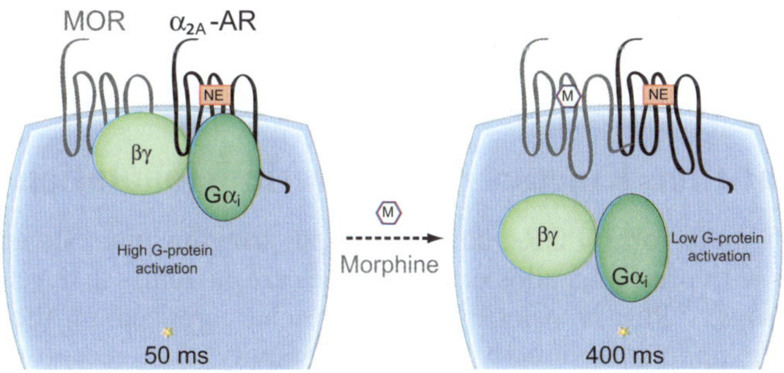

Figure 9.4 Molecular mechanism of the *trans*-conformational switches occurring in the MOR/α_{2A}-AR heterodimer. Norepinephrine (NE) binding to the α_{2A}-AR receptor leads to α_{2A}-AR activation within ~50 ms. Morphine-induced activation of the neighbouring MOR triggers a MOR conformational change which is directly propagated to the norepinephrine-activated α_{2A}-AR receptor and alters its structural conformation within ~400 ms. This event causes a reduction of the α_{2A}-AR-dependent G_i-protein activation.

~400 ms. Because this event is somewhat faster than the α_{2A}-AR-dependent G_i activation (~500 ms), it is unlikely to be mediated *via* G_i or *via* other indirect mechanisms. Rather, it is compatible with a conformational change directly induced by the MOR on the interacting α_{2A}-AR.[43] This hypothesis is also supported by the observation that the trans-conformational switch was not altered by the pertussis toxin-induced inhibition of G_i, and thus does not seem to be dependent on a receptor/G protein interaction.[43] An interesting observation is that the cross-conformational switch was visible only when both agonists were present, suggesting that morphine may stabilize an inactive state of the α_{2A}-AR, thus compromising its norepinephrine-dependent activation. One receptor, when occupied by its agonist, therefore inactivates the other receptor in such a heterodimer. This idea is supported by data reporting morphine to significantly decrease the level of G_i protein activation in the presence of norepinephrine instead of producing the expected additive effect.[43]

The observation that morphine binding to MOR regulates the conformational state of the norepinephrine-occupied α_{2A}-AR and inhibits G_i protein activation is in agreement with numerous studies suggesting—negative or positive—cooperative conformational changes within receptor homo- and heterodimers.[44–47] Recent reports suggest that activation of G proteins by GPCR dimers derives from the interaction of a single G protein molecule binding two receptors. In this view, one receptor protomer in a dimer is activated, whereas the other provides a docking support to G protein binding and selectivity.[48–50] In such a situation, only one protomer per dimer would be active. Taken together, all these studies indicate that different GPCRs may directly inhibit each other in receptor heterodimers and may thus create receptor pharmacologies that are clearly distinct from those of the parent receptors. It remains to be seen how general this phenomenon is, and to what extent such regulation occurs *in vivo*. An exception to this negative cooperativity appears to be the case of the $GABA_B$-receptor, where only one of the two subunits binds agonists, whereas the other one triggers the G protein signal.[51,52,71]

9.5 Real-time Kinetics of Allosteric Modulation

Allosteric ligands for GPCRs have attracted growing interest, and several compounds with presumed allosteric properties are currently in development as potential drugs.[53] Pure allosteric ligands are characterized by the fact that, despite being unable to directly affect the activity of a receptor, they can interact with binding sites on it that are topographically distinct from the orthosteric site recognized by the receptor's endogenous agonists. Thereby, they may modulate either the binding or the efficacy (or both) of orthosteric ligands, either in a positive or in a negative manner. Mixed-type allosteric ligands may, in addition, have direct effects on receptor function (*i.e.* independent of an orthosteric ligand).[54–56]

Examples for such allosteric regulation are the muscarinic acetylcholine receptors, and in particular, the M_2 muscarinic receptor subtype has been studied with regard to such modulation. Numerous allosteric ligands for muscarinic receptors have been identified, some of which show some subtype selectivity for the M_2 muscarinic receptor (*i.e.* the negative allosteric modulators gallamine and dimethyl-W84), whereas others are rather unselective.[57] Looking at the kinetics of allosteric modulations, studies using radioligand binding assays indicate that the effects of some muscarinic allosteric ligands might be quite rapid.[58,59] However, even though most studies in the field of allosteric modulation have been performed with radioligand binding assays, this approach does not allow a precise analysis of the effects on receptors, which require methods that can detect effects in the millisecond time range.

Therefore, we have developed a FlAsH/CFP FRET-based M_2 muscarinic receptor sensor in order to monitor receptor modulation triggered by orthosteric and allosteric ligands in real-time.[34] As described above, this technique has the great advantage of allowing the study of effects caused by various compounds on the receptor protein itself, rather than on downstream processes, thus representing a unique tool to monitor the effects of allosteric ligands on the receptors themselves. As in other FRET-based receptor sensors, the exposure of the M_2 muscarinic receptor sensor to full agonists (*i.e.* acetylcholine, carbachol) resulted in a rapid decrease of the FRET signal, with a half-time of again below 100 ms. This signal can be completely reverted by application of saturating concentrations of antagonists (*i.e.* atropine).

As expected for pure allosteric ligands, the stimulation of the sensor with the specific allosteric ligands gallamine or dimethyl-W84 did not affect the FRET signal. However, when given in the presence of the agonists acetylcholine or carbachol, the allosteric ligands caused a concentration-dependent reversal of the agonist-induced FRET signal as expected for a negative allosteric modulator. This is compatible with a switch of the receptor back into an inactive state. A reversal can obviously also be obtained by washout of agonists or by addition of high concentrations of orthosteric antagonists; the latter method has the advantage of preventing rebinding of dissociated agonist.

With regard to the kinetics of this reversal of agonist-induced FRET signals, the rates obtained for gallamine and dimethyl-W84 were significantly faster ($\sim$80–200 ms) than those observed with saturating concentrations of orthosteric antagonists ($\sim$500–800 ms) or with the simple washout of the agonist ($\sim$1100 ms).[34] Overall, these observations suggest that the allosteric ligands actively change the receptor conformation and do not need to 'wait' until the orthosteric agonist has dissociated. They are in agreement with the characterization of gallamine and dimethyl-W84 as negative allosteric modulators of muscarinic receptors, which reduce the equilibrium affinity of orthosteric ligands and slow down both the association and dissociation of antagonist radioligands.[60–63] Taken together, these data suggest that allosteric ligands induce an inactivation process that is faster than the usual dissociation of agonists and is presumably due to the transition of the receptor into an inactive conformation. Similar studies with positive allosteric modulators have, to our knowledge, not yet been performed.

As mentioned above, several lines of experimental evidence indicate that some GPCRs may exist and function as dimers and/or high order oligomers.

This raises the possibility that allosteric ligands might act across a receptor dimer. Data obtained with early studies on the effects of gallamine on antagonist radioligand binding reported that the allosteric effects persist in solubilized muscarinic receptors characterized by an apparent molecular weight of ~80 kDa, a value lying between the actual values of a mono- and a dimer.[63] Since the apparent molecular weight might include a contribution by detergent, and since proteins often have a higher apparent molecular weight than predicted from their sequence, this supports the notion that allosteric effects occur in a single muscarinic receptor.

Furthermore, the kinetic data obtained with our FRET-based approach argue for an action of allosteric and orthosteric ligands on the same protein. In fact, we measured switch times of allosteric ligands down to <100 ms, a value significantly smaller than the switch times we observed across a receptor dimer (~400 ms) or between a GPCR and its cognate G protein (~500 ms). Thus, the switch times we observed for allosteric ligands are well below the switch times that have so far been observed between GPCR and other proteins, and for this reason seem to be more compatible with intramolecular switch times in GPCRs. This suggests that both allosteric and orthosteric ligands bind to the same receptor molecule, which integrates different signals into a receptor output.

9.6 How Fast are GPCRs in Living Cells?

Our data about the 'true' speed of the agonist-induced conformational changes that represent activation of a GPCR are still quite uncertain. The currently available data from several methods used to investigate the switch times of GPCRs are quite different. FRET signals detected in intact cells after agonist stimulation with the FRET-based receptor constructs occur in a mono-exponential manner, with switch times of 50–100 ms for most cases (Table 9.1).[27]

Thus, the kinetics of GPCR activation in living cells are much faster than the fluorescence signals obtained in isolated membranes or reconstituted systems. Full agonist-dependent receptor activation occurs with a similar rate constants τ of 50–100 ms for α_{2A}-adrenergic, β_1-adrenergic, β_2-adrenergic, muscarinic and A_{2A}-adenosine receptors.[26,27,30,31,34] The fast kinetics of the initial steps of the signalling pathway are in accordance with the rapid control of ion channels by G protein mentioned above (Figure 9.5). One exception encountered so far is the parathyroid hormone receptor (PTHR), which shows a significantly slower activation half-life ($\tau \approx 1$ s). This correlates well with agonist binding studies suggesting a biphasic interaction of the PTH with its receptor: after a first rapid interaction ($\tau \approx 100$ ms, dependent on agonist concentration) a second slower component follows ($\tau \approx 1$ s), which presumably represents the interaction of the ligand with the transmembrane core of the receptor and may coincide with receptor activation. Similar slow activation kinetics have been detected for an analogous bradykinin B2 receptor sensor.[32,64]

The kinetic studies described above suggest that the rate constant of GPCR activation depends on the nature of both the receptor and the ligand applied.

Table 9.1 Kinetics of different receptor activation measured by FRET in living cells after stimulation with saturating concentrations of the respective full agonists.

Receptor	$t_{1/2}$, *ms*	
PTHR[27]	1000	Slow receptor activation correlates with the slow interaction of the PTH with the transmembrane core of the receptor.
α_{2A}-AR[27,36]	50	
β_1-AR[30]	60	
β_2-AR[31]	47	The speed of the agonist-induced conformational changes detected in GPCRs activated by small full agonists (*i.e.* norepinephrine, epinephrine, carbachol, adenosine) is similar.
M_1-AChR[35,72]	< 100	
M_2-AChR[34]	50	
M_3-AChR[35]	62	
M_5-AChR[35]	65	
A_{2A}-R[26]	66	
mGluR1β[65,66]	10 ~ 5000	Activation kinetics denote inter-subunit conformational changes occurring in a class C GPCR; the slower kinetics were recorded by TIRF microscopy.[65,66]

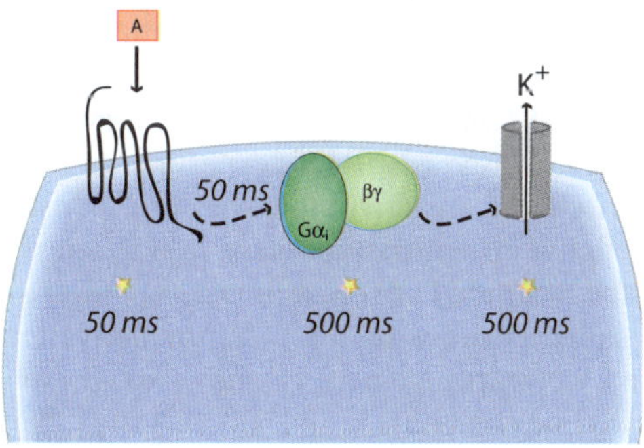

Figure 9.5 Kinetics of the steps occurring during receptor activation and signalling in living cells. From the left: stimulation of the receptor with a full agonist (**A**) induces a conformational change of the receptor that results in its full activation within ~ 50 ms.[26,28,30,31,34,35] The recruitment of the cognate G protein by the activated receptor is reported to be as fast as the activation of the receptor itself. It leads to activation of the G protein within ~ 500 ms.[68–70] The activated G protein couples to downstream effectors (*i.e.* adenylyl cyclases or ion channels) and modulates their activity. As reported by FRET and whole-cell electrophysiological recordings, there is a tight temporal correlation between the active state of the G_i-protein and the GIRK channel activity (~ 500 ms).[69]

Kinetic studies with the α_{2A}-AR have shown that different types of ligands determine changes in FRET at very different rates.[28,31,36] The signals caused by conformational changes induced by partial agonists were not only smaller but also slower ($\tau \approx 1$ s) than those generated by full agonists ($\tau \approx 60$ ms). This observation is in agreement with data obtained very recently with the β_2-AR;

in this FRET-based sensor, norepinephrine induces only partial (~50% amplitude) and slower (rate constants ~50 *vs.* ~130 ms) signals compared to those triggered by epinephrine.[31]

In addition to these receptor- and ligand-specific values, there is recent evidence that conformational changes may also differ at various parts of a receptor. These data suggest that ligands of different efficacies induce different conformational changes of the α_{2A}-AR that imply a different orientation of the third intracellular loop.[37] Using three FRET-based α_{2A}-AR sensors labelled with FlAsH at three distinct positions of the third intracellular loop (close to TM5, in the middle of iCL3 or close to TM6), we observed that full agonists (*i.e.* norepinephrine) trigger conformational rearrangements which are detectable in all the three reference points. In contrast, partial agonists (*i.e.* dopamine) induce stronger movements of the region of the third intracellular loop near to TM6, but weaker movements in the regions closed to TM5. Those latter movements are not only smaller but also slower than those induced by full agonists (the τ values for the two compounds differ by a factor of ~3). This supports the hypothesis that different agonists may induce distinct conformations with different speeds of GPCRs activation. As highlighted by these findings, the activation rate does not appear to be an intrinsic feature of an individual receptor but rather to depend largely on the type of ligand. It remains to be investigated whether the class of receptor and the downstream signalling pathway also may influence the speed of activation of individual receptors.

Figure 9.5 summarizes what we know about the individual steps of a GPCR cascade from analogous FRET experiments in intact cells. Most GPCRs appear to switch into an active state with time constants on the order of 50–100 ms, and the active receptors then very rapidly engage G proteins such that the kinetics of the receptor/G protein interaction after addition of agonists are essentially the same as those of the receptor activation itself. G protein activation, measured *via* rearrangement of the α- *vs.* $\beta\gamma$-subunits, however, is ten times slower with time constants in the 500 ms range. Again, this activation step appears to be very closely linked to effector activation and the latter (at least in the case of ion channels, which can be measured electrophysiologically) display activation kinetics that are again very similar to those of the G proteins.

In contrast to these FRET data, fluorescence data obtained with isolated reconstituted receptors suggest much slower conformational changes in response to ligands; these signals were observed to occur on the second to minute scale, depending on the reconstitution procedure and the type of experiment and ligand (see above; ref. 20). While such differences may be attributed to various technical factors, an alternative explanation might be that these experiments actually measure a process that is distinct from those measured by FRET experiments in intact cells. For example, agonists might first induce an active (*i.e.* G protein activating) state of receptors, followed by other states, such as β-arrestin interacting states, and ultimately a desensitized state (where desensitization lies in the receptor protein itself, as is the case with ligand-gated ion channels[6]). The recent indications that GPCRs might indeed adopt multiple 'active' conformations suggest that such sequential conformational changes may indeed be a possibility.

The detailed analysis of the various states that rhodopsin assumes after light activation suggest that such schemes could also occur in GPCRs.

On the other hand, rhodopsin switching is much more rapid than the conformational changes observed in any other GPCR, with rate constants for the active metarhodopsin II state of $\sim$1 ms, and much faster initial steps. While this rapid speed of activation may be related to the particular biological function of rhodopsin, there may be also many technical reasons why such rates can be measured much more precisely for rhodopsin. These reasons include the fact that the ligand (retinal) is already pre-bound in inactive rhodopsin, that activation can be initiated by light-induced isomerization of retinal with extreme temporal precision by light flashes and that it can also be recorded with high accuracy. In contrast, experiments with FRET-based sensors in intact cells represent ensemble responses that may be confounded by asynchronous switching of a large number of receptors (caused for example by unequal diffusion of ligands to the receptors' binding sites).

Another interesting question is whether inter-subunit switching in GPCR dimers occurs with the same speed as intra-molecular switches. In the metabotropic glutamate receptor type 1, this has been investigated by two groups.[65,66] While the first group reported signals of a large amplitude but slow kinetics, the second group observed signals that were small ($\sim$0.5 % change) but very rapid (see Table 9.1). Thus, it remains to be seen whether inter-subunit conformational changes are slower or faster than those measured within a single GPCR.

In summary, much remains to be done to unravel the switching behaviour of GPCRs. However, many technologies have recently been developed that allow us to approach these questions and to address the possibility that this switching entails a number of distinct processes, which may occur with distinct speeds ranging from the millisecond range to several seconds, and that each of these processes is a regulated step with a defined biological function.

References

1. K. L. Pierce, R. T. Premont and R. J. Lefkowitz, *Nat. Rev. Mol. Cell Biol.*, 2002, **3**, 639–650.
2. M. C. Lagerstrom and H. B. Schioth, *Nat. Rev. Drug Discov.*, 2008, **7**, 339–357.
3. K. Palczewski, T. Kumasaka, T. Hori, C. A. Behnke, H. Motoshima, B. A. Fox, I. Le Trong, D. C. Teller, T. Okada, R. E. Stenkamp, M. Yamamoto and M. Miyano, *Science*, 2000, **289**, 739–745.
4. D. M. Rosenbaum, S. G. Rasmussen and B. K. Kobilka, *Nature*, 2009, **459**, 356–363.
5. M. J. Marinissen and J. S. Gutkind, *Trends Pharmacol. Sci.*, 2001, **22**, 368–376.
6. M. J. Lohse, *Biochim. Biophys. Acta*, 1993, **1179**, 171–188.
7. J. P. Pin, R. Neubig, M. Bouvier, L. Devi, M. Filizola, J. A. Javitch, M. J. Lohse, G. Milligan, K. Palczewski, M. Parmentier and M. Spedding, *Pharmacol. Rev.*, 2007, **59**, 5–13.
8. M. J. Lohse, P. Hein, C. Hoffmann, V. O. Nikolaev, J. P. Vilardaga and M. Bünemann, *Br. J. Pharmacol.*, 2008, **153**(Suppl 1), S125–132.

9. A. G. Gilman, *Annu. Rev. Biochem.*, 1987, **56**, 615–649.
10. H. R. Bourne, D. A. Sanders and F. McCormick, *Nature*, 1991, **349**, 117–127.
11. C. J. Lingle, D. Maconochie and J. H. Steinbach, *J. Membr. Biol.*, 1992, **126**, 195–217.
12. R. L. MacDonald and R. E. Twyman, *Ion Channels*, 1992, **3**, 315–343.
13. M. A. Lemmon and J. Schlessinger, *Cell*, 2010, **141**, 1117–1134.
14. P. J. Pfaffinger, J. M. Martin, D. D. Hunter, N. M. Nathanson and B. Hille, *Nature*, 1985, **317**, 536–538.
15. M. Bünemann, M. M. Bücheler, M. Philipp, M. J. Lohse and L. Hein, *J. Biol. Chem.*, 2001, **276**, 47512–47517.
16. W. L. Hubbell, C. Altenbach, C. M. Hubbell and H. G. Khorana, *Adv. Protein Chem.*, 2003, **63**, 243–290.
17. C. L. Makino, X. H. Wen and J. Lem, *Curr. Opin. Neurobiol.*, 2003, **13**, 404–412.
18. U. Gether, S. Lin and B. K. Kobilka, *J. Biol. Chem.*, 1995, **270**, 28268–28275.
19. M. J. Lohse, C. Hoffmann, V. O. Nikolaev, J. P. Vilardaga and M. Bünemann, *Adv. Protein Chem.*, 2007, **74**, 167–188.
20. B. K. Kobilka and X. Deupi, *Trends Pharmacol. Sci.*, 2007, **28**, 397–406.
21. J. H. Park, P. Scheerer, K. P. Hofmann, H. W. Choe and O. P. Ernst, *Nature*, 2008, **454**, 183–187.
22. P. Ghanouni, J. J. Steenhuis, D. L. Farrens and B. K. Kobilka, *Proc. Natl. Acad. Sci. U. S. A.*, 2001, **98**, 5997–6002.
23. P. Ghanouni, Z. Gryczynski, J. J. Steenhuis, T. W. Lee, D. L. Farrens, J. R. Lakowicz and B. K. Kobilka, *J. Biol. Chem.*, 2001, **276**, 24433–24436.
24. G. Swaminath, Y. Xiang, T. W. Lee, J. Steenhuis, C. Parnot and B. K. Kobilka, *J. Biol. Chem.*, 2004, **279**, 686–691.
25. X. Yao, C. Parnot, X. Deupi, V. R. Ratnala, G. Swaminath, D. Farrens and B. Kobilka, *Nat. Chem. Biol.*, 2006, **2**, 417–422.
26. C. Hoffmann, G. Gaietta, M. Bünemann, S. R. Adams, S. Oberdorff-Maass, B. Behr, J. P. Vilardaga, R. Y. Tsien, M. H. Ellisman and M. J. Lohse, *Nat. Methods*, 2005, **2**, 171–176.
27. J. P. Vilardaga, M. Bünemann, C. Krasel, M. Castro and M. J. Lohse, *Nat. Biotechnol.*, 2003, **21**, 807–812.
28. J. P. Vilardaga, R. Steinmeyer, G. S. Harms and M. J. Lohse, *Nat. Chem. Biol.*, 2005, **1**, 25–28.
29. S. Marullo and M. Bouvier, *Trends Pharmacol. Sci.*, 2007, **28**, 362–365.
30. F. Rochais, J. P. Vilardaga, V. O. Nikolaev, M. Bünemann, M. J. Lohse and S. Engelhardt, *J. Clin. Invest.*, 2007, **117**, 229–235.
31. S. Reiner, M. Ambrosio, C. Hoffmann and M. J. Lohse, *J. Biol. Chem.*, 2010, **285**, 36188–36198.
32. M. Chachisvilis, Y. L. Zhang and J. A. Frangos, *Proc. Natl. Acad. Sci. U. S. A.*, 2006, **103**, 15463–15468.
33. S. R. Adams, R. E. Campbell, L. A. Gross, B. R. Martin, G. K. Walkup, Y. Yao, J. Llopis and R. Y. Tsien, *J. Am. Chem. Soc.*, 2002, **124**, 6063–6076.
34. M. Maier-Peuschel, N. Frölich, C. Dees, L. G. Hommers, C. Hoffmann, V. O. Nikolaev and M. J. Lohse, *J. Biol. Chem.*, 2010, **285**, 8793–8800.

35. N. Ziegler, J. Bätz, U. Zabel, M. J. Lohse and C. Hoffmann, *Bioorg. Med. Chem.*, 2010.
36. V. O. Nikolaev, C. Hoffmann, M. Bünemann, M. J. Lohse and J. P. Vilardaga, *J. Biol. Chem.*, 2006, **281**, 24506–24511.
37. A. Zürn, U. Zabel, J. P. Vilardaga, H. Schindelin, M. J. Lohse and C. Hoffmann, *Mol. Pharmacol.*, 2009, **75**, 534–541.
38. A. Zürn, C. Klenk, U. Zabel, S. Reiner, M. J. Lohse and C. Hoffmann, *Bioconjug. Chem.*, 2010, **21**, 853–859.
39. S. Dorsch, K. N. Klotz, S. Engelhardt, M. J. Lohse and M. Bünemann, *Nat. Methods*, 2009, **6**, 225–230.
40. B. A. Jordan, I. Gomes, C. Rios, J. Filipovska and L. A. Devi, *Mol. Pharmacol.*, 2003, **64**, 1317–1324.
41. T. C. Spaulding, S. Fielding, J. J. Venafro and H. Lal, *Eur. J. Pharmacol.*, 1979, **58**, 19–25.
42. G. L. Wilcox, K. H. Carlsson, A. Jochim and I. Jurna, *Brain Res.*, 1987, **405**, 84–93.
43. J. P. Vilardaga, V. O. Nikolaev, K. Lorenz, S. Ferrandon, Z. Zhuang and M. J. Lohse, *Nat. Chem. Biol.*, 2008, **4**, 126–131.
44. L. Albizu, M. N. Balestre, C. Breton, J. P. Pin, M. Manning, B. Mouillac, C. Barberis and T. Durroux, *Mol. Pharmacol.*, 2006, **70**, 1783–1791.
45. J. Y. Springael, E. Urizar, S. Costagliola, G. Vassart and M. Parmentier, *Pharmacol. Ther.*, 2007, **115**, 410–418.
46. D. Mesnier and J. L. Baneres, *J. Biol. Chem.*, 2004, **279**, 49664–49670.
47. E. Urizar, L. Montanelli, T. Loy, M. Bonomi, S. Swillens, C. Gales, M. Bouvier, G. Smits, G. Vassart and S. Costagliola, *EMBO J.*, 2005, **24**, 1954–1964.
48. J. F. White, J. Grodnitzky, J. M. Louis, L. B. Trinh, J. Shiloach, J. Gutierrez, J. K. Northup and R. Grisshammer, *Proc. Natl. Acad. Sci. U. S. A.*, 2007, **104**, 12199–12204.
49. J. L. Baneres and J. Parello, *J. Mol. Biol.*, 2003, **329**, 815–829.
50. M. Damian, A. Martin, D. Mesnier, J. P. Pin and J. L. Baneres, *EMBO J.*, 2006, **25**, 5693–5702.
51. J. P. Pin, L. Comps-Agrar, D. Maurel, M. L. Rives, E. Trinquet, J. Kniazeff, P. Rondard and L. Prezeau, *J. Physiol.*, 2009.
52. A. Pinard, R. Seddik and B. Bettler, *Adv. Pharmacol.*, 2010, **58**, 231–255.
53. P. J. Conn, A. Christopoulos and C. W. Lindsley, *Nat. Rev. Drug Discov.*, 2009, **8**, 41–54.
54. L. T. May and A. Christopoulos, *Curr. Opin. Pharmacol.*, 2003, **3**, 551–556.
55. J. H. Li, S. J. Han, F. F. Hamdan, S. K. Kim, K. A. Jacobson, L. M. Bloodworth, X. Zhang and J. Wess, *J. Biol. Chem.*, 2007, **282**, 26284–26293.
56. A. Christopoulos, L. T. May, V. A. Avlani and P. M. Sexton, *Biochem. Soc. Trans.*, 2004, **32**, 873–877.
57. L. T. May, V. A. Avlani, C. J. Langmead, H. J. Herdon, M. D. Wood, P. M. Sexton and A. Christopoulos, *Mol. Pharmacol.*, 2007, **72**, 463–476.
58. C. Tränkle, A. Dittmann, U. Schulz, O. Weyand, S. Buller, K. Johren, E. Heller, N. J. Birdsall, U. Holzgrabe, J. Ellis, H. D. Holtje and K. Mohr, *Mol. Pharmacol.*, 2005, **68**, 1597–1610.

59. C. Tränkle, O. Weyand, U. Voigtlander, A. Mynett, S. Lazareno, N. J. Birdsall and K. Mohr, *Mol. Pharmacol.*, 2003, **64**, 180–190.
60. K. Mohr, C. Tränkle and U. Holzgrabe, *Recept. Channels*, 2003, **9**, 229–240.
61. T. K. Narayanan and R. S. Aronstam, *Neurochem. Res.*, 1986, **11**, 1397–1406.
62. B. D. Pearce and L. T. Potter, *Neurosci. Lett.*, 1988, **88**, 281–285.
63. J. M. Stockton, N. J. Birdsall, A. S. Burgen and E. C. Hulme, *Mol. Pharmacol.*, 1983, **23**, 551–557.
64. M. Castro, V. O. Nikolaev, D. Palm, M. J. Lohse and J. P. Vilardaga, *Proc. Natl. Acad. Sci. U. S. A.*, 2005, **102**, 16084–16089.
65. M. Tateyama, H. Abe, H. Nakata, O. Saito and Y. Kubo, *Nat. Struct. Mol. Biol.*, 2004, **11**, 637–642.
66. P. Marcaggi, H. Mutoh, D. Dimitrov, M. Beato and T. Knopfel, *Proc. Natl. Acad. Sci. U. S. A.*, 2009, **106**, 11388–11393.
67. C. Hoffmann, G. Gaietta, A. Zürn, S. R. Adams, S. Terrillon, M. H. Ellisman, R. Y. Tsien and M. J. Lohse, *Nat. Protoc.*, 2010, **5**, 1666–1677.
68. P. Hein, F. Rochais, C. Hoffmann, S. Dorsch, V. O. Nikolaev, S. Engelhardt, C. H. Berlot, M. J. Lohse and M. Bünemann, *J. Biol. Chem.*, 2006, **281**, 33345–33351.
69. M. Bünemann, M. Frank and M. J. Lohse, *Proc. Natl. Acad. Sci. U. S. A.*, 2003, **100**, 16077–16082.
70. M. Frank, L. Thümer, M. J. Lohse and M. Bünemann, *J. Biol. Chem.*, 2005, **280**, 24584–24590.
71. C. Monnier, H. Tu, E. Bourrier, C. Vol, L. Lamarque, E. Trinquet, J. P. Pin and P. Rondard, *EMBO J.*, 2011, **30**, 32–42.
72. J. B. Jensen, J. S. Lyssand, C. Hague and B. Hille, *J. Gen. Physiol.*, 2009, **133**, 347–359.

CHAPTER 10

Emerging Signalling Properties of the PTH Receptor

JEAN-PIERRE VILARDAGA

Laboratory for GPCR Biology, Department of Pharmacology and Chemical Biology, University of Pittsburgh, School of Medicine, Pittsburgh, PA 15261, USA, and Endocrine Unit, Department of Medicine, Massachusetts General Hospital and Harvard Medical School, MA 02114, USA

10.1 Introduction

Parathyroid hormone (PTH) receptor type 1 (PTHR),[1] a paradoxical G protein-coupled receptor (GPCR) of the B family, transmits signals *via* two distinct ligand systems:[2,3]

- PTH is a circulating and homeostatic hormone secreted in a pulsatile fashion from the parathyroid gland with bone (chondrocytes, osteoblasts and osteocytes) and kidney (distal and proximal tubule cells) as the main targets. It regulates concentrations of calcium and phosphate ions, and 1,25-dihydroxy-vitamin-D (the active form of vitamin D) in blood and extracellular fluids.
- PTH-related peptide (PTHrP) is a paracrine, autocrine or intacrine hormone found in developmental tissues such as bone, lung, heart and mammary glands among other tissues.

Despite the activation of shared G_S- and G_q proteins, which trigger adenylyl cyclases/cAMP/PKA and phospholipase C/inositol phosphates/Ca^{2+}/PKC

RSC Drug Discovery Series No. 8
G Protein-Coupled Receptors: From Structure to Function
Edited by Jesús Giraldo and Jean-Philippe Pin

Published by the Royal Society of Chemistry, www.rsc.org

signalling cascades, respectively, it is unclear how a single receptor discriminates between the two ligand signalling systems.[1] It is also unclear why PTH stimulates more prolonged increases in serum levels of 1,25-dihydroxyvitamin-D, Ca^{2+} and bone resorption markers than PTHrP when the ligands are administered by continuous infusion.[4,5] This chapter discusses recent findings suggesting that the capacity of PTH to stabilize a high affinity PTHR conformation able to sustain G_S signalling from intracellular domain is a key determinant that differentiates the action of PTH *vs.* PTHrP.[6–8]

10.2 Kinetics of the PTHR Signalling System

10.2.1 GPCR Studies in Live Cells

The transfer of information through GPCR signalling systems involves sequential and reversible events that initially take place at the cell membrane and modulate the production and propagation of second messenger molecules inside the cell. These events begin when ligand (L) binding converts an inactive receptor (R) into its active conformation (R*) through intramolecular rearrangement of its transmembrane helices and intracellular loops.[9–12] These conformational changes are thought to expose receptor epitopes at the intracellular side that permit their interaction with G proteins and promote a rapid GDP/GTP exchange on the $G\alpha$ subunit.[13,14]

The application of fluorescent resonance energy transfer (FRET) and related fluorescent-based methods—recently reviewed[15–17] and illustrated in Figure 10.1 for the PTHR system—allows us to monitor the kinetics of individual reactions involved in GPCR signalling in real-time and in live cells.[18] These new quantitative FRET-based approaches have increased our understanding of the basic molecular mechanisms by which ligands interact with several GPCRs[8,19–24] and modulate G protein-mediated signalling processes in both cells[8,25–27] and tissues.[28,29] They have also allowed us to determine the complete sequence of early steps in the PTHR signalling cascade, revealing rate-limiting reactions for PTH and PTHrP, usually studied with their respective *N*-terminal synthetic fragments PTH(1-34) and PTHrP(1-36), which are fully functional[30] (Figure 10.1).

10.2.2 Hormone–PTHR Interaction

The hormone–receptor interaction begins with the association of the *C*-terminal domain of PTH(1-34) to the amino-terminal extracellular domain of PTHR (N-domain).[31–34] The kinetics of this interaction reflect a simple bimolecular interaction between the ligand and the receptor, and are strictly dependent on ligand concentrations.[35] The observed rate constant (k_{obs}) of this interaction is thus defined by $k_{obs} = k_{off} + (k_{on} \times L)$ under pseudo-first-order conditions, where k_{on} and k_{off} represent the binding and dissociation rate constants, respectively.[36,37] This initial *N*-binding event occurs with a time

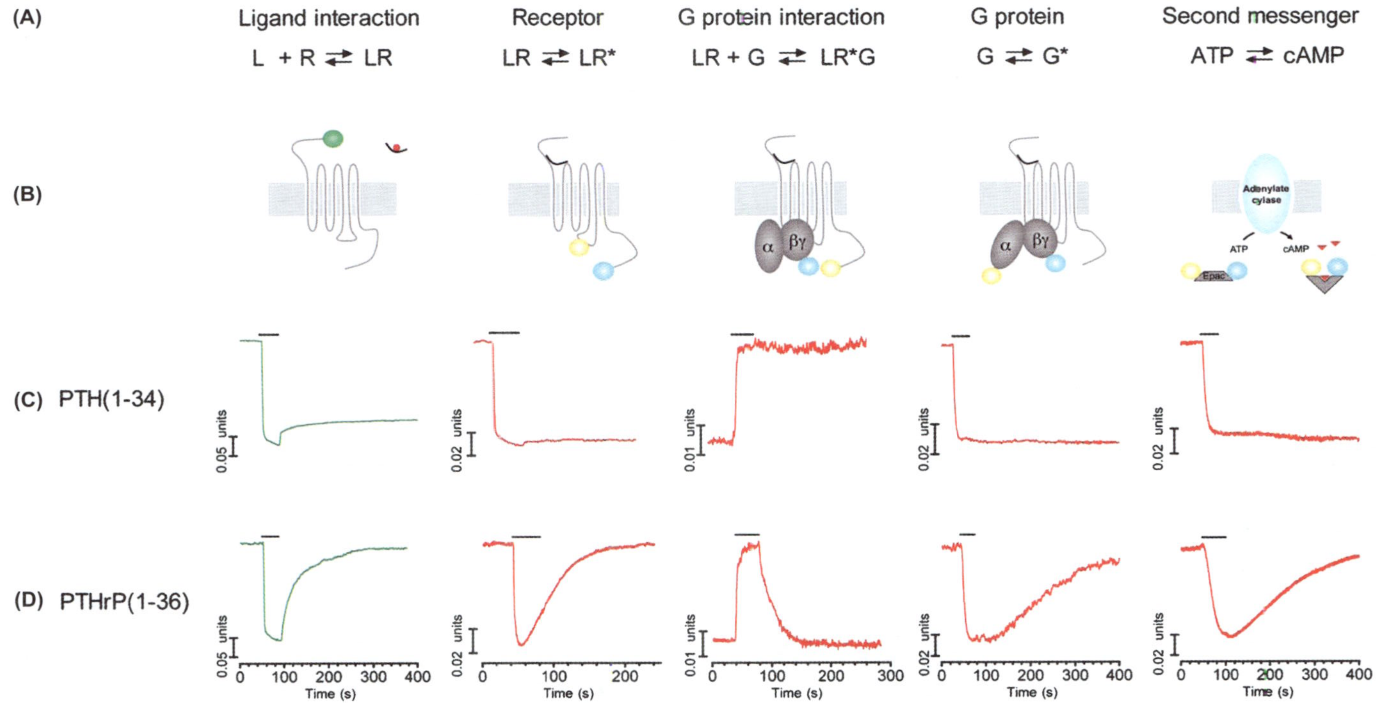

Figure 10.1 Time courses of early reactions in the signalling cascade of PTHR. (**A**) Biochemical reactions. (**B**) Experimental approaches for FRET measurements of the different kinetic events: ligand/receptor interactions are measured by FRET between GFP-tagged PTH-receptor and tetramethylrhodamine (TMR) labelled peptides; PTHR activation is monitored by a decrease in FRET between CFP and YFP inserted in the third intracellular loop and *C*-terminus of the receptor, respectively; PTHR and G protein interaction is measured as an increase in FRET between YFP-labelled PTHR and CFP-labelled $G\gamma_2$ in combination with $G\alpha_S$ and $G\beta_1$ proteins; G protein activation is monitored by recording FRET between YFP-labeled Gα and CFP-labelled $G\gamma_2$-subunits; ligand-mediated cAMP response upon PTHR activation in cells is measured as a reduction of FRET in the Epac-based cAMP biosensor, $Epac^{CFP/YFP}$. (**C** and **D**) Time courses of distinct reactions mediated by PTH(1-34) or PTHrP(1-36) at a saturating concentration. Measurements are performed in single cells continuously perfused with buffer or with ligand for the time indicated by the horizontal bar. For binding, the trace represents changes in emission of GFP fluorescence normalized to the initial value. For the other events, traces represent the normalized FRET ratio F_{YFP}/F_{CFP} after correction for bleedthrough (CFP emission into the YFP channel), cross-talk (direct excitation of YFP by the CFP excitation light) and photobleaching. Adapted from ref. 8.

constant ($\tau_{obs} = 1/k_{obs}$) of ~150 ms at a high concentration (10 μM) of PTH(1-34).[35] Then the *N*-terminal domain of the hormone interacts with the juxtamembrane region of the PTHR (J-domain; encompassing the transmembrane helical domain and connecting extracellular loops). This interaction has more complex and slower kinetics ($\tau \geq 1$ s) involving not only ligand and receptor interactions, but also the conformational changes of the ligand/receptor complex required for PTHR activation.[35] The time constants of this second and slower binding step are temporally coupled to the isomerization of PTHR into its active conformation and are rate limiting for the activation of PTHR.

PTHrP(1-36), like PTH(1-34), associates with PTHR by a similar two-step process, though with slightly slower kinetics.[8] Each hormone dissociates by a distinct mechanism, however. PTHrP(1-36) dissociates fully from the receptor, whereas PTH(1-34) forms a persistent complex with PTHR. The dissociation of PTHrP(1-36) follows a two-step mechanism involving fast and slow components. The rapid dissociation process ($\tau = 1.3$ s) corresponds to a minor fraction (<15%) of ligand-bound receptor and is a G protein independent process, whereas the slower dissociation process ($\tau = 28$ s) is dependent on the release of G proteins from the receptor and reflects ligand dissociation from a ligand–receptor–G protein complex.[8]

10.2.3 PTHR Activation/Deactivation

Following ligand binding, PTHR reaches its active state at a relatively slow speed ($\tau \approx 1$ s) compared with the kinetics of activation of other GPCRs such as the adenosine A_{2A} receptor, α- and β-adrenergic receptors and muscarinic receptors (<100 ms).[19,20,24,38–41] Differences in the time course of receptor activation presumably reflect differences in the structure of ligands and their mode of binding (peptide *vs.* small molecules). Additional allosteric modulation mediated by PTHR interactions to single transmembrane proteins[42–44] (*i.e.,* receptor activity-modifying proteins and LRP5) and cytosolic adaptor proteins[25,45,46] (dishevelled, and Na^+/H^+ exchanger regulatory factor adaptor proteins) may also influence to the kinetics of PTHR activation.

The time course of PTHR deactivation measured after washout of PTHrP(1-36) proceeds with a time constant $\tau \approx 58$ s, which is much slower than that measured for adrenergic receptors such as the α_{2A}AR after norepinephrine (NE) washout (~2 s).[21,27] This difference is most likely dependent on the faster dissociation of α_{2A}AR ligands, but also on the intrinsic capacity of individual receptors to switch back to their inactive states after ligand dissociation. This point is illustrated in the case of PTHrP(1-36), which dissociates from PTHR with a half-time $t_{1/2} \approx 19$ s whereas PTHR deactivates with a $t_{1/2} \approx 40$ s after PTHrP(1-36) washout (Figure 10.3 and Table 10.1). This suggests that ligand dissociation is completed before receptor deactivation. The active state of PTHR thus persists even after PTHrP(1-36) has dissociated. As a result of the slow receptor deactivation compared with the faster ligand dissociation, the receptor remains in its active state for an extended

Table 10.1 Kinetics of ligand association and dissociation (reaction 1), PTHR activation and deactivation (reaction 2), PTHR and G_S interaction (reaction 3), G_S activation and deactivation (reaction 4), cAMP accumulation and degradation (reaction 5) (ref. 8). Reactions were recorded from single HEK-293 cells at a saturating concentration of ligand.[8] Values represent the mean ± s.e.m. of the rate constant (τ).

	Switch on (s)		*Turn off (s)*	
	PTH(1-34)	*PTHrP(1-36)*	*PTH(1-34)*	*PTHrP(1-36)*
(1) $L + R \rightleftarrows LR$	$\tau^{fast} = 0.14 \pm 0.01$ $\tau^{slow} = 1.15 \pm 0.10$	$\tau^{fast} = 0.17 \pm 0.05$ $\tau^{slow} = 1.54 \pm 0.15$	$\tau^{fast} = 1.50 \pm 0.27$ τ^{slow} NA	$\tau^{fast} = 1.38 \pm 0.23$ $\tau^{slow} = 28.12 \pm 0.60$
(2) $LR \rightleftarrows LR^*$	$\tau = 0.95 \pm 0.15$	$\tau = 1.59 \pm 0.11$	NA	$\tau = 58.54 \pm 6.42$
(3) $LR^* + G \rightleftarrows LR^*G$	$\tau = 0.96 \pm 0.13$	$\tau = 1.58 \pm 0.19$	NA	$\tau = 48.14 \pm 5.29$
(4) $G \rightleftarrows G^*$	$\tau = 1.58 \pm 0.13$	$\tau = 2.04 \pm 0.14$	NA	$\tau = 121.50 \pm 6.35$
(5) cAMP	$\tau = 10.89 \pm 2.26$	$\tau = 12.66 \pm 1.06$	NA	$\tau = 296.70 \pm 17.47$

time, which presumably permits additional cycles of G protein activation/deactivation even in the absence of the ligand (Table 10.1). The isomerization of the active receptor state R* into its inactive state R, as opposed to ligand dissociation, is therefore the limiting step for the termination of PTHR activation mediated by PTHrP.

10.2.4 PTHR–G protein Interaction

The next step, which involves receptor–G protein interaction, proceeds with kinetics that can be as fast as receptor activation itself ($\tau = 40$–1000 ms) depending on the receptor and the expression level of G proteins.[27] For PTHR, the maximal time constants obtained at a high level of G_S expression are 0.96 s for PTH(1-34) and 1.58 s for PTHrP(1-36), similar to those obtained for the corresponding PTHR activation switch.[8,19] At low expression levels of G_S, however, the time constant of the PTHR/G_S interaction can take >1 s whereas the kinetic of receptor activation remains unchanged.[8] The speed of PTHR/G protein interaction, as seen for other GPCRs such as α_{2A}-AR and the adenosine A_2-receptor, is not therefore determined by the time course of receptor activation, but rather by a diffusion-controlled collision process which is dependent on the relative expression levels of receptors and G proteins.

10.2.5 G Protein Activation/Deactivation

Conformational rearrangements and/or disassembly events between Gα and G$\beta\gamma$ subunits are associated with activation of G proteins[26,47] and proceed with time constants of $\tau \approx 1-2$ s at a saturating concentration of PTHrP(1-36) or PTH(1-34).[8] After PTHrP(1-36) washout, G_S deactivates with approximately two-fold slower kinetics than those observed for PTHR deactivation, which may be associated with the slow intrinsic GTPase activity of Gα-subunits, as well as a limited action of GTPase-activating proteins (GAPs) known to accelerate the Gα-mediated GTP hydrolysis.[48] Although PTHR activation and G protein activation mediated by PTH(1-34) and PTHrP(1-36) follow similar mechanisms, the corresponding deactivation processes are quite divergent. Thus, for both ligands, receptor-ligand binding, as opposed to receptor conformational change, is the rate-limiting step for receptor activation (Table 10.1). In addition, the conformational/dissociational event between the Gα and G$\beta\gamma$ subunits—as opposed to receptor and G protein interaction—is the rate-limiting step for G protein activation. In contrast, the limiting step for termination of PTHR signalling diverges markedly for PTH and PTHrP; PTH(1-34) induces a 'locked on' active-state PTHR conformation that can mediate persistent Gs activation, whereas PTHrP(1-36) dissociates rapidly from the PTHR, prompting rapid signal termination.

10.3 Persistent cAMP Production Induced by Internalized PTHR: A New Concept for GPCR

10.3.1 PTHR Conformations

Membrane-based competition and kinetic binding assays[6,49] demonstrate that PTH(1-34) and PTHrP(1-36) can stabilize distinct high-affinity conformational states of the PTHR—a state independent of G protein coupling that is GTPγS-resistant and stabilized by PTH(1-34), and another state dependent on G protein coupling that is GTPγS-sensitive and stabilized preferentially by PTHrP(1-36) (Figure 10.2).

The high affinity observed in the absence of GTPγS, which is expected from the classical ternary complex model of GPCR binding mechanisms,[50] indicates binding to a G protein coupled PTHR conformation. However, the high affinity binding observed for PTH(1-34) in the presence of GTPγS is not predicted by this classical model. Earlier pharmacological studies on the secretin receptor[51] and the corticotropin-releasing factor (CRF) receptor[52–54] (two other family B GPCRs) also showed high affinity agonist binding in the presence of non-hydrolysable GTP analogs [GTPγS or Gpp(NH)p], suggesting that this binding is mediated by the stabilization of a receptor conformation presumably not

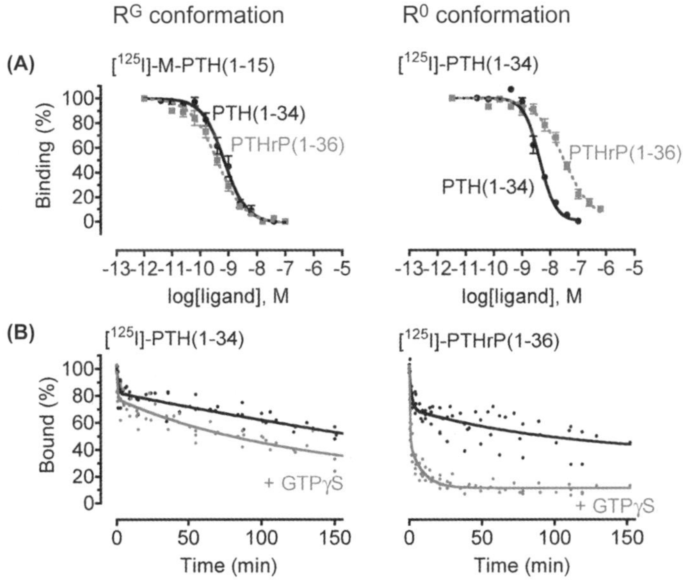

Figure 10.2 PTHR conformations: cell membrane binding assays. Binding to the R^0 and R^G conformations of the PTHR are determined by competition reactions. For R^0, [^{125}I]-PTH(1-34) is used as a tracer radioligand and including GTPγS in the reaction. For RG, [^{125}I]-PTH(1-15) is used as a radioligand in the presence of a high-affinity, negative-dominant $G\alpha_S$ subunit ($G\alpha_S{}^{ND}$). Adapted from ref*s* 6 and 49.

coupled to G proteins, which was initially termed R^0 by Hoare *et al.*[53,54] By analogy with these studies, Gardella and colleagues[6,49] postulated that PTH(1-34) stabilizes a putative high affinity R^0 PTHR conformation, which is uncoupled from G proteins, and is thus distinct from the high affinity G protein-coupled conformation, termed R^G (Figure 10.2). In support of this hypothesis, Dean *et al.* showed that a modified *N*-terminal PTH fragment analog, M-PTH(1-15) (where M = Ala/Aib1, Aib3, Gln10, Har11, Ala12, Trp14, Arg19), which binds weakly in the presence of GTPγS, does not bind to the PTHR in membranes extracted from cells genetically lacking the Gα_S subunit, whereas PTH(1-34) binds to these membranes as well as it does to membranes prepared from wild-type cells.[6,49]

In complementary FRET experiments,[8] Ferrandon *et al.* confirmed that PTH and PTHrP stabilize distinct PTHR conformations, and further showed that the bimolecular ligand/receptor (LR) complex induced by PTH(1-34) is highly stable whereas that induced by PTHrP(1-36) or M-PTH(1-14) is reversible. In concordance with the radioligand binding studies performed in membranes, FRET data showed that a dominant-negative form of Gα_S (DN-G_S), which is inactive and binds the receptor in an irreversible fashion, has no effect on PTH(1-34) release, but blocks instead the dissociation of PTHrP(1-36) and M-PTH(1-14) (Figure 10.3).

These results imply that the slow dissociation component seen with these two ligands depends on the release of G proteins from the receptor, a process inhibited by DN-G_S. Taken together, these studies suggest that PTHR forms either an unusually high-affinity complex with PTH(1-34) that is not dependent on classical G protein coupling, but which can nevertheless be associated with G proteins or a more conventional high-affinity complex with other ligands such PTHrP(1-36) and M-PTH(1-14), that are transient and depend on coupling to G proteins.

10.3.2 Sustained PTHR Signalling

PTH(1-34) differentiates itself from PTHrP(1-36) by inducing prolonged cAMP responses in cultured cells.[6,8] Additional PTH variants that bound to the R^0 PTHR conformation with even higher affinity than does PTH(1-34) have been recently identified.[7] This enhanced affinity and selectivity for the R^0 conformation is accompanied by prolonged cAMP responses in cultured cells expressing the PTHR and also mediating protracted signalling responses in bone and/or kidney PTH target cells as demonstrated by prolonged hypercalcemic (elevated blood Ca^{2+} concentrations) and hypophosphatemic (decrease in circulating phosphate) responses in mice.[7] Importantly, these *in vivo* effects are mediated at the receptor level and not by extended bioavailability of ligands.[7] The selectivity and affinity with which a given PTH ligand binds to the R^0 receptor conformation seems to determine the duration of the signal and thus contributes to the biological properties of that ligand. The R^0 PTHR conformation is viewed as an intermediary state between the classically defined inactive receptor state, R, and the active state, R*, such that a ligand that binds

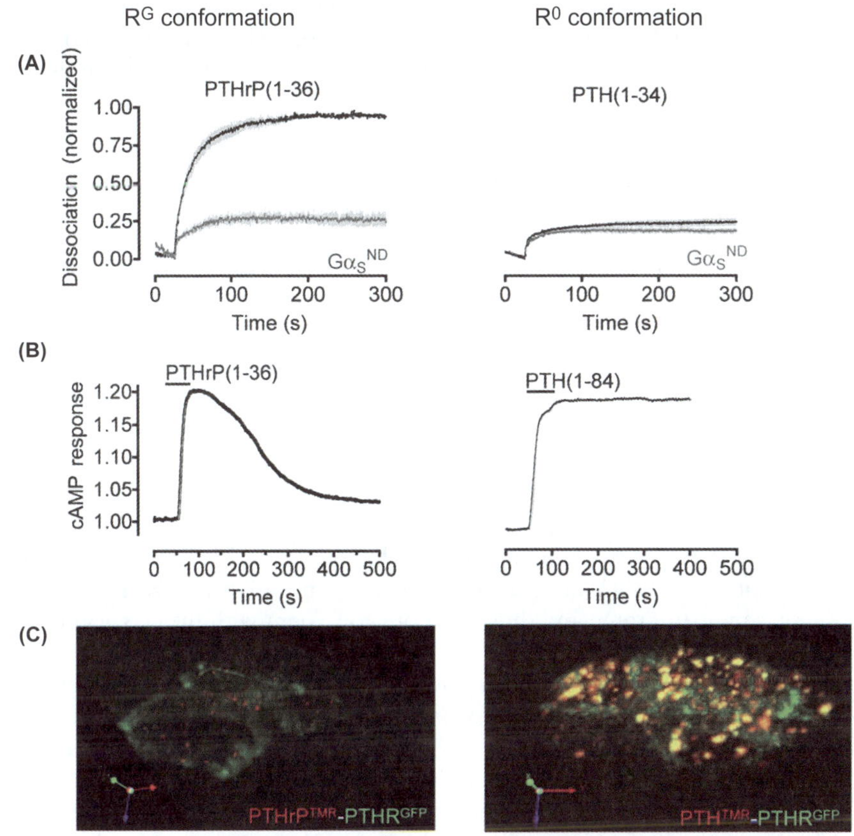

Figure 10.3 PTHR conformations: life cells FRET assays. (**A**) Averaged dissociation time courses of TMR-labelled ligands, PTH(1–34)^{TMR} (right panel) and PTHrP(1–36)^{TMR} (left panel), from GFP-tagged PTHR, ^{GFP}N-PTHR, are shown in the absence or presence of a dominant negative Gαs (GαsND). FRET recordings from HEK-293 cells are shown as normalized ratios. (**B**) Average time-courses of cAMP production in response to PTHrP(1–36) (left) and PTH(1-84) (right) in HEK-293 cells stably expressing PTHR and co-transfected with the cAMP biosensor, Epac$^{CFP/YFP}$. Individual cells were continuously perfused with buffer or with the hormone for the time indicated by the horizontal bar. (**C**) A 3D view of TMR-labelled hormones, and ^{GFP}N-PTHR in live HEK-293 cells by spinning disc confocal microscopy 30 min after ligand wash out. PTH(1–34)^{TMR} (red) and ^{GFP}N-PTHR (green) colocalized within endocytic compartments (right). In contrast, PTHrP(1–36)^{TMR} alone is detected as small puntae at internalized sites (left). Adapted from ref. 8.

stably to R^0 will, *via* its R$^0 \rightarrow$ R^{0*} isomerization, have the capacity to induce prolonged signalling responses in cells, whereas a ligand that binds only weakly to R^0 will induce only transient signalling responses.[55,56]

One prediction, previously discussed,[56] is that R^0-selective ligands, due to prolonged cAMP elevation in bone cells expressing PTHR, may favour net bone-resorption responses with sustained calcium release, and thus be candidate therapies for hypoparathyroidism,[57,58] whereas R^G-selective ligands, due to transient action at the receptor, may favour bone anabolic responses and thus be candidate therapies for osteoporosis.

10.3.3 Endosomal G Protein Signalling: Emerging Paradigm for PTHR

By monitoring the effects of brief stimuli of PTH(1-34) or PTHrP(1-36) on cAMP responses and PTHR internalization in cultured HEK-293 cells, Ferrandon *et al.* obtained a series of new findings that may explain the distinct modes of action of PTH and PTHrP.[8] First, the complex PTH(1-34)/PTHR, which can be tracked using tetramethylrhodamine-labelled PTH and GFP-tagged PTHR, internalizes rapidly into Rab5-postive endosomes in association with $G\alpha_S$ and adenylyl cyclases (Figure 10.3). Second, the internalization of PTH(1-34)/PTHR is not associated with desensitization of the cAMP response. Thirdly, blocking PTHR internalization reduces markedly the time course of the cAMP response. In contrast, PTHrP(1-36) or M-PTH(1-14), which promote short-lived signalling responses, are completely reversible and limited to the plasma membrane, and thus are more consistent with the classical model for GPCR desensitization.

These data provide support for the hypothesis that PTHR can generate cAMP not only from the plasma membrane but also from intracellular membranes, and that early endosomes serve as a platform for PTHR-mediated sustained cAMP production, as can happen with arrestin-dependent ERK signalling pathway[59] and the epidermal growth factor receptor signalling cascade[60] (Figure 10.4). This hypothesis, which represents a major paradigm shift in our understanding of mechanisms of GPCR signalling, is supported by recent studies on the TSH receptor[28] and the sphingosine-1-phosphate type 1 receptor.[61]

10.4 Perspectives and Conclusions

The mechanism by which PTH mediate prolonged signalling from endosomes needs to be further elucidated. One possibility is that a stable PTH/receptor complex, formed at the plasma membrane, internalizes and persists at endosomal membranes to mediate sustained cAMP production either through multiple activation/deactivation cycles of G_S proteins or through sustained activation of a single G_S protein.

The well-known capacity of PTH to recruit β-arrestins to PTHR[62–65] and to stabilize a persistent ternary PTH–PTHR–arrestin complex raises the intriguing question as to how β-arrestin and G_S can bind PTHR at the same time. A second possibility is that complexes formed through interactions between β-arrestins and $G\beta\gamma$ subunits known to provide a mechanism for scaffolding signalling complexes[66] regulate the sustained cAMP signalling mediated by

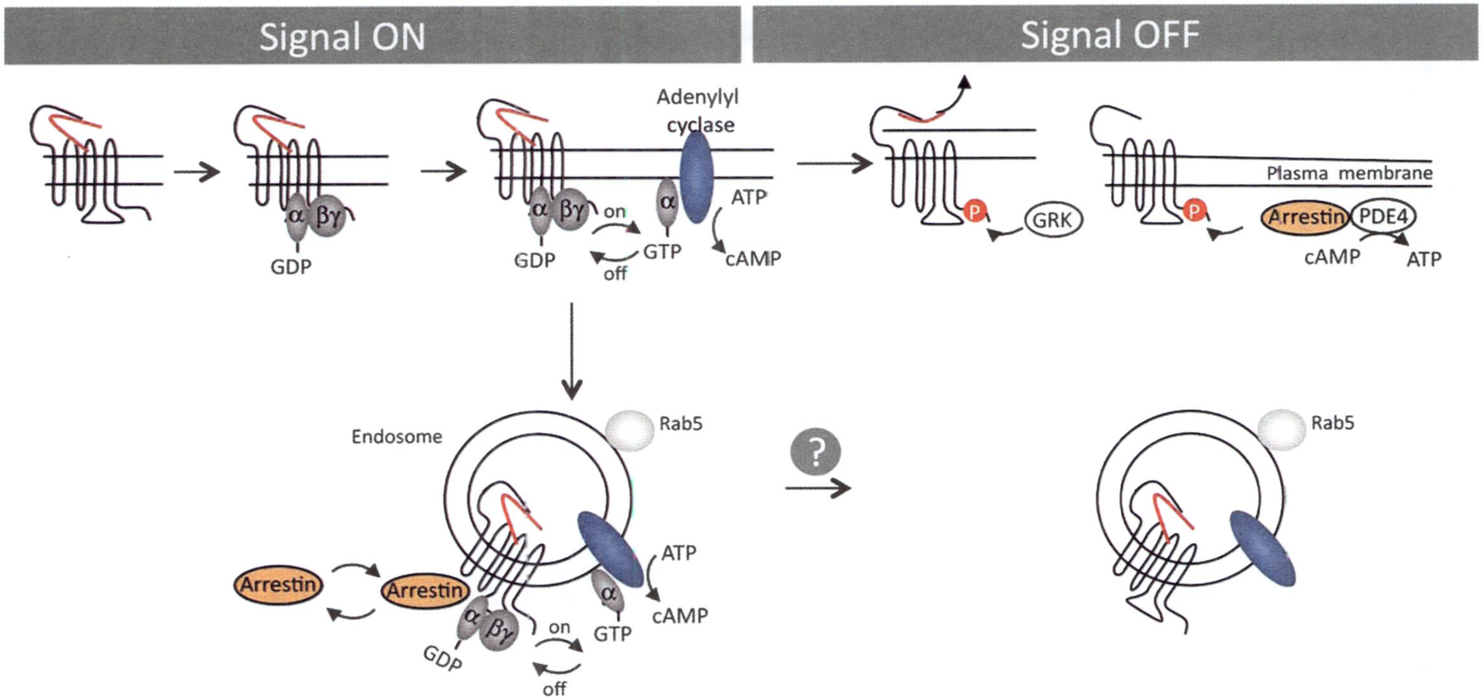

Figure 10.4 Protein-exchange dynamic at PTHR microdomains as a putative model for sustained signalling. Cyclic AMP production mediated by ligand-activated G protein-coupled receptors (GPCRs) is thought to originate at the plasma membrane and is rapidly extinguished within minutes by mechanisms involving receptor endocytosis, with the bound ligands being either released at the cell surface or internalized separately from receptors and G proteins, and by PKA activation of phosphodiesterase PDE4, which can be recruited by arrestins at the plasma membrane to degrade rapidly cAMP. This model cannot explain recent observation that certain GPCRs such as the PTHR continue to stimulate cAMP production even after receptor internalization. The molecular origin of sustained cAMP mediated by internalized PTHR, for example, may depend on the capacity of PTH to form a high-order signalling complex in endosomes that contains PTH, PTHR, G_S, and arrestins, and where PTHR can assemble/disassemble alternatively with either arrestin or Gs through dynamic interactions to regulate cAMP levels in magnitude and duration.

PTH. Another possibility, which is illustrated in Figure 10.4, is that PTH triggers the formation of a high-order signalling complex that contains PTH, PTHR, G_S and arrestins, and where PTHR can assemble/disassemble alternatively with either arrestin or Gs through dynamic interactions to regulate cAMP levels in magnitude and duration.

Acknowledgements

This work was supported by National Institutes of Health (NIH) grant award DK087688.

References

1. H. Juppner, A. B. Abou-Samra, M. Freeman, X. F. Kong, E. Schipani, J. Richards, L. F. Kolakowski, Jr., J. Hock, J. T. Potts, Jr. and H. M. Kronenberg *et al.*, *Science*, 1991, **254**, 1024–1026.
2. D. Goltzman, *Trends Endocrinol. Metab.*, **21**, 512–518.
3. J. J. Wysolmerski, in *Primer on the Metabolic Bone Diseases and Disorders of Mineral Metabolism*, ed. C. J. Rosen, ASBMR, Washington, DC, 2008, pp. 127–133.
4. M. J. Horwitz, M. B. Tedesco, S. M. Sereika, B. W. Hollis, A. Garcia-Ocana and A. F. Stewart, *J. Clin. Endocrinol. Metab.*, 2003, **88**, 1603–1609.
5. M. J. Horwitz, M. B. Tedesco, S. M. Sereika, M. A. Syed, A. Garcia-Ocana, A. Bisello, B. W. Hollis, C. J. Rosen, J. J. Wysolmerski, P. Dann, C. Gundberg and A. F. Stewart, *J. Bone Miner. Res.*, 2005, **20**, 1792–1803.
6. T. Dean, J. P. Vilardaga, J. T. Potts, Jr. and T. J. Gardella, *Mol. Endocrinol.*, 2008, **22**, 156–166.
7. M. Okazaki, S. Ferrandon, J. P. Vilardaga, M. L. Bouxsein, J. T. Potts, Jr. and T. J. Gardella, *Proc. Natl. Acad. Sci. U. S. A.*, 2008, **105**, 16525–16530.
8. S. Ferrandon, T. N. Feinstein, M. Castro, B. Wang, R. Bouley, J. T. Potts, T. J. Gardella and J. P. Vilardaga, *Nat. Chem. Biol.*, 2009, **5**, 734–742.
9. S. P. Sheikh, T. A. Zvyaga, O. Lichtarge, T. P. Sakmar and H. R. Bourne, *Nature*, 1996, **383**, 347–350.
10. D. L. Farrens, C. Altenbach, K. Yang, W. L. Hubbell and H. G. Khorana, *Science*, 1996, **274**, 768–770.
11. U. Gether, *Endocr. Rev.*, 2000, **21**, 90–113.
12. J. Wess, S. J. Han, S. K. Kim, K. A. Jacobson and J. H. Li, *Trends Pharmacol. Sci.*, 2008, **29**, 616–625.
13. H. R. Bourne, *Curr. Opin. Cell Biol.*, 1997, **9**, 134–142.
14. J. Hu, Y. Wang, X. Zhang, J. R. Lloyd, J. H. Li, J. Karpiak, S. Costanzi and J. Wess, *Nat. Chem. Biol.*, **6**, 541–548.
15. J. P. Vilardaga, M. Bunemann, T. N. Feinstein, N. Lambert, V. O. Nikolaev, S. Engelhardt, M. J. Lohse and C. Hoffmann, *Mol. Endocrinol.*, 2009, **23**, 590–599.

16. M. J. Lohse, V. O. Nikolaev, P. Hein, C. Hoffmann, J. P. Vilardaga and M. Bunemann, *Trends Pharmacol. Sci.*, 2008, **29**, 159–165.
17. F. Ciruela, J. P. Vilardaga and V. Fernandez-Duenas, *Trends Biotechnol.*, **28**, 407–415.
18. J. P. Vilardaga *J. Recept. Signal Transduction*, 2010, **30**, 304–312.
19. J. P. Vilardaga, M. Bunemann, C. Krasel, M. Castro and M. J. Lohse, *Nat Biotechnol.*, 2003, **21**, 807–812.
20. C. Hoffmann, G. Gaietta, M. Bunemann, S. R. Adams, S. Oberdorff-Maass, B. Behr, J. P. Vilardaga, R. Y. Tsien, M. H. Ellisman and M. J. Lohse, *Nat. Methods*, 2005, **2**, 171–176.
21. J. P. Vilardaga, R. Steinmeyer, G. S. Harms and M. J. Lohse, *Nat. Chem. Biol.*, 2005, **1**, 25–28.
22. J. P. Vilardaga, V. O. Nikolaev, K. Lorenz, S. Ferrandon, Z. Zhuang and M. J. Lohse, *Nat. Chem. Biol.*, 2008, **4**, 126–131.
23. V. O. Nikolaev, C. Hoffmann, M. Bunemann, M. J. Lohse and J. P. Vilardaga, *J. Biol. Chem.*, 2006, **281**, 24506–24511.
24. M. Maier-Peuschel, N. Frolich, C. Dees, L. G. Hommers, C. Hoffmann, V. O. Nikolaev and M. J. Lohse, *J. Biol. Chem.*, **285**, 8793–8800.
25. D. Wheeler, J. L. Garrido, A. Bisello, Y. K. Kim, P. A. Friedman and G. Romero, *Mol. Endocrinol.*, 2008, **22**, 1163–1170.
26. M. Bunemann, M. Frank and M. J. Lohse, *Proc. Natl. Acad. Sci. U. S. A.*, 2003, **100**, 16077–16082.
27. P. Hein, M. Frank, C. Hoffmann, M. J. Lohse and M. Bunemann, *EMBO J.*, 2005, **24**, 4106–4114.
28. D. Calebiro, V. O. Nikolaev, M. C. Gagliani, T. de Filippis, C. Dees, C. Tacchetti, L. Persani and M. J. Lohse, *PLoS Biol.*, 2009, **7**, e1000172.
29. V. O. Nikolaev, A. Moshkov, A. R. Lyon, M. Miragoli, P. Novak, H. Paur, M. J. Lohse, Y. E. Korchev, S. E. Harding and J. Gorelik, *Science*, **327**, 1653–1657.
30. G. W. Tregear, J. Van Rietschoten, E. Greene, H. T. Keutmann, H. D. Niall, B. Reit, J. A. Parsons and J. T. Potts, Jr., *Endocrinology*, 1973, **93**, 1349–1353.
31. T. J. Gardella, A. K. Wilson, H. T. Keutmann, R. Oberstein, J. T. Potts, Jr., M. Kronenberg and S. R. Nussbaum, *Endocrinology*, 1993, **132**, 2024–2030.
32. H. Juppner, E. Schipani, F. R. Bringhurst, I. McClure, H. T. Keutmann, J. T. Potts, Jr., H. M. Kronenberg, A. B. Abou-Samra, G. V. Segre and T. J. Gardella, *Endocrinology*, 1994, **134**, 879–884.
33. C. Lee, T. J. Gardella, A. B. Abou-Samra, S. R. Nussbaum, G. V. Segre, J. T. Potts, Jr., H. M. Kronenberg and H. Juppner, *Endocrinology*, 1994, **135**, 1488–1495.
34. T. J. Gardella, M. D. Luck, A. K. Wilson, H. T. Keutmann, S. R. Nussbaum, J. T. Potts, Jr. and H. M. Kronenberg, *J. Biol. Chem.*, 1995, **270**, 6584–6588.
35. M. Castro, V. O. Nikolaev, D. Palm, M. J. Lohse and J. P. Vilardaga, *Proc. Natl. Acad. Sci. U. S. A.*, 2005, **102**, 16084–16089.
36. G. A. Weiland and P. B. Molinoff, *Life Sci.*, 1981, **29**, 313–330.

37. P. B. Molinoff, B. B. Wolfe and G. A. Weiland, *Life Sci.*, 1981, **29**, 427–443.
38. F. Rochais, J. P. Vilardaga, V. O. Nikolaev, M. Bunemann, M. J. Lohse and S. Engelhardt, *J. Clin. Invest.*, 2007, **117**, 229–235.
39. P. Hein, F. Rochais, C. Hoffmann, S. Dorsch, V. O. Nikolaev, S. Engelhardt, C. H. Berlot, M. J. Lohse and M. Bunemann, *J. Biol. Chem.*, 2006, **281**, 33345–33351.
40. J. B. Jensen, J. S. Lyssand, C. Hague and B. Hille, *J. Gen. Physiol.*, 2009, **133**, 347–359.
41. B. H. Falkenburger, J. B. Jensen and B. Hille, *J. Gen. Physiol.*, **135**, 81–97.
42. L. M. Ittner, D. Koller, R. Muff, J. A. Fischer and W. Born, *Biochemistry*, 2005, **44**, 5749–5754.
43. P. M. Sexton, M. Morfis, N. Tilakaratne, D. L. Hay, M. Udawela, G. Christopoulos and A. Christopoulos, *Ann. N. Y. Acad. Sci.*, 2006, **1070**, 90–104.
44. M. Wan, C. Yang, J. Li, X. Wu, H. Yuan, H. Ma, X. He, S. Nie, C. Chang and X. Cao, *Genes Dev.*, 2008, **22**, 2968–2979.
45. W. B. Sneddon, C. A. Syme, A. Bisello, C. E. Magyar, M. D. Rochdi, J. L. Parent, E. J. Weinman, A. B. Abou-Samra and P. A. Friedman, *J. Biol. Chem.*, 2003, **278**, 43787–43796.
46. G. Romero, W. B. Sneddon, Y. Yang, D. Wheeler, H. C. Blair and P. A. Friedman, *J. Biol. Chem.*, **285**, 14756–14763.
47. N. A. Lambert, *Sci. Signal.*, 2008, **1**, re5.
48. G. X. Xie and P. P. Palmer, *J. Mol. Biol.*, 2007, **366**, 349–365.
49. T. Dean, A. Linglart, M. J. Mahon, M. Bastepe, H. Juppner, J. T. Potts, Jr. and T. J. Gardella, *Mol. Endocrinol.*, 2006, **20**, 931–943.
50. P. Samama, S. Cotecchia, T. Costa and R. J. Lefkowitz, *J. Biol. Chem.*, 1993, **268**, 4625–4636.
51. J. P. Vilardaga, E. Ciccarelli, C. Dubeaux, P. De Neef, A. Bollen and P. Robberecht, *Mol. Pharmacol.*, 1994, **45**, 1022–1028.
52. S. R. Hoare, S. K. Sullivan, N. Ling, P. D. Crowe and D. E. Grigoriadis, *Mol. Pharmacol.*, 2003, **63**, 751–765.
53. S. R. Hoare, S. K. Sullivan, A. Pahuja, N. Ling, P. D. Crowe and D. E. Grigoriadis, *Peptides*, 2003, **24**, 1881–1897.
54. S. R. Hoare, S. K. Sullivan, D. A. Schwarz, N. Ling, W. W. Vale, P. D. Crowe and D. E. Grigoriadis, *Biochemistry*, 2004, **43**, 3996–4011.
55. J. T. Potts and T. J. Gardella, *Ann. N. Y. Acad. Sci.*, 2007, **1117**, 196–208.
56. J. P. Vilardaga, G. Romero, P. A. Friedman and T. J. Gardella, *Cell. Mol. Life Sci.*, 2011, **68**, 1–13.
57. K. K. Winer, C. W. Ko, J. C. Reynolds, K. Dowdy, M. Keil, D. Peterson, L. H. Gerber, C. McGarvey and G. B. Cutler, Jr., *J. Clin. Endocrinol. Metab.*, 2003, **88**, 4214–4220.
58. S. D. Mittelman, G. N. Hendy, R. A. Fefferman, L. Canaff, I. Mosesova, D. E. Cole, L. Burkett and M. E. Geffner, *J. Clin. Endocrinol. Metab.*, 2006, **91**, 2474–2479.

59. A. C. Hanyaloglu and M. von Zastrow, *Annu. Rev. Pharmacol. Toxicol.*, 2008, **48**, 537–568.
60. M. Miaczynska, L. Pelkmans and M. Zerial, *Curr. Opin. Cell Biol.*, 2004, **16**, 400–406.
61. F. Mullershausen, F. Zecri, C. Cetin, A. Billich, D. Guerini and K. Seuwen, *Nat. Chem. Biol.*, 2009, **5**, 428–434.
62. J. P. Vilardaga, M. Frank, C. Krasel, C. Dees, R. A. Nissenson and M. J. Lohse, *J. Biol. Chem.*, 2001, **276**, 33435–33443.
63. J. P. Vilardaga, C. Krasel, S. Chauvin, T. Bambino, M. J. Lohse and R. A. Nissenson, *J. Biol. Chem.*, 2002, **277**, 8121–8129.
64. M. Castro, F. Dicker, J. P. Vilardaga, C. Krasel, M. Bernhardt and M. J. Lohse, *Endocrinology*, 2002, **143**, 3854–3865.
65. S. L. Ferrari, V. Behar, M. Chorev, M. Rosenblatt and A. Bisello, *J. Biol. Chem.*, 1999, **274**, 29968–29975.
66. M. Yang, R. L. He, J. L. Benovic and R. D. Ye, *Biochem. J.*, 2009, **417**, 287–296.

CHAPTER 11

Metabotropic Glutamate Receptors: A Paradigm of Structural and Functional Receptor Complexity

JEAN-PHILIPPE PIN,* JULIE KNIAZEFF, CYRIL GOUDET, THIERRY DURROUX, PHILIPPE RONDARD AND LAURENT PRÉZEAU

Centre National de la Recherche Scientifique (CNRS, UMR5203), Institut National de la Santé et de la Recherche Médicale (INSERM, U661), Universités de Montpellier I and II, Institut de Génomique Fonctionnelle, Montpellier, France

11.1 Introduction

The L-amino acid glutamate is the neurotransmitter involved in most fast excitatory synapses in the central nervous system. To rapidly depolarize the postsynaptic neurons, it activates various types of ionotropic receptors known as the AMPA, NMDA and kainate receptors. In the mid-1980s, glutamate was also discovered to activate G protein-coupled receptors (GPCRs) now named the metabotropic glutamate (mGlu) receptors. First identified as receptors activating phospholipase C (PLC) through the Gq subtype of G proteins,[1,2] other mGlu receptors were expected, as revealed by specific action of the glutamate analog L-AP4,[3,4] and the observation that glutamate can inhibit adenylyl cyclase.[5]

RSC Drug Discovery Series No. 8
G Protein-Coupled Receptors: From Structure to Function
Edited by Jesús Giraldo and Jean-Philippe Pin

Published by the Royal Society of Chemistry, www.rsc.org

The multiplicity of mGlu receptor subtypes was confirmed by the identification of eight genes encoding such receptors in the mid-1990s.[6–13] These receptors are sub-classified into three groups based on their sequence similarity, their pharmacological properties and coupling specificity.[14–16] The mGlu1 and mGlu5 receptors composed group I receptors. They both activate the PLC pathway through Gq and can be selectively activated by dihydroxyphenylglycine (DHPG). These two receptors are mostly distributed postsynaptically, from where they regulate the activity of the ionotropic receptors. Group II is composed of mGlu2 and mGlu3, which both inhibit adenylyl cyclase through Gi and are selectively activated by agonist LY354740. These receptors are often located on pre-synaptic elements where they control the release of various neurotransmitters, but are also found on some postsynaptic elements. The mGlu3 subtype, together with mGlu5, is also found in astrocytes. The four other mGlu subtypes (mGlu4, 6, 7 and 8) make up group III. Group III mGluRs couple to Gi/o proteins and regulate neurotransmitter release at pre-synaptic sites, except mGlu6 which is specifically located in ON bipolar cells in the retina and responsible for the 'ON' response.[17]

Not surprisingly, these eight mGlu receptors represent promising targets for the development of new therapeutic drugs for the treatment of various diseases, mostly neurologic and psychiatric diseases.[14,18,19] These include anxiety, depression, schizophrenia and epilepsy, as well as the neurodegenerative diseases such as Parkinson's and Alzheimer's. Although no molecules acting on mGlu receptors have entered the market yet, several clinical studies have validated their use for the treatment of anxiety, schizophrenia, Parkinson's disease, migraine and gastro-oesophageal reflux disease. There is therefore much interest in better elucidating the structure and activation mechanism of these receptors to open new avenues for the discovery of molecules regulating their activity.

Not only mGlu receptors are considerable interest in drug development, but they also represent a fascinating model to decipher the functioning of many other GPCRs—those sharing structural similarity like the other class C GPCRs such as the $GABA_B$, the Calcium-sensing and the sweet and umami taste receptors,[20] but also the class A, rhodopsin-like GPCRs. Furthermore, unlike class A receptors, mGlu receptors are covalently linked dimers[21] and therefore represent an excellent model system to understand the physiological meaning of GPCR dimerization, which despite a number of years of extensive studies, remains elusive.[22–28]

This review concentrates on the general structure of mGlu receptors and the current hypothesis regarding their activation mechanism. We show that mGlu receptors are complex proteins made up of several domains that confer to the receptor a number of possibilities to modulate their activity. We also see that some key aspects of their functioning as dimers help to propose a functional role for class A GPCR dimerization.

11.2 Structural Organization of mGlu Receptors

Like most other class C GPCRs, mGlu receptors are multidomain proteins composed of a ligand-binding domain linked through a cysteine-rich domain

(CRD) to a typical heptahelical domain common to all GPCRs, followed by a *C*-terminal intracellular tail variable in length (Figure 11.1).

We first review our current knowledge of the structure of each of these domains before considering how the functioning of each of these is coordinated to allow G protein activation.

11.2.1 The Agonist Binding Domain: A Venus Flytrap Domain

The cDNA sequences of the first mGlu receptor revealed a large extracellular domain composed of about 580 residues,[6,12] much larger than most extracellular

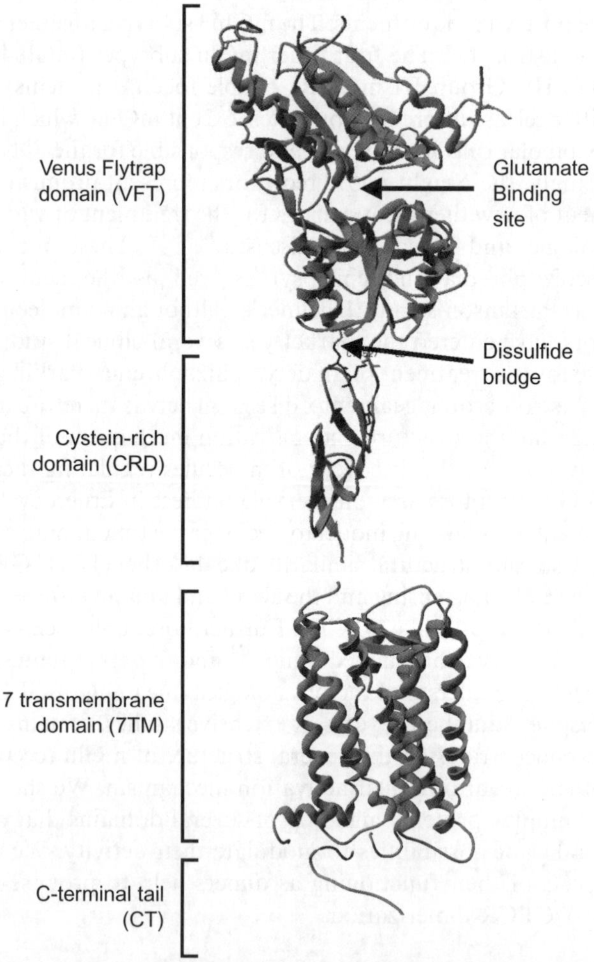

Figure 11.1 Structural organization of a mGlu receptor as illustrated with the Venus flytrap domain (VFT) associated with the cysteine-rich domain (CRD) from mGlu3 (pdb 2E4U), and a 3-D model of a mGlu seven-transmembrane (7TM) generated from the rhodopsin structure using the class C GPCR 7TM sequence alignment and the method previously reported.[20,46]

domains of GPCRs cloned at that time. Pharmacological properties of chimeric receptors soon revealed that the first two-thirds of the extracellular domain composed the ligand recognition domain.[29] This was supported by bioinformatics studies revealing structural similarities of the extracellular domain of mGlu receptors with a bacterial periplasmic amino acid binding proteins involved in the transport of amino acids. Of note is the leucine–isoleucine–valine binding protein (LIVBP),[30] known to contain a Venus flytrap domain (VFT) (Figure 11.1). Indeed, not only does the succession of predicted secondary elements perfectly match those of LIVBP and other similar proteins, but five residues important for the binding of the amino acid moiety of the ligand are conserved in mGlu receptors. This prediction was eventually confirmed by the resolution of the crystal structure of the VFT of mGluR1 with and without glutamate,[31] followed by those of mGlu3, mGlu7[32,33] and mGlu5.[34]

The domain is composed of two globular lobes, each made of a central beta sheet surrounded with alpha helices (Figure 11.1). The ligand binding site is located within the cleft that separates both lobes and engages the five residues predicted to bind the alpha amino acid part of glutamate, while the gamma carboxylic group interacts with three conserved basic residues (together with water molecules) (Figure 11.2). Analysis of the binding pocket within the eight mGlu receptor subtypes revealed a high degree of conservation of most residues, especially between mGlu receptors from the same group, making the development of highly subtype-selective orthosteric ligands very difficult.[35,36] It is not surprising that, although a number of compounds have been identified that are either selective agonist or antagonists for a specific group of mGlu receptors, only very few show a preferential affinity for a single mGlu subtype.[37] Only ligands that extend out of the strict glutamate binding pocket were found to display a clear mGlu receptor subtype selectivity.[38,39]

11.2.2 The Cysteine-Rich Domain

For a long time, the structure of the cysteine-rich domain (CRD) was unknown. It was first reported that this domain shares some sequence similarity with a cysteine-rich of tumour necrosis factor receptor,[40] and this was further refined and found consistent with mutagenesis data.[41] Only in 2007 was the structure of the first CRD solved, together with the VFT domain of mGlu3.[32] As illustrated in Figure 11.1, this domain is composed of seven beta strands, organized into two modules and stabilized by four disulfide bonds, making this domain likely to be highly rigid. Of note, one of the nine cysteines (Cys527 in mGlu3) is involved in a disulfide bridge with another highly conserved Cys residue of the VFT, thus making the connection between the VFT and the CRD more rigid (Figure 11.1). However, the contact area between the CRD and the VFT is still very limited, positioning the CRD in a very extended fashion compared with the VFT. Such interconnection between these two domains is indeed limited to the polypeptide chain and the disulfide bridge, offering the possibility that the CRD can still

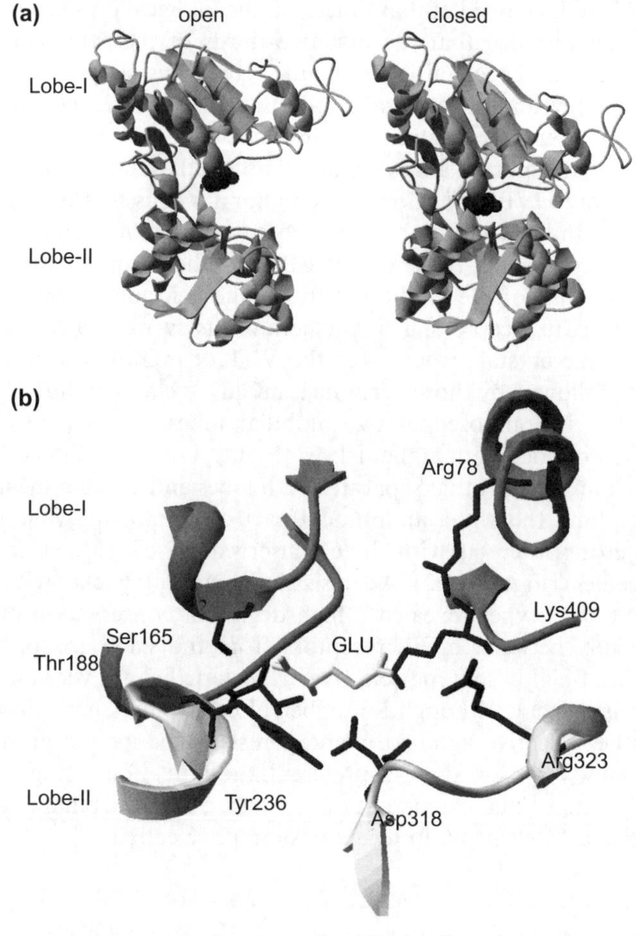

Figure 11.2 Glutamate binds in the VFT cleft while in the open form (**a**, left), and stabilizes a closed conformation (**a**, right) (ribbon representations generated using Swiss-PdbViewer software, and the B (left) and A (right) protomer of the Protein Data Bank (PDB) file 1EWK). Lobe I is in dark grey and lobe II is in light grey. (**b**) A detailed view of the glutamate binding site of the closed form of the mGlu1 VFT (generated using Swiss-PdbViewer and the A protomer of the PDB file 1EWK).[31]

move compared to the VFT. As discussed later, this could possibly be involved in the activation process.

11.2.3 The 7TM Domain

After cloning of the first mGlu receptors, it was evident that there was very low sequence similarity with the other class A GPCRs in the seven-transmembrane (7TM) region, suggesting that these receptors may not necessarily share overall structural similarity. This idea was supported by the observation that, whereas the third intracellular loop of the class A receptors plays a critical role in G protein coupling selectivity, mGluRs utilize the second loop.[42,43] Most importantly, it was even reported that replacement of the third intracellular loop of rhodopsin by the second intracellular loop of the mGlu6 receptor could lead to a functional receptor.[44] However, such apparent differential roles of the i2 and i3 loops could well be explained by the fact that both loops contact the G protein alpha-subunits such that any of these can play a critical role in G protein coupling selectivity.

In contrast, recent studies are more in favour of the 7TM domain of mGlu and other class C GPCRs being structurally similar to rhodopsin-like receptors.[20] First, like in rhodopsin, the third transmembrane (TM) of mGlu receptors is not highly hydrophobic and is one of the longest TM, making it likely to be central in the 7TM core domain. Secondly, class C GPCR 7TM domains also possess a putative eighth amphipatic helix after TM7 interconnected to TM7 *via* a NPxxY-like motif.[42] Thirdly, the disulfide bridge interconnecting the top of TM3 with that of TM5, that is highly conserved in class A GPCRs, is also conserved in class C receptors. Fourthly, the conserved Trp residue of TM6, which plays a critical role in GPCR activation, is conserved in mGlu receptors. Fifthly, whereas the DRY motif of the class A GPCRs is not found in class C receptors, an ionic interaction between TM3 and TM6, known to play a critical role in stabilizing the inactive state of the receptors,[45] is also found in class C GPCRs, where it appears to have a similar role.[46] Sixthly, synthetic compounds have been identified that bind in the 7TM domains of mGlu receptors, and act as positive and negative allosteric modulators. When characterizing their binding sites, these were found to perfectly match three-dimensional (3-D) models of the mGlu 7TM core based on the structure of rhodopsin.[47–50] Finally, the 7TM domain of mGlu receptors can be expressed alone, can reach the cell surface and can couple to G proteins when activated by positive allosteric modulators (PAMs).[51]

11.2.4 The *C*-terminal Tail

The *C*-terminal tails of mGlu receptors are quite variable, not only at the sequence level between receptors of the same group, but also in length. Indeed, this part of the mGlu receptor can vary from a few residues to up to 350 residues. The *C*-termini of mGlu receptors are also the subject of alternative splicing, including mGlu1, mGlu5, mGlu4 and mGlu7.[15,16] Very likely unstructured, such intracellular tails contain a number of sites that allow for the interaction with various intracellular proteins such as SH3 or PDZ domain containing proteins. It is now well accepted that these interacting proteins control the targeting as well as the function of these receptors. As these interacting proteins have not yet been shown to modify either the pharmacological properties of the receptors or the general activation process, and because comprehensive reviews focusing on the function of these proteins have recently been published[52,53] (see chapter 13), this aspect of mGlu receptors is not discussed further here.

11.3 mGlu Receptors Are Constitutive Dimers

11.3.1 Dimeric Organization of mGlu Receptors

In 1995, Romano and colleagues[21] were the first to demonstrate that the mGlu5 receptor was a dimer, not only in transfected cells, but most importantly in the brain.[21] In the absence of a high concentration of reducing agents, the mGlu5 receptor was only detected as a dimer in acrylamide gels; this was supported by co-immunoprecipitation experiments using epitope tagged receptors. It was then further demonstrated that the mGlu subunits are indeed covalently linked through a disulfide bond, and the dimeric organization of mGlu receptors was firmly confirmed by the solved crystal structures of a number of mGlu VFTs.[31,32,54] These structures revealed a conserved and large hydrophobic area responsible for the association of two VFTs.

GPCRs have been proposed to form even larger complexes than just dimers, as based on AFM images and by a number of crosslinking and fluorescence resonance energy transfer (FRET) studies.[55–59] When the oligomer stoichiometry was examined for mGlu receptors, only strict dimers were observed.[60–63] Such a conclusion was reached thanks to the use of the quality control system of the $GABA_B$ receptor which enabled control of the subunit composition of a cell surface class C receptor dimer and then allowed introduction of a single tagged subunit. Under such conditions, no FRET could be detected between dimers. The formation of strict dimers at the cell surface was further supported by quantitative FRET measurements.[63] This is in contrast with the $GABA_B$ receptor, which spontaneously forms tetramers in transfected cells and probably also in neurons.[62,64,65]

11.3.2 Can mGlu Subunits Form Heterodimeric Receptors?

Early on mGlu receptors were considered to assemble exclusively into homodimers. This was because no co-immunoprecipitation was observed in cells expressing both mGlu1 and mGlu5 receptors,[21] and also because each mGlu receptor subtype has a rather specific distribution in the brain.[66] However, this is in contrast with other class C GPCRs such as the $GABA_B$ and the sweet and umami taste receptors, where heteromeric entities are clearly formed. By using snap and clip tags to specifically label cell surface mGlu subunits, we recently demonstrated that mGlu receptors can indeed exist as heterodimeric entities, at least in transfected cells.[63] When looking at possible combinations, it was found that group I mGlu subunits can form heterodimers, but not with the other subunits from groups II and III. However, heterodimers composed of any of the group II and III subunits can form. Such complexes correspond to strict heterodimers and not to the assembly of two homodimeric entities. Finally, such heterodimers have the same capacity to form as the homodimers.

It is interesting to note that mGlu subunits that localize postsynaptically can heterodimerize, but not with those mainly targeted to the pre-synaptic element; this is consistent with such heterodimers possibly existing *in vivo*. Indeed, careful examination of the localization of mGlu subunits revealed a number of postsynaptic elements in which both mGlu1 and mGlu5 are found,[67,68] as well as many terminals coexpressing two group III mGlu subtypes.[69]

These observations also raise the possibility that mGlu subunits can assemble with other class C receptor subunits such as the CaS receptor or GPRC6a. In agreement with such possibility, heteromers composed of mGlu1 and CaS subunits have been reported.[70]

Further studies are necessary to identify the specific functional properties of these heterodimeric mGlu receptors and to validate their existence *in vivo*. But if their significance can be confirmed, such receptor entities offer new opportunities to more precisely target synapses where these are specifically located, thus limiting a number of side effects.

Taken together, these data reveal that mGlu receptors are complex multi-domain proteins assembled into dimeric entities. These observations raise the question as to how agonist binding into the VFT cleft can activate the 7TM domains to allow G protein activation.

11.4 Functioning of the Venus Flytrap Dimer

11.4.1 VFT Closure: A Key Step in Agonist-induced Activation of mGlu Receptors

As observed with many VFT domains (including bacterial periplasmic-binding proteins), the mGlu VFT is in an open conformation in the absence of agonist, but in a closed conformation in the presence of an agonist (Figure 11.2).[31–32,54] It was soon proposed that the agonists act by stabilizing a closed form of the VFT. However, an open form of the mGlu1 VFT was also observed with glutamate bound. This may reflect the capacity of this domain to oscillate between a closed and an open form in the presence of agonist, as observed with other VFTs. This mechanism would likely be required to allow the agonist to escape the cleft.

Indeed, a number of studies support this initial step in class C receptor activation. First, all structures solved with a bound antagonist are in the open form,[33,54,71] and either steric or ionic hindrance have been identified that prevent domain closure when an antagonist is bound in the cleft. Secondly, removing the residues from the mGlu8 VFT responsible for such steric or ionic hindrance converted two distinct antagonists, MAP4 and ACPT-II, into full agonists.[72] Thirdly, although not reported for mGlu receptors, locking the closely related $GABA_B1$ VFT in a closed form by introducing a disulfide bridge resulted into a dithiothreitol (DTT) sensitive, constitutively active receptor.[73]

But how can the closure of the VFT lead to 7TM activation?

11.4.2 Dimeric Functioning of MGlu VFTs

The initial mGlu1 VFT structures solved either in an apo or antagonist bound form, or in the presence of glutamate, revealed an important difference in the relative position of the VFTs (Figure 11.3).[31,54] This suggested a possible agonist-induced movement such that two lobe IIs of the VFT domain are distal and do not contact each other in the so-called 'resting' state—as observed in the absence of agonist, or in the presence of antagonist, when both VFTs are open (Roo state, see Figure 11.3). However, the lobes II may contact each other in the 'active' form when bound to agonists. When analysing the residues at the lobe II dimer interface in the 'active' form, it was proposed that due to the relative position of positively and negatively charged residues, the 'active' conformation of the dimer would only be stable if one at least of the VFT is closed (the Aco and Acc states, Figure 11.3).[31,54,74] These observation lead to the proposal that mGlu receptor activation resulted from a relative movement

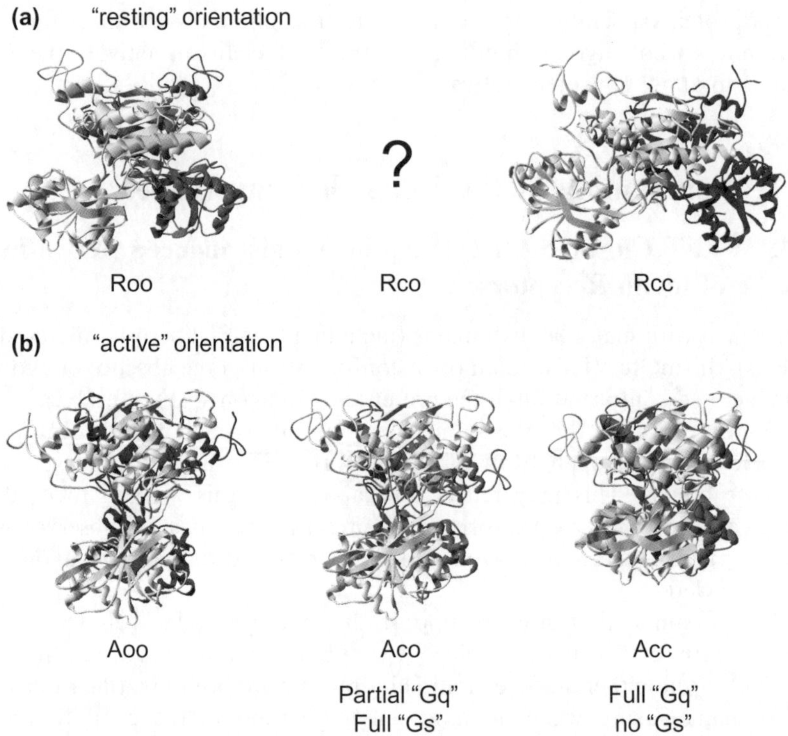

Figure 11.3 Structures of mGlu VFT dimers. (**a**) Structures in the 'resting' orientation, with either both VFTs in the open form (Roo), as observed with the empty or antagonist MCPG bound form of mGlu1 (PDB 1ISS), or with both in the closed form (Rcc) as observed with the agonist bound mGlu3 VFTs (PDB 2E4U). No structures are available in the resting orientation with only one VFT closed (Rco). (**b**) Structures in the 'active' orientation with both VFTs open (Aoo), as observed with mGlu1 VFT bound to antagonist LY341495 (PDB 3KS9), with only one VFT closed (Aco) (glutamate-bound mGlu1, PDB 1EWK), or with both VFT closed (Acc) (glutamate- and Gd^{3+}-bound form of mGlu1, PDB 1ISR). In mGlu1, the Aco form has been proposed to preferentially activates Gs, and partially activate Gq, while the Acc form fully activates Gq.[77]

of one subunit compared to the other, as reported for other membrane receptor dimers like the cytokine receptors or the atrial natriuretic peptide receptors.

However, recent structural evidence with additional mGlu VFT dimers structures casts some doubt on this proposal. Indeed, any of the mGlu3 VFT structures occupied by five different agonists all remain in the so-called 'resting' orientation, even though both VFTs are in the closed state (the Rcc state, Figure 11.3).[32] Furthermore, the structure of an mGlu1 VFT dimer bound to the antagonist LY341795 in each VFT was solved and revealed that both VFTs are in the open form with the relative VFT orientation corresponding to the 'active' state (the Aoo state).

Such observations may well be the consequence of the conditions necessary to obtain the crystals and because only a portion of the receptor is being crystallized. Indeed, a number of studies support the idea that a relative movement of the VFTs is critical for receptor activation. First, a dimeric organization is important for function since mutations that prevent the association of the VFTs suppress function.[75] Secondly, agonist binding in only one VFT activates one 7TM equally as well as the other within the dimer.[61] Thirdly, preventing the relative movement of the $GABA_B$ VFTs using a glycan wedge approach suppresses function.[76] Finally, preliminary FRET data with fluorophores inserted in the mGlu VFTs revealed a movement during receptor activation (Doumazane, Scholler, Zwier, Trinquet, Rondard and Pin, in preparation). Although the exact amplitude of the relative movement cannot be deduced from the available crystal structures, these data are consistent with such a movement being required for the signal transduction from the binding site to the G protein activating site.

11.4.3 Symmetric *versus* Asymmetric Functioning of the VFT Dimer

Since mGlu receptors are dimers (either homo or heterodimers), they all have two agonists binding sites, raising the question of the stoichiometry of binding site occupancy for function. The first agonist-bound structure solved revealed an asymmetric dimer in which only one VFT is in the closed form while the other remained open (Aco state). A second agonist occupied structure was solved in the presence of Gd^{3+}, a cation that binds at the lobe II interface, and revealed a fully symmetrical closed-closed structure (Acc state) (Figure 11.3).

To examine whether both symmetric and asymmetric conformations lead to a functional receptor, Kniazeff *et al.* generated a mGlu1 dimer with two point mutations in the agonist binding site largely decreasing agonist affinity.[60] It was found that a dimer with a single agonist-bound dimer was able to activate G proteins, but with a three-fold lower efficacy. Indeed, only when both binding sites are occupied can the receptor be fully active. In this same study, the authors demonstrate that a dimeric receptor with two distinct VFTs allows specific agonist binding to one VFT with the simultaneous binding of a specific antagonist to the other VFT. Again, this leads to a partially activated receptor, as observed with a single activated VFT, and further documents that this partially active state corresponds to an asymmetric VFT dimer in which one VFT is closed while the other is open (Aco state) (Figure 11.3).

Tateyama and Yubo have proposed that distribution between asymmetric and symmetric states of the VFT dimers is related to the activation of signalling-specific cascades.[77] In the absence of Gd^{3+}, glutamate was found to effectively activate the Gs pathway, but to a lesser extend the PLC pathway, suggesting that the asymmetric dimer (as observed in the crystal structure without Gd^{3+}) couples preferentially to Gs. In contrast, in the presence of Gd^{3+}, which stabilizes the symmetric VFT dimer, a full phospholipase C (PLC) response was observed while the Gs coupling disappeared (Figure 11.3). The concept of VFT structural symmetry is consistent with data obtained with the CaS receptor, for which an antibody isolated from patients with acquired hypocalciuric hypercalcemia which targets the extracellular domain of this receptor, enhances Gq coupling while it inhibits the extracellular signal-regulated kinase (ERK) pathway.[78] This further indicates that different conformations of the VFT dimer leads to the activation of different signalling pathways.

11.5 Cysteine-Rich Domain and Intramolecular Transduction in mGluRs

As mentioned above, the VFT is interconnected to the 7TM domain *via* the CRD, and this connection is crucial for the allosteric coupling between the VFT and the 7TM domains. Indeed, altering the disulfide bond that interconnects the VFT and the CRD through cysteine mutagenesis (Figure 11.1) is sufficient to suppress this allosteric coupling, since agonist-induced activation of the receptor is impaired. In addition, alterations in the disulfide bond impaired the enhanced agonist affinity offered by the positive allosteric modulators in the 7TM domain.[41] This indicates that a tight, at least semi-rigid, association of the VFT with the CRD is required for activation.

According to the X-ray crystal structure of the mGlu3 extracellular domain with both VFT and CRD, a different scenario can be proposed for how the CRD transmits the information from the VFT dimeric state to the 7TM domains (Figures 11.4 and 11.5).[32] Although the authors of this study propose the involvement of oligomerization in the activation process, no physical proximity between mGlu dimers could be detected in transfected cells,[61,62] such that this possibility is not considered here.

One possibility, as suggested by the crystal structure, is that the inactive orientation of the VFT maintains the 7TM domains away from one another (Figure 11.4), while the active conformation orients the two 7TMs in close contact with each other, cooperatively stabilizing their active state (Figures 11.4 and 11.5). However, other biophysical data are inconsistent with this model. When CFP and YFP are inserted in the intracellular loops of the 7TM domains, an efficient FRET is observed under resting condition,[79,80] inconsistent with the rather large distance (>100 Å) between the 7TM domains expected according to the structure of the mGlu3 extracellular domain dimer. Moreover, this model does not provide a good explanation on how the VFT

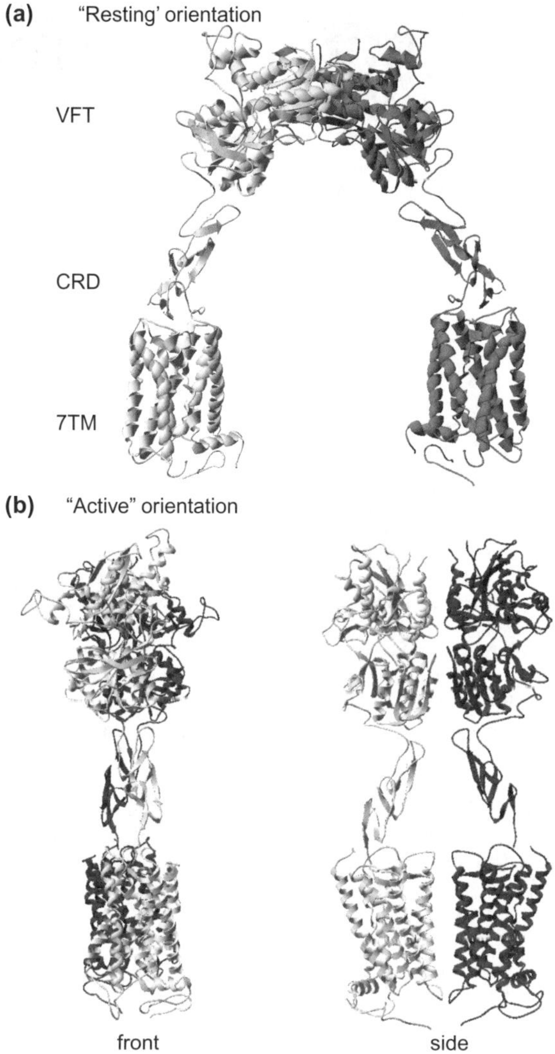

Figure 11.4 General model of a mGlu receptor dimer when the VFT dimer is in the 'resting' orientation (**a**) or in an 'active' orientation (**b**). In (**b**) front and side views are shown.

dimer prevents direct activation of the 7TMs by positive allosteric modulators, since these effectively activate isolated 7TMs.[41,51]

A second possibility is that even when the VFT dimer is in the resting orientation, the 7TMs are still in contact (Figure 11.5). This can be achieved only if the CRD can move relative to the VFT, a possibility that can be considered because of the limited contact area between these two domains interconnected only by a disulfide bridge. Activation of the VFT dimer would then force the two 7TMs to interact in a manner in which the angle between the

(a)

(b)

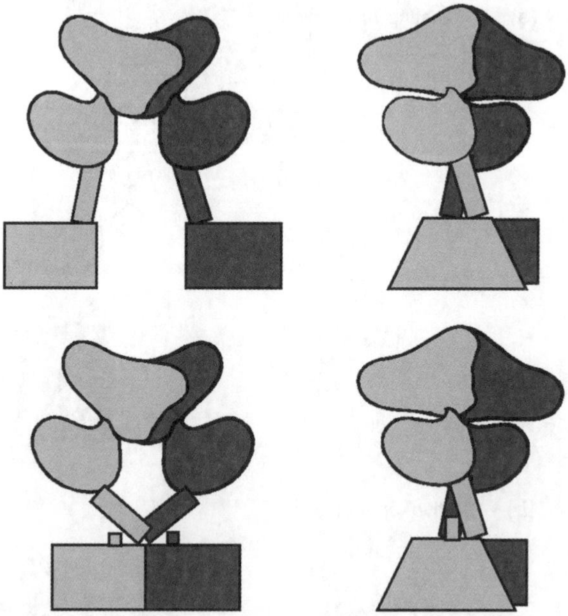

Figure 11.5 Possible activation mechanisms of mGluR dimers. (**a**) The model based on the structure of the mGlu3 extracellular dimer with the CRD predicts that the 7TMs are not in contact in the inactive state while they come into closed proximity in the active state. The contact between the two 7TMs leads to the activation of only one subunit. (**b**) A second model predicts that both 7TMs are in contact whether the VFT dimer is in the 'resting' or 'active' orientation. This model is possible only if the CRD can move relative to the VFT. Accordingly, the angle between the CRD and the 7TM is expected to vary, affecting the conformation of the extracellular loops of the 7TMs and leading to activation of one of them.

CRD and the 7TM would be altered (Figure 11.5). Such a possible movement of the CRD could then affect the conformation of the extracellular loops within the 7TM domain, leading to its activation. This second model would be more consistent with the data showing a minimal change in distance between the 7TMs as measured by FRET.[79,80] It is also consistent with the fact that modulators do not efficiently activate the 7TMs if the VFT dimer remains in the resting state,[51] as expected in the absence of agonist.

Obviously, other possibilities exist and only when the crystal structure of the full-length mGlu dimer is solved will it be possible to test these different models easily.

11.6 Contributions of the 7TM Domain

11.6.1 Active and Inactive States of the 7TM Domain

As mentioned above, the 7TM domain of mGluRs shares structural similarities with the class A rhodopsin-like receptors. But does this domain activate the

G protein in a similar way? Although the intracellular loops of the class C 7TM are short, the *C*-terminal end of the G protein alpha subunit appears to bind in a cavity between the intracellular loops 2 and 3,[43] likely as observed in rhodopsin.[81] In addition, the amphipatic helix 8 also plays a role in G protein coupling.[42]

To better understand how the 7TM domain of mGlu receptors can be activated, the isolated domain of mGlu5 was expressed in cells. It appeared to fold correctly as revealed by its correct targeting to the cell surface and its ability to bind molecules known to interact in this domain.[51] Moreover, levels of basal activity, similar to that observed with the full-length receptor was observed with the isolated mGlu5 7TM, which could be inhibited by the inverse agonist MPEP that directly binds in the 7TM domain. Of most interest, molecules acting as positive allosteric modulators of mGlu receptors are incapable of activating the full-length receptor but enhancing agonist potency and/or efficacy, were found to fully activate the isolated 7TM domain.[51] Similar observations were made with other mGlu subtypes tested, including mGlu5, mGlu1, mGlu2,[41,51,82] mGlu4 and mGlu7 (Goudet and Pin, unpublished data), as well as with the related $GABA_B$ receptor.[83] These data indicate that a change in conformation in the 7TM domain of mGlu receptors plays a critical role in G protein activation.

To further document this conclusion, our laboratory examined whether an isolated, monomeric 7TM domain of mGlu receptors, such as the monomeric class A GPCRs, could also activate G proteins.[84] We first noticed that G protein activation was observed in cells under conditions where no FRET between the isolated 7TM domain could be measured, suggesting that a monomeric mGlu 7TM could activate G proteins. This was firmly demonstrated by isolating a single 7TM into a membrane nanodiscs.[84]

Although no information is available today regarding the conformational changes of mGlu 7TM required for G protein activation, it is possible that, as observed in rhodopsin, an opening of the intracellular side of the receptor occurs during activation, resulting from an increased distance between the intracellular sides of TM3 and TM6. Indeed, an ionic interaction between these two TM regions appears to be conserved in all class C GPCRs. Furthermore, mutating residues that participate in the ionic lock in the $GABA_B2$ receptor subunit results in an increase in agonist affinity, consistent with an active state of this domain.[46]

Taken together, these different data indicate that a change in conformation, likely similar to that observed in rhodopsin, is involved in mGlu 7TM activation.

11.6.2 Asymmetric Functioning of mGlu 7TM Dimer

As indicated above, a single isolated mGlu 7TM is sufficient to activate G proteins. But can both 7TMs in the dimer be activated simultaneously, and does this symmetric activation lead to specific properties of the receptor? To date this question is only partly solved. Using both positive and negative allosteric ligands to control the conformation of a single 7TM domain within a dimer, our laboratory has been able to demonstrate that indeed, activation of a single 7TM within the dimer is needed for G protein activation.[85,86] To reach this conclusion, we adapted the mGlu receptors to an optimized quality control system that allowed the targeting to the cell surface of dimers composed of well-defined subunits. This allowed us to introduce mutations in a single subunit of a mGlu dimer. We showed that the G protein coupling efficacy of one subunit A is dependent of the state of the associated subunit B. Maintaining this associated B 7TM in an active conformation with a positive allosteric modulator largely decreased the coupling efficacy of the A 7TM.[85] In contrast, maintaining the B 7TM in an inactive conformation with an inverse agonist increased the coupling efficacy of the A.[86]

Such an asymmetric functioning of the mGlu 7TM dimer (Figure 11.5) is consistent with the observation that only a single subunit is responsible for G protein activation in the heterodimeric $GABA_B$,[87,88] sweet and umami taste receptors.[89]

But how can a symmetrical VFT dimer of the homodimeric mGlu receptors lead to an asymmetrical change in conformation in the 7TM dimer? One possibility is that the G protein itself is indeed responsible for this asymmetry. This is consistent with the G protein alpha subunit being required to stabilize 7TM proteins in their active state and with the general agreement that only a single G protein heterotrimer can interact with a GPCR dimer at one time (Figure 11.6).[90]

Although a single 7TM in a mGlu dimer appears sufficient for G protein activation, an active role of the associated 7TM in the agonist-induced activation process cannot be excluded. It is also possible that the associated 7TM domain, although not directly involved in G protein activation, plays additional roles such as those involving associated intracellular regulatory proteins. More work is required to clarify this issue.

11.7 Implications for Other GPCR Dimers

These recent studies reveal interesting features of the activation mechanism of mGlu and other class C GPCRs:

(i) These receptors are dimers.
(ii) Dimerization is required for agonist binding in the VFT to activate the 7TM domains through inter-subunit rearrangement.
(iii) Agonist binding leads to the activation of a single 7TM within the dimer, leading to an asymmetric receptor dimer.
(iv) The minimal unit required for G protein activation is a single 7TM domain (Figure 11.5).

Such information is of major interest to better understand the significance of the dimerization of other GPCRs such as the class A receptors. As recently

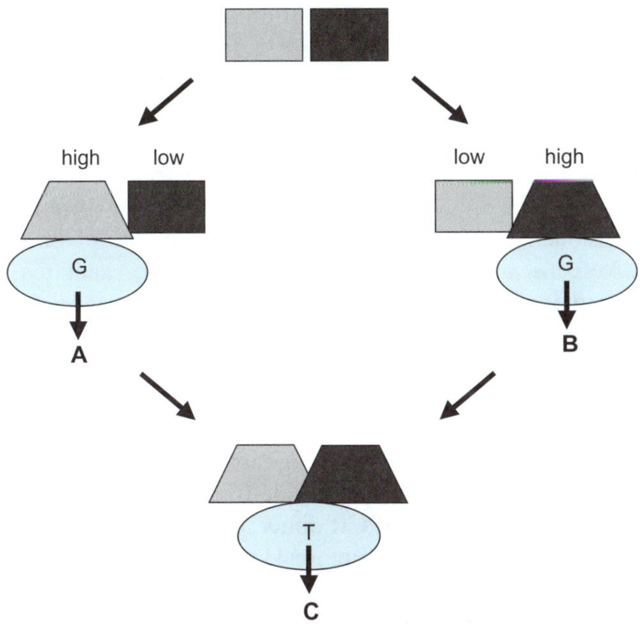

Figure 11.6 Asymmetrical functioning of GPCR dimers. Upon agonist binding to one subunit, G protein association stabilizes a high affinity state of that agonist bound subunit, resulting in an asymmetric GPCR dimer, in which the second subunit remains in a low agonist affinity state. It is still possible that interacting proteins other than G proteins are mostly recruited by a receptor dimer in which both subunits are in an active state, although one cannot exclude the possibility that either one of the active 7TMs can still activate G protein in some cases. The model is presented with a heterodimeric receptor able to activate a first pathway (**A**) when the one receptor (light grey) is activated, a second pathway (**B**) when the other receptor (black) is activated, and a third pathway (**C**), possibly resulting from a G protein independent mechanism, when both receptors are activated.

observed with class C 7TMs, it is clear that a monomeric class A GPCR represents the minimal unit required for G protein activation,[91–97] but this does not exclude the existence of GPCR dimers or oligomers in native cells. Not only have GPCRs dimers (or larger oligomers) been observed under physiological conditions,[57,98] but oxytocin receptor dimers have recently been reported in native tissues using high-affinity fluorescent ligands.[99]

Of interest, the functional asymmetric dimers reported for class C GPCRs have also been observed in several cases with class A GPCRs. Indeed, G protein coupling efficacy was found to be equivalent for a receptor monomer and a receptor dimer,[91,94,100,101] consistent with a single protomer within a dimeric receptor being sufficient for full G protein activation. By analysing conformational changes in each subunit of a reconstituted leukotriene B4 receptor (BLT1 receptor), it was clearly shown that ligand binding in one subunit does not lead to the same change in conformation of the associated subunit in the presence of the heterotrimeric G protein.[101]

An asymmetric state of GPCR dimers had already been suggested as being based on negative or positive cooperativity in ligand binding, and this was among the early data to suggest that GPCR dimers (or oligomers) in native tissue preparations.[102] In many cases, a negative cooperativity was observed when using radioactive agonists as a ligand. This was best demonstrated using excess non-radioactive agonist to enhance the OFF rate of bound agonists, both in native cells and in native membranes.[103] The exact reason for this observation remained unclear, until Vassart's group put it in the context of receptor dimerization and provided the first evidence that a single agonist may bind with high affinity to a receptor dimer.[104] This was first studied using the glycoprotein hormone receptors and then extended to the chemokine receptors.[103,105] This phenomenon was found to depend on G proteins and was assumed to result from the increase in agonist affinity in a receptor directly stabilized in its active state by a nucleotide-free G protein. Since a single G protein is likely able to bind per GPCR dimer, only one site per dimer is capable of reaching the high affinity state (Figure 11.6).

However, this proposal has recently been challenged. It is possible, if both the number of G proteins and the amount of GTP are limited, that only a fraction of the agonist-occupied GPCRs are stabilized in their high affinity state by G protein alpha subunits in their nucleotide-free form; this would explain the apparent negative cooperativity without the need for GPCR dimerization.[106] In a recent study, fluorescent ligands were used to examine whether agonist binding cooperativity could be related to a proximity between binding sites.[99] It had been previously reported that both agonists and antagonists of the vasopressin and oxytocin receptors display negative and positive binding cooperativity, both in transfected cells and in native tissues.[107] By using a combination of ligands carrying either a fluorophore donor or an acceptor, close proximity between antagonists binding sites were revealed by large FRET signals between these bound fluorescent ligands. This is consistent with two antagonists being able to bind per dimer.[99] In contrast, a very low but significant FRET signal was observed when fluorescent agonists were used, indicating that within a receptor dimer, only a single agonist is able to bind with high affinity.[99] This is consistent with only a single receptor being able to reach a G protein-stabilized high agonist affinity state per receptor dimer. It is also consistent with the observation that a single G protein heterotrimer associates with a GPCR dimer and stabilizes a specific conformation of one subunit only.[90,101]

Such a model predicts that simultaneous activation of both subunits of a receptor dimer may decrease G protein-mediated signalling.[108] In contrast, locking one subunit in an inactive state should favour signalling by the associated subunit. This fits nicely with recent data obtained with the dopamine D2 receptor homodimer,[109] as well as with the a2A-AR–muOR heterodimer.[110] Of major interest, is the possibility that other signalling cascades are specifically activated when both subunits are simultaneously in the active state. Accordingly, GPCR dimerization offers a much larger range of possible conformational states, thus providing various possibilities for different ligands to activate specific signalling cascades, depending on the cooperativity between the two binding sites. It is then tempting to speculate that dimerization, by amplifying minor conformational changes generated by different ligands, may play a key role in ligand-biased agonism.

11.8 Conclusions

The mGluRs, like the other class C GPCRs, are complex multidomain, dimeric receptors. The elucidation of the activation mechanism of these receptors reveals a number of possibilities to develop new and original compounds acting at several positions, stabilizing either the active or the inactive state of the complex, and acting as negative or positive allosteric modulators respectively. The identification of such ligands will certainly have important clinical applications in various therapeutic areas.

Our pursuit to understand the activation mechanism of mGlu receptors also brought important information on the functioning of other GPCR dimers, including class A rhodopsin-like receptors. These studies revealed an asymmetric component of the functioning GPCR dimers, with one subunit being involved in G protein activation, while the other is likely to be involved in other roles. Putative roles may be to help scaffold and recruit a number of regulatory proteins in the close vicinity of the receptor complex. Perhaps a more exciting possibility would suggest that dimerization offers a way to amplify minimal conformational changes through the relative movement of the two protomers, therefore offering a basis for ligand-biased efficacy.

Although GPCRs are certainly among the most studied receptors, being the most abundant, and representing a major target for drug development, there remains many unanswered questions yet to be answered regarding the GPCR functioning and regulation.

Acknowledgements

The work in J. P. Pin's group is supported by the CNRS, INSERM, the French Ministry of Research, grants from the Agence Nationale pour la Recherche (ANR-05-NEUR-0121-04, ANR-05-PRIB-02502, ANR-06-BLAN-0087, ANR-09-PIRI-0014, ANR-09-BLANC-0272-01), Foundation Marato TV3, Eranet Neuron (ANR-08-NEUR-006-02), CisBio and an unrestricted grant from Senomyx.

References

1. F. Sladeczek, J.-P. Pin, M. Recasens, J. Bockaert and S. Weiss, *Nature*, 1985, **317**, 717.
2. F. Nicoletti, J. T. Wroblewski, A. Novelli, H. Alho, A. Guidotti and E. Costa, *J. Neurosci.*, 1986, **6**, 1905.

3. M. M. Slaughter and R. F. Miller, *Nature*, 1985, **314**, 96.
4. G. E. Fagg and T. H. Lanthorm, *Br. J. Pharmacol.*, 1985, **86**, 743.
5. L. Prézeau, O. Manzoni, V. Homburger, F. Sladeczek, K. Curry and J. Bockaert, *Proc. Natl. Acad. Sci. U. S. A.*, 1992, **89**, 8040.
6. M. Masu, Y. Tanabe, K. Tsuchida, R. Shigemoto and S. Nakanishi, *Nature*, 1991, **349**, 760.
7. T. Abe, H. Sugihara, H. Nawa, R. Shigemoto, N. Mizuno and S. Nakanishi, *J. Biol. Chem.*, 1992, **267**, 13361.
8. I. Aramori and S. Nakanishi, *Neuron*, 1992, **8**, 757.
9. Y. Tanabe, M. Masu, T. Ishii, R. Shigemoto and S. Nakanishi, *Neuron*, 1992, **8**, 169.
10. Y. Nakajima, H. Iwakabe, C. Akazawa, H. Nawa, R. Shigemoto, N. Mizuno and S. Nakanishi, *J. Biol. Chem.*, 1993, **268**, 11868.
11. N. Okamoto, S. Hori, C. Akazawa, Y. Hayashi, R. Shigemoto, N. Mizuno and S. Nakanishi, *J. Biol. Chem.*, 1994, **269**, 1231.
12. K. M. Houamed, J. L. Kuijper, T. L. Gilbert, B. A. Haldeman, P. J. O'Hara, E. R. Mulvihill, W. Almers and F. S. Hagen, *Science*, 1991, **252**, 1318.
13. R. M. Duvoisin, C. Zhang and K. Ramonell, *J. Neurosci.*, 1995, **15**, 3075.
14. C. M. Niswender and P. J. Conn, *Annu. Rev. Pharmacol. Toxicol.*, 2010, **50**, 295.
15. J.-P. Pin and R. Duvoisin, *Neuropharmacology*, 1995, **34**, 1.
16. P. Conn and J.-P. Pin, *Ann. Rev. Pharmacol. Toxicol.*, 1997, **37**, 205.
17. M. Masu, H. Iwakabe, Y. Tagawa, T. Miyoshi, M. Yamashita, Y. Fukuda, H. Sasaki, K. Hiroi, Y. Nakamura, R. Shigemoto, M. Takada, K. Nakamura, K. Nakao, M. Katsuki and S. Nakanishi, *Cell*, 1995, **80**, 757.
18. P. J. Conn, G. Battaglia, M. J. Marino and F. Nicoletti, *Nat. Rev. Neurosci.*, 2005, **6**, 787.
19. S. T. Patil, L. Zhang, F. Martenyi, S. L. Lowe, K. A. Jackson, B. V. Andreev, A. S. Avedisova, L. M. Bardenstein, I. Y. Gurovich, M. A. Morozova, S. N. Mosolov, N. G. Neznanov, A. M. Reznik, A. B. Smulevich, V. A. Tochilov, B. G. Johnson, J. A. Monn and D. D. Schoepp, *Nat. Med.*, 2007, **13**, 1102.
20. J.-P. Pin, T. Galvez and L. Prézeau, *Pharmacol. Ther.*, 2003, **98**, 325.
21. C. Romano, W.-L. Yang and K. L. O'Malley, *J. Biol. Chem.*, 1996, **271**, 28612.
22. J.-P. Pin, R. R. Neubig, M. Bouvier, L. Devi, M. Filizola, J. A. Javitch, M. J. Lohse, G. Milligan, K. Palczewski, M. Parmentier and M. Spedding, *Pharmacol. Rev.*, 2007, **59**, 5.
23. S. Angers, A. Salahpour and M. Bouvier, *Annu. Rev. Pharmacol. Toxicol.*, 2002, **42**, 409.
24. S. Ferre, R. Baler, M. Bouvier, M. G. Caron, L. A. Devi, T. Durroux, K. Fuxe, S. R. George, J. A. Javitch, M. J. Lohse, K. Mackie, G. Milligan, K. D. Pfleger, J.-P. Pin, N. D. Volkow, M. Waldhoer, A. S. Woods and R. Franco, *Nat. Chem. Biol.*, 2009, **5**, 131.

25. M. Bouvier, *Nat. Rev. Neurosci.*, 2001, **2**, 274.
26. S. Bulenger, S. Marullo and M. Bouvier, *Trends Pharmacol. Sci.*, 2005, **26**, 131.
27. G. Milligan, *Br. J. Pharmacol.*, 2009, **158**, 5.
28. G. Milligan, *Curr. Opin. Pharmacol.*, 2010, **10**, 23.
29. K. Takahashi, K. Tsuchida, Y. Tanabe, M. Masu and S. Nakanishi, *J. Biol. Chem.*, 1993, **268**, 19341.
30. P. J. O'Hara, P. O. Sheppard, H. Thøgersen, D. Venezia, B. A. Haldeman, V. McGrane, K. M. Houamed, C. Thomsen, T. L. Gilbert and E. R. Mulvihill, *Neuron*, 1993, **11**, 41.
31. N. Kunishima, Y. Shimada, Y. Tsuji, T. Sato, M. Yamamoto, T. Kumasaka, S. Nakanishi, H. Jingami and K. Morikawa, *Nature*, 2000, **407**, 971.
32. T. Muto, D. Tsuchiya, K. Morikawa and H. Jingami, *Proc. Natl. Acad. Sci. U. S. A.*, 2007, **104**, 3759.
33. E. Dobrovetsky, G. Khutoreskaya, A. Seitova, H. He, J. Weigelt, A. M. Edwards, C. H. Arrowsmith, C. Bountra, D. Cossar, A. Bochkarev and S. G. Consortium, Protein Data Bank, 2010, DOI: 10.2210/pdb 3MQ4/pdb.
34. E. Dobrovetsky, G. Khutoreskaya, A. Seitova, D. Cossar, A. M. Edwards, C. H. Arrowsmith, C. Bountra, J. Weigelt, A. Bochkarev and S. G. Consortium, Protein Data Bank, 2010, DOI: 10.2210/pdb 3LMK/pdb.
35. F. C. Acher and H. O. Bertrand, *Biopolymers*, 2005, **80**, 357.
36. H. O. Bertrand, A. S. Bessis, J.-P. Pin and F. C. Acher, *J. Med. Chem.*, 2002, **45**, 3171.
37. J.-P. Pin and F. Acher, *Curr. Drug Targets CNS Neurol. Disord.*, 2002, **1**, 297.
38. C. Selvam, N. Oueslati, I. Lemasson, I. Brabet, D. Rigault, T. Courtiol, S. Cesarini, N. Triballeau, H.-O. Bertrand, C. Goudet, J.-P. Pin and F. C. Acher, *J. Med. Chem.*, 2010, **53**, 2797.
39. C. Beurrier, S. Lopez, D. Révy, C. Selvam, C. Goudet, M. Lhérondel, P. Gubellini, L. Kerkerian-LeGoff, F. Acher, J.-P. Pin and M. Amalric, *FASEB J.*, 2009, **23**, 3619.
40. X. Liu, Q. He, D. J. Studholme, Q. Wu, S. Liang and L. Yu, *Trends Biochem. Sci.*, 2004, **29**, 458.
41. P. Rondard, J. Liu, S. Huang, F. Malhaire, C. Vol, A. Pinault, G. Labesse and J.-P. Pin, *J. Biol. Chem.*, 2006, **281**, 24653.
42. J.-P. Pin, C. Joly, S. F. Heinemann and J. Bockaert, *EMBO J.*, 1994, **13**, 342.
43. M. Havlickova, J. Blahos, I. Brabet, J. Liu, B. Hruskova, L. Prezeau and J.-P. Pin, *J. Biol. Chem.*, 2003, **278**, 35063.
44. T. Yamashita, A. Terakita and Y. Shichida, *J. Biochem.*, 2001, **130**, 149.
45. J. H. Park, P. Scheerer, K. P. Hofmann, H. W. Choe and O. P. Ernst, *Nature*, 2008, **454**, 183.
46. V. Binet, B. Duthey, J. Lecaillon, C. Vol, J. Quoyer, G. Labesse, J.-P. Pin and L. Prézeau, *J. Biol. Chem.*, 2007, **282**, 12154.
47. F. Knoflach, V. Mutel, S. Jolidon, J. N. Kew, P. Malherbe, E. Vieira, J. Wichmann and J. A. Kemp, *Proc. Natl. Acad. Sci. U. S. A.*, 2001, **98**, 13402.

48. S. Litschig, F. Gasparini, D. Rueegg, N. Munier, P. J. Flor, I.-T. Vranesic, L. Prézeau, J.-P. Pin, C. Thomsen and R. Kuhn, *Mol. Pharmacol.*, 1999, **55**, 453.
49. P. Malherbe, N. Kratochwil, F. Knoflach, M. T. Zenner, J. N. Kew, C. Kratzeisen, H. P. Maerki, G. Adam and V. Mutel, *J. Biol. Chem.*, 2003, **278**, 8340.
50. A. Pagano, D. Rüegg, S. Litschig, N. Stoehr, C. Stierlin, M. Heinrich, P. Floersheim, L. Prézeau, F. Carroll, J.-P. Pin, A. Cambria, I. Vranesic, P. J. Flor, F. Gasparini and R. Kuhn, *J. Biol. Chem.*, 2000, **275**, 33750.
51. C. Goudet, F. Gaven, J. Kniazeff, C. Vol, J. Liu, M. Cohen-Gonsaud, F. Acher, L. Prezeau and J.-P. Pin, *Proc. Natl. Acad. Sci. U. S. A.*, 2004, **101**, 378.
52. J. Bockaert, L. Fagni, A. Dumuis and P. Marin, *Pharmacol. Ther.*, 2004, **103**, 203.
53. J. Bockaert, J. Perroy, C. Becamel, P. Marin and L. Fagni, *Annu. Rev. Pharmacol. Toxicol.*, 2010, **50**, 89.
54. D. Tsuchiya, N. Kunishima, N. Kamiya, H. Jingami and K. Morikawa, *Proc. Natl. Acad. Sci. U. S. A.*, 2002, **99**, 2660.
55. D. Fotiadis, Y. Liang, S. Filipek, D. A. Saperstein, A. Engel and K. Palczewski, *Nature*, 2003, **421**, 127.
56. P. Carriba, G. Navarro, F. Ciruela, S. Ferre, V. Casado, L. Agnati, A. Cortes, J. Mallol, K. Fuxe, E. I. Canela, C. Lluis and R. Franco, *Nat. Methods*, 2008, **5**, 727.
57. W. Guo, E. Urizar, M. Kralikova, J. C. Mobarec, L. Shi, M. Filizola and J. A. Javitch, *EMBO J.*, 2008, **27**, 2293.
58. L. F. Pisterzi, D. B. Jansma, J. Georgiou, M. J. Woodside, J. T. Chou, S. Angers, V. Raicu and J. W. Wells, *J. Biol. Chem.*, 2010, **285**, 16723.
59. J. J. Fung, X. Deupi, L. Pardo, X. J. Yao, G. A. Velez-Ruiz, B. T. Devree, R. K. Sunahara and B. K. Kobilka, *EMBO J.*, 2009, **28**, 3315.
60. J. Kniazeff, A.-S. Bessis, D. Maurel, H. Ansanay, L. Prezeau and J.-P. Pin, *Nat. Struct. Mol. Biol.*, 2004, **11**, 706.
61. C. Brock, N. Oueslati, S. Soler, L. Boudier, P. Rondard and J.-P. Pin, *J. Biol Chem*, 2007, **282**, 33000.
62. D. Maurel, L. Comps-Agrar, C. Brock, M.-L. Rives, E. Bourrier, M. A. Ayoub, H. Bazin, N. Tinel, T. Durroux, L. Prézeau, E. Trinquet and J.-P. Pin, *Nat. Methods*, 2008, **5**, 561.
63. E. Doumazane, P. Scholler, J. Zwier, E. Trinquet, P. Rondard and J.-P. Pin, *FASEB J.*, 2011, **25**, 66.
64. L. Comps-Agrar, J. Kniazeff, L. Nørskov-Lauritsen, D. Maurel, M. Gassmann, N. Gregor, L. Prézeau, B. Bettler, T. Durroux, E. Trinquet, and J.-P. Pin, *EMBO J.*, 2011, in press.
65. J. Schwenk, M. Metz, G. Zolles, R. Turecek, T. Fritzius, W. Bildl, E. Tarusawa, A. Kulik, A. Unger, K. Ivankova, R. Seddik, J. Y. Tiao, M. Rajalu, J. Trojanova, V. Rohde, M. Gassmann, U. Schulte, B. Fakler and B. Bettler, *Nature*, 2010, **465**, 231.
66. F. Ferraguti and R. Shigemoto, *Cell Tissue Res.*, 2006, **326**, 483.

67. S. J. Tallaksen-Greene, K. W. Kaatz, C. Romano and R. L. Albin, *Brain Res.*, 1998, **780**, 210.
68. L. J. Volk, C. A. Daly and K. M. Huber, *J. Neurophysiol.*, 2006, **95**, 2427.
69. F. Ferraguti, T. Klausberger, P. Cobden, A. Baude, J. D. Roberts, P. Szucs, A. Kinoshita, R. Shigemoto, P. Somogyi and Y. Dalezios, *J. Neurosci.*, 2005, **25**, 10520.
70. L. Gama, S. G. Wilt and G. E. Breitwieser, *J. Biol. Chem.*, 2001, **276**, 39053.
71. E. Dobrovetsky, G. Khutoreskaya, A. Seitova, D. Cossar, A. M. Edwards, C. H. Arrowsmith, C. Bountra, J. Weigelt, A. Bochkarev and S. G. Consortium, *unpublished*, Protein Data Bank, 2009, DOI: 10.2210/pdb3KS9/pdb.
72. A. S. Bessis, P. Rondard, F. Gaven, I. Brabet, N. Triballeau, L. Prezeau, F. Acher and J.-P. Pin, *Proc. Natl. Acad. Sci. U. S. A.*, 2002, **99**, 11097.
73. J. Kniazeff, P.-P. Saintot, C. Goudet, J. Liu, A. Charnet, G. Guillon and J.-P. Pin, *J. Neurosci.*, 2004, **24**, 370.
74. H. Jingami, S. Nakanishi and K. Morikawa, *Curr. Opin. Neurobiol.*, 2003, **13**, 271.
75. T. Sato, Y. Shimada, N. Nagasawa, S. Nakanishi and H. Jingami, *J. Biol. Chem.*, 2003, **278**, 4314.
76. P. Rondard, S. Huang, C. Monnier, H. Tu, B. Blanchard, N. Oueslati, F. Malhaire, Y. Li, D. Maurel, E. Trinquet, G. Labesse, J.-P. Pin* and J. Liu, *EMBO J.*, 2008, **27**, 1321.
77. M. Tateyama and Y. Kubo, *Proc. Natl. Acad. Sci. U. S. A.*, 2006, **103**, 1124.
78. N. Makita, J. Sato, K. Manaka, Y. Shoji, A. Oishi, M. Hashimoto, T. Fujita and T. Iiri, *Proc. Natl. Acad. Sci. U. S. A.*, 2007, **104**, 5443.
79. P. Marcaggi, H. Mutoh, D. Dimitrov, M. Beato and T. Knopfel, *Proc. Natl. Acad. Sci. U. S. A.*, 2009, **106**, 11388.
80. M. Tateyama, H. Abe, H. Nakata, O. Saito and Y. Kubo, *Nat. Struct. Mol. Biol.*, 2004, **11**, 637.
81. P. Scheerer, J. H. Park, P. W. Hildebrand, Y. J. Kim, N. Krauss, H. W. Choe, K. P. Hofmann and O. P. Ernst, *Nature*, 2008, **455**, 497.
82. Y. Chen, C. Goudet, J.-P. Pin and P. J. Conn, *Mol. Pharmacol.*, 2008, **73**, 909.
83. V. Binet, C. Brajon, L. Le Corre, F. Acher, J.-P. Pin and L. Prezeau, *J. Biol. Chem.*, 2004, **279**, 29085.
84. S. Granier, E. Doumazane, P. Scholler, D. El-Moustaine, R. Rahmeh, P. Bronb, P. Rondard, J.-L. Banères and J.-P. Pin, submitted.
85. C. Goudet, J. Kniazeff, V. Hlavackova, F. Malhaire, D. Maurel, F. Acher, J. Blahos, L. Prezeau and J.-P. Pin, *J. Biol. Chem.*, 2005, **280**, 24380.
86. V. Hlavackova, C. Goudet, J. Kniazeff, A. Zikova, D. Maurel, C. Vol, J. Trojanova, L. Prézeau, J.-P. Pin and J. Blahos, *EMBO J.*, 2005, **24**, 499.
87. B. Duthey, S. Caudron, J. Perroy, B. Bettler, L. Fagni, J.-P. Pin and L. Prézeau, *J. Biol. Chem.*, 2002, **277**, 3236.
88. T. Galvez, B. Duthey, J. Kniazeff, J. Blahos, G. Rovelli, B. Bettler, L. Prézeau and J.-P. Pin, *EMBO J.*, 2001, **20**, 2152.

89. H. Xu, L. Staszewski, H. Tang, E. Adler, M. Zoller and X. Li, *Proc. Natl. Acad. Sci. U. S. A.*, 2004, **101**, 14258.
90. J. L. Baneres and J. Parello, *J. Mol. Biol.*, 2003, **329**, 815.
91. L. Arcemisbehere, T. Sen, L. Boudier, M. N. Balestre, G. Gaibelet, E. Detouillon, H. Orcel, C. Mendre, R. Rahmeh, S. Granier, C. Vives, F. Fieschi, M. Damian, T. Durroux, J. L. Baneres and B. Mouillac, *J. Biol. Chem.*, 2010, **285**, 6337.
92. T. H. Bayburt, A. J. Leitz, G. Xie, D. D. Oprian and S. G. Sligar, *J. Biol. Chem.*, 2007, **282**, 14875.
93. O. P. Ernst, V. Gramse, M. Kolbe, K. P. Hofmann and M. Heck, *Proc. Natl. Acad. Sci. U. S. A.*, 2007, **104**, 10859.
94. J. F. White, J. Grodnitzky, J. M. Louis, L. B. Trinh, J. Shiloach, J. Gutierrez, J. K. Northup and R. Grisshammer, *Proc. Natl. Acad. Sci. U. S. A.*, 2007, **104**, 12199.
95. M. R. Whorton, M. P. Bokoch, S. G. Rasmussen, B. Huang, R. N. Zare, B. Kobilka and R. K. Sunahara, *Proc. Natl. Acad. Sci. U. S. A.*, 2007, **104**, 7682.
96. M. Chabre and M. le Maire, *Biochemistry*, 2005, **44**, 9395.
97. A. J. Kuszak, S. Pitchiaya, J. P. Anand, H. I. Mosberg, N. G. Walter and R. K. Sunahara, *J. Biol. Chem.*, 2009, **284**, 26732.
98. J. F. Mercier, A. Salahpour, S. Angers, A. Breit and M. Bouvier, *J. Biol. Chem.*, 2002, **277**, 44925.
99. L. Albizu, M. Cottet, S. Stoev, R. Seyer, I. Brabet, T. Roux, H. Bazin, E. Bourrier, L. Lamarque, C. Breton, M. Rives, M. Kralikova, A. H. Newman, J. A. Javitch, E. Trinquet, M. Manning, J.-P. Pin, B. Mouillac and T. Durroux, *Nat. Chem. Biol.*, 2010, **6**, 587.
100. S. L. Chinault, M. C. Overton and K. J. Blumer, *J. Biol. Chem.*, 2004, **279**, 16091.
101. M. Damian, A. Martin, D. Mesnier, J.-P. Pin and J. L. Banères, *EMBO J.*, 2006, **25**, 5693.
102. K. A. Wreggett and J. W. Wells, *J. Biol. Chem.*, 1995, **270**, 22488.
103. J. Y. Springael, E. Urizar, S. Costagliola, G. Vassart and M. Parmentier, *Pharmacol. Ther.*, 2007, **115**, 410.
104. E. Urizar, L. Montanelli, T. Loy, M. Bonomi, S. Swillens, C. Gales, M. Bouvier, G. Smits, G. Vassart and S. Costagliola, *EMBO J.*, 2005, **24**, 1954.
105. J. Y. Springael, P. N. Le Minh, E. Urizar, S. Costagliola, G. Vassart and M. Parmentier, *Mol. Pharmacol.*, 2006, **69**, 1652.
106. M. Chabre, P. Deterre and B. Antonny, *Trends Pharmacol. Sci.*, 2009, **30**, 182.
107. L. Albizu, M.-N. Balestre, C. Breton, J.-P. Pin, M. Manning, B. Mouillac, C. Barberis and T. Durroux, *Mol. Pharmacol.*, 2006, **70**, 1783.
108. X. Rovira, J.-P. Pin and J. Giraldo, *Trends Pharmacol. Sci.*, 2010, **31**, 15.
109. Y. Han, I. S. Moreira, E. Urizar, H. Weinstein and J. A. Javitch, *Nat. Chem. Biol.*, 2009, **5**, 688.
110. J. P. Vilardaga, V. O. Nikolaev, K. Lorenz, S. Ferrandon, Z. Zhuang and M. J. Lohse, *Nat. Chem. Biol.*, 2008, **4**, 126.

CHAPTER 12

Crosstalk Between Receptors: Challenges of Distinguishing Upstream from Downstream Mechanisms

MAHALAXMI ABURI,[a] MARIE-LAURE RIVES,[a] YANG HAN,[a] MICHAELA KRALIKOVA,[b] ENEKO URIZAR,[a] HIDEAKI YANO[a] AND JONATHAN A. JAVITCH*[a]

[a] Center for Molecular Recognition and Departments of Psychiatry and Pharmacology, College of Physicians and Surgeons, Columbia University, 630 W. 168th Street, New York, NY 10032, USA; [b] Department of Auditory Neuroscience, Czech Academy of Science, Videnska 1083, Prague 4, 14220 Czech Republic

12.1 Introduction

There is substantial evidence indicating that G-protein coupled receptors (GPCRs) exist as dimers or oligomers.[1,2] Dimerization can occur between identical GPCRs, resulting in the formation of homomers, or between close family members, or even GPCRs from different families, resulting in the formation of heteromers. During the past several years, GPCR dimerization has been supported by a large number of biochemical, pharmacological and biophysical approaches used *in vitro* as well as, in selected cases, *in vivo*.[2,3] Receptor heteromerization has been proposed to generate novel pharmacological profiles and/or activate specific signalling pathways, with significant

RSC Drug Discovery Series No. 8
G Protein-Coupled Receptors: From Structure to Function
Edited by Jesús Giraldo and Jean-Philippe Pin

Published by the Royal Society of Chemistry, www.rsc.org

implications for drug discovery.[4,5] In this chapter we discuss the emerging concept of 'crosstalk' that arises from receptor coexpression and how it may contribute to the diversity and complexity of GPCR signalling. We discuss recently developed approaches designed to unravel the functional importance of dimerization in a signalling complex in Family A GPCRs.

12.2 Signalling Complexities

Each cell expresses multiple GPCRs with downstream signalling pathways that are organized as coordinated communication networks. Upon ligand activation, these multiprotein complexes integrate signals both temporally and spatially to produce a cellular response. In many cases, two or more simultaneous signals induce a response different from the simple sum of the two independent signals. Such coincidence detection is ubiquitous in biology,[6] but the cellular and molecular mechanisms that underlie signal integration are not fully understood. One example of such a coincidence detector is phospholipase C (PLC), which can integrate signalling from different GPCRs.[7,8] Thus, crosstalk could occur at the level of the signalling pathways downstream of the receptors themselves or directly at the receptor level through physical interactions between receptors (Figure 12.1).

One of the great challenges in GPCR biology today is to understand the mechanistic link between the physical interaction of receptors in the plasma membrane and the signalling crosstalk of receptor heteromers. Numerous studies have demonstrated signalling crosstalk between coexpressed family A GPCRs,[4,9,10] but in almost all cases the mechanistic link between heteromerization and signalling is not clear. Although the presence of two coexpressed receptors may be essential, signalling crosstalk could nonetheless take

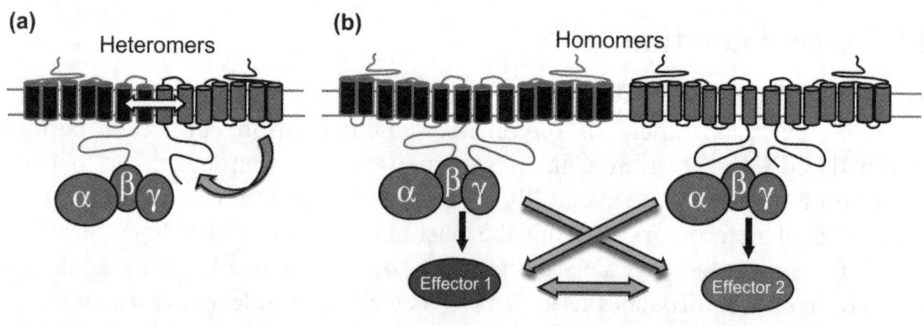

Figure 12.1 Schematic representation of alternate mechanisms for functional crosstalk between two GPCRs. The receptors are represented here as black and grey 7-transmembrane proteins, but each receptor could be a monomer or a homomer or a heteromer. (**a**) Functional crosstalk from heteromerization, *i.e.* physical interaction between two receptors. (**b**) The receptors could either form a heteromer or homomer, but the functional crosstalk may nonetheless occur at the level of signalling. This crosstalk can also occur between two receptors without dimerization.

place downstream of parallel monomeric or homomeric receptor-mediated G protein and/or G protein-independent signalling, and in such a case would not be a direct result of signalling by a heteromeric receptor unit. Such a downstream crosstalk mechanism, while often ignored, is very difficult to rule out due to the great complexity and overlap of signalling systems.

Further complicating the situation, a number of findings now support the existence of higher order complexes,[11,12] raising the possibility that GPCR heteromers may interact not as heterodimers *per se* but rather as higher order hetero-oligomers composed of homomeric subunits. In addition, the stability and specificity of GPCR interactions has also become a matter of intense debate, fuelling the ongoing controversy regarding the physiological/functional role of receptor oligomerization.[5,13,14]

A recent single-molecule study using total internal reflection fluorescence microscopy has enabled researchers to track the position of putative individual molecules of a family A GPCR, M1 muscarinic acetylcholine receptor, in living cells over a period of several seconds.[15] The results of this study point to a transient formation of M1 receptor dimers (lifetime of ~0.5 s at 23 °C) and to an estimate that ~30% of the total receptor molecules exist as dimers at a given point in time. Whether these findings can be generalized to other family A GPCRs and whether they reflect alterations of dimerization status by the high-affinity fluorescent antagonist molecules used to visualize the receptors is not yet known.

Recent fluorescence recovery after photobleaching (FRAP) studies of β1-adrenoceptors[16] and dopamine D2 receptor[17] also suggested that receptor interactions are transient and dynamic. Although the lifetime of the M1 dimers appears to be relatively short, it is longer than expected for a random collisional event and is sufficient, in principle, for the emergence of unique pharmacological function from homomeric and/or heteromeric units. Thus, it is essential to explore the various kinds of crosstalk and verify that GPCR heteromers are actual signalling units in which the two receptors interact directly to modulate G protein activation. While much progress has been achieved in these efforts for family C receptors (see chapter 11), investigation of the molecular mechanism of family A GPCRs has proven more challenging for a number of technical reasons, as discussed below.

12.2.1 Potential Crosstalk due to Dimerization

Coexpression of GPCRs can lead to the emergence of signalling properties that differ from those resulting from expression of individual receptors, and this has been proposed to originate from signalling by a GPCR heteromer. One example of this, and of the associated complexity, is the proposed dopamine D1–D2 receptor heteromer.

Dopamine D1 and D2 receptors are distinguished by their opposing effects on adenylate cyclase (AC). While dopamine D1 receptor (D1R) signals *via* Gs/olf coupling, leading to stimulation of AC and increased cAMP (cyclic adenosine monophosphate), the D2 receptor (D2R) signals *via* Gi/o proteins, leading to inhibition of AC and decreased cAMP.[18] Although the two receptors belong to distinct subfamilies of dopamine receptors, a number of physiological functions are mediated by activation of both receptors. For example, behavioural studies suggest that co-stimulation of D1R is essential for D2R agonists to exert maximal locomotor effects[19] and that co-activation of both D1R and D2R is required to cause cocaine-induced hyperlocomotion in rats.[20] While the mechanism underlying this synergy is not clear and may reflect simultaneous regionally distinct effects, the colocalization of the two receptors in a small subset of striatal neurons[21,22] raises the possibility of their direct association.

Several studies[23–25] have suggested that coexpressed D1R and D2R can interact to form a heteromeric unit that activates Gq/11 and PLC upon co-activation of both receptors, thereby triggering intracellular Ca^{2+} release.[22] In these studies, neither D1R nor D2R when expressed alone activated Gq, which was taken as evidence that this signalling requires the presence of a D1R–D2R heteromer. Consistent with this argument, Ca^{2+} signalling was blocked by antagonists of either D1R or D2R and was independent of Gi/o and Gs activation. A subset of D1R agonists (SKF81297, SKF83959) was reported to target the heteromer specifically with no effect on Gq/11 pathways when either of the receptors was expressed alone. This switch in coupling of D2R from Gi/o to Gq/11 proteins upon interaction with D1R has been attributed to heteromerization.[24] However, these results are in contrast with reports demonstrating that antagonism of D2R does not affect the ability of D1R agonists to activate Gq/11 in neuronal preparations, suggesting the existence of an uncharacterized mechanism by which D1-like receptors can activate Gq/11, perhaps independently of D2R.[26,27] Both D1R and the related D5R have been implicated in Gq activation in the brain and this remains a subject of some controversy.[18,28] Furthermore, studies using scintillation proximity assays (SPA)[29] in striatal membranes and in HEK293 cells have shown that dopamine and other D1R agonists (*e.g.* SKF83959) activate Gs and Gq with similar potencies and efficacies—in the absence of D2R in HEK293 cells. Finally, stimulation of D2R when expressed alone can lead to elevation of Ca^{2+} levels through a $G\beta/\gamma$ mediated pathway requiring Gi/o activation.[30] How these disparate pieces of evidence fit together in a molecular mechanism of signalling is a topic of ongoing investigation, and many questions remain unanswered.

The proposed adenosine A2A receptor (A2AR)–D2R heteromer is another example of the complexity of signal integration. A2AR activation is associated with stimulation of cAMP, whereas D2R activation leads to inhibition of cAMP. The A2AR and D2R are coexpressed within GABAergic neurons of the striatum where they play a key role in the modulation of neuronal activity.[31] A2AR–D2R heteromerization was first demonstrated in transfected mammalian cells using co-immunoprecipitation and fluorescence and bioluminescence resonance energy transfer (FRET and BRET) techniques respectively.[32] Furthermore, an A2AR agonist was shown to decrease the ability of a D2R agonist to displace the binding of a selective D2R radioligand in striatal tissue.[33,34] This phenomenon of modulation of binding, together with the biophysical data in heterologous cells, was taken to infer that the modulation is mediated through a conformational change within a receptor heteromer.[35]

A2AR–D2R functional interactions have also been reported at the second messenger level, both in transfected cells and in the brain.[36] In this case, stimulation of D2R prevents the effects of A2AR stimulation. Specifically, D2R-mediated activation of the Gi/o protein counteracts the ability of the Gs/Golf coupled A2AR to activate the protein kinase A (PKA) pathway.[37] Such a modulation of PKA is expected based on concurrent activation of Gi/o and Gs by D2R and A2AR, respectively. However, physical interactions between the receptors have also been proposed to play an important role in this crosstalk.[31] An arginine-rich epitope at the *N*-terminal end of the third intracellular loop (IL3) of D2R is necessary for its coupling to Gi/o.[38] Curiously, the same epitope has been implicated in binding to the *C*-terminus of A2AR.[39,40] In GABAergic striatal medium spiny neurons (MSNs), a peptide corresponding to the *C*-terminus of A2AR disrupted the ability of A2AR activation to block the inhibitory effects of D2R activation on *N*-methyl-D-aspartic acid (NMDA) induced spike firing,[41] possibly by binding to IL3 of D2R and altering D2R–A2AR interaction. These findings led to the proposal that D2R signalling to Gi/o can be inhibited by interaction of this region with the activated A2AR in a heteromer.[31,36] Notably, the long *C*-terminal chain of A2AR has been proposed to be a coincidence detector since it can bind several accessory proteins, including GPCR kinases and arrestins.[42] Therefore, some of the altered signalling observed with the A2AR *C*-terminal peptide might result from altered interactions of the A2AR with interacting proteins other than D2R.

The existence of additional partners in this interaction has also been proposed.[43–45] A2AR forms heteromers with metabotropic glutamate receptor (mGlu5R) in transfected cells,[43] and both receptors have been co-immunoprecipitated from striatal membranes. Co-stimulation of mGlu5R (which activates Gq) in the striatum abolished the inhibitory effect of D2R on A2AR signalling, suggesting that the interaction is more complex than simple off-setting effects on Gi/o and Gs. Recent studies have proposed an association between these three receptors based on the co-distribution of D2R, A2AR and mGluR5 within the extra synaptic plasma membrane of dendritic spines at glutamatergic striatal synapses.[45] In addition, D2R has also been inferred to be involved in an oligomeric interaction with A2AR and cannabinoid CB1 receptor in HEK293 cells.[44] Although these data suggest receptor colocalization, evidence for direct physical interaction between D2R–A2AR *in vivo* is still lacking.

In summary, although these results are intriguing and suggest the possibility of an untapped level of pharmacological diversity for new compound development, it cannot yet be definitively argued that receptor heteromerization is responsible for the observed crosstalk, as we cannot exclude downstream signal integration triggered by the independent activation of each receptor as the cause of the functional regulation.

12.2.2 Potential Crosstalk due to Signal Integration

It is important to recognize that not all crosstalk results from receptor dimerization. For example, in cerebellar Purkinje cells, activation of the $GABA_B$ receptor increases the Ca^{2+} responses generated by the metabotropic glutamate 1a receptor (mGlu1aR),[46,47] thereby facilitating the generation of long-term depression (LTD) at the parallel fibre-Purkinje cell synapses.[48] Both receptors have been shown to colocalize perisynaptically at postsynaptic densities of Purkinje cell dendritic spines[49] and have been co-immunoprecipitated from brain preparations.[46,47]

Like the arguments discussed above, these results suggested the possibility that mGlu1aR–$GABA_B$R heteromers could be responsible for the observed synergy. However, although the functional interaction was reproduced in transfected HEK293 cells, no evidence for direct physical interaction could be demonstrated,[49] suggesting that the observed crosstalk arises at the level of the downstream signalling pathways and not from receptor heteromers. Moreover, the crosstalk was completely eliminated when a mutant of $GABA_B$R that is unable to couple to the Gi/o protein was used. It has been established that the observed potentiation of the Gq-mediated mGlu1aR calcium responses is modulated by the $G\beta/\gamma$ subunits of the Gi/o G proteins activated by the $GABA_B$R, possibly due to the activation of a specific PLC.[48,49] Indeed, Gi–Gq synergism in some systems has been explained by their actions on phospholipase C beta 3 (PLC-β3), which acts as a coincidence detector upon stimulation by $G\beta/\gamma$ released by Gi activation and by Gq[50] in a mechanism that does not require receptor dimerization.

Interestingly, Gi–Gq signalling synergy has been extended to other pairs of Gi/Gq-coupled receptors such as adenosine A1 receptors (A1R) and mGlu1aR in HEK293 cells and mGlu3R/mGlu5R in cortical astrocytes[49,51,52] for which no evidence of direct physical association has been detected. Although this crosstalk also appears to be independent of receptor oligomerization, in contrast to the findings in HEK293 cells, no cross-regulation between A1R and mGlu1aR was observed in Purkinje cells. One possible explanation is that these two receptors have different subcellular compartmentalization *in vivo,* whereas they may be colocalized at the surface of HEK293 cells, thereby leading to sufficient proximity between the effectors of the signalling pathways to generate signal integration. Differences in the expression and localization of PLC isoforms may also contribute to differences in synergy.[7,8]

These examples indicate that observation of molecular proximity between two receptors at the plasma membrane and differential receptor pharmacology or modified intracellular signalling properties does not mean that a receptor heteromer is directly responsible for the differential response. Indeed, even though there is substantial evidence for direct physical interaction between several GPCRs,[32,53,54] it has not been possible to rule out downstream signalling crosstalk and to establish incontrovertibly that cooperative interactions within receptor heteromers are responsible for the crosstalk.

Such a mechanistic interrogation of heteromeric signalling has been particularly difficult for family A GPCRs. In contrast, our mechanistic understanding of the functional role of GPCR dimerization is more advanced for the family C receptors, due in large part to the clever adaptation of the endoplasmic reticulum (ER) retention signal from the $GABA_BR$ to enable controlled cell surface expression and signalling by defined mGluR heteromers (see chapter 11).[55] This approach has allowed a detailed exploration of the mechanistic roles of both protomers in the signalling unit. Unfortunately, related approaches with ER retention signals have been unsuccessful in family A receptors.

12.3 New Methodologies to Control the Identity of the Components Comprising the Signalling Unit

Described below are novel complementation and energy transfer-based biosensor approaches that our lab has used to monitor receptor activation and explore the functional stoichiometry of the signalling unit of D2R, a representative family A GPCR, and to differentiate the functional role of individual protomers of D2R homo- and heteromers.

12.3.1 Individual Components Involved in a Signalling Unit

For class A GPCRs, a major obstruction to expanding our understanding of dimerization has been methodological, caused by the limited ability to control the identity of components of the G protein signalling unit. Although both rhodopsin[56] and the β_2-adrenergic receptor (β2AR)[57] have been shown to signal efficiently to G proteins when reconstituted into lipid nanodiscs containing a single receptor, such studies cannot determine whether these receptors function alone *in vivo*. To date, the native functional signalling unit in class A GPCRs remains unclear. Han *et al.* developed a functional complementation assay that ensures tight control of the components of the D2R signalling unit, allowing exploration of individual contributions from each GPCR protomer to G protein signalling within the dimeric functional unit.[58] This system reports directly on receptor–G protein interactions, which rules out downstream crosstalk as a potential confound.

Han *et al.* engineered Flp-In T-Rex-293 cells to stably express aequorin (AQ), which produces luminescence in response to Gq activation and subsequent Ca^{2+} release (Figure 12.2).[58] Activation of D2R in these cells did not generate luminescence, as expected due to the lack of D2R coupling to Gq. Stable expression of a chimeric Gα, Gqi5, in which the last five residues of Gq were substituted with those from Gi, led to robust signalling of D2R through the PLC pathway. Moreover, agonist-mediated activation was observed when Gqi5 was fused to the *C*-terminus of D2R *via* a flexible linker (D2R–L–Gqi5). In contrast, when Gqi5 was directly fused to the D2R *C*-terminus (D2R–Gqi5), the fusion construct expressed at the surface and bound ligand, but agonist treatment did not lead to luminescence—probably because the absence of a

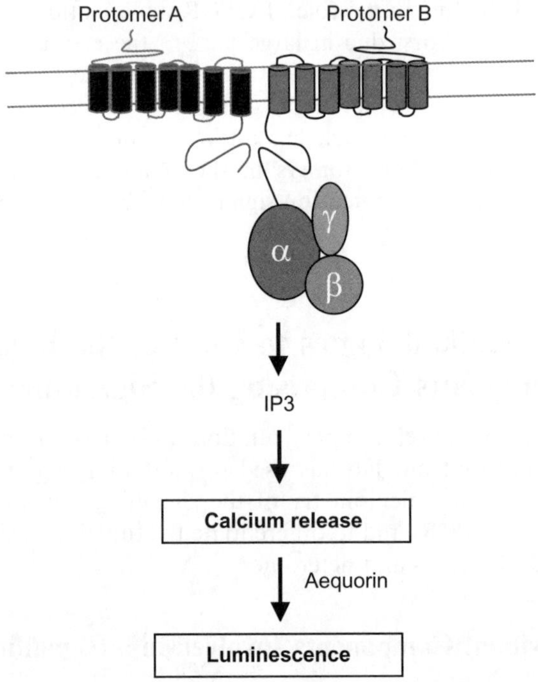

Figure 12.2 Schematic representation of the functional complementation paradigm and receptor-activated luminescence in the aequorin system.

linker prevented positioning of the fused G protein where it could form a productive interaction with the cytoplasmic loops of the fused receptor or another fused receptor. Furthermore, signalling of D2R–Gqi5 was not rescued by free Gqi5—probably because the fusion construct sterically blocks free Gqi5 from a productive interaction. Remarkably, however, coexpression of D2R (protomer A) and D2R–Gqi5 (protomer B), neither of which could function alone, reconstituted robust D2R agonist-mediated activation. In this scenario, the fused Gqi5 from protomer B can be activated by agonist binding to protomer A in which the cytoplasmic surface is accessible due to absence of the steric constraint imposed by the G protein fusion. The reconstitution of such a signalling unit provided a unique opportunity to manipulate each protomer independently and to determine its role in signalling.

Based on this method, Han *et al.* proposed a mechanistic explanation for the reciprocal modulation of protomer functions in a dimeric signalling complex. The minimal signalling unit, consisting of two GPCRs and a single heterotrimeric G protein, was maximally activated by agonist binding to a single protomer, which suggested an asymmetrical activated dimer. Indeed, agonist binding to the second protomer, involved in the dimer formation led to decrease in signalling, whereas inverse agonist binding to the second protomer enhanced signalling. The fact that a constitutively active receptor, which was also unable to bind agonist, blunted signalling of a coexpressed wild-type (WT) receptor indicated the importance of the conformational state of the second protomer. Thus, GPCR heteromer function is modulated not only by ligand binding to the second protomer, but also by its activation state. Although similar inferences of asymmetric activation of receptors in a dimer have been drawn in family C receptors (see chapter 11) and for other family A receptors,[59] the extent to which this mechanism is shared by other GPCRs must be evaluated experimentally. For example, it has been proposed that the serotonin 5-hydroxytryptamine 2C receptor functions more efficiently when both protomers can bind ligand,[60] suggesting a different mode of interaction in this dimer.

Although Han *et al.* took advantage of the unique situation made possible by the very short length of the D2R *C*-terminus to prevent promiscuous delivery of the G protein to other receptors or receptor pairs; the system is unlikely to be suited for use as a universal biosensor. This is due to the fact that most other GPCR's, unlike D2R, have a substantially longer *C*-terminus that can function as a flexible linker, allowing the G-protein to be activated by the protomer to which it is fused. This scenario complicates interpretation of the data, as the fused G protein can be provided to the same protomer or to other potentially unrelated receptors, as well as to a *bona fide* receptor partner. Thus, other approaches to differentiate homomeric and heteromeric signalling are necessary.

12.3.2 Differentiating Signalling of Heteromers from Homomers

Ideally, one would like to monitor interaction of the receptor signalling unit with a variety of non-chimeric unfused G proteins to mimic native signalling more faithfully. To achieve this goal, Urizar *et al.*[61] combined luminescence and/or fluorescence complementation with BRET to devise a novel biophysical method termed Complemented Donor Acceptor resonance energy transfer (CODA-RET) (Figure 12.3).

The method is based on quantifying the BRET between a receptor heteromer or homomer, and a G protein subunit. Specific receptor–receptor interaction reconstitutes a luminescent signal that serves as the energy donor of the biosensor. This establishes tight control over the receptor species engaged in the energy transfer to the G protein heterotrimer, allowing measurement of the communication between a defined receptor pair and the G protein of choice, without contribution from homomeric signalling complexes. Furthermore, this approach excludes downstream crosstalk or signalling integration since it is based on the conformational changes that occur upon ligand binding in a very confined environment.

Urizar *et al.* coexpressed mVenus labelled G protein fusions with split *Renilla luciferase 8* (RLuc8) such that the *C*-terminus of protomer A was fused to an *N*-terminal fragment (L1) and the *C*-terminus of protomer B to the *C*-terminal fragment (L2) of RLuc8. Specific energy transfer between the complemented receptors used as donor and the acceptor m-Venus-fused G protein was

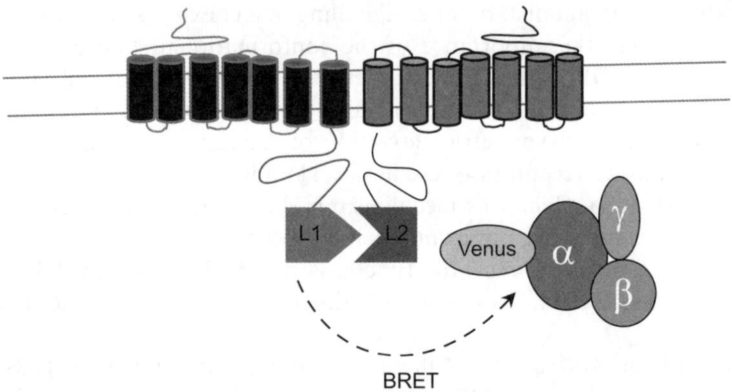

Figure 12.3 A novel biosensor for differentiating homomeric and heteromeric signalling. BRET is observed in cells coexpressing receptor (represented as black and grey 7-transmembrane proteins) fused to L1 and L2 (split RLuc8) as the complemented donor and mVenus as the acceptor inserted into the helical domain of the Gα subunit.

detected upon agonist stimulation, serving as a proximity-based receptor–receptor–G protein functional readout. This strategy eliminated the signal stemming from homomeric receptor complexes, which do not luminesce. Furthermore, no significant differences in agonist-induced BRET changes were observed when comparing the full-length receptor–luciferase fusions and the split luciferase receptor constructs for several GPCR pairs. This not only demonstrated that the splits express and function normally but also that the orientation of the complemented donor does not introduce a detectable alteration of the receptor–G protein interface.

Urizar *et al.* applied this biosensor to study the putative D1R–D2R heteromer and demonstrated that receptor heteromerization can lead to functional selectivity, in which a drug acts differently at the same GPCR protomer depending on the identity of the partner protomer that participates in forming the signalling unit. As the luminescence signal resulted from complementation of the split luciferase fused to two defined receptors in molecular proximity and the BRET monitored direct receptor heteromer–G protein conformational rearrangement, the authors concluded that this particular case of functional selectivity results from a direct interaction of the heteromer with G protein and not due to downstream signalling. Importantly, this approach can also be applied to explore the ability of defined receptor heteromers to recruit arrestin. It could also be extended to different interacting partners, as well as to FRET-based assays by substituting split green fluorescent protein variants for the split luciferase constructs used in these studies.

In summary, receptor heteromerization is an important factor that must be considered in drug design and pharmacological analysis. Whether this is a general mechanism that can be exploited to develop new compounds selectively targeting the various GPCR heteromers remains to be established. Resonance energy transfer is an extremely versatile and highly sensitive technique that has made substantial contribution to our understanding of GPCRs. The next generation of this technique integrated with other labelling techniques has the potential to significantly increase our understanding of cellular protein networks. It promises to play an important role in future research and drug discovery by allowing analysis of drug effects on defined GPCR heteromers with very fine control over the entire biosensor system.

12.4 Conclusions

Understanding the prevalence and relevance of receptor dimerization is crucial for the pharmacological analysis and design of new drugs that could target specific aspects of cellular signalling. Development of heteromer-specific compounds could result in the generation of new drugs inducing fewer side effects due to their more cell-type specific action based on the coexpression pattern of receptors composed of the targeted heteromeric complex.

It is clear from the evidence described above that much more research is required to answer the questions concerning GPCR oligomerization and its consequences on cell signalling. GPCR di/oligo-merization has been implicated in heterologous cell systems and recently also in native tissues, but it is still a major challenge to prove the functional relevance of such receptor complexes. Therefore, development of new approaches and their adaptation to *in vivo* studies are extremely crucial to advancing our knowledge of GPCR biology.

Acknowledgements

This work was supported in part by NIH grants DA022413 and MH54137 (J.A.J.) and by the Lieber Center for Schizophrenia Research and Treatment.

References

1. J.-P. Pin, R. Neubig, M. Bouvier, L. Devi, M. Filizola, J. A. Javitch, M. J. Lohse, G. Milligan, K. Palczewski, M. Parmentier and M. Spedding, *Pharmacol. Rev.*, 2007, **59**, 5–13.
2. S. Ferre, R. Baler, M. Bouvier, M. G. Caron, L. A. Devi, T. Durroux, K. Fuxe, S. R. George, J. A. Javitch, M. J. Lohse, K. Mackie, G. Milligan, K. D. G. Pfleger, J.-P. Pin, N. D. Volkow, M. Waldhoer, A. S. Woods and R. Franco, *Nat. Chem. Biol.*, 2009, **5**, 131–134.
3. L. Albizu, M. Cottet, M. Kralikova, S. Stoev, R. Seyer, I. Brabet, T. Roux, H. Bazin, E. Bourrier, L. Lamarque, C. Breton, M. L. Rives, A. Newman, J. Javitch, E. Trinquet, M. Manning, J.-P. Pin, B. Mouillac and T. Durroux, *Nat. Chem. Biol.*, 2010, **6**, 587–594.
4. G. Milligan, *Mol. Pharmacol.*, 2004, **66**, 1–7.

5. M. Bouvier, N. Heveker, R. Jockers, S. Marullo and G. Milligan, *Nat. Methods*, 2007, **4**, 3–4.
6. H. R. Bourne and R. Nicoll, *Cell*, 1993, **72**(Suppl), 65–75.
7. F. Philip, G. Kadamur, R. G. Silos, J. Woodson and E. M. Ross, *Curr. Biol.*, 2010, **20**, 1327–1335.
8. R. A. Rebres, T. I. Roach, I. D. Fraser, F. Philip, C. Moon, K. M. Lin, J. Liu, L. Santat, L. Cheadle, E. M. Ross, M. I. Simon and W. E. Seaman, *J. Biol. Chem.*, 2011, **286**, 942–951.
9. M. Bouvier, *Nat. Rev. Neurosci.*, 2001, **2**, 274–286.
10. S. R. George, T. Fan, Z. Xie, R. Tse, V. Tam, G. Varghese and B. F. O'Dowd, *J. Biol. Chem.*, 2000, **275**, 26128–26135.
11. P. S. H. Park and J. W. Wells, *J. Neurochem.*, 2004, **90**, 537–548.
12. F. Philip, P. Sengupta and S. Scarlata, *J. Biol. Chem.*, 2007, **282**(26), 19203–19216.
13. M. Chabre and M. le Maire, *Biochemistry*, 2005, **44**(27), 9395–9403.
14. J. R. James, M. I. Oliveira, A. M. Carmo, A. Iaboni and S. J. Davis, *Nat. Methods*, 2006, **3**, 1001–1006.
15. J. A. Hern, A. H. Baig, G. I. Mashanov, B. Birdsall, J. E. T. Corrie, S. Lazareno, J. E. Molloy and N. J. M. Birdsall, *Proc. Natl. Acad. Sci. U. S. A.*, 2010, **107**, 2693–2698.
16. S. Dorsch, K. N. Klotz, S. Engelhardt, M. J. Lohse and M. Bunemann, *Nat. Methods*, 2009, **6**, 225–230.
17. J. M. Fonseca and N. A. Lambert, *Mol. Pharmacol.*, 2009, **75**, 1296–1299.
18. A. Hasbi, B. F. O'Dowd and S. R. George, *Curr. Opin. Pharmacol.*, 2010, **10**, 93–99.
19. J. K. Dreher and D. M. Jackson, *Brain Res.*, 1989, **487**, 267–277.
20. K. Kita, T. Shiratani, K. Takenouchi, H. Fukuzako and M. Takigawa, *Eur. Neuropsychopharmacol.*, 1999, **9**, 1–7.
21. O. Aizman, H. Brismar, P. Uhlen, E. Zettergren, A. I. Levey, H. Forssberg, P. Greengard and A. Aperia, *Nat. Neurosci.*, 2000, **3**, 226–230.
22. S. P. Lee, C. H. So, A. J. Rashid, G. Varghese, R. Cheng, A. J. Lanca, B. F. O'Dowd and S. R. George, *J. Biol. Chem.*, 2004, **279**, 35671–35678.
23. C. H. So, G. Varghese, K. J. Curley, M. M. Kong, M. Alijaniaram, X. Ji, T. Nguyen, B. F. O'Dowd and S. R. George, *Mol. Pharmacol.*, 2005, **68**, 568–578.
24. A. J. Rashid, C. H. So, M. M. Kong, T. Furtak, M. El-Ghundi, R. Cheng, B. F. O'Dowd and S. R. George, *Proc. Natl. Acad. Sci. U. S. A.*, 2007, **104**, 654–659.
25. C. H. So, V. Verma, M. Alijaniaram, R. Cheng, A. J. Rashid, B. F. O'Dowd and S. R. George, *Mol. Pharmacol.*, 2009, **75**, 843–854.
26. H. Y. Wang, A. S. Undie and E. Friedman, *Mol. Pharmacol.*, 1995, **48**, 988–994.
27. L. Q. Jin, S. Goswami, G. Cai, X. Zhen and E. Friedman, *J. Neurochem.*, 2003, **85**, 378–386.
28. A. Sahu, K. R. Tyeryar, H. O. Vongtau, D. R. Sibley and A. S. Undieh, *Mol. Pharmacol.*, 2009, **75**, 447–453.

29. C. C. la Mannoury, S. Vidal, V. Pasteau, D. Cussac and M. J. Millan, *Neuropharmacology*, 2007, **52**, 1003–1014.
30. D. J. Surmeier, J. Ding, M. Day, Z. Wang and W. Shen, *Trends Neurosci.*, 2007, **30**, 228–235.
31. S. Ferre, A. S. Woods, G. Navarro, M. Aymerich, C. Lluis and R. Franco, *Curr. Opin. Pharmacol.*, 2010, **10**, 67–72.
32. J. Hillion, M. Canals, M. Torvinen, V. Casado, R. Scott, A. Terasmaa, A. Hansson, S. Watson, M. E. Olah, J. Mallol, E. I. Canela, M. Zoli, L. F. Agnati, C. F. Ibanez, C. Lluis, R. Franco, S. Ferre and K. Fuxe, *J. Biol. Chem.*, 2002, **277**, 18091–18097.
33. S. Ferre, G. von Euler, B. Johansson, B. B. Fredholm and K. Fuxe, *Proc. Natl. Acad. Sci. U. S. A*, 1991, **88**, 7238–7241.
34. Z. az-Cabiale, Y. Hurd, D. Guidolin, U. B. Finnman, M. Zoli, L. F. Agnati, J. J. Vanderhaeghen, K. Fuxe and S. Ferre, *Neuroreport*, 2001, **12**, 1831–1834.
35. R. Franco, V. Casado, A. Cortes, J. Mallol, F. Ciruela, S. Ferre, C. Lluis and E. I. Canela, *Br. J. Pharmacol.*, 2008, **153**(Suppl 1), S90–S98.
36. S. Ferre, C. Quiroz, A. S. Woods, R. Cunha, P. Popoli, F. Ciruela, C. Lluis, R. Franco, K. Azdad and S. N. Schiffmann, *Curr. Pharm. Des.*, 2008, **14**, 1468–1474.
37. B. Kull, S. Ferre, G. Arslan, P. Svenningsson, K. Fuxe, C. Owman and B. B. Fredholm, *Biochem. Pharmacol.*, 1999, **58**, 1035–1045.
38. T. Voss, E. Wallner, A. P. Czernilofsky and M. Freissmuth, *J. Biol. Chem.*, 1993, **268**, 4637–4642.
39. F. Ciruela, J. Burgueno, V. Casado, M. Canals, D. Marcellino, S. R. Goldberg, M. Bader, K. Fuxe, L. F. Agnati, C. Lluis, R. Franco, S. Ferre and A. S. Woods, *Anal. Chem.*, 2004, **76**, 5354–5363.
40. A. S. Woods and S. Ferre, *J. Proteome. Res.*, 2005, **4**, 1397–1402.
41. K. Azdad, D. Gall, A. S. Woods, C. Ledent, S. Ferre and S. N. Schiffmann, *Neuropsychopharmacology*, 2009, **34**, 972–986.
42. I. Gsandtner and M. Freissmuth, *Mol. Pharmacol.*, 2006, **70**, 447–449.
43. S. Ferre, M. Karcz-Kubicha, B. T. Hope, P. Popoli, J. Burgueno, M. A. Gutierrez, V. Casado, K. Fuxe, S. R. Goldberg, C. Lluis, R. Franco and F. Ciruela, *Proc. Natl. Acad. Sci. U. S. A.*, 2002, **99**, 11940–11945.
44. P. Carriba, G. Navarro, F. Ciruela, S. Ferre, V. Casado, L. Agnati, A. Cortes, J. Mallol, K. Fuxe, E. I. Canela, C. Lluis and R. Franco, *Nat. Methods*, 2008, **5**, 727–733.
45. N. Cabello, J. Gandia, D. C. Bertarelli, M. Watanabe, C. Lluis, R. Franco, S. Ferre, R. Lujan and F. Ciruela, *J. Neurochem.*, 2009, **109**, 1497–1507.
46. M. Hirono, T. Yoshioka and S. Konishi, *Nat. Neurosci.*, 2001, **4**, 1207–1216.
47. T. Tabata, K. Araishi, K. Hashimoto, Y. Hashimotodani, H. van der Putten, B. Bettler and M. Kano, *Proc. Natl. Acad. Sci. U. S. A.*, 2004, **101**, 16952–16957.
48. Y. Kamikubo, T. Tabata, S. Kakizawa, D. Kawakami, M. Watanabe, A. Ogura, M. Iino and M. Kano, *J. Physiol.*, 2007, **585**, 549–563.

49. M. L. Rives, C. Vol, Y. Fukazawa, N. Tinel, E. Trinquet, M. A. Ayoub, R. Shigemoto, J.-P. Pin and L. Prezeau, *EMBO J.*, 2009, **28**, 2195–2208.
50. F. Philip, G. Kadamur, R. G. Silos, J. Woodson and E. M. Ross, *Curr. Biol.*, 2010, **20**, 1327–1335.
51. E. M. Hur and K. T. Kim, *Cell Signal.*, 2002, **14**, 397–405.
52. T. D. Werry, G. F. Wilkinson and G. B. Willars, *Biochem. J.*, 2003, **374**, 281–296.
53. M. Rocheville, D. C. Lange, U. Kumar, S. C. Patel, R. C. Patel and Y. C. Patel, *Science*, 2000, **288**, 154–157.
54. C. S. Kearn, K. Blake-Palmer, E. Daniel, K. Mackie and M. Glass, *Mol. Pharmacol.*, 2005, **67**, 1697–1704.
55. J.-P. Pin, T. Galvez and L. Prezeau, *Pharmacol. Therapeut.*, 2003, **98**, 325–354.
56. T. H. Bayburt, A. J. Leitz, G. F. Xie, D. D. Oprian and S. G. Sligar, *J. Biol. Chem.*, 2007, **282**, 14875–14881.
57. M. R. Whorton, M. P. Bokoch, S. G. F. Rasmussen, B. Huang, R. N. Zare, B. Kobilka and R. K. Sunahara, *Proc. Natl. Acad. Sci. U. S. A.*, 2007, **104**, 7682–7687.
58. Y. Han, I. S. Moreira, E. Urizar, H. Weinstein and J. A. Javitch, *Nat. Chem. Biol.*, 2009, **5**, 688–695.
59. J. P. Vilardaga, V. O. Nikolaev, K. Lorenz, S. Ferrandon, Z. Zhuang and M. J. Lohse, *Nat. Chem. Biol.*, 2008, **4**, 126–131.
60. K. Herrick-Davis, E. Grinde, T. J. Harrigan and J. E. Mazurkiewicz, *J. Biol. Chem.*, 2005, **280**, 40144–40151.
61. E. Urizar, H. Yano, R. Kolster, C. Galés, N. Lambert and J. A. Javitch, *Nat. Chem. Biol.*, 2011, In press.

CHAPTER 13

Functional Crosstalk between Group I Metabotropic Glutamate Receptors and Ionotropic Glutamate Receptors Controls Synaptic Transmission

JOEL BOCKAERT,* LAURENT FAGNI AND JULIE PERROY

Institute of Functional Genomics, CNRS UMR5203, INSERM U661, Universities I and II of Montpellier, 141 Rue de la Cardonille, 34000 Montpellier, France

13.1 Introduction

Half century ago, the discovery of the neuro-excitant properties of glutamate opened a new area in the understanding of brain physiology and pathologies. Ionotropic glutamate (iGlu) receptors (channel receptors permeable to Na^+, K^+ and eventually Ca^{2+}) are the main actors of fast excitatory synaptic transmission.[1,2] In the early 1980s, these receptors were classified into three groups based on their sensitivity to three exogenous agonists: *N*-methyl-D-aspartate (NMDA), quisqualate and kainate.[3] It is remarkable that this first classification was later entirely confirmed by the cloning of these receptor subunits. Quisqualate receptors are also activated by α-amino-3-hydroxy-5-methyl-4-isoxazolepropionic acid

RSC Drug Discovery Series No. 8
G Protein-Coupled Receptors: From Structure to Function
Edited by Jesús Giraldo and Jean-Philippe Pin

Published by the Royal Society of Chemistry, www.rsc.org

(AMPA) and were renamed AMPA receptors after the discovery that, in contrast to AMPA, quisqualate also activated metabotropic glutamate (mGlu) receptors coupled to the Gq protein/phospholipase C pathway.[4,5]

The first member of G protein-coupled glutamate receptor family to be cloned was the metabotropic glutamate receptor coupled to Gq, mGlu$_1$.[6] Seven other mGlu receptors were subsequently identified (Table 13.1).[7] All belong to the G-protein-coupled receptor (GPCR) superfamily, thus constituting the first members of the so-called GPCR class C. This GPCR family also includes taste receptors that are sensitive to sweet molecules and umami substances, $GABA_B$ receptors, Ca^{2+}-receptors and pheromone receptors.[8,9] Thus, as for most of neurotransmitter systems, the mammalian glutamatergic system displays both ionotropic and metabotropic receptors. Typical iGlu receptors act in a fast time (millisecond range) whereas mGlu receptors are slowly acting receptors (0.1–1 second range).

Table 13.1 Nomenclature for glutamate receptors. For each table cell are listed the names of the human gene (*italic*), the IUPHAR subunit nomenclature (**bold**) and commonly used names of the receptors.

NMDA receptors	*AMPA receptors*	*Kainate receptors*	*Orphan glutamate receptors*	*Metabotropic glutamate receptors*
GRIN1 **GluN1** NR1 GluRζ1	*GRIA1* **GluA1** GluR1 GluRA	*GRIK1* **GluK1** GluR5	*GRID1* **GluD1** GluRδ1	*GRM1* **mGlu$_1$** mGluR1
GRIN2A **GluN2A** NR2A GluRε1	*GRIA2* **GluA2** GluR2 GluRB	*GRIK2* **GluK2** GluR6	*GRID2* **GluD2** GluRδ2	*GRM2* **mGlu$_2$** mGluR2
GRNI2B **GluN2B** NR2B GluRε2	*GRIA3* **GluA3** GluR3 GluRC	*GRIK3* **GluK3** GluR7		*GRM3* **mGlu$_3$** mGluR3
GRIN2C **GluN2C** NR2C GluRε3	*GRIA4* **GluA4** GluR4 GluRD	*GRIK4* **GluK4** KA-1		*GRM4* **mGlu$_4$** mGluR4
GRIN2D **GluN2D** NR2D GluRε4		*GRIK5* **GluK5** KA-2		*GRM5* **mGlu$_5$** mGuR5
GRIN3A **GluN3A** NR3A				*GRM6* **mGlu$_6$** mGuR6
GRIN3B **GluN3B** NR3B				*GRM7* **mGlu$_7$** mGluR7
				GRM8 **mGlu$_8$** mGluR8

NMDA receptors are tetramers composed of two GluN1 subunits (eight different GluN1 subunits are generated by alternative splicing of a single gene) and two GluN2 subunits (Table 13.1). GluN2 subunits (GluN2A, GluN2B, GluN2C and GluN2D) are generated by four different genes. Two genes encode for two different GluN3 subunits (GluN3A and GluN3B), which are associated with GluN1 and probably GluN2A/B when present. NMDA receptors are permeable to Na^{2+}, K^{+} and Ca^{2+}. Activation of NMDA receptors requires coincident binding of glutamate and glycine to GluN2 and GluN1/GluN3 subunits, respectively, and a third concomitant event (membrane depolarization) to release the voltage-dependent channel block exerted by extracellular Mg^{2+} on the pore of the receptor channel.[10] Under physiological conditions, these receptors have an essential role in the induction of some forms of synaptic plasticity, whereas in pathological conditions such as stroke, amyotrophic lateral sclerosis and other neurodegenerative diseases, NMDA receptors mediate neuronal death.

AMPA receptors are tetramers essentially composed of two GluA1, two GluA3 or two GluA4 subunits, associated with two GluA2 subunits (Table 13.1).[11–13] However, endogenous GluA1 or GluA3 homotetramers do exist. The lack of GluA2 subunit makes the channel permeable to Ca^{2+}, whereas GluA2-containing channels are virtually impermeable to Ca^{2+}. GluA2 lacking AMPA receptors display large single-channel conductance, an inwardly rectifying current–voltage (I-V) relationship and are blocked by the Joro-toxin (JST) polyamine.[11,14–16] The lack of permeability to Ca^{2+} of AMPA receptors is determined by post-transcriptional modification (*i.e.* RNA editing) of the uncharged amino acid glutamine (Q) to the positively charged arginine (R) in the pore lining region of the GluA2 subunit. Indeed the positively charged amino acid at the critical site makes it energetically unfavourable for Ca^{2+} to enter the pore.[11,14–16] Functional kainate (KA) receptors are pentameric assemblies of five subunits (GluK1–5) in various combinations (Table 13.1).[17] Two additional members of the iGlu receptor superfamily, GluD1 and GluD2, have been isolated by homology screening from rat and mouse brain cDNA libraries (Table 13.1). These are orphan receptors and functional modality of their channel pore remains to be determined.[18,19]

mGlu receptors are classified into three groups.[9] Group I is composed of two receptors, $mGlu_1$ and $mGlu_5$. The $mGlu_1$ receptor subtype is expressed as four splice variants: a long *C*-terminal form ($mGlu_{1a}$, 350 residues) and three shorter *C*-terminal forms ($mGlu_{1b,c,d}$), whereas the $mGlu_5$ receptor is expressed as two long *C*-terminal splice variants ($mGlu_{5a,b}$). These receptors are coupled to the Gq protein-dependent pathway. They are mainly postsynaptic and localized in the forebrain, hippocampus and striatum for $mGlu_5$, and in the mid-brain, hippocampus, striatum and cerebellum for $mGlu_1$. Group II mGlu receptors include $mGlu_2$, $mGlu_3$ and the *Drosophila* $DmGlu_A$. They are coupled to the Gi/Go pathways and so inhibit adenylyl cyclase and Ca^{2+} channels whereas they activate K^{+} channels. They are localized both pre- and postsynaptically. Group III mGlu receptors include $mGlu_4$, $mGlu_6$, $mGlu_7$ and $mGlu_8$. The $mGlu_7$ and $mGlu_8$ subtypes display two splice variants ($mGlu_{7a,b}$ and $mGlu_{8a,b}$) of roughly similar length.[9] They are mainly presynaptic and down-regulate synaptic glutamate and ã-aminobutyric acid (GABA) release.[20–22]

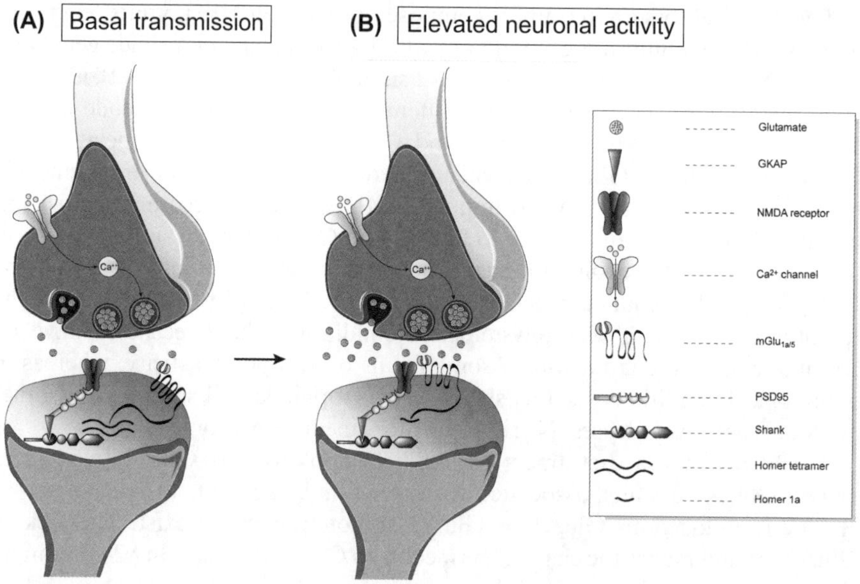

Figure 13.1 Dynamics of postsynaptic glutamate receptor complex. (**A**) Under normal conditions, the $mGlu_{1a/5}$ receptor is physically linked to the NMDA receptor *via* a multiprotein complex composed of Homer dimer, shank, GKAP and PSD95 proteins. (**B**) Following elevated neuronal activity, induced monomeric Homer1a competes with Homer dimers and disrupts the physical link between $mGlu_{1a/5}$ and NMDA receptor complex, allowing physical interaction between the receptors and their mutual inhibition.

At the postsynaptic membrane, iGlu receptors are localized within the postsynaptic density, whereas group I mGlu receptor subtypes are localized in the perisynaptic zone surrounding the postsynaptic density.[23,24] This specific organization makes direct physical interaction between these receptors very unlikely, except under certain conditions as described below (see also Figure 13.1). However, $mGlu_{1a/5}$ receptors are indirectly linked to NMDA receptors *via* a large sub-synaptic multiprotein network that includes in sequence, tetrameric Homer proteins (four subunits linked head-to-tail by their coiled-coil *C*-terminal domain), guanylate kinase-associated protein (GKAP), multimeric shank PDZ protein complex and the PDZ postsynaptic density protein of 95 kDa (PSD-95) (Figure 13.1A).[25–28] The $mGlu_{1a/5}$ receptor subtypes are also linked to inositol-1,4,5-trisphosphate (IP3)/ryanodine receptors, transient receptor potential cation channels 1 and 4 (TRPC1, TRPC4) and the phosphoinositide 3-kinase (PI3K) enhancer-long (PIKE-L) *via* the Homer tetrameric structure.[25–27] This network is probably dynamic since the monomeric Homer1a/ania3 splice variants, which lack a coiled-coil domain and compete with the binding of tetrameric forms of Homer, are supposed to disrupt the physical link between $mGlu_{1a/5}$ receptor and the Shank complex (Figure 13.1B).[25–27] Those variants are immediate early genes that are rapidly synthesized following elevated neuronal activity, or acute or chronic use of drugs of abuse such as cocaine, amphetamine, phencyclidines and nicotine.[29–33]

The dynamic properties of the postsynaptic glutamate receptor complex prompted us to revisit the reductionist view of independent actions of iGlu *vs.* mGlu receptors, and to rather consider these receptors as functionally interconnected entities that can act in concert in given physiological and/or pathological conditions. Our knowledge of functional interplay between postsynaptic iGlu and mGlu receptors covers the group I mGlu and both AMPA and NMDA receptors. Here we review the functional interplay between these receptors, focusing on synaptic plasticity as well as neurological and psychiatric disorders.

13.2 Functional Crosstalk Between mGlu and NMDA Receptors

13.2.1 Potentiation of NMDA Receptor-mediated Responses by group I mGlu receptors

A prominent effect of group I mGlu receptor activation on iGlu receptors is to selectively potentiate agonist-induced NMDA receptor responses in a number of brain structures.[34,35] In cultured striatal neurons, for instance, stimulation of NMDA receptors rapidly induces phosphorylation/activation of cAMP response element-binding (CREB). Co-incubation of the neurons with NMDA and group I mGluR agonists, at concentrations which alone do not activate CREB, induces CREB phosphorylation. This synergistic effect is blocked by protein kinase C (PKC) inhibitors, suggesting that coincident detection processes between NMDA and group I mGlu receptors activate PKC. Interestingly, the synergistic effect between group I mGlu and NMDA receptors is blocked by selective $mGlu_5$ but not $mGlu_1$ receptor antagonists.[36–39] Recent studies using the selective mGlu5 receptor positive allosteric modulator ADX-47273 corroborated these results. ADX-47273 potentiates the NMDA receptor-mediated Ca^{2+} response induced by subthreshold concentrations of agonists *R*, *S*-dihydrophenylglycine (DHPG) and NMDA) in primary neuronal cultures, enhances NMDA receptor-dependent long-term potentiation (LTP) induced by tetanic electrical stimulation of afferents in rat hippocampal slices, and reverses NMDA receptor antagonist-induced hyperlocomotion in rats.[40]

What is the mechanism underlying facilitation of NMDA receptor-mediated responses by $mGlu_5$ receptors? Studies using the *Xenopus* oocyte expression system show that stimulation of group I mGlu receptors triggers the recruitment of new NMDA channels to the plasma membrane *via* regulation of exocytosis processes.[41] Although most of the studies indicate a prevalent modulatory effect of $mGlu_5$ rather than mGlu1 receptors on NMDA responses, it seems that both group I mGlu receptor subtypes can modulate NMDA receptors, but *via* two distinct pathways. In hippocampal and cortical neurons, activation of $mGlu_5$ receptors potentiates NMDA receptor-mediated current *via* G protein and PKC-dependent activation of the Src family of kinases.[42] Activation of $mGlu_1$ receptor also potentiates NMDA-induced currents, but *via* a G protein-independent pathway that involves Pyk2/CAKβ in addition to Src family kinases.[43,44] Tyrosine phosphorylation of GluN2A/B subunits by Src *via* either pathway stabilizes NMDA receptors at the cell surface, thereby increasing the receptor responses.[42] Reciprocally, NMDA receptor activation can potentiate group I mGlu receptor responses. Stimulation of NMDA receptors induces dephosphorylation of $mGlu_5$ receptors by the Ca^{2+}-activated protein phosphatase 2B/calcineurin at sites responsible for PKC-dependent desensitization, thus potentiating mGlu5 receptor responses.[45,46]

As mentioned above, $mGlu_{1a/5}$ subunits are physically linked to NMDA receptors at the postsynaptic membrane *via* scaffolding proteins (Figure 13.1A). $mGlu_{1a}$ and $mGlu_5$ bind directly to Homer tetramers, which in turn interact with the Shank–GKAP–PSD95–NMDA receptor complex.[28] Several findings suggest that such a physical association may support functional interplay between iGlu and mGlu receptors. Co-application of NMDA and the non selective group I mGlu receptor agonist DHPG synergistically increases ERK1/2 phosphorylation (ERK: extracellular regulated kinase). This effect requires the integrity of PSD95–NMDA receptor binding and is blocked by small interfering RNAs that selectively reduce cellular level of multimeric Homer1b/c proteins. Extracellular regulated kinase (ERK) activation through this pathway results in phosphorylation of Elk-1 and CREB transcription factors, thereby regulating gene expression through synapse to nucleus communication.[47]

13.2.2 Inhibition of NMDA Receptor-mediated Responses by Group I mGlu Receptors

In contrast to the findings described above, studies indicate that group I mGlu receptor agonists inhibit NMDA-evoked responses in cultured neurons.[48–50] This effect is not well characterized, but recent findings suggest that it may result from direct interaction between $mGlu_{1a/5}$ and NMDA receptors (NMDARs). Direct physical association between $mGlu_5$ and GluN1A/2B subunits has been shown using bioluminescence resonance energy transfer (BRET) technology in a heterologous HEK293 expression system. This interaction triggers reciprocal inhibition of the mGlu and NMDA receptors, independently of activation of both receptors by agonists.[51]

We propose the following hypothesis. NMDA receptors are localized within the postsynaptic density (PSD) whereas group I mGlu receptor subtypes are found outside the PSD. The Shank–GKAP–PSD95 complex might be responsible for this peculiar postsynaptic organization. The constitutively expressed multimeric Homer complex binds to $mGlu_{1a/5}$ subunits and physically links them to the GKAP–PSD95–NMDAR complex. Elevated neuronal activity induces expression of Homer1a, which acts as a dominant negative endogenous protein by disrupting the physical link between $mGlu_{1a/5}$ receptors and Shank. The released $mGlu_{1a/5}$ receptors may then move towards and interact with/inhibit NMDA receptors. This model needs to be confirmed and could provide new molecular process by which $mGlu_{1a/5}$ receptors would dynamically downregulate synaptic NMDA receptor functions (Figure 13.1B).

13.2.3 Role of Group I mGlu Receptors in the Plasticity of NMDA Receptor-mediated Synaptic Responses

An important aspect of functional mGlu-NMDA receptor crosstalk is to control the long-term potentiation (LTP) of the NMDA receptor-mediated component of glutamatergic EPSC - excitatory post-synaptic current (NMDAR-EPSC). Short bursts of perforant path stimulation, or application of group I mGlu receptor agonist, induce a rapid onset LTP of NMDAR-EPSC in granule cells of the rat dentate gyrus *in vitro* that depends on co-activation of mGlu and NMDA receptors, as well as mobilization of a PKC-dependent pathway.[52] Indeed recent studies performed at the mossy fibre to CA3 pyramidal synapse suggest that this LTP most likely occurs *via* mobilization of postsynaptic SNARE-dependent insertion of vesicle-associated NMDA receptors.[38] A similar selective LTP of NMDAR-EPSC can also be induced at the hippocampal mossy fibre synapse *via* activation of postsynaptic adenosine A_{2A} receptors.[53]

Neurons of mammalian brain can also undergo group I mGlu receptor-dependent long-term depression (LTD) of NMDAR-EPSC (in addition to LTD of AMPAR-EPSC) following low frequency stimulation of excitatory afferents.[54] In the Schaffer/commissural afferents to CA1 hippocampal pyramidal cell synapse, this form of LTD results from NMDA receptor internalization.[55] The actin stabilizer, jasplakinolide, as well as activation of extrasynaptic NMDA receptors following blockade of glutamate uptake, abolishes this mGlu receptor-induced LTD.[56] The authors proposed that this effect results from lateral movement of NMDA receptors towards extra-synaptic pools. These findings suggest that both local endocytosis and migration of NMDA receptors within the membrane could participate to the mGlu receptor-induced LTD of NMDAR-EPSC. This LTD does not involve tyrosine kinases or phosphatases, or protein synthesis, and therefore differs from the tyrosine kinase dependent LTD of NMDAR-EPSC and LTD of AMPAR-EPSC which relies on protein tyrosine phosphatase activity.[57–60] It also differs from LTD of NMDAR-EPSC which involves actin depolymerisation, but not endocytosis.[61,62]

The switch between the facilitatory and inhibitory effects of group I mGlu receptors on NMDA receptors critically depends on intracellular Ca^{2+} concentration. Group I mGlu receptor agonist-induced depression of NMDA currents can be converted into potentiation following strong buffering of intracellular Ca^{2+} by BAPTA (1,2bis(o-aminophenoxy)ethane-*N*,*N*,*N*′-tetraacetic acid) in hippocampal CA3 pyramidal cells.[63] This is consistent with the well-known depressive effect of elevated intracellular Ca^{2+} concentration on NMDA receptor responses. A number of Ca^{2+}-dependent proteins including calmodulin, calcineurin, PKC and α-actinin-2 are implicated in NMDA receptor regulation. Of particular interest is the striatal-enriched protein tyrosine phosphatase (STEP) that is also expressed in various brain neuronal cell types and which is activated by the Ca^{2+}-dependent protein phosphatase calcineurin upon Ca^{2+} entry through NMDA receptors.[64]

In light of these observations, it has been proposed that low intracellular concentration of Ca^{2+} favours facilitation of NMDA receptors by a Src-dependent pathway, whereas highly increased intracellular Ca^{2+} favours inhibition of the receptors by a STEP-dependent pathway.[63] At the perforant path–granule cell synapse of the hippocampal formation, a similar Ca^{2+}-dependent bidirectional modulation of NMDA receptors would determine whether the receptors undergo mGlu receptor-dependent LTP or LTD.[65]

13.2.4 Inter-dependent Pathological Action of NMDA and Group I mGlu Receptors

Several studies have shown a synergistic effect of $mGlu_5$ and NMDA receptor antagonists on instrumental learning, working memory, prepulse inhibition and stereotypy.[66,67] Thus in the condition of NMDA receptor hypofunction, potentiation of $mGlu_5$ receptor responses may be beneficial to improve cognitive functions.[67–69] It has also been shown that co-activation of NMDA and group I mGlu receptors could trigger hyperalgesia after inflammation.[70] $mGlu_5$ receptor antagonists improve parkinsonian symptomatology in 6-hydroxydopamine (6-OHDA) lesioned rats.[67] Finally, $mGlu_5$ receptor activity-dependent loss of GluN1 immunoreactivity from the synaptic membrane is observed following long-lasting application of soluble oligomers of amyloid β synthetic peptide (Aβo).[71] Together these observations suggest functional interaction between NMDA and group I mGlu receptors (namely mGu_5) in pathological conditions.

Down-regulation of NMDA receptor-mediated responses by group I mGlu receptors is consistent with the well-known neuroprotective action of this class of mGlu receptors against NMDA receptor-mediated toxicity.[72] This neuroprotective effect seems to involve both $mGlu_1$ and $mGlu_5$ receptor subtypes, depending on the cell type, as well as the presence of the GluN2C-containing NMDA receptors.[73]

13.2.5 Endocannabinoid-dependent Modulation of Synaptic Strength by Synergistic Action of NMDA and Group I mGlu Receptors

The endocannabinoid system contributes to activity-dependent synaptic modulation in the central nervous system (CNS), namely short- and long-term suppression of transmitter release.[74] Indeed application of NMDA facilitates endocannabinoid release in cultured hippocampal neurons and this effect is enhanced by co-application of a group I mGlu receptor agonist.[75] Similarly to hippocampus, the release of endocannabinoids by cerebellar inhibitory interneurons regulates the strength of excitatory parallel fibre inputs to Purkinje cells. This regulation is also mediated by both NMDA and group I mGlu receptor activation.[76]

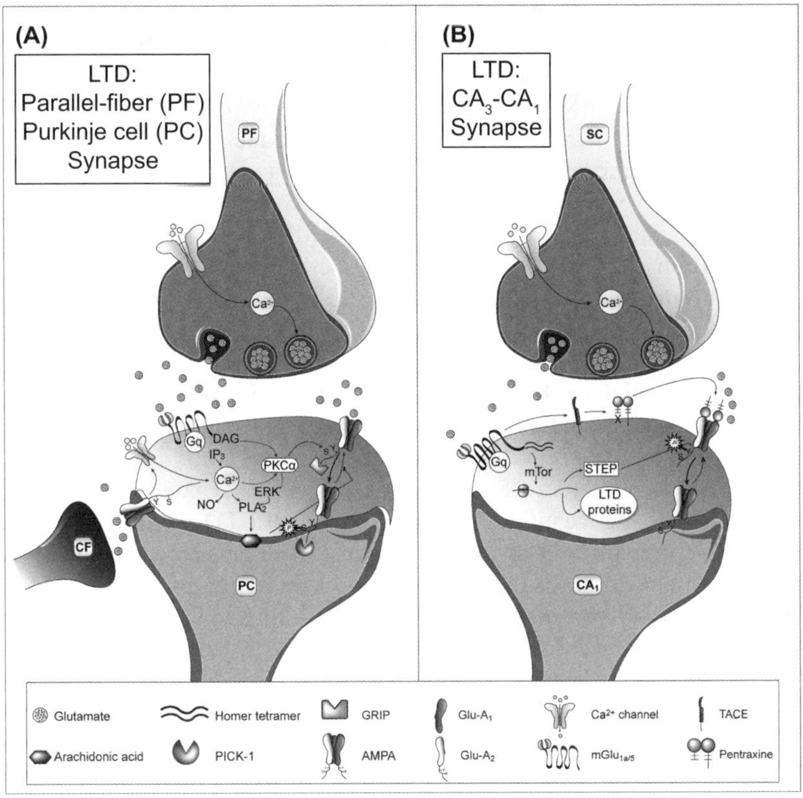

Figure 13.2 Models of mGluR-mediated LTD at three different types of synapse. (**A**) cerebellar mGlu-LTD. This type of LTD occurs in Purkinje neurons (PC) following coincident activity of parallel (PF) and climbing (CF) afferents. It results from co-activation of post-synaptic mGlu1 and AMPA receptor. This leads to PCKα activation, phosphorylation of GluA2 subunits and internalization of AMPA receptors *via* PICK1 and GRIP adaptor proteins. (**B**) Hippocampal mGlu-LTD. This type of LTD occurs at the synapse between Schaffer-collateral afferents (SC) and CA1 pyramidal cells. It results from internalization of AMPA receptors following STEP-mediated dephosphorylation of the GluA2 subunit, $mGlu_5$-induced activation of the metalloproteinase TACE and hydrolysis of pentraxin extracellular domain.

13.3 Group I mGlu Receptor-mediated Long-term Depression

The best-studied synaptic plasticity induced by $mGlu_1$ receptors is LTD of glutamatergic excitatory synaptic strength ($mGlu_1$-LTD; Figure 13.2A).[77–79] $mGlu_1$-LTD was first described in the granule cell parallel fibre (PF)/Purkinje cell (PC) synapse in cerebellum.[79] In PCs, coincident activation of postsynaptic $mGlu_1$ receptor and membrane depolarisation *via* climbing fibre (CF) mediated activation of AMPA receptors leads to mobilization of intracellular Ca^{2+} and opening of voltage-dependent Ca^{2+} channels, followed by PKCα activation. Other signalling events such as phospholipase A2 activation—likely *via* its phosphorylation by the PKC-activated ERK pathway or mobilization of the Na^{+}/Ca^{2+} exchanger and nitrous oxide (NO) production—also participate in $mGlu_1$-LTD.[80–82] Expression of this LTD is the result of internalization of GluA2-containing AMPA receptors. In PCs, AMPA receptors are mainly composed of GluA2/GluA3 subunits. Internalization of AMPA receptors correlates with phosphorylation of GluA2 at *C*-terminal Ser880 by PKCα̃. Interaction of unphosphorylated GluA2 with the PDZ domain glutamate receptor interacting protein (GRIP) stabilizes AMPA receptors at the plasma membrane, whereas phosphorylated GluA2 preferentially interacts with protein interacting with PKC (PICK1), thus leading to clathrin-dependent endocytosis of AMPA receptors. In parallel, arachidonic acid production following phospholipase A2 activation inhibits AMPA receptor recycling and promotes the receptor degradation.[80,83]

At the Schaffer collateral (SC)/CA1 pyramidal cell synapses, two protocols have been used to induce $mGlu_5$ receptor-dependent LTD ($mGlu_5$-LTD; Figure 13.2B) in slice preparations or awake rodents.[78] The first protocol consists of a brief (5 min) application of the specific group I mGlu receptor agonist DHPG, whereas the second protocol consists of long-lasting low frequency electrical stimulation (1–3 Hz, 5–15 min) of SC afferents.[77,78] In the second protocol, both group I mGlu and muscarinic receptors cooperate to induce LTD. The mechanisms of this LTD are postsynaptic, at least in adult rodents.[77] Surprisingly, the signalling pathway triggered by the pharmacologically induced $mGlu_5$-LTD does not necessarily involve Gαq/PLC/PKC/Ca^{2+}.

As in PF–PC synapse LTD, the depression of synaptic strength results from AMPA receptor internalization, but what are the mechanisms at the SC–CA1 pyramidal cell synapses? The first is a tyrosine dephosphorylation (Y848, Y852, Y855) of the *C*-terminal GluA2 subunit, which modifies the GluA2–GRIP/PICK1 interactions.[84,85] The second is a group I mGlu receptor-mediated activation of the matrix metalloproteinase tumour necrosis factor-a-converting enzyme (TACE), which hydrolyses the extracellular domain of neuronal pentraxin.[86] The released peptide then clusters and stimulates endocytosis of AMPA receptors. The group I mGlu receptor-mediated tyrosine dephosphorylation involves a phosphatase (STEP) that is rapidly synthesized during the induction phase of the LTD.

It has been proposed that hippocampal mGlu-LTD contributes to Fragile X mental retardation syndrome. The syndrome is the most common inherited disease that causes mental retardation and it has been associated to mutation of the *fmr1* gene. The larger mGlu-LTD that characterizes the Fragile X syndrome in *fmr1* KO mice can be corrected by reducing the level of $mGlu_5$ subunit expression, supporting the theory that the Fragile X syndrome results from unbalanced activation of group I mGlu (namely $mGlu_5$) receptors and impaired synaptic maturation.[26,87,88]

It is admitted that long-term synaptic plasticity in the brain reward system, including neurons of ventral tegmental area (VTA), nucleus accumbens and pre-frontal cortex, constitutes the cellular basis of addiction induced by drugs of abuse such as cocaine, morphine, nicotine, ethanol and amphetamines. The drugs generate a pathological reinforcement of the natural rewarding process until they become excessive and compulsive. Recently in both the VTA and nucleus accumbens, it has been shown that cocaine potentiates glutamatergic synaptic transmission onto GABAergic interneurons and this correlates with a change in subunit composition of AMPA receptors. In naïve rodents, AMPA receptors typically consist of GluA2-containing heteromers, as in most brain areas. After cocaine treatments (hours in the VTA and days in the accumbens), a fraction of AMPA receptors becomes homomeric GluA1 (GluA2-lacking) Ca^{2+} permeable rectifying receptors.[79,89,90] This modification can be reversed by activation of $mGlu_1$, which induces mammalian target of rapamycin (mTOR) dependent GluA2 synthesis and substitution of GluA1 homomeric AMPA receptors by GluA1/GluA2 heteromeric AMPA receptors.[90,91] Such a reversal of cocaine-induced plasticity by $mGlu_1$ receptor is in line with behavioural sensitization without drug exposure, when $mGlu_1$ is hypofunctioning.[92]

The mGlu-LTD, which involves a switch from rectifying Ca^{2+} permeable homomeric GluA1-containing AMPA receptors to Ca^{2+} impermeable heteromeric GluA1/GluA2 AMPA receptors is similar to the synaptic plasticity observed at cerebellar PF–stellate cell synapses. At this synapse, activation of both group I mGlu receptors and Ca^{2+} permeable rectifying AMPA receptors drives a switch in AMPA receptor subunit composition, leading to protein-synthesis dependent synaptic expression of Ca^{2+} impermeable (likely GluA2-containing) AMPA receptors. Interestingly, $GABA_B$ receptors promote the effect of group I mGlu receptors. Both receptors are tonically activated and control this form of synaptic plasticity.[93]

13.4 Conclusions

We have provided several examples showing that functional crosstalk between iGlu and mGlu receptors plays an important role in synaptic function. This seems largely supported by scaffolding proteins that physically link these receptors at the postsynaptic membrane. Such scaffold multiprotein complexes are clearly dynamically regulated, allowing the receptors to be included or released from the complex in order to laterally move outside of the synapse and/or to be internalized. Such an exquisite regulation provides fine-tuning of synaptic efficacy, including long-term plasticity. Here we provided several examples showing that alteration of such regulatory mechanisms can lead to severe neurological and psychiatric disorders. As such, these processes and underlying protein–protein interactions should be considered valuable targets for the development of future therapeutic strategies.

Acknowledgements

We would like to thank Muriel Gien-Asari for art work. This work was supported by DIATRAL (a EuroBioMed and Medicen programme financed by

Fonds Unique Interministeriel), REPLACES (EC Contract LSHM-CT-2008-222918), NewTagAE and SYNGEN (ANR-06-NEUR-035-01 and ANR-08-MNPS-037-01 projects), CNRS, Inserm and the Universities of Montpellier I and II.

References

1. T. Hayashi, *Jpn. J. Physiol.*, 1952, **3**, 46–64.
2. D. R. Curtis, J. W. Phillis and J. C. Watkins, *Nature*, 1959, **183**, 611–612.
3. R. Dingledine, K. Borges, D. Bowie and S. F. Traynelis, *Pharmacol. Rev.*, 1999, **51**, 7–61.
4. F. Sladeczek, J. P. Pin, M. Recasens, J. Bockaert and S. Weiss, *Nature*, 1985, **317**, 717–719.
5. F. Nicoletti, J. L. Meek, M. J. Iadarola, D. M. Chuang, B. L. Roth and E. Costa, *J. Neurochem.*, 1986, **46**, 40–46.
6. M. Masu, Y. Tanabe, K. Tsuchida, R. Shigemoto and S. Nakanishi, *Nature*, 1991, **349**, 760–765.
7. C. M. Niswender and P. J. Conn, *Annu. Rev. Pharmacol. Toxicol.*, 2010, **50**, 295–322.
8. J. P. Pin, J. Kniazeff, C. Goudet, A. S. Bessis, J. Liu, T. Galvez, F. Acher, P. Rondard and L. Prezeau, *Biol. Cell*, 2004, **96**, 335–342.
9. P. J. Conn and J. P. Pin, *Annu. Rev. Pharmacol. Toxicol.*, 1997, **37**, 205–237.
10. P. Paoletti and J. Neyton, *Curr. Opin. Pharmacol.*, 2007, **7**, 39–47.
11. I. H. Greger, E. B. Ziff and A. C. Penn, *Trends Neurosci.*, 2007, **30**, 407–416.
12. D. Bleakman and D. Lodge, *Neuropharmacology*, 1998, **37**, 1187–1204.
13. A. I. Sobolevsky, M. P. Rosconi and E. Gouaux, *Nature*, 2009, **462**, 745–756.
14. G. T. Swanson, S. K. Kamboj and S. G. Cull-Candy, *J. Neurosci.*, 1997, **17**, 58–69.
15. C. D. Aizenman, G. Munoz-Elias and H. T. Cline, *Neuron*, 2002, **34**, 623–634.
16. D. Soto, I. D. Coombs, L. Kelly, M. Farrant and S. G. Cull-Candy, *Nat. Neurosci.*, 2007, **10**, 1260–1267.
17. G. L. Collingridge, R. W. Olsen, J. Peters and M. Spedding, *Neuropharmacology*, 2009, **56**, 2–5.
18. M. Yamazaki, K. Araki, A. Shibata and M. Mishina, *Biochem. Biophys. Res. Commun.*, 1992, **183**, 886–892.
19. H. Lomeli, R. Sprengel, D. J. Laurie, G. Kohr, A. Herb, P. H. Seeburg and W. Wisden, *FEBS Lett.*, 1993, **315**, 318–322.
20. J. Perroy, O. El Far, F. Bertaso, J. P. Pin, H. Betz, J. Bockaert and L. Fagni, *EMBO J.*, 2002, **21**, 2990–2999.
21. N. Kogo, Y. Dalezios, M. Capogna, F. Ferraguti, R. Shigemoto and P. Somogyi, *Eur J. Neurosci.*, 2004, **19**, 2727–2740.

22. S. R. Bradley, D. G. Standaert, K. J. Rhodes, H. D. Rees, C. M. Testa, A. I. Levey and P. J. Conn, *J. Comp. Neurol.*, 1999, **407**, 33–46.
23. F. Ferraguti and R. Shigemoto, *Cell Tissue Res.*, 2006, **326**, 483–504.
24. M. M. Bolton, T. A. Blanpied and M. D. Ehlers, *Cell. Mol. Life Sci.*, 2000, **57**, 1517–1525.
25. L. Fagni, F. Ango, J. Perroy and J. Bockaert, *Semin. Cell Dev. Biol.*, 2004, **15**, 289–298.
26. J. Bockaert, J. Perroy, C. Becamel, P. Marin and L. Fagni, *Annu. Rev. Pharmacol. Toxicol.*, 2010, **50**, 89–109.
27. J. Bockaert, L. Fagni, A. Dumuis and P. Marin, *Pharmacol. Ther.*, 2004, **103**, 203–221.
28. M. D. Ehlers, *Curr. Biol.*, 1999, **9**, R848–850.
29. P. R. Brakeman, A. A. Lanahan, R. O'Brien, K. Roche, C. A. Barnes, R. L. Huganir and P. F. Worley, *Nature*, 1997, **386**, 284–288.
30. G. C. Zhang, L. M. Mao, X. Y. Liu, N. K. Parelkar, A. Arora, L. Yang, M. Hains, E. E. Fibuch and J. Q. Wang, *Mol. Pharmacol.*, 2007, **71**, 1148–1158.
31. K. Hashimoto, T. Nakahara, H. Yamada, M. Hirano, T. Kuroki and S. Kanba, *Neurochem. Int.*, 2007, **51**, 227–232.
32. S. M. Cochran, M. Fujimura, B. J. Morris and J. A. Pratt, *Synapse*, 2002, **46**, 206–214.
33. K. Rezvani, Y. Teng, D. Shim and M. De Biasi, *J. Neurosci.*, 2007, **27**, 10508–10519.
34. A. Pisani, P. Calabresi, D. Centonze and G. Bernardi, *Br. J. Pharmacol.*, 1997, **120**, 1007–1014.
35. H. Awad, G. W. Hubert, Y. Smith, A. I. Levey and P. J. Conn, *J. Neurosci.*, 2000, **20**, 7871–7879.
36. L. Mao and J. Q. Wang, *Neuroscience*, 2002, **115**, 395–402.
37. F. Codazzi, A. Di Cesare, N. Chiulli, A. Albanese, T. Meyer, D. Zacchetti and F. Grohovaz, *J. Neurosci.*, 2006, **26**, 3404–3411.
38. H. B. Kwon and P. E. Castillo, *Neuron*, 2008, **57**, 108–120.
39. Z. Jia, Y. Lu, J. Henderson, F. Taverna, C. Romano, W. Abramow-Newerly, J. M. Wojtowicz and J. Roder, *Learn. Mem.*, 1998, **5**, 331–343.
40. H. Rosenbrock, G. Kramer, S. Hobson, E. Koros, M. Grundl, M. Grauert, K. G. Reymann and U. H. Schroder, *Eur. J. Pharmacol.*, 2010.
41. J. Y. Lan, V. A. Skeberdis, T. Jover, X. Zheng, M. V. Bennett and R. S. Zukin, *J. Neurosci.*, 2001, **21**, 6058–6068.
42. M. W. Salter and L. V. Kalia, *Nat. Rev. Neurosci.*, 2004, **5**, 317–328.
43. P. Benquet, C. E. Gee and U. Gerber, *J. Neurosci.*, 2002, **22**, 9679–9686.
44. V. Heidinger, P. Manzerra, X. Q. Wang, U. Strasser, S. P. Yu, D. W. Choi and M. M. Behrens, *J. Neurosci.*, 2002, **22**, 5452–5461.
45. S. Alagarsamy, M. J. Marino, S. T. Rouse, R. W. t. Gereau, S. F. Heinemann and P. J. Conn, *Nat. Neurosci.*, 1999, **2**, 234–240.
46. S. Alagarsamy, J. Saugstad, L. Warren, I. M. Mansuy, R. W. T. Gereau and P. J. Conn, *Neuropharmacology*, 2005, **49**(Suppl 1), 135–145.
47. L. Yang, L. Mao, Q. Tang, S. Samdani, Z. Liu and J. Q. Wang, *J. Neurosci.*, 2004, **24**, 10846–10857.

48. C. S. Colwell and M. S. Levine, *Neuroscience*, 1994, **61**, 497–507.
49. S. P. Yu, S. L. Sensi, L. M. Canzoniero, A. Buisson and D. W. Choi, *J. Physiol.*, 1997, **499**(Pt 3), 721–732.
50. M. Blaabjerg, L. Fang, J. Zimmer and A. Baskys, *Exp. Neurol.*, 2003, **183**, 573–580.
51. J. Perroy, F. Raynaud, V. Homburger, M. C. Rousset, L. Telley, J. Bockaert and L. Fagni, *J. Biol. Chem*, 2008, **283**, 6799–6805.
52. J. J. O'Connor, M. J. Rowan and R. Anwyl, *J. Neurosci.*, 1995, **15**, 2013–2020.
53. N. Rebola, R. Lujan, R. A. Cunha and C. Mulle, *Neuron*, 2008, **57**, 121–134.
54. P. L. Yi, F. C. Chang, J. J. Tsai, C. R. Hung and P. W. Gean, *Neurosci. Lett.*, 1995, **185**, 207–210.
55. E. M. Snyder, B. D. Philpot, K. M. Huber, X. Dong, J. R. Fallon and M. F. Bear, *Nat. Neurosci.*, 2001, **4**, 1079–1085.
56. D. R. Ireland and W. C. Abraham, *J. Neurophysiol*, 2009, **101**, 1375–1385.
57. C. M. Coussens, J. M. Williams, D. R. Ireland and W. C. Abraham, *Neuropharmacology*, 2000, **39**, 2267–2277.
58. C. C. Huang and K. S. Hsu, *J. Neurochem.*, 2006, **96**, 179–194.
59. A. Kumar and T. C. Foster, *J. Neurophysiol.*, 2007, **98**, 2729–2736.
60. P. R. Moult, C. M. Gladding, T. M. Sanderson, S. M. Fitzjohn, Z. I. Bashir, E. Molnar and G. L. Collingridge, *J. Neurosci.*, 2006, **26**, 2544–2554.
61. W. Morishita, H. Marie and R. C. Malenka, *Nat. Neurosci.*, 2005, **8**, 1043–1050.
62. Y. Peng, J. Zhao, Q. H. Gu, R. Q. Chen, Z. Xu, J. Z. Yan, S. H. Wang, S. Y. Liu, Z. Chen and W. Lu, *Hippocampus*, 2010, **20**, 646–658.
63. A. A. Grishin, C. E. Gee, U. Gerber and P. Benquet, *J. Neurosci.*, 2004, **24**, 350–355.
64. S. Paul, A. C. Nairn, P. Wang and P. J. Lombroso, *Nat. Neurosci.*, 2003, **6**, 34–42.
65. S. C. Harney, M. Rowan and R. Anwyl, *J. Neurosci.*, 2006, **26**, 1128–1132.
66. H. Homayoun and B. Moghaddam, *Eur. J. Pharmacol.*, 2010.
67. N. Turle–Lorenzo, N. Breysse, C. Baunez and M. Amalric, *Psychopharmacology*, 2005, **179**, 117–127.
68. M. J. Marino and P. J. Conn, *Curr. Opin. Pharmacol.*, 2006, **6**, 98–102.
69. B. Moghaddam, *Psychopharmacology*, 2004, **174**, 39–44.
70. W. Guo, F. Wei, S. Zou, M. T. Robbins, S. Sugiyo, T. Ikeda, J. C. Tu, P. F. Worley, R. Dubner and K. Ren, *J. Neurosci.*, 2004, **24**, 9161–9173.
71. M. Renner, P. N. Lacor, P. T. Velasco, J. Xu, A. Contractor, W. L. Klein and A. Triller, *Neuron*, 2010, **66**, 739–754.
72. F. Nicoletti, V. Bruno, M. V. Catania, G. Battaglia, A. Copani, G. Barbagallo, V. Cena, J. Sanchez-Prieto, P. F. Spano and M. Pizzi, *Neuropharmacology*, 1999, **38**, 1477–1484.
73. M. Pizzi, F. Boroni, K. M. Bianchetti, M. Memo and P. Spano, *Eur. J. Neurosci.*, 1999, **11**, 2489–2496.

74. V. Chevaleyre, K. A. Takahashi and P. E. Castillo, *Annu. Rev. Neurosci.*, 2006, **29**, 37–76.
75. T. Ohno-Shosaku, Y. Hashimotodani, M. Ano, S. Takeda, H. Tsubokawa and M. Kano, *J. Physiol.*, 2007, **584**, 407–418.
76. M. Beierlein and W. G. Regehr, *J. Neurosci.*, 2006, **26**, 9935–9943.
77. C. Bellone, C. Luscher and M. Mameli, *Cell. Mol. Life Sci.*, 2008, **65**, 2913–2923.
78. C. M. Gladding, S. M. Fitzjohn and E. Molnar, *Pharmacol. Rev.*, 2009, **61**, 395–412.
79. C. Luscher and K. M. Huber, *Neuron*, 2010, **65**, 445–459.
80. T. D. Le, Y. Shirai, T. Okamoto, T. Tatsukawa, S. Nagao, T. Shimizu and M. Ito, *Proc. Natl. Acad. Sci. U. S. A.*, 2010, **107**, 3198–3203.
81. M. Mashimo, T. Hirabayashi, T. Murayama and T. Shimizu, *J. Cell Sci.*, 2008, **121**, 3015–3024.
82. A. Dumuis, J. P. Pin, K. Oomagari, M. Sebben and J. Bockaert, *Nature*, 1990, **347**, 182–184.
83. J. P. Steinberg, K. Takamiya, Y. Shen, J. Xia, M. E. Rubio, S. Yu, W. Jin, G. M. Thomas, D. J. Linden and R. L. Huganir, *Neuron*, 2006, **49**, 845–860.
84. Y. Zhang, D. V. Venkitaramani, C. M. Gladding, Y. Zhang, P. Kurup, E. Molnar, G. L. Collingridge and P. J. Lombroso, *J. Neurosci.*, 2008, **28**, 10561–10566.
85. C. M. Gladding, V. J. Collett, Z. Jia, Z. I. Bashir, G. L. Collingridge and E. Molnar, *Mol. Cell. Neurosci.*, 2009, **40**, 267–279.
86. R. W. Cho, J. M. Park, S. B. Wolff, D. Xu, C. Hopf, J. A. Kim, R. C. Reddy, R. S. Petralia, M. S. Perin, D. J. Linden and P. F. Worley, *Neuron*, 2008, **57**, 858–871.
87. M. F. Bear, G. Dolen, E. Osterweil and N. Nagarajan, *Neuropsychopharmacology*, 2008, **33**, 84–87.
88. M. W. Waung and K. M. Huber, *Curr. Opin. Neurobiol.*, 2009, **19**, 319–326.
89. K. L. Conrad, K. Y. Tseng, J. L. Uejima, J. M. Reimers, L. J. Heng, Y. Shaham, M. Marinelli and M. E. Wolf, *Nature*, 2008, **454**, 118–121.
90. C. Bellone and C. Luscher, *Nat. Neurosci.*, 2006, **9**, 636–641.
91. M. Mameli, B. Balland, R. Lujan and C. Luscher, *Science*, 2007, **317**, 530–533.
92. K. K. Szumlinski, S. Toda, L. D. Middaugh, P. F. Worley and P. W. Kalivas, *Ann. N. Y. Acad. Sci.*, 2003, **1003**, 468–471.
93. L. Kelly, M. Farrant and S. G. Cull-Candy, *Nat. Neurosci.*, 2009, **12**, 593–601.

CHAPTER 14

Modulating Receptor Function through RAMPs

JOSEPH J. GINGELL AND DEBBIE L. HAY*

School of Biological Sciences and Maurice Wilkins Centre for Molecular Biodiscovery, University of Auckland, Auckland, New Zealand

14.1 Introduction

Receptor activity-modifying proteins (RAMPs) are a family of three single-transmembrane spanning proteins that are capable of forming complexes with G protein-coupled receptors (GPCRs), altering their trafficking, glycosylation, pharmacology and signalling properties.

Their interactions with the family B GPCRs—the calcitonin receptor (CTR) and the calcitonin receptor-like receptor (CLR)—are the most extensively studied. RAMPs have been demonstrated to modulate the pharmacological profile of these receptors, generating the calcitonin gene-related peptide (CGRP) receptor and adrenomedullin (AM) receptor phenotypes with CLR and the amylin receptors with CTR. There is strong evidence that RAMPs are involved in ligand binding, although it is unclear whether this is a direct action or if RAMPs allosterically modulate the CLR or CTR to modulate ligand affinity. RAMPs are also capable of interacting with other family B GPCRs and the family C calcium-sensing receptor, suggesting they may have a broader role.

RSC Drug Discovery Series No. 8
G Protein-Coupled Receptors: From Structure to Function
Edited by Jesús Giraldo and Jean-Philippe Pin

Published by the Royal Society of Chemistry, www.rsc.org

14.2 Discovery

RAMPs were discovered whilst trying to clone the receptor for CGRP.[1] There was evidence that CGRP acted through a GPCR, as stimulation of certain tissues and cell lines by CGRP led to the accumulation of cAMP. The orphan receptor CLR (with close homology to the CTR) was activated by CGRP when transfected into HEK293 cells but did not respond to CGRP in Cos-7 cells.[2–4] A rat homologue of CLR was also found to be responsive to CGRP in HEK293 cells, suggesting a factor present in HEK293 cells conferred high affinity for CGRP on the receptor.[5]

This factor, RAMP1, was identified by McLatchie and colleagues in 1998. It was found that a family of single transmembrane domain proteins (RAMPs) were required for functional expression of the CLR at the cell surface, explaining the lack of response to CGRP in cells that lacked RAMP expression.[1,4,5] A single cDNA encoding the 148 amino acid protein RAMP1 was isolated from SK-N-MC cells (cells known to express CGRP receptors). When the cRNA was injected into *Xenopus* oocytes, a cAMP response to CGRP was observed. The structure of the protein was unexpected as it was not a GPCR; co-transfection with CLR resulted in a cAMP response and CGRP binding in HEK293T cells. Neither RAMP1 nor CLR alone are efficiently expressed at the cell surface; both components are required for co-transport to the cell surface. Some cell lines express endogenous RAMPs, explaining the earlier findings.

The discovery of an accessory protein that promotes expression of GPCRs was not unprecedented as, in *Caenorhabditis elegans*, ODR4 had been shown to control expression of olfactory receptors,[6] while the cyclophilin NinaA is involved in the expression of opsins in *Drosophila*.[7] The unique finding of McLatchie and colleagues was that two RAMP1 related sequences were found which could also interact with CLR, but instead of forming CGRP receptors, formed high affinity receptors for the related peptide AM.

14.3 Pharmacology

The key feature of RAMPs is the ability to modulate the pharmacology of CLR and CTR. RAMP1/CLR is a high affinity CGRP receptor; RAMP2/CLR has high affinity for AM but low affinity for CGRP, while RAMP3/CLR has high affinity for AM and moderate affinity for CGRP. The pharmacology of these receptors is generally consistent across several mammalian species—human, rat, mouse and pig.[8–10]It was initially thought that differential glycosylation was responsible for the differences in pharmacology observed at the CLR with the different RAMPs, but this was later discounted.[11,12]

RAMPs also modulate the pharmacology of CTR[13–17] to form high affinity receptors for amylin. The properties of receptors for this peptide, denoted AMY receptors, are variable and depend on the host cell environment and endogenous CLR/RAMP levels.[14,18] The International Union of Basic and Clinical Pharmacology (IUPHAR) consensus is that there are three AMY receptors formed by CTR with RAMPs 1–3, but this can be further subdivided

Table 14.1 Summary of pharmacological profiles for calcitonin, CGRP, amylin and adrenomedullin receptors.[21]

Receptor type	*Molecular composition*	*Order of agonist potency*
Amylin receptor (AMY_1)	RAMP1 + CTR	rAMY ≥ αCGRP > hCT > AM
Amylin receptor (AMY_2)	RAMP2 + CTR	Poorly defined pharmacology
Amylin receptor (AMY_3)	RAMP3 + CTR	rAMY > αCGRP > AM
Adrenomedullin receptor (AM_1)	RAMP2 + CLR	AM > αCGRP > AMY
Adrenomedullin receptor (AM_2)	RAMP3 + CLR	AM > αCGRP > AMY
CGRP receptor	RAMP1 + CLR	αCGRP > AM ≥ AMY
Calcitonin receptor (CTR)	CTR	CT > rAMY,αCGRP > AM

Abbreviations: AMY = amylin; AM = adrenomedullin; CT = calcitonin; r = rat.

based on the splice variant of CTR present. The most highly expressed form is designated $CTR_{(a)}$, while a splice variant which contains 16 additional amino acids in the first intracellular loop is designated $CTR_{(b)}$.[19] Amylin binds only with high affinity to RAMP-complexed CTR, while CGRP binds with high affinity to the RAMP1/CTR complex as well as the RAMP1/CLR complex.

The interactions with AM are inconsistent; in HEK293 cells AM sensitivity with all three RAMPs at the CTR has been demonstrated,[20] but in Cos-7 cells, AM only weakly binds and activates AMY receptors.[14,17] The influence of cellular background may make a significant contribution to the pharmacology observed by different researchers as Cos-7 cells have weaker coupling to Gα compared to HEK293 cells (see Section 14.6.2). The pharmacological profiles of the receptors are summarized in Table 14.1.

14.4 Distribution

RAMPs, CLR and CTR have a widespread distribution, with mRNA found in many cell lines and tissues.[22] Studies in rat have found CLR with RAMP1 and RAMP2 located in the atria and ventricle, and CLR with all three RAMPs in the cerebellum, spinal cord, liver, spleen, vas deferens and lungs.[23] In the rodent brain, RAMPs were detected in several regions, and overlap with areas expressing CLR and CTR; this correlating with sites known to bind CGRP and amylin.[24,25] In human CLR and RAMP2 have been identified in aortic endothelial cells and umbilical vascular endothelial cells.[26]

Immunostaining found CLR and RAMP1 colocalized in the trigeminal nucleus of the rat[27] and similar findings were recently also observed in humans.[28] Overall the distribution of the RAMPs, CLR and CTR account for the observed pharmacology present in the cell lines and tissues studied.[23,29] However, it is important to point out that there is still much work to do in terms of colocalizing RAMPs with GPCR partners due to the paucity of high quality antibodies.

14.5 Structure of RAMPs

All three RAMPs have a common structure of a single transmembrane domain α-helix, an extracellular *N*-terminus and a short intracellular *C*-terminus. The largest part is the extracellular *N*-terminus which is approximately 90 amino acids in RAMP1 and RAMP3, depending on the precise cleavage site of the predicted signal peptide. RAMP2 has a longer predicted signal peptide and is also extended by around 13 residues at its *N*-terminus (Figure 14.1).[1] There is about 31% amino acid sequence identity between the three human RAMPs and 56% sequence similarity to each other;[1] the transmembrane domains are conserved across species while the *N*-terminal domains are less conserved. All RAMPs across species contain four cysteine residues which are predicted to form disulphide bonds. This suggests a common secondary structure.

A crystal structure of the human RAMP1 *N*-terminal domain was reported in 2008 by Kusano and colleagues.[30] The structure is tri-helical (Figure 14.2) with three disulphide bonds between C27–C82, C40–C72 and C57–C104. Helices 1 and 2 are mostly hydrophobic in character and interaction between the two is important for the structure, while helices 2 and 3 were identified as those most likely to be involved in interaction with the receptor and ligand binding. The disulphide bond between C40 and C72 in helices 1 and 2 was suggested to be important for maintaining the structure of the ligand binding site.

A hydrophobic patch was identified which was suggested to form the receptor and ligand interaction sites. The residues R67, D71, E78, W74 and W84 were proposed to form part of the ligand binding site, as the side chains are predicted to be exposed to solvent. The residues predicted to form the receptor interaction site were F93, H97 and F101; these are highly conserved across RAMPs.

Despite the low sequence identity, the three RAMPs are believed to share a similar structure as the two disulphide bonds important for stabilizing the RAMP1

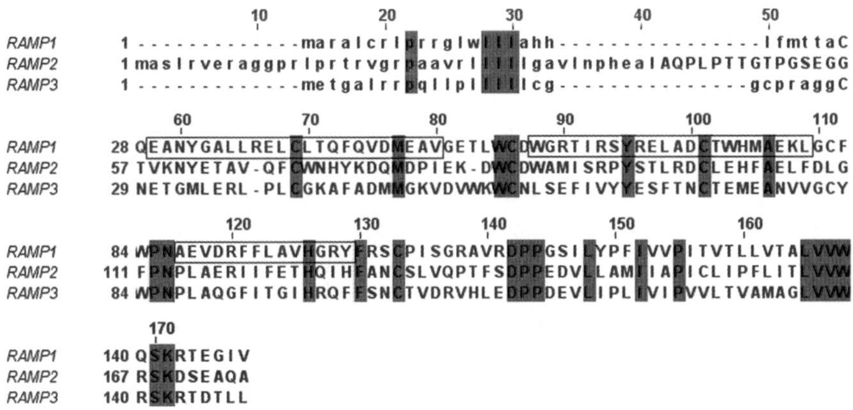

Figure 14.1 Sequence alignment of the three human RAMPs. The predicted signal sequence is shown in lower case and the boxes denote the positions of the three helices in RAMP1. Conserved residues are shaded. Alignment performed with ClustalW and edited with Jalview.

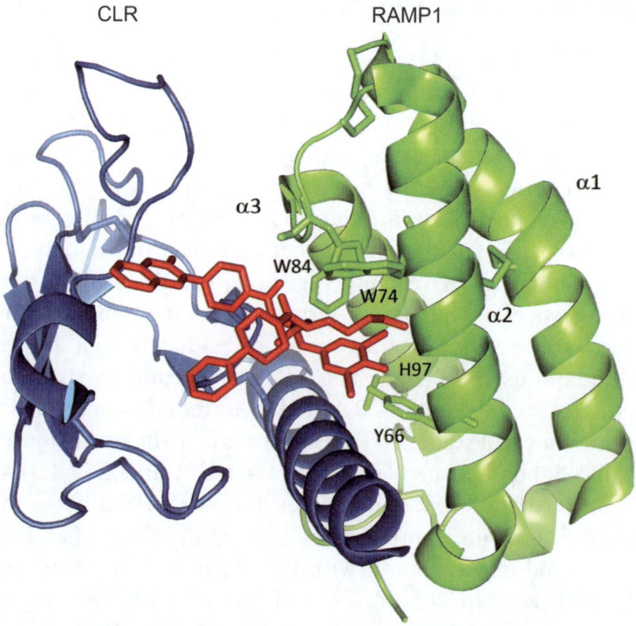

Figure 14.2 Ribbon diagram of the crystal structure of the CLR and RAMP1 ectodomains with olcegepant bound.[31] RAMP1 helices are labelled α1, α2 and α3. RAMP1 residues W74 and W84 important for olcegepant and telcagepant interactions and Y66 and H97 important for CLR interaction are shown in stick form, as are disulphide bonds.[47,54,56]

structure are conserved. The equivalent residues to those identified in RAMP1 predicted to form the receptor and ligand interaction sites are also hydrophobic.[30]

14.5.1 Structure of the CGRP Receptor Ectodomain

Recently the crystal structure of the CGRP receptor ectodomain (CLR + RAMP1) has been solved, with the small molecule antagonists telcagepant and olcegepant bound.[31] The ectodomains form a 1 : 1 complex; CLR has a similar fold to other family B GPCR ectodomain structures[32] and RAMP1 closely matched the previously solved structure. This complex was capable of binding CGRP and truncated CGRP analogues, and also small molecule antagonists—though with lower affinity than the full-length receptor complex.[33] The interaction between CLR and RAMP1 extracellular domains is stabilized by electrostatic and hydrophobic interactions; these are discussed in more detail in Section 14.6.3.

14.6 RAMP Interactions with Receptors

Although only RAMP interactions with CLR and CTR are reportedly capable of altering the pharmacological phenotype, RAMPs can associate with other family B GPCRs including:

- parathyroid hormone (PTH-1) receptor—with RAMP2;
- vasoactive intestinal polypeptide/pituitary adenylate cyclase-activating peptide (VPAC1) receptor—with RAMP1, RAMP2 and RAMP3;[34]
- secretin receptor—with RAMP3;[35]
- calcium sensing family C GPCR—with RAMP1 and RAMP3.[36]

The stoichiometry of the CLR–RAMP1 interaction is still unclear. It has been reported to be two CLR to one RAMP1 molecule,[37] though the data did not rule out two RAMP1 monomers interacting with a CLR dimer. As mentioned in Section 14.5.1, the crystal structure of the CLR/RAMP1 ectodomain complex shows 1 : 1 stoichiometry, although whether this is the case for the intact receptor is not known. Thus, there is insufficient data to be confident of the physiologically relevant stoichiometry of RAMP/GPCR complexes. No studies have been performed on the interactions with RAMP2 and RAMP3 or the CTR-RAMP interactions, so it is unclear whether the stoichiometry is the same.

14.6.1 Trafficking

A key function for RAMPs is the ability to promote the cell surface expression of CLR; RAMPs associate with the receptor in the endoplasmic reticulum and promote terminal glycosylation.[1] RAMPs 1 and 3 are also important for the trafficking of the calcium sensing receptor, as they deliver the receptor from the endoplasmic reticulum to the Golgi allowing glycosylation to occur.[36] On the other hand, RAMPs do not appear to be needed for CTR expression at the cell surface.

RAMPs can also modulate trafficking in receptor down-regulation where the RAMP/CLR complex is internalized with β-arrestin.[38] In HEK293 cells, RAMP2-linked AM_1 receptors show greater internalization than RAMP3-linked receptors.[39] The *C*-terminal residues of RAMP3 (DTLL) form a type-1 PDZ recognition site, while in rat and mouse, the residues RLL form a PDZ domain. In cells with *N*-ethylmaleimide-sensitive factor (NSF), the PDZ domain can bring the receptor back to the cell surface but there is no effect on CGRP or AM_1 receptors lacking PDZ domains.[40] Na^+/H^+ exchanger regulatory factor-1 (NHERF-1), which is also capable of binding to the PDZ domain, blocks internalization of the AM_2 receptor but not that of CGRP or AM_1 receptors.[41]

14.6.2 Signalling

RAMPs have been found to play a role in the signalling of AMY receptors. There are receptor isoform and cell background dependent differences in the ability of RAMP2 to create high affinity AMY receptors; in Cos-7 cells, RAMP2 only weakly induces the AMY receptor phenotype, but strongly induces AMY receptors when transfected with the $CTR_{(b)}$ isoform.[14,15,18] However, expression of either receptor isoform in CHO-P cells leads to strong induction of AMY receptor phenotype suggesting that the cellular background is important.[18] *C*-terminal chimeras of RAMP1 and RAMP2 showed that the RAMP1 *C*-terminus could increase cAMP production in response to CGRP compared to $CTR_{(a)}$ alone, indicating a signalling role for this domain.[42] *C*-terminal truncation of RAMP reduced amylin potency, although this was partially reversed through overexpression of Gαs.[43] Preferential coupling of AMY_1 and AMY_3 receptors to Gαs *vs.* Gαq compared to $CTR_{(a)}$ expressed alone has been demonstrated.[44] This supports a direct role for RAMPs *via* the *C*-terminal domain in G protein interactions.

RAMPs do not appear to be involved in signalling when in complex with CLR. This may be due to the presence of receptor component protein (RCP) in CGRP and AM receptor signalling.[45] RCP is an intracellular peripheral membrane protein, essential for Gαs- mediated signalling. There is currently no evidence of a role for RCP in other RAMP–GPCR complexes.

Alteration of phosphoinositide synthesis without altering cAMP formation has been observed for the interaction of the VPAC1 receptor with RAMP2;[34] however this could be due to a change in subcellular location rather than alteration of the signalling pathway. For several receptors with which RAMPs in interact, the impact of RAMPs on signalling has yet to be formally tested.

14.6.3 Mechanisms of Receptor Interaction

The best characterized molecular interactions between GPCRs and RAMPs are those between RAMP1 and CLR. The *N*-terminus alone of RAMP1 has found to be capable of inducing trafficking of CLR to the cell surface and signalling in response to CGRP, although with reduced potency, suggesting roles for multiple domains in the CLR–RAMP1 interaction.[46] The structure of the ectodomain of CLR with RAMP1 shows an interaction of the *N*-terminal α-helix of CLR with helices 2 and 3 of RAMP1; numerous hydrophobic and hydrogen bonding interactions hold the complex together. Of particular importance is Q54 of CLR which interacts with the backbone amides of RAMP1 residues R102 and C104, while CLR Q50 interacts with H97, F101 and P105.[31] Many of these observations are supported by previous findings from mutagenesis work as described below.

A study looking at the role of the conserved human RAMP residues in RAMP1 by mutating them to Ala found that the mutations Y66A and H97A significantly reduced cell surface expression,[47] suggesting these residues are of particular importance for RAMP1 interaction with CLR (Figure 14.2). F93A and F101A had a smaller effect on cell surface expression, but are also found to interact with CLR.[31,47] The residues Y66 and H97 are conserved across all three RAMPs, suggesting that they may form a common in interaction interface with all GPCRs that interact with RAMPs.[47] It is difficult to speculate as to how a CLR or other receptor dimer might affect this proposition.

Residue swapping between RAMP1 and RAMP3 found that RAMP1 with the substitutions E88L, V89A and F93I induced a significant reduction in cell surface expression when expressed with CTR with subsequent loss of potency for CGRP and rat amylin; however, no loss of cell surface expression was observed when expressed with CLR. This indicates that there may be differences in how RAMP1 interacts with CLR and CTR. Given that both RAMP1 and RAMP3 interact with the same receptors, it is difficult to draw firm conclusions on whether these residues form part of the receptor interaction site.[48]

Peptide competition experiments have indicated that the interaction between the secretin receptor and RAMP3 occurs through transmembrane domains six and seven. In contrast to interactions between RAMP1 and 3 with the CLR and CTR, the *N*-terminus does not appear to be important for this interaction.[35] There is around 30% sequence identity between RAMPs in the helical region, possibly explaining the different receptor interaction profiles between the three RAMPs.

14.6.4 RAMP Domains/Residues Involved in Ligand Interactions

The *N*-terminus of RAMPs has been found to be the key determinant of receptor pharmacology when in complex with CLR or CTR.[11,49,50] The role of domains within the *N*-termini has been further explored using chimeras between the three RAMPs, domain deletions and point mutations.

An initial study using domain deletion mutants to look at the role that RAMP1 *N*-terminal domains play in ligand binding found that residues 91–103 of RAMP1 were required for the binding of CGRP; however, using site-directed mutagenesis, no individual residues were identified that interacted with CGRP.[51] It should be noted that this could have been the result of reduced cell surface expression.

A small chimera swapping residues 86–89 of RAMP1 with the equivalent residues in RAMP3 resulted in a reduction in potency (six-fold) for hαCGRP at the CGRP receptor, but no reduction in potency with hβCGRP; at the AMY_1 receptor there was large reduction in potency for hβCGRP (350-fold), although this could be due to reduced surface expression which was not determined. Interestingly, the reciprocal chimera in RAMP3 did not affect agonist (hαCGRP and AM) potency at the AM_2 receptor.[52] These residues are present in the loop between helices 2 and 3, suggesting that this part of RAMP1 is involved in ligand binding; however there must be differences between the CGRP and AMY_1 complexes.

The RAMP1 residue W74 was found to be important for high affinity binding of the small molecule CGRP receptor antagonists olcegepant (BIBN4096BS) and telcagepant (MK-0974);[31,33,53–55] however, it is not important for CGRP binding.[44,52,53] Interestingly, the analogous position of E74 in RAMP3 has been demonstrated to be important for the binding of AM;[48,56] as the two RAMPs are predicted to share a similar structure.[57] This provides evidence that this region of helix 2 in RAMPs may be involved in forming the ligand binding site.

Residue swapping between RAMP1 and RAMP3 at position 74 increased the affinity of the CGRP receptor for AM; the opposite effect with equal magnitude was observed with the reciprocal mutation E74W in the AM_2 receptor with respect to AM potency, but no effect was observed on hαCGRP potency. W74E slightly enhanced the potency of rAMY at the AMY_1 receptor but did not have an effect on βCGRP potency,[48] providing additional evidence that W74 is not important for CGRP binding.

Mutation of the RAMP1 residues P85 and N86 (located in the loop between helices 2 and 3) to Ala resulted in substantial reductions in potency (200-fold and 10-fold respectively) in response to hαCGRP at the CGRP receptor. However, there were also large decreases in cell surface expression, making it difficult to attribute the loss in potency to disruption of ligand binding or disruption in the structure of RAMP1. The mutations L69A and T73A (on helix 2) were found to have a small effect on the potency of CGRP with CLR.[47] These residues are not exposed,[30,31] suggesting the mutations may have an indirect effect by altering the position of helix 2.

Based on the crystal structure of the RAMP1 *N*-terminus, the residues R67, D71, E78, W74 and W84 were postulated to form part of the CGRP binding site.[30] Mutation of these residues to Ala in the CGRP receptor revealed that W84 was important for binding telcagepant and compound 3 (an olcegepant analogue) as well as CGRP and CGRP(8-37).[5] R67A was found to reduce telcagepant potency and binding, but was not important for compound 3.[55] W84A was also found to significantly reduce human αCGRP and rat amylin potency at the AMY_1 receptor, whereas substitution to Phe restored some of this function suggesting a bulky residue is important at this position; both these mutations also reduced cell surface expression to a greater or lesser degree.[58] W74 and W84 are in close proximity and solvent exposed, suggesting that ligands could directly interact with these residues. The crystal structure of the CGRP receptor ectodomain reveals that these two residues form the ceiling and back of a hydrophobic binding pocket which is important for both olcegepant and telcagepant binding.[31]

NMR analysis with olcegepant bound supports these findings with interactions with the residues W74 and W84 indicated, which is also consistent with the results of mutagenesis studies.[54–56]

Overall the results of chimera and mutagenesis studies suggest that the upper region of helix 2 and the loop between helices 2 and 3 in RAMPs are important for ligand interactions, potentially by directly forming part of the binding site. Differences in the position and composition of amino acid residues in this region between RAMPs may give rise to the different pharmacological phenotypes observed. Although the structure of the CGRP receptor ectodomain does not have the peptide bound, orientation of the peptide consistent with that observed in the family B receptors, glucose-dependent insulinotropic polypeptide receptor (GIPR) and the PTH-1 receptor would likely bring the *C*-terminus of the peptide in close proximity to the loop between helices 2 and 3 of RAMP1, enabling CGRP selectivity and would be blocked by the presence of olcegepant and telcagepant.[31]

14.7 RAMPs as Drug Targets

The CGRP receptor has been identified as a target for the treatment of migraine. Currently four small molecule CGRP receptor antagonists have made it to clinical trials—olcegepant, telcagepant, BI 44370 and MK-3207.[59] Telcagepant and olcegepant are the best characterized; as previously discussed RAMP1 is important for the selectivity of these molecules for the CGRP receptor, suggesting that RAMPs themselves may represent drug targets although this could lead to non-specific effects.

Most data suggest that, for CLR and CTR complexes at least, the interface between receptor and RAMP should be targeted to generate selective drugs. SYMLIN® (pramlintide acetate), an amylin analogue, is marketed for the treatment of diabetes but small molecule amylin receptor agonists could be therapeutically superior. There is interest in pursuing both AM agonists and antagonists that could be useful in cardiovascular disease, lymphoedema and cancer.

14.8 Conclusions

RAMPs modulate the pharmacology, trafficking and signalling of CLR and CTR, although not all aspects occur with every combination of receptor and RAMP. There are a growing number of related GPCRs with which RAMPs have been reported to interact, but in many cases, the consequences of these interactions have not been studied in depth. RAMPs are an important class of protein which enhance the complexity and intricacy of GPCR function.

References

1. L. M. McLatchie, N. J. Fraser, M. J. Main, A. Wise, J. Brown, N. Thompson, R. Solari, M. G. Lee and S. M. Foord, *Nature*, 1998, **393**, 333–339.
2. N. Aiyar, K. Rand, N. A. Elshourbagy, Z. Zeng, J. E. Adamou, D. J. Bergsma and Y. Li, *J. Biol. Chem.*, 1996, **271**, 11325–11329.
3. F. Njuki, C. G. Nicholl, A. Howard, J. C. Mak, P. J. Barnes, S. I. Girgis and S. Legon, *Clin. Sci. (Lond.)*, 1993, **85**, 385–388.
4. B. Fluhmann, R. Muff, W. Hunziker, J. A. Fischer and W. Born, *Biochem. Biophys. Res. Commun.*, 1995, **206**, 341–347.
5. Z. Q. Han, H. A. Coppock, D. M. Smith, S. Van Noorden, M. W. Makgoba, C. G. Nicholl and S. Legon, *J. Mol. Endocrinol.*, 1997, **18**, 267–272.
6. N. D. Dwyer, E. R. Troemel, P. Sengupta and C. I. Bargmann, *Cell*, 1998, **93**, 455–466.
7. E. K. Baker, N. J. Colley and C. S. Zuker, *EMBO J.*, 1994, **13**, 4886–4895.
8. K. Husmann, P. M. Sexton, J. A. Fischer and W. Born, *Mol. Cell. Endocrinol.*, 2000, **162**, 35–43.
9. K. R. Oliver, S. A. Kane, C. A. Salvatore, J. J. Mallee, A. M. Kinsey, K. S. Koblan, N. Keyvan-Fouladi, R. P. Heavens, A. Wainwright, M. Jacobson, I. M. Dickerson and R. G. Hill, *Eur. J. Neurosci.*, 2001, **14**, 618–628.

10. K. Kikumoto, T. Katafuchi and N. Minamino, *Hypertens. Res.*, 2003, **26** (Suppl), S15–23.
11. S. Hilairet, S. M. Foord, F. H. Marshall and M. Bouvier, *J. Biol. Chem.*, 2001, **276**, 29575–29581.
12. A. Aldecoa, R. Gujer, J. A. Fischer and W. Born, *FEBS Lett.*, 2000, **471**, 156–160.
13. S. L. Armour, S. Foord, T. Kenakin and W. J. Chen, *J. Pharmacol. Toxicol. Methods*, 1999, **42**, 217–224.
14. G. Christopoulos, K. J. Perry, M. Morfis, N. Tilakaratne, Y. Gao, N. J. Fraser, M. J. Main, S. M. Foord and P. M. Sexton, *Mol. Pharmacol.*, 1999, **56**, 235–242.
15. R. Muff, K. Leuthauser, N. Buhlmann, S. M. Foord, J. A. Fischer and W. Born, *FEBS Lett.*, 1998, **441**, 366–368.
16. K. Leuthauser, R. Gujer, A. Aldecoa, R. A. McKinney, R. Muff, J. A. Fischer and W. Born, *Biochem. J.*, 2000, **351**(Pt 2), 347–351.
17. D. L. Hay, G. Christopoulos, A. Christopoulos, D. R. Poyner and P. M. Sexton, *Mol. Pharmacol.*, 2005, **67**, 1655–1665.
18. N. Tilakaratne, G. Christopoulos, E. T. Zumpe, S. M. Foord and P. M. Sexton, *J. Pharmacol. Exp. Ther.*, 2000, **294**, 61–72.
19. D. R. Poyner, P. M. Sexton, I. Marshall, D. M. Smith, R. Quirion, W. Born, R. Muff, J. A. Fischer and S. M. Foord, *Pharmacol. Rev.*, 2002, **54**, 233–246.
20. K. Kuwasako, K. Kitamura, Y. Nagoshi and T. Eto, *Biochem. Biophys. Res. Commun.*, 2003, **301**, 460–464.
21. D. L. Hay, D. R. Poyner and P. M. Sexton, *Pharmacol. Ther.*, 2006, **109**, 173–197.
22. P. M. Sexton, A. Albiston, M. Morfis and N. Tilakaratne, *Cell Signal.*, 2001, **13**, 73–83.
23. P. Chakravarty, T. P. Suthar, H. A. Coppock, C. G. Nicholl, S. R. Bloom, S. Legon and D. M. Smith, *Br. J. Pharmacol.*, 2000, **130**, 189–195.
24. P. M. Sexton, J. S. McKenzie, R. T. Mason, J. M. Moseley, T. J. Martin and F. A. Mendelsohn, *Neuroscience*, 1986, **19**, 1235–1245.
25. P. M. Sexton, G. Paxinos, M. A. Kenney, P. J. Wookey and K. Beaumont, *Neuroscience*, 1994, **62**, 553–567.
26. S. Kamitani, M. Asakawa, Y. Shimekake, K. Kuwasako, K. Nakahara and T. Sakata, *FEBS Lett.*, 1999, **448**, 111–114.
27. J. K. Lennerz, V. Ruhle, E. P. Ceppa, W. L. Neuhuber, N. W. Bunnett, E. F. Grady and K. Messlinger, *J. Comp. Neurol.*, 2008, **507**, 1277–1299.
28. S. Eftekhari, C. A. Salvatore, A. Calamari, S. A. Kane, J. Tajti and L. Edvinsson, *Neuroscience*, 2010, **169**, 683–696.
29. T. Choksi, D. L. Hay, S. Legon, D. R. Poyner, S. Hagner, S. R. Bloom and D. M. Smith, *Br. J. Pharmacol.*, 2002, **136**, 784–792.
30. S. Kusano, M. Kukimoto-Niino, R. Akasaka, M. Toyama, T. Terada, M. Shirouzu, T. Shindo and S. Yokoyama, *Protein Sci.*, 2008, **17**, 1907–1914.

31. E. ter Haar, C. M. Koth, N. Abdul-Manan, L. Swenson, J. T. Coll, J. A. Lippke, C. A. Lepre, M. Garcia-Guzman and J. M. Moore, *Structure*, 2010, **18**, 1083–1093.
32. F. Y. Siu and R. C. Stevens, *Structure*, 2010, **18**, 1067–1068.
33. C. M. Koth, N. Abdul-Manan, C. A. Lepre, P. J. Connolly, S. Yoo, A. K. Mohanty, J. A. Lippke, J. Zwahlen, J. T. Coll, J. D. Doran, M. Garcia-Guzman and J. M. Moore, *Biochemistry*, 2010, **49**, 1862–1872.
34. A. Christopoulos, G. Christopoulos, M. Morfis, M. Udawela, M. Laburthe, A. Couvineau, K. Kuwasako, N. Tilakaratne and P. M. Sexton, *J. Biol. Chem.*, 2003, **278**, 3293–3297.
35. K. G. Harikumar, J. Simms, G. Christopoulos, P. M. Sexton and L. J. Miller, *Biochemistry*, 2009, **48**, 11773–11785.
36. T. Bouschet, S. Martin and J. M. Henley, *J. Cell Sci.*, 2005, **118**, 4709–4720.
37. M. Heroux, M. Hogue, S. Lemieux and M. Bouvier, *J. Biol. Chem.*, 2007, **282**, 31610–31620.
38. K. Kuwasako, Y. N. Cao, C. P. Chu, S. Iwatsubo, T. Eto and K. Kitamura, *J. Biol. Chem.*, 2006, **281**, 7205–7213.
39. K. Kuwasako, Y. Shimekake, M. Masuda, K. Nakahara, T. Yoshida, M. Kitaura, K. Kitamura, T. Eto and T. Sakata, *J. Biol. Chem.*, 2000, **275**, 29602–29609.
40. J. M. Bomberger, N. Parameswaran, C. S. Hall, N. Aiyar and W. S. Spielman, *J. Biol. Chem.*, 2005, **280**, 9297–9307.
41. J. M. Bomberger, W. S. Spielman, C. S. Hall, E. J. Weinman and N. Parameswaran, *J. Biol. Chem.*, 2005, **280**, 23926–23935.
42. M. Udawela, G. Christopoulos, M. Morfis, A. Christopoulos, S. Ye, N. Tilakaratne and P. M. Sexton, *Mol. Pharmacol.*, 2006, **70**, 1750–1760.
43. M. Udawela, G. Christopoulos, N. Tilakaratne, A. Christopoulos, A. Albiston and P. M. Sexton, *Mol. Pharmacol.*, 2006, **69**, 1984–1989.
44. M. Morfis, N. Tilakaratne, S. G. Furness, G. Christopoulos, T. D. Werry, A. Christopoulos and P. M. Sexton, *Endocrinology*, 2008, **149**, 5423–5431.
45. M. A. Prado, B. Evans-Bain, K. R. Oliver and I. M. Dickerson, *Peptides*, 2001, **22**, 1773–1781.
46. T. J. Fitzsimmons, X. Zhao and S. A. Wank, *J. Biol. Chem.*, 2003, **278**, 14313–14320.
47. J. Simms, D. L. Hay, R. J. Bailey, G. Konycheva, G. Bailey, M. Wheatley and D. R. Poyner, *Biochemistry*, 2009, **48**, 198–205.
48. T. Qi, G. Christopoulos, R. J. Bailey, A. Christopoulos, P. M. Sexton and D. L. Hay, *Mol. Pharmacol.*, 2008, **74**, 1059–1071.
49. N. J. Fraser, A. Wise, J. Brown, L. M. McLatchie, M. J. Main and S. M. Foord, *Mol. Pharmacol.*, 1999, **55**, 1054–1059.
50. E. T. Zumpe, N. Tilakaratne, N. J. Fraser, G. Christopoulos, S. M. Foord and P. M. Sexton, *Biochem. Biophys. Res. Commun.*, 2000, **267**, 368–372.
51. K. Kuwasako, K. Kitamura, Y. Nagoshi, Y. N. Cao and T. Eto, *J. Biol. Chem.*, 2003, **278**, 22623–22630.

52. T. Qi, J. Simms, R. J. Bailey, M. Wheatley, D. L. Rathbone, D. L. Hay and D. R. Poyner, *Biochemistry*, 2010, **49**, 522–531.
53. J. J. Mallee, C. A. Salvatore, B. LeBourdelles, K. R. Oliver, J. Longmore, K. S. Koblan and S. A. Kane, *J. Biol. Chem.*, 2002, **277**, 14294–14298.
54. P. S. Miller, J. Barwell, D. R. Poyner, M. J. Wigglesworth, S. L. Garland and D. Donnelly, *Biochem. Biophys. Res. Commun.*, 2010, **391**, 437–442.
55. E. L. Moore, J. J. Gingell, S. A. Kane, D. L. Hay and C. A. Salvatore, *Biochem. Biophys. Res. Commun.*, 2010, **394**, 141–145.
56. D. L. Hay, G. Christopoulos, A. Christopoulos and P. M. Sexton, *Mol. Pharmacol.*, 2006, **70**, 1984–1991.
57. R. J. Bailey, J. W. Bradley, D. R. Poyner, D. L. Rathbone and D. L. Hay, *Peptides*, 2010, **31**, 579–584.
58. J. J. Gingell, T. Qi, R. J. Bailey and D. L. Hay, *Peptides*, 2010, **31**, 1400–1404.
59. M. J. Fischer, *Expert Opin. Investig. Drugs*, 2010, **19**, 815–823.

CHAPTER 15

Activation of G Protein-Coupled Receptor (GPCR) Kinases by GPCRs

JOHN J. G. TESMER

Life Sciences Institute and the Department of Pharmacology, University of Michigan, 210 Washtenaw Avenue, Ann Arbor, MI, 48104-2216, USA

15.1 Introduction

Under normal physiological conditions, signal transduction pathways are subject to multiple levels of desensitization that ensure termination of the signal after an appropriate period of time. Such processes enable cells to adapt to rapidly changing extracellular conditions and to avoid harmful metabolic consequences when they are not absolutely necessary. This is particularly true in the case of G protein coupled receptor (GPCR) signalling where trace amounts of extracellular signals, such as a single photon of light, can lead to detectable and profound physiological change.

In GPCR signalling, desensitization occurs at the level of the heterotrimeric G protein α subunit (Gα), which hydrolyses GTP to return to its deactivated GDP-bound state. For some classes of Gα subunit, GTP hydrolysis can be accelerated by regulator of G protein signalling (RGS) proteins, or by the effector proteins that they interact with (*e.g.* phospholipase Cβ enzymes or RGS homology (RH) domain-containing RhoGEFs).[1] Desensitization can also occur downstream of Gα subunits through the action of enzymes that counteract the change in concentration of second messengers (*e.g.* cAMP

RSC Drug Discovery Series No. 8
G Protein-Coupled Receptors: From Structure to Function
Edited by Jesús Giraldo and Jean-Philippe Pin

Published by the Royal Society of Chemistry, www.rsc.org

phosphodiesterases). However, if an activated GPCR remains at the cell membrane and accessible to heterotrimeric G proteins, then these compensatory mechanisms will be insufficient to fully terminate signal transduction because the receptor will catalyse the formation of additional activated G proteins.

In metazoans, an important pathway for GPCR desensitization is initiated by a family of protein kinases that phosphorylate the exposed cytoplasmic loops and/or *C*-terminal tails of activated receptors. Phosphorylation tags the receptors for sequestration by arrestin and, in most cases, endocytosis. These kinases, known as GPCR kinases (GRKs),[2] share an important quality with heterotrimeric G proteins in that they both selectively recognize receptors in their active configuration. This property also distinguishes them from second-messenger regulated protein kinases, such as protein kinase A (PKA), which can phosphorylate receptors whether they are active or not.

In vertebrates, seven GRKs (GRK1–7) are responsible for activation-dependent phosphorylation of most, if not all, of the receptors in the human genome.[3] Of these GRKs, only four are broadly expressed (GRK2, 3, 5 and 6) and thus are expected to be responsible for desensitization of the balance of GPCRs in the human genome. This is an impressive feat, especially considering the limited sequence conservation among GPCRs. Thus, GRKs must be able to distinguish and respond to the unique topology presented by the cytoplasmic surface of an activated GPCR, as opposed to specific residues presented at the cytoplasmic surface by activated receptors.[4] The activity of most GRKs is also dependent on the presence of anionic phospholipids, which are abundant in cell membranes.[5,6]

The vertebrate GRKs are grouped into three subfamilies based on their gene structure.[7] The GRK1 subfamily is vertebrate specific and includes GRK1 (rhodopsin kinase) and GRK7, which are expressed primarily in the rod and cone cells of the vertebrate retina, respectively. The GRK2 subfamily includes GRK2 and GRK3 (β-adrenergic receptor kinases 1 and 2, respectively), which have a characteristic *C*-terminal pleckstrin homology (PH) domain that binds Gβγ subunits and anionic phospholipids. The GRK4 subfamily includes GRK4, 5 and 6. GRK4 and 6 are *C*-terminally palmitoylated, whereas GRK5 appears to be constitutively bound to the membrane without any post-translational modification. All metazoans appear to have at least one GRK2-like and GRK4-like enzyme in their genome. The chief structural difference in these three GRK subfamilies is found in their *C*-terminal regions, which contain various elements that contribute to membrane targeting and, in some cases, activation. The domain organization of GRK6 is shown in Figure 15.1A.

Because of easy access to ample quantities of bovine rhodopsin, the simplicity of measuring rhodopsin activation in rod outer segments and the identification of mutations in GRK1 that cause human disease, GRK1 is arguably the best understood GRK. The activity of GRK1 was first identified in the early 1970s as being responsible for the light-induced phosphorylation of rhodopsin in rod outer segments (ROS),[8,9] where the combined action of GRK1 and

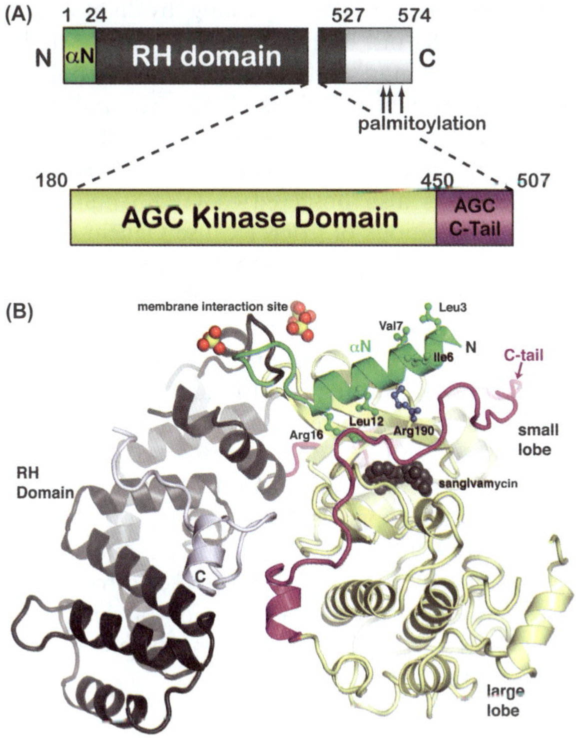

Figure 15.1 Structure of GRK6, a typical GRK. (**A**) Primary structure of GRK6. GRKs contain a characteristic RH domain which plays a scaffolding role. Preceding the RH domain is an *N*-terminal region of helical propensity (αN) that is important for receptor interactions. Inserted into the RH domain is an AGC kinase domain, including its characteristic regulatory C-tail. The sequence numbering at domain boundaries corresponds to that of human GRK6. (**B**) Atomic structure of GRK6 in a relatively closed and active conformation (PDB entry 3NYN).[57] The colour of each domain matches that shown in panel A. In this structure, the N-terminal region of GRK6 forms a single helix (αN) that packs in a cleft formed between a conserved patch on the small lobe (demarked by Arg190) and the kinase C-tail. Hydrophobic residues at the *N*-terminus of αN (Leu3, Ile6 and Val7) are positioned away from the kinase domain and are available to dock with activated GPCRs, whereas the side chains of residues at the *C*-terminus of αN (Leu12 and Arg16) pack in the cleft formed between the small lobe and the C-tail, just above the hinge joining the large and small lobes. Adjacent to αN, sulfate anions (sulfurs are coloured yellow and oxygens red) bind to a region that is responsible for binding to anionic phospholipids,[5] suggesting that this region will be in close proximity to the cell membrane in the GRK–GPCR complex. The structure was determined in complex with the adenosine analog sangivamycin (grey spheres), which occupies the ATP-binding site of the kinase domain.

visual arrestin is able to terminate signalling by light-activated rhodopsin (Rho*) in the span of ~80 ms, although this is not the rate-limiting step for termination of visual signalling.[10] In transgenic mice, deletion of the *C*-terminal 15 residues of rhodopsin, which include all the Ser and Thr residues phosphorylated by GRK1,[11] leads to prolonged decay of the photoresponse and aberrant rod cell morphology,[12] as do knockouts of GRK1.[13] Light-induced apoptosis of the retina is also observed in GRK1-null mice,[13] whereas in humans, inactivating mutations of GRK1 cause Oguchi disease, a stationary form of night blindness.[14,15] These examples demonstrate the negative consequences that can emerge from the loss of GRK function in a physiological setting. Elevated GRK function also plays a role in disease and leads to abnormal suppression of GPCR signalling. The best-characterized example is the overexpression of GRK2 during heart failure.[16]

This review explores what is currently known about the functional interactions between GPCRs and GRKs, and attempts to evaluate older studies in light of more recent atomic structures of GPCRs and GRKs. We start with biochemical evidence that supports the existence of an allosteric receptor docking site on GRKs. Next, we survey studies aimed at identifying the structural elements of GPCRs that likely interact directly with GRKs, and *vice versa*. The review ends with the presentation of a mechanistic model for the activation of GRKs by GPCRs based chiefly on recent structural and functional analyses of GRK6 and GRK1.

15.2 Biochemical Evidence for a Receptor Docking Site

Many mechanistic rules regarding how kinases are activated and interact with their protein substrates were derived through structural and functional studies of PKA—the first protein kinase to be structurally characterized.[17,18] However, PKA is a somewhat of an outlier among kinases because it can phosphorylate peptide substrates with high efficiency and its targets have a well-defined consensus sequence. Many other protein kinases instead bind to target sequences in their substrates with low affinity, and therefore must rely on additional docking interactions that provide not only selectivity but also allosteric control of kinase activity.[19,20]

This is also true for GRKs. GRKs phosphorylate peptides with orders of magnitude less efficiency than receptors, and despite efforts to define ideal peptide substrates for the various GRK subfamilies,[21,22] there is no GRK consensus sequence evident in the ~800 GPCRs of the human genome. However, the various GRK subfamilies do seem to utilize a unique pattern of phosphorylation sites within a given receptor.[23] Not all of these sites are required for GRK-mediated desensitization,[24,25] but they may underlie the observed differences in signalling outcomes that occur as a consequence of receptor phosphorylation by different GRKs.[26–28]

Kinetic data provided the first hints of receptor-mediated allosteric control of GRK activity. The estimated K_M for GRK1-mediated phosphorylation of synthetic peptides derived from the *C*-terminus of rhodopsin is orders of magnitude higher than for Rho*.[29–31] Likewise, the catalytic efficiency of GRK2 for receptor-derived peptide substrates is 10^6-fold lower than for activated β_2-adrenergic receptors (β_2ARs).[32] The fact that GRK1 can efficiently use detergent-solubilized Rho* as a substrate all but eliminates the possibility that other factors in ROS membranes could be involved in regulation of its kinase activity.[29,33] Allosteric control of GRK activity is thus a property imbued in its interaction with the activated receptor.

Direct proof for allosteric modulation of kinase activity came with the observation that Rho*, or Rho* treated with endopeptidase Asp-N to remove its *C*-terminal phosphorylation sites, stimulates GRK1 phosphorylation of peptides derived from the *C*-terminus of rhodopsin by several orders of magnitude.[11,31] The ability of GRK2 to phosphorylate peptide substrates is likewise dramatically enhanced by Rho*, Asp-N treated Rho*, activated β_2ARs,[34] and by agonist-occupied m_2 muscarinic receptors engineered to lack GRK phosphorylation sites.[35] The ability of GRK5 to phosphorylate peptides is also stimulated in the presence of Rho* or Asp-N treated Rho*.[36] These observations suggest that receptor mediated activation is a property shared by all GRKs and, conversely, that the ability to activate GRKs is a property shared by many, if not all GPCRs.

15.3 Structural Elements of GPCRs Involved in Binding GRKs

Although the phosphoacceptor sites of receptors such as the *C*-terminal tail of Rho* must obviously bind in the phosphotransfer site of GRKs, the kinetic data suggest that this interface is of relatively low affinity. Furthermore, if GRKs preferentially recognize the active form of GPCRs, then they would have to form significant interactions with the transmembrane bundle of the receptor, which is expected to undergo the largest conformational change upon activation.[37] Indeed, crystal structures of Ops* (which presumably has a Rho*-like conformation)[38] confirmed that transmembrane helices 5 and 6 (TM5 and TM6), and consequently the IL3 loop that joins them, undergo a large conformational change as the receptor approaches its activated state.[39,40] Attempts to define the GRK binding site on receptors include the use of receptor-derived peptides as inhibitors or bait, cleavage of rhodopsin with specific proteases, and functional analysis of recombinant mutants.

15.3.1 Studies using Receptor-derived Peptides

Peptides derived from the cytoplasmic loops and tails of rhodopsin and β_2AR inhibit GRK activity. Most potent are peptides derived from IL3 and

IL1.[29,32] Remarkably, peptides from IL1 of β_2AR can be modified to achieve IC_{50} values as low as 0.6 μM.[41] Sequences from IL2 and IL3, but not IL1, of the α_2 adrenergic receptor (α_2AR), expressed as GST-fusion proteins, reportedly bind GRK2.[42] However, some caution must be used in interpreting such studies, as it is not known if peptides behave similarly to their corresponding elements in the context of a full-length, integral membrane receptor. For example, peptides derived from the intradiscal loops of rhodopsin (*i.e.* inaccessible to GRKs) inhibit GRK1 with similar potencies to those from the intracellular loops,[29] and the IL1 β_2AR peptides are noncompetitive against receptor,[41] which is inconsistent with their role as a direct binding site for GRKs. Peptides derived from the *N*-termini of GRKs[43,44] and from α9 of the RH domain[45] also inhibit kinase activity with similar potencies and likewise exhibit non-competitive behaviour. Taken together, these results suggest that such peptides interact with GRKs in a nonspecific manner.

15.3.2 Studies using Modified Receptors

Collectively, these studies point to the IL2 and IL3 loops, in particular the regions adjacent to the transmembrane spans, as being the most important for GRK interactions with receptors. This is consistent with the fact that these loops are more accessible to cytoplasmic proteins than IL1. Cleavage of IL3 of Rho* with V8 protease greatly reduces GRK1 activity, even though this loop contains no phosphorylation sites.[11] However, recombinant deletion of a large segment of IL3 of the m_2 muscarinic receptor (residues 233–380) has no significant impact on the ability of this receptor to activate GRK2, indicating that if IL3 is involved in interactions with GRKs, then the membrane proximal regions of the loop are the most important.[35] This conclusion is consistent with the lack of strong sequence conservation of IL3 among GPCRs, but inconsistent with results reported for the interactions of GRK2 with the α_2AR, where mutation of basic residues within its extended IL3 loop diminish GRK2 activity and its ability to stimulate phosphorylation of peptides by GRK2.[42] This may indicate that different receptors have distinct mechanisms for maintaining high affinity interactions with GRKs. Deletion of the IL2 and IL3 loops of rhodopsin, including portions of their membrane proximal helices, abrogate GRK1 phosphorylation and binding,[33] and the introduction of disulfide bonds between TM6 and TM7 or TM3 and TM6 of rhodopsin blocks phosphorylation by GRK1.[46] Based on subsequent crystal structures, such modifications probably restrict the light-induced structural changes in the cytoplasmic face of the receptor and thereby limit the access of GRK1. Even so, such studies confirm that GRKs must have access to the cytoplasmic surface of the transmembrane bundle of Rho* in order to efficiently phosphorylate the *C*-terminal tail of the receptor.

Alteration of pairs or triplets of residues to alanine or glycine in recombinantly expressed rhodopsin revealed that some substitutions in IL1 increase activity, while others in IL2 and IL3 decrease it.[47] Based on current atomic structures of rhodopsin and Ops*, the residues mutated in IL1 that increase activity (T62A/V63A/Q64A) are not situated so that they could directly interact with cytoplasmic proteins. However, the A233/234/235G triple mutant, which reduces activity by 80%, is located at the cytoplasmic end of TM5 is readily accessible. Replacement of the IL2 and IL3 loops of rhodopsin with unrelated sequences abrogates GRK1 activity, whereas mutations in IL2 and IL3 can either increase or decrease catalytic efficiency, although by no more than 5.5-fold or 0.4-fold, respectively.[33] The relatively subtle effects of such point mutants suggests that the affinity of GPCRs for GRKs is dictated by many elements on the cytoplasmic surface of the receptor and is not heavily reliant on any single specific interaction.

15.4 Structural Elements of GRKs Involved in Binding GPCRs

By 1991, it was established that GRK1, GRK2 and GRK3 are closely related in primary structure as well as function,[48–50] and constitute a novel family of kinases that selectively phosphorylate activated GPCRs.[2] Where then are the sites on these GRKs required for recognition of activated receptors? Unlike GPCRs, GRKs are soluble proteins and consequently there is much more real estate to consider. Furthermore, interpretation of how different regions of GRKs affect kinase activity is complicated because these regions could play a role in either (a) direct interactions with a GPCR, (b) direct interactions with the cell surface, (c) intramolecular contacts that stabilize the activated state of the GRK, or (d) a combination of some or all of the above.

The structural element most strongly implicated in receptor recognition in GRKs is the highly conserved *N*-terminal region (residues 1 to ~20, Figure 15.1A). Truncation of this region leads to nearly complete loss of receptor phosphorylation.[43,51–54] Furthermore, the binding of recoverin[53,55] or an antibody[56] to this region in GRK1 blocks phosphorylation of Rho*, and point mutants of conserved residues in the *N*-terminal region exhibit defects in catalytic activity.[43,44,51,57,58] The *N*-terminal region is disordered in nearly all crystal structures of GRKs published to date and is susceptible to limited proteolysis,[54,59] indicating that it is not stably associated with the rest of the enzyme in solution. Peptides derived from the *N*-terminal region are unstructured in solution,[44] but assume a helical structure when in complex with other proteins such as recoverin.[60] When residues from the *N*-terminal region have been observed in crystal structures of GRKs, they are held in place by crystal contacts and likewise demonstrate helical character.[54,57] Interestingly, peptides derived from the *N*-terminal region can inhibit GRK activity against the β_2AR or Rho*[43,44] with μM potencies, but their mechanism of action is unclear.

The second structural element known to be functionally important for receptor phosphorylation is the AGC kinase *C*-terminal tail (C-tail), an extension of the canonical protein kinase fold (Figure 15.1A) that plays a regulatory role in AGC kinases and that becomes increasingly well-ordered as the kinase approaches its closed, active state.[61,62] Like the *N*-terminal region, the C-tail is poorly ordered in most GRK crystal structures,[54,63,64] as might be expected because the kinase domains in these structures adopt a relatively open, therefore inactive conformation. The central segment of the C-tail, known as the 'active site tether' (AST), passes over the active site of the AGC kinase domain and contains residues known to be important for GRK activity.[52,65] The beginning of the AST contains a proline rich motif that mutagenesis studies of *E. coli* expressed GRK2 have indicated are important for GPCR binding and activity.[66] However, the prolines in this motif play a structural role in GRK2,[63] and thus their relationship to receptor binding is probably indirect.

The third, most recently recognized region important for receptor interactions is a small surface on the small lobe of the kinase domain that is uniquely conserved in GRKs compared with other AGC kinases.[45,52] In the centre of this conserved patch is an invariant arginine (Arg191 in GRK1) that is solvent exposed in all but one of the reported GRK structures. Strikingly, substitution of GRK1-Arg191 with alanine or lysine is almost as detrimental for receptor phosphorylation as is truncation of the entire *N*-terminal region. Similar results are obtained upon mutation of the analogous arginine residues in GRK2 and GRK6.[52] However, these mutations also affect phosphorylation of the rhodopsin *C*-terminal peptide, indicating that the conserved surface on the small lobe demarked by Arg191 is most likely involved in stabilizing the active conformation of the kinase domain. However, the proximity of this conserved surface to both the *N*-terminal region and to the C-tail suggests that these three elements functionally interact in the active, receptor-bound conformation of GRKs.[52]

The RH domains of GRKs have also been implicated in direct interactions with receptors. The most compelling story is the phosphorylation-independent inhibition of metabotropic glutamate receptors by GRK2. One difficulty in evaluating the molecular basis for phosphorylation-independent desensitization mediated by GRK2 and GRK3 is that they can bind $G\alpha_q$ family members with high affinity *via* their RH domain, which blocks the binding of $G\alpha_q$ to downstream effectors.[67] In many cases, sequestration of $G\alpha_q$ may be entirely responsible for the observed desensitization of $G\alpha_q$-coupled receptors by GRK2 (for example, see ref. 68). However, the *N*-terminal RH domain of GRK2 (residues 45–185) co-immunoprecipitates with the metabotropic glutamate receptor,[69] and this apparent interaction is unaffected by mutations in the GRK2 RH domain that abolish $G\alpha_q$ binding. Furthermore, a D527A mutation in the RH domain of full-length GRK2 eliminates association with the receptor and blocks the ability of GRK2 to regulate $G\alpha_q$ signaling.[70] Other examples of direct, presumably functional interactions of GRK RH domains with GPCRs include the GRK4 RH domain with the $GABA_B$ receptor[71] and GRK2 RH domain-mediated inhibition of the endothelin A receptor.[72] However, a caveat that should be considered in such studies is that RH domain fragments, such as GRK2 45–185, have unfulfilled domain interfaces and also lack the last two helices of the domain, which occur after the kinase domain in the primary sequence (Figure 15.1A). These properties may confer non-physiological behaviour. Interestingly, the Q41L polymorphism in GRK5 enhances agonist-promoted phosphorylation and agonist-promoted desensitization of the β_2AR in cells.[73] GRK5 Gln41 occurs in the α1 helix of the RH domain, once again suggesting the involvement of the RH domain in receptor recognition. However, the mechanism by which this polymorphism promotes activity is unclear. For example, it may simply reflect enhanced stability of the protein *in vivo* due to favourable packing interactions with the adjacent Trp175 side chain.

15.5 Molecular Basis for GRK Activation

15.5.1 Structure of GRK6 in a Closed Conformation

In a recently reported crystal structure of GRK6,[57] the enzyme is fortuitously trapped in a conformation believed to be similar to its activated, receptor-bound conformation (Figure 15.1B). Several observations support this claim. First, the kinase domain adopts a relatively closed conformation compared to all the prior structures of GRKs and one that is similar to the structure of PKA in its active state.[74] Secondly and as would be expected for an active AGC kinase, the C-tail of the GRK6 kinase domain is relatively well-ordered and the AST region of its C-tail follows a path similar to those observed in other active AGC kinases, although it contains a deletion of four residues that creates a small cleft between the small lobe and the C-tail. Thirdly, and perhaps most importantly, the *N*-terminal region is completely ordered and forms an extended α helix (αN) that interacts with this cleft, forming interfaces with both the C-tail and the conserved patch on the small lobe of the kinase domain. This intramolecular assembly involves nearly all of the residues remote from the kinase active site that are known to be important for phosphorylation of activated receptors.

If the new GRK6 structure represents a GRK in a conformation close to its active state, then where is the GPCR docking site? Examination of the interactions between GRK6 molecules in the crystal lattice provides one clue. Exposed, highly conserved hydrophobic residues at the *N*-terminal end of αN form a short coiled-coil interaction with symmetry-related αN helices in the lattice. These interactions likely stabilize the helical conformation of the *N*-terminal region, which may explain how a relatively active state of GRK6 could be crystallized in the absence of an activated GPCR. Thus, it was hypothesized that activated GPCRs interact with this same region of αN, stabilizing the αN helix and consequently its interactions with the C-tail and the small lobe.[57]

To test this model, residues in the αN helix of GRK6 were altered by site-directed mutagenesis and the activities of the resulting proteins towards either Rho* or a peptide substrate derived from the *C*-terminal tail of rhodopsin were compared.[57] It was hypothesized that, if a mutant was deficient in both receptor and peptide phosphorylation, then it would indicate that the residue is important for stabilizing the catalytically active state of the kinase domain (although an additional role in direct binding to GPCRs cannot be ruled out). If only rhodopsin phosphorylation were affected, then that residue is only involved in direct interactions with receptors. Indeed, mutation of the residues on αN that face away from the kinase domain in the GRK6 structure (*i.e.* Leu3, Ile6, and Val7) negatively affect the efficiency of only GPCR phosphorylation, whereas those of residues that contact the kinase domain negatively affect both GPCR and peptide phosphorylation (*e.g.* Leu12), as do those of residues on the kinase domain that contact αN.[57] An analogous study of mutations in the *N*-terminal region of GRK1 yielded a similar activity profile[58] and analysis of mutations in the more distantly related *N*-terminus of GRK2 suggests a similar story.[44] Based on these results, the *N*-terminal end of αN likely represents the principal receptor docking site of GRKs.

15.5.2 An Alternative role for the *N*-terminal Region?

An alternative role proposed for the *N*-terminal region is to facilitate the interactions of GRKs with the inner leaflet of the phospholipid bilayer. Fuelling this hypothesis is the observation that mutations in the *N*-terminal region of GRK5 and GRK2 decrease the association of these enzymes with phospholipid vesicles[43] or mixed micelles,[44] and decrease the phospholipid-dependent autophosphorylation of GRK5.[43] Peptides corresponding to the *N*-terminal region also appear to drive GRK2 into the pellet fraction in mixed-micelle pull-down assays and to inhibit GRK2 noncompetitively, suggesting that they are involved in binding lipids and not in direct interactions with receptors.[44]

However, a model in which the αN helix interacts directly with the membrane is inconsistent with the closed GRK6 structure, as this would require the insertion of several negatively charged residues (Glu2 and Glu4) into the hydrophobic phase of the membrane. It is also difficult to interpret the results derived from *N*-terminal peptides because their mechanism of inhibition is not understood. Furthermore, if these peptides bind to the same site on a GRK as αN, then one might expect them to be activating rather than inhibitory, especially as they seem to promote GRK association with membranes, where their substrates are localized.[44]

Some of the phospholipid binding data, at least in studies of full-length GRKs, might be explained as an indirect effect of stabilizing the closed state of the kinase domain. The interactions of αN with the kinase domain, as observed in the closed GRK6 structure, may help configure the phospholipid-binding determinants of the GRK so that they have enhanced affinity for lipid bilayers. In line with this idea, a region known to be required for the binding of phosphatidylinositol-4,5-bisphosphate in the GRK4 subfamily[5] is positioned immediately adjacent to the αN helix in the GRK6 structure (Figure 15.1B).

15.5.3 A Receptor Docking Model and a Proposed Mechanism of Activation

Structures of Ops* alone and in complex with a peptide derived from the *C*-terminus of Gα_t (Gα_t-CT K341L) provide images of a GPCR in what is proposed to be a relatively active conformation, and a model for how the *C*-termini of Gα subunits interact with GPCRs.[39,40] Although opsin is far less active than Rho*, low pH and the transducin peptide can stabilize a Rho*-like conformation of the receptor.[38] Their reported structures are indeed similar to that of the β_2AR in an agonist-bound state (B. Kobilka, personal communication) or a structure of constitutively active mutant of rhodopsin (G. Schertler, personal communication). In these structures, the most prominent conformational change from dark state rhodopsin is the reorganization of TM5, TM6 and the end of TM7 to create a shallow pocket on the cytoplasmic face of the receptor, which allows the Gα_t-CT K341L peptide to bind as an amphipathic helix a hydrophobic surface formed by TM3, 5 and 6 (Figure 15.2A and 15.2B). Given the importance of the IL3 region for GRK activity, it is easy to hypothesize that the αN helix of GRKs binds in a similar manner as the Gα_t-CT K341L peptide, or by extension, the *C*-terminal helix of transducin.[57] Both helices have amphipathic character, protrude conspicuously from the surface of their respective proteins, and are disordered in crystal structures when not in contact with another protein surface.[58] A GRK–GPCR docking model can thus be generated by superimposing the hydrophobic patch in the GRK6 αN helix with hydrophobic residues in the Gα_t-CT K341L helix in the Ops* structure (Figure 15.2C). The resulting model positions the proposed membrane interacting region of GRK6 (and by extension, those of other GRKs) adjacent to the lipid bilayer and the *C*-terminus of opsin in an ideal position to enter the active site cleft of the kinase domain (Figure 15.3). In other GPCRs, the IL3 loop can be quite long. However, variability in IL3 is readily accommodated by the docking model because there are no obvious steric collisions between IL3 and the rest of the GRK.

Although no equivalent structures are available, homology modelling suggests that the *N*-terminal regions of other GRK subfamilies (GRK1 and GRK2) can form a similar helix that forms analogous complementary interactions with their respective small lobes and C-tails (J. Tesmer, unpublished data). In these homology models, the GRK1 and GRK2 subfamilies would have three and one additional amino acids at their amino termini, respectively. It is not known if these extra residues would extend the helix, but due to the lack of sequence conservation in the first several residues of GRK1 and GRK2, which both efficiently phosphorylate Rho*, these residues are unlikely to form functionally important interactions with the receptor. In the docking model, the *N*-terminus of GRK6 is positioned such that any additional residues at the *N*-termini of GRK1 and GRK2 can exit the pocket through a gap formed between TM6 and the H8 helix of the receptor (Figure 15.2C).

The balance of the structural and biochemical data presented above suggests the following model for the activation of GRKs (Figure 15.3). GRKs identify activated GPCRs through the interaction of their *N*-terminal regions with the shallow cavity formed on the cytoplasmic surface of these receptors. This docking event stabilizes the αN helix, which then (or simultaneously) interacts with the GRK kinase domain. Because interactions between αN and the small lobe and C-tail of the kinase domain are only likely to occur when the kinase domain assumes a relatively closed state, receptor docking consequently stabilizes a more active conformation of the kinase domain. A recent study in which GRK1 was engineered to have a disulfide linkage between αN and the C-tail seems to support the proposed mechanism, as the presence of the disulfide linkage enhances activity towards peptide substrates ~10-fold.[58] Favourable interactions of GRKs with anionic phospholipids in the cell membrane adjacent to the receptor would likewise enhance the affinity of this docking complex.

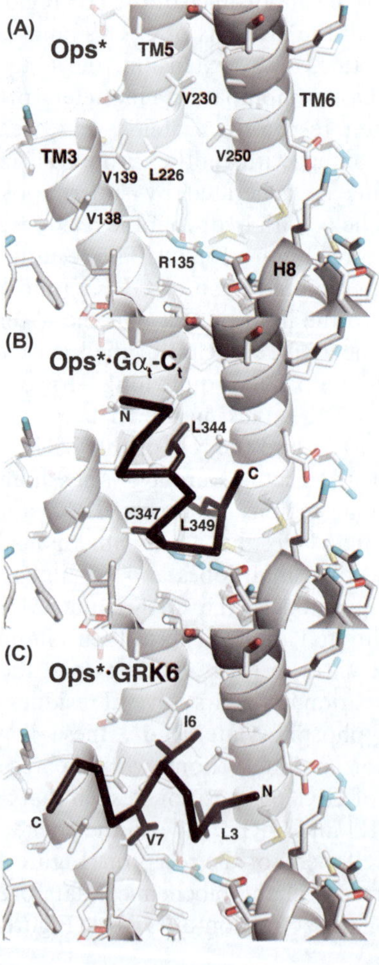

Does the *N*-terminal region play a functional role in GRKs that are not docked with receptors? Although some reports indicate that loss of the *N*-terminal region has no affect on phosphorylation of peptides or soluble substrates,[43,51] recent work with GRK1 demonstrated a ~10-fold reduction in peptide phosphorylation in a GRK1 variant that lacks the amino terminus.[52] Mutation of residues in the *N*-terminal regions of GRK1 and GRK6 that are predicted to interact with the small lobe or the C-tail of the kinase domain also exhibit up to 20-fold decreases in catalytic efficiency against peptides.[57,58] This implies that αN must dock with the kinase domain in order for GRK to phosphorylate peptide substrates with maximum efficiency. The role of αN thus appears to be two-fold. The first is to recognize the fact that a GPCR is in an activated state and the second is to bias the conformational equilibrium of the kinase domain towards its most active state. When bound to a GPCR, αN is clearly more efficient at this process, perhaps because the receptor reinforces the helical conformation of the *N*-terminal region. It should be noted, however, that receptors probably form additional interactions with GRKs besides αN, such as with residues in the RH domain or in the C-tail of the kinase domain. In some GRK–receptor pairs, these additional interactions may contribute to selectivity or provide further stabilization of the closed state of the kinase domain. Contacts with the RH domain would most easily be accomplished by oligomeric receptors such as the metabotropic glutamate receptor. However, such a mechanism may not be pertinent for class A GPCRs, because monomeric rhodopsin has been shown to be an efficient substrate for GRK1.[76]

Figure 15.2 A shallow pocket is formed on the surface of the cytoplasmic domain of activated GPCRs and is the primary site of interaction for Gα subunits and likely GRKs. (**A**) The cytoplasmic surface of Ops*,[40] a receptor presumed to be in an activated conformation. The labelled side chains are conserved as hydrophobic residues in all GPCR structures reported to date (β_2AR,[79] β_1AR,[80] adenosine A2A,[81] squid rhodopsin,[82] CXCR4,[83] and dopamine D3)[84] and form a continuous wall on one side of the pocket. These receptors couple to Gα subunits representing three of the four Gα subfamilies, and thus these conserved residues likely interact with residues common to the αC helices of all Gα subunits. Arg135 is part of the conserved (E/D)RY motif found in most class A GPCRs.[85] (**B**) The $G\alpha_t$–C_t peptide bound to Ops* (PDB entry 3DQB). The peptide is shown as a black Cα trace. The side chains of three residues that are almost invariant in Gα subunits directly interact with the conserved hydrophobic surface shown in panel A. The amino and carboxyl termini of the peptide are labelled as 'N' and 'C', respectively. (**C**) Conceptual model of how the *N*-terminal 11 amino acids of GRK6 might dock in the same pocket. There are three highly conserved hydrophobic residues at the *N*-termini of GRKs that could interact with the same sites as the conserved $G\alpha_t$–C_t side chains in panel B. Alternatively, GRK6 can be modelled with its *N*-terminal helix essentially superimposed on that of $G\alpha_t$–C_t using a different set of resides on the $G\alpha_t$–C_t peptide.[57] In either case, the complex positions the *C*-terminal tail of Ops* close to the active site cleft of the kinase domain (Figure 15.3). A high resolution structure of the complex is needed to clarify the GRK–GPCR interaction, which will likely include other residues of the GRK.

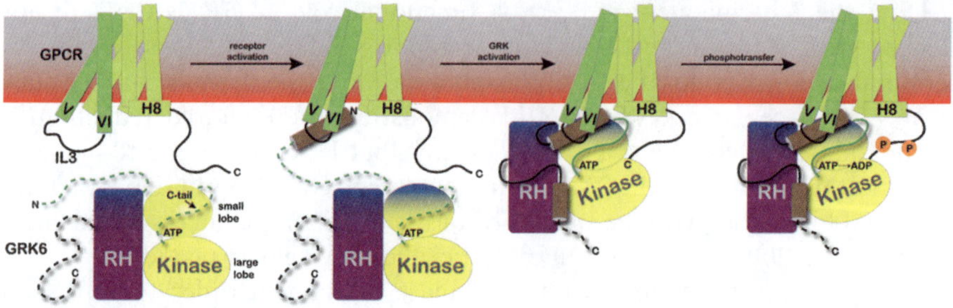

Figure 15.3 Proposed mechanism for the activation of GRKs by activated GPCRs. The GRK model shown corresponds to that of GRK6, which is presumed to have disordered *N*-terminal, C-tail and *C*-terminal regions (dashed lines, left panel) in its basal, cytoplasmic state. Receptor activation reorganizes the cytoplasmic surface of the receptor, most notably TM5 and TM6 (labelled V and VI), to create a small pocket, in which the *N*-terminal region of GRK6 docks as a helix. This *N*-terminal helix in turn (and perhaps simultaneously) interacts with the kinase domain in a manner that stabilizes the small and large lobes in a closed state and orders the kinase C-tail, thereby activating the enzyme and allowing basic regions in the GRK (blue) to form favourable electrostatic interactions with the negatively charged plasma membrane (red). The *C*-terminus of GRK6 also becomes ordered as the kinase domain closes and packs against the RH domain, although the physiological importance of this is not yet understood. The GRK can then catalyze multiple rounds of phosphotransfer into the *C*-terminal tail of the receptor, which is positioned close to the active site of the kinase domain, or into IL3 of receptors that contain longer IL3 loops (right panel).[57]

15.6 Conclusions

In 1997 Palczewski wrote that 'the GPCR*·GRK complex involves geometrical complementaries of the *N*-terminal domain of the kinase and cytoplasmic loops of the receptor'.[4] The balance of the current biochemical and structural evidence still supports this hypothesis. The primary interface between GRKs and GPCRs is most likely formed between one side of the αN helix, as observed in the closed GRK6 structure, and the inner hydrophobic surface of the pocket formed on the cytoplasmic face of the activated GPCR (Figure 15.2C). Mutation of specific residues in αN that are proposed to interact directly with GPCRs diminish, but do not abrogate activity nearly to the same extent as deletion of the entire αN helix. Substitution of individual side chains in IL2 and IL3 also has more subtle phenotypes than cleavage, substitution or deletion of the entire loop.[33,47] These results indeed reinforce the idea that it is complementary shape and electrostatic character– rather than specific interactions, that matter most for the formation of the activated GRK–GPCR complex. This is, after all, consistent with the fact GRKs must be able to recognize hundreds of different sequence-diverse receptors solely on the basis of their conformation. Non-specific (but conserved) hydrophobic interactions such as those predicted in Figure 15.2C without accompanying hydrogen bonds or salt bridges may contribute to the lack of receptor specificity. Consequently the GRK–GPCR complex is of relatively low affinity (~10 μM), but of high selectivity for the activated state of the receptor. This relatively low affinity presents an additional challenge to determining the atomic structure of the GRK–GPCR complex, which is essential for verifying and elaborating upon the proposed model for GRK activation.

Based on their ability to selectively regulate receptor function, GRKs represent important drug targets, particularly with respect to the treatment of cardiovascular disease.[16] Do the new structural insights into their mechanisms of activation open the door to any new avenues for the development of drugs that selectively target GRKs? The GRK6 αN helix protrudes from the surface of the kinase domain and most likely interacts with the same pocket of an activated receptor as do heterotrimeric G proteins. Targeting this pocket on GPCRs by small molecules would block GRK function, but it would also inhibit heterotrimeric G protein coupling, and consequently would likely reinforce any phenotype resulting from high GRK activity (*i.e.* loss of receptor function). The *C*-terminal end of the αN helix uses a pair of large side chains (Leu12 and Arg16 in GRK6, Tyr13 and Met17 in GRK2, and Phe15 and Arg19 in GRK1) to dock into the cleft formed between the small lobe, the C-tail and the hinge of the kinase domain (Figure 15.1B). Comparison with other AGC kinases suggests that this pocket is unique to GRKs.[61] Thus, it represents an interesting allosteric site that could be exploited by small molecules to block the docking of GPCRs with GRKs. However, its small size and the fact that it is formed between mobile structural elements of the kinase domain suggest that it would be challenging for small molecules not only to achieve high affinity, but also to compete efficiently with the interaction of αN, which by virtue of its covalent linkage would be at high local concentration relative to any compounds presented in *trans*. Thus, the most likely strategy remains to develop selective GRK inhibitors that target the active site cavity of the kinase domain while it is in a unique, inactive conformation,[77,78] as suggested by comparing the crystal structures of GRK1, GRK2 and GRK6 in their various unique, open states.[54] Perhaps in this manner, drugs can be identified that are selective for individual GRK subfamilies.

Acknowledgements

This work was supported by National Institutes of Health grants HL086865 and HL071818. The author is grateful to Dr. Rachel Sterne-Marr (Siena College) for her helpful comments and suggestions.

References

1. S. R. Sprang, Z. Chen and X. Du, *Adv. Protein Chem.*, 2007, **74**, 1–65.
2. K. Palczewski and J. L. Benovic, *Trends Biochem. Sci.*, 1991, **16**, 387–391.

3. J. A. Pitcher, N. J. Freedman and R. J. Lefkowitz, *Annu. Rev. Biochem.*, 1998, **67**, 653–692.
4. K. Palczewski, *Eur. J. Biochem.*, 1997, **248**, 261–269.
5. J. A. Pitcher, Z. L. Fredericks, W. C. Stone, R. T. Premont, R. H. Stoffel, W. J. Koch and R. J. Lefkowitz, *J. Biol. Chem.*, 1996, **271**, 24907–24913.
6. C. V. Carman, L. S. Barak, C. Chen, L. Y. Liu-Chen, J. J. Onorato, S. P. Kennedy, M. G. Caron and J. L. Benovic, *J. Biol. Chem.*, 2000, **275**, 10443–10452.
7. R. T. Premont, A. D. Macrae, S. A. Aparicio, H. E. Kendall, J. E. Welch and R. J. Lefkowitz, *J. Biol. Chem.*, 1999, **274**, 29381–29389.
8. H. Kuhn and W. J. Dreyer, *FEBS Lett.*, 1972, **20**, 1–6.
9. D. Bownds, J. Dawes, J. Miller and M. Stahlman, *Nat. New Biol.*, 1972, **237**, 125–127.
10. C. M. Krispel, D. Chen, N. Melling, Y. J. Chen, K. A. Martemyanov, N. Quillinan, V. Y. Arshavsky, T. G. Wensel, C. K. Chen and M. E. Burns, *Neuron*, 2006, **51**, 409–416.
11. K. Palczewski, J. Buczylko, M. W. Kaplan, A. S. Polans and J. W. Crabb, *J. Biol. Chem.*, 1991, **266**, 12949–12955.
12. J. Chen, C. L. Makino, N. S. Peachey, D. A. Baylor and M. I. Simon, *Science*, 1995, **267**, 374–377.
13. C. K. Chen, M. E. Burns, M. Spencer, G. A. Niemi, J. Chen, J. B. Hurley, D. A. Baylor and M. I. Simon, *Proc. Natl. Acad. Sci. U. S. A.*, 1999, **96**, 3718–3722.
14. S. Yamamoto, K. C. Sippel, E. L. Berson and T. P. Dryja, *Nat. Genet.*, 1997, **15**, 175–178.
15. A. V. Cideciyan, X. Zhao, L. Nielsen, S. C. Khani, S. G. Jacobson and K. Palczewski, *Proc. Natl. Acad. Sci. U. S. A.*, 1998, **95**, 328–333.
16. G. W. Dorn, 2nd, *J. Mol. Med.*, 2009, **87**, 455–463.
17. D. R. Knighton, J. H. Zheng, L. F. Ten Eyck, N. H. Xuong, S. S. Taylor and J. M. Sowadski, *Science*, 1991, **253**, 414–420.
18. S. S. Taylor, J. Yang, J. Wu, N. M. Haste, E. Radzio-Andzelm and G. Anand, *Biochim. Biophys. Acta*, 2004, **1697**, 259–269.
19. E. J. Goldsmith, R. Akella, X. Min, T. Zhou and J. M. Humphreys, *Chem. Rev.*, 2007, **107**, 5065–5081.
20. A. Remenyi, M. C. Good and W. A. Lim, *Curr. Opin. Struct. Biol.*, 2006, **16**, 676–685.
21. J. J. Onorato, K. Palczewski, J. W. Regan, M. G. Caron, R. J. Lefkowitz and J. L. Benovic, *Biochemistry*, 1991, **30**, 5118–5125.
22. P. Kunapuli, J. J. Onorato, M. M. Hosey and J. L. Benovic, *J. Biol. Chem.*, 1994, **269**, 1099–1105.
23. Z. L. Fredericks, J. A. Pitcher and R. J. Lefkowitz, *J. Biol. Chem.*, 1996, **271**, 13796–13803.
24. A. Seibold, B. G. January, J. Friedman, R. W. Hipkin and R. B. Clark, *J. Biol. Chem.*, 1998, **273**, 7637–7642.
25. A. Seibold, B. Williams, Z. F. Huang, J. Friedman, R. H. Moore, B. J. Knoll and R. B. Clark, *Mol. Pharmacol.*, 2000, **58**, 1162–1173.

26. D. A. Zidar, J. D. Violin, E. J. Whalen and R. J. Lefkowitz, *Proc. Natl. Acad. Sci. U. S. A.*, 2009, **106**, 9649–9654.
27. J. Kim, S. Ahn, X. R. Ren, E. J. Whalen, E. Reiter, H. Wei and R. J. Lefkowitz, *Proc. Natl. Acad. Sci. U. S. A.*, 2005, **102**, 1442–1447.
28. X. R. Ren, E. Reiter, S. Ahn, J. Kim, W. Chen and R. J. Lefkowitz, *Proc. Natl. Acad. Sci. U. S. A.*, 2005, **102**, 1448–1453.
29. K. Palczewski, J. H. McDowell and P. A. Hargrave, *Biochemistry*, 1988, **27**, 2306–2313.
30. D. J. Kelleher and G. L. Johnson, *J. Biol. Chem.*, 1990, **265**, 2632–2639.
31. C. Fowles, R. Sharma and M. Akhtar, *FEBS Lett.*, 1988, **238**, 56–60.
32. J. L. Benovic, J. Onorato, M. J. Lohse, H. G. Dohlman, C. Staniszewski, M. G. Caron and R. J. Lefkowitz, *Br. J. Clin Pharmacol.*, 1990, **30**(Suppl 1), 3S–12S.
33. R. L. Thurmond, C. Creuzenet, P. J. Reeves and H. G. Khorana, *Proc. Natl. Acad. Sci. U. S. A.*, 1997, **94**, 1715–1720.
34. C. Y. Chen, S. B. Dion, C. M. Kim and J. L. Benovic, *J. Biol. Chem.*, 1993, **268**, 7825–7831.
35. K. Kameyama, K. Haga, T. Haga, O. Moro and W. Sadee, *Eur. J. Biochem.*, 1994, **226**, 267–276.
36. R. T. Premont, W. J. Koch, J. Inglese and R. J. Lefkowitz, *J. Biol. Chem.*, 1994, **269**, 6832–6841.
37. D. L. Farrens, C. Altenbach, K. Yang, W. L. Hubbell and H. G. Khorana, *Science*, 1996, **274**, 768–770.
38. R. Vogel and F. Siebert, *J. Biol. Chem.*, 2001, **276**, 38487–38493.
39. J. H. Park, P. Scheerer, K. P. Hofmann, H. W. Choe and O. P. Ernst, *Nature*, 2008, **454**, 183–187.
40. P. Scheerer, J. H. Park, P. W. Hildebrand, Y. J. Kim, N. Krauss, H. W. Choe, K. P. Hofmann and O. P. Ernst, *Nature*, 2008, **455**, 497–502.
41. R. Winstel, H. G. Ihlenfeldt, G. Jung, C. Krasel and M. J. Lohse, *Biochem. Pharmacol.*, 2005, **70**, 1001–1008.
42. C. S. Pao and J. L. Benovic, *J. Biol. Chem.*, 2005, **280**, 11052–11058.
43. B. Noble, L. A. Kallal, M. H. Pausch and J. L. Benovic, *J. Biol. Chem.*, 2003, **278**, 47466–47476.
44. C. S. Pao, B. L. Barker and J. L. Benovic, *Biochemistry*, 2009, **48**, 7325–7333.
45. F. Baameur, D. H. Morgan, H. Yao, T. M. Tran, R. A. Hammitt, S. Sabui, J. S. McMurray, O. Lichtarge and R. B. Clark, *Mol. Pharmacol.*, 2010, **77**, 405–415.
46. K. Cai, J. Klein-Seetharaman, J. Hwa, W. L. Hubbell and H. G. Khorana, *Biochemistry*, 1999, **38**, 12893–12898.
47. W. Shi, S. Osawa, C. D. Dickerson and E. R. Weiss, *J. Biol. Chem.*, 1995, **270**, 2112–2119.
48. J. L. Benovic, A. De Blasi, W. C. Stone, M. G. Caron and R. J. Lefkowitz, *Science*, 1989, **246**, 235–240.
49. W. Lorenz, J. Inglese, K. Palczewski, J. J. Onorato, M. G. Caron and R. J. Lefkowitz, *Proc. Natl. Acad. Sci. U. S. A.*, 1991, **88**, 8715–8719.

50. J. L. Benovic, J. J. Onorato, J. L. Arriza, W. C. Stone, M. Lohse, N. A. Jenkins, D. J. Gilbert, N. G. Copeland, M. G. Caron and R. J. Lefkowitz, *J. Biol. Chem.*, 1991, **266**, 14939–14946.
51. Q. M. Yu, Z. J. Cheng, X. Q. Gan, G. B. Bao, L. Li and G. Pei, *J. Neurochem.*, 1999, **73**, 1222–1227.
52. C. C. Huang, K. Yoshino-Koh and J. J. Tesmer, *J. Biol. Chem.*, 2009, **284**, 17206–17215.
53. M. K. Higgins, D. D. Oprian and G. F. Schertler, *J. Biol. Chem.*, 2006, **281**, 19426–19432.
54. P. Singh, B. Wang, T. Maeda, K. Palczewski and J. J. Tesmer, *J. Biol. Chem.*, 2008, **283**, 14053–14062.
55. C. K. Chen, J. Inglese, R. J. Lefkowitz and J. B. Hurley, *J. Biol. Chem.*, 1995, **270**, 18060–18066.
56. K. Palczewski, J. Buczylko, L. Lebioda, J. W. Crabb and A. S. Polans, *J. Biol. Chem.*, 1993, **268**, 6004–6013.
57. C. A. Boguth, P. Singh, C. C. Huang and J. J. Tesmer, *EMBO J.*, 2010, **29**, 3249–3259.
58. C. C. Huang, T. Orban, B. Jastrzebska, K. Palczewski and J. J. Tesmer, *Biochemistry*, 2011, **50**, 1940–1949.
59. S. Sarnago, R. Roca, A. de Blasi, A. Valencia and F. Mayor, Jr. and C. Murga, *Mol. Pharmacol.*, 2003, **64**, 629–639.
60. J. B. Ames, K. Levay, J. N. Wingard, J. D. Lusin and V. Z. Slepak, *J. Biol. Chem.*, 2006, **281**, 37237–37245.
61. N. Kannan, N. Haste, S. S. Taylor and A. F. Neuwald, *Proc. Natl. Acad. Sci. U. S. A.*, 2007, **104**, 1272–1277.
62. C. Hauge, T. L. Antal, D. Hirschberg, U. Doehn, K. Thorup, L. Idrissova, K. Hansen, O. N. Jensen, T. J. Jorgensen, R. M. Biondi and M. Frodin, *EMBO J.*, 2007, **26**, 2251–2261.
63. D. T. Lodowski, J. A. Pitcher, W. D. Capel, R. J. Lefkowitz and J. J. Tesmer, *Science*, 2003, **300**, 1256–1262.
64. D. T. Lodowski, V. M. Tesmer, J. L. Benovic and J. J. Tesmer, *J. Biol. Chem.*, 2006, **281**, 16785–16793.
65. R. Sterne-Marr, P. A. Leahey, J. E. Bresee, H. M. Dickson, W. Ho, M. J. Ragusa, R. M. Donnelly, S. M. Amie, J. A. Krywy, E. D. Brookins-Danz, S. C. Orakwue, M. J. Carr, K. Yoshino-Koh, Q. Li and J. J. Tesmer, *Biochemistry*, 2009, **48**, 4285–4293.
66. X. Gan, Z. Ma, N. Deng, J. Wang, J. Ding and L. Li, *J. Biol. Chem.*, 2004, **279**, 49741–49746.
67. J. M. Willets, R. A. Challiss and S. R. Nahorski, *Trends Pharmacol. Sci.*, 2003, **24**, 626–633.
68. J. M. Willets, S. R. Nahorski and R. A. Challiss, *J. Biol. Chem.*, 2005, **280**, 18950–18958.
69. G. K. Dhami, P. H. Anborgh, L. B. Dale, R. Sterne-Marr and S. S. Ferguson, *J. Biol. Chem.*, 2002, **277**, 25266–25272.
70. G. K. Dhami, L. B. Dale, P. H. Anborgh, K. E. O'Connor-Halligan, R. Sterne-Marr and S. S. Ferguson, *J. Biol. Chem.*, 2004, **279**, 16614–16620.

71. J. Perroy, L. Adam, R. Qanbar, S. Chenier and M. Bouvier, *EMBO J.*, 2003, **22**, 3816–3824.
72. N. J. Freedman, A. S. Ament, M. Oppermann, R. H. Stoffel, S. T. Exum and R. J. Lefkowitz, *J. Biol. Chem.*, 1997, **272**, 17734–17743.
73. W. C. Wang, K. A. Mihlbachler, E. R. Bleecker, S. T. Weiss and S. B. Liggett, *Pharmacogenet. Genomics*, 2008, **18**, 729–732.
74. Madhusudan, P. Akamine, N. H. Xuong and S. S. Taylor, *Nat. Struct. Biol.*, 2002, **9**, 273–277.
75. C. C. Huang and J. J. Tesmer, *J. Biol. Chem.*, 2011, **286**, 7715–7721.
76. T. H. Bayburt, S. A. Vishnivetskiy, M. A. McLean, T. Morizumi, C. C. Huang, J. J. Tesmer, O. P. Ernst, S. G. Sligar and V. V. Gurevich, *J. Biol. Chem.*, 2011, **286**, 7715–7721.
77. L. N. Johnson, *Q. Rev. Biophys.*, 2009, **42**, 1–40.
78. M. Rabiller, M. Getlik, S. Kluter, A. Richters, S. Tuckmantel, J. R. Simard and D. Rauh, *Arch. Pharm. (Weinheim)*, 2010, **343**, 193–206.
79. V. Cherezov, D. M. Rosenbaum, M. A. Hanson, S. G. Rasmussen, F. S. Thian, T. S. Kobilka, H. J. Choi, P. Kuhn, W. I. Weis, B. K. Kobilka and R. C. Stevens, *Science*, 2007, **318**, 1258–1265.
80. T. Warne, M. J. Serrano-Vega, J. G. Baker, R. Moukhametzianov, P. C. Edwards, R. Henderson, A. G. Leslie, C. G. Tate and G. F. Schertler, *Nature*, 2008, **454**, 486–491.
81. V. P. Jaakola, M. T. Griffith, M. A. Hanson, V. Cherezov, E. Y. Chien, J. R. Lane, A. P. Ijzerman and R. C. Stevens, *Science*, 2008, **322**, 1211–1217.
82. M. Murakami and T. Kouyama, *Nature*, 2008, **453**, 363–367.
83. B. Wu, E. Y. Chien, C. D. Mol, G. Fenalti, W. Liu, V. Katritch, R. Abagyan, A. Brooun, P. Wells, F. C. Bi, D. J. Hamel, P. Kuhn, T. M. Handel, V. Cherezov and R. C. Stevens, *Science*, 2010, **330**, 1066–1071.
84. E. Y. Chien, W. Liu, Q. Zhao, V. Katritch, G. W. Han, M. A. Hanson, L. Shi, A. H. Newman, J. A. Javitch, V. Cherezov and R. C. Stevens, *Science*, 2010, **330**, 1091–1095.
85. G. E. Rovati, V. Capra and R. R. Neubig, *Mol. Pharmacol.*, 2007, **71**, 959–964.

CHAPTER 16

The Complex Role of G Protein-coupled Receptor Kinase 2 (GRK2) in Cell Signalling: Beyond GPCR Desensitization

FEDERICO MAYOR JR.,* PETRONILA PENELA, CATALINA RIBAS AND CRISTINA MURGA

Departamento de Biología Molecular and Centro de Biología Molecular 'Severo Ochoa', (Consejo Superior de Investigaciones Científicas- Universidad Autónoma de Madrid), Universidad Autónoma, 28049 Madrid, Spain

16.1 Introduction

G protein-coupled receptor kinases (GRKs) were first described as serine/threonine kinases that participate together with arrestins in the regulation of multiple G protein-coupled receptors (GPCRs). Once activated, these seven-transmembrane (7TM) receptors become specifically phosphorylated by GRKs in their intracellular domains.[1,2] This promotes the association of cytosolic proteins termed arrestins, leading to uncoupling from heterotrimeric G proteins and thus abrogating further signal propagation. As a result of β-arrestin binding, phosphorylated receptors are also targeted for clathrin-mediated internalization.[3,4]

RSC Drug Discovery Series No. 8
G Protein-Coupled Receptors: From Structure to Function
Edited by Jesús Giraldo and Jean-Philippe Pin

Published by the Royal Society of Chemistry, www.rsc.org

Besides such a 'negative' role of GRKs/arrestins in GPCR signalling, recent data indicate that, in addition to signal termination, GRKs and arrestins can initiate alternative signalling pathways and participate in a variety of cellular processes. In the case of arrestins, this is achieved by serving as scaffold proteins for several signalling mediators such as c-Src, JNK-3, components of the Raf/MEK/ERK cascade, the cAMP phosphodiesterase PDE4, the Ral-GDS regulator of the cytoskeleton, components of the NfκB signalling pathway and the Mdm2 ubiquitin ligase, among others.[2,3,5–7] Therefore, arrestin recruitment is critical for triggering the modulation of important intracellular signalling cascades by GPCRs and other plasma membrane receptors, contributing to the overall cellular response to the presence of a messenger.

On the other hand, the cellular role of GRKs is not limited to promoting β-arrestin binding to activated GPCRs. GRKs are multidomain proteins with diverse cellular functions. In particular, the ubiquitous and prototypic GRK isoform, GRK2, is emerging as a key node in signal transduction pathways, displaying a complex network of functional interactions ('interactome'). GRKs can regulate other receptor families, such as tyrosine kinase receptors,[8] and are able to modulate signals downstream of G proteins[9] and to functionally interact with many non-receptors partners.[10] Interestingly, the expression and function of GRK2 is tightly regulated, and its levels and functionality altered in several physiological and pathological situations, suggesting that this protein is involved in a variety of physiological processes and that its changes may contribute to the onset and/or development of different pathologies.[1,2,11,12]

16.2 GRK2 is a Multidomain Protein

Seven GRK genes are known in mammals, subdivided into three different groups: visual GRK subfamily (GRK1 and GRK7); the β-adrenergic receptor kinase subfamily (GRK2/GRK3); and the GRK4 subfamily (GRK4, GRK5 and GRK6).[1,2,10] Except for visual GRKs and GRK4, GRKs are almost ubiquitously expressed, showing differences in their levels depending on the tissue and cell type. GRK2 appears to be the more functionally important isoform as judged by the lethality of homozygous GRK2-deficient mice.[13]

A common structural architecture is shared by all GRK isoforms, with a conserved central catalytic domain ($\sim$270 amino acids), similar to that of the AGC family of serine–threonine kinases, flanked by an *N*-terminal domain ($\sim$185 amino acids) and a variable length *C*-terminal domain ($\sim$105–230 amino acids). The *N*-terminal domain is important for receptor recognition, intracellular membrane anchoring and allosteric activation. It also contains an RH (regulator of G protein signalling homology) domain. In the case of GRK2/3, the RH domain interacts with Gαq family members, thus hampering phospholipase C beta (PLCβ) activation by GPCR. The *C*-terminal region of GRK2/3 contains a pleckstrin homology domain (PH) with binding sites for the membrane phospholipid PIP2 and free Gβγ subunits, which mediate agonist-dependent translocation to the plasma membrane of this mainly cytoplasmic protein.[1,12]

The crystal structures of GRK2 in complex with G protein β1γ2 subunits and Gαq and, more recently, of human GRK2 in complex with a general kinase inhibitor[14–16] have indicated that the three distinct GRK2 regions (RH, kinase and PH domains) are located at the vertices of a triangle—a clear example of how multiple modular domains are integrated in a single molecule to transduce and modulate different signalling events. Moreover, these data have provided insights into GRK regulation of different effectors and activators and on possible mechanisms of intramolecular regulation, suggesting that receptor recognition, plasma membrane targeting and GRK activation are intimately coupled processes.[17] In addition to this general domain architecture, ongoing research is unveiling the localization of regions involved in interaction with different cellular proteins and of regulatory phosphorylation sites in the GRK2 structure [reviewed in refs. 1,10 and 12].

16.3 GPCR Phosphorylation by GRK2 or Other GRKs can Differentially Trigger Downstream Signalling Cascades

Recent studies have suggested that the precise sets of residues phosphorylated in the intracytoplasmic domains of the receptor by one or another GRK or by the same GRK in response to different receptor agonists (what has been termed the GRK-specific 'bar code'[18]) can engage distinct signalosomes downstream of the receptor, probably by recruiting β-arrestin in different conformations, leading to distinct β-arrestin-scaffolded complexes. For instance, phosphorylation of the CXCR4 receptor varies in kinetics, location, extent and nature of the precise residues that become phosphorylated depending on the kinase that is engaged.[19] In addition, different pools of β-arrestin are mobilized when the CCR7 receptor is stimulated by different ligands by activating GRK3/6 or GRK6 alone; CCL19-mediated GRK3/6 phosphorylation causes β-arrestin to redistribute in endocytic vesicles, while CCL21-dependent GRK6 activation recruits β-arrestin to the cell membrane.[20]

16.4 GRK2 Phosphorylates non-GPCR Substrates and Displays a Complex Network of Functional Interactions

Adding to its classical role as a GPCR desensitizing enzyme, GRK2 is also able to control the activity and function of other types of membrane receptors. In fact, several data demonstrate that GRK2 can act as a regulator of members of the tyrosine kinase receptor family. For instance, beta-type platelet-derived growth factor receptor (PDGF-Rβ) is phosphorylated by GRK2, which reduces receptor activation in vascular smooth muscle cells.[21] On the other hand, stimulation of epidermal growth factor (EGF) receptor triggers the recruitment of GRK2, leading to its phosphorylation at tyrosine residues.

This leads to an increase in GRK2 catalytic activity towards nearby membrane GPCR, thus promoting opioid receptor trans-regulation.[22]

Moreover, a growing number of non-receptor substrates have been identified for GRK2[10,12] (see Figure 16.1). This kinase phosphorylates tubulin, synucleins, phosducin, ribosomal protein P2, the inhibitory γ subunit of the type 6 retinal cyclic guanosine monophosphate (cGMP) phosphodiesterase, a subunit of the epithelial Na+ channel, the ERM family protein ezrin, the calcium-binding protein DREAM, IKappaBalpha (IκBα), and the p38 mitogen-activated protein kinase (MAPK) (see refs. 12,23 and 24, and references therein). Some of these phosphorylation events appear to be inhibitory. For instance, GRK2-mediated phosphorylation of p38MAPK blocks its activation and impairs cytokine secretion,[24] thus contributing to signal termination downstream of receptor proteins. GRK2 also inhibits transforming growth factor beta (TGF-β) mediated cell growth arrest and apoptosis by inducing Smad phosphorylation in hepatoma cells.[25] On the contrary, GRK2/3 or orthologue proteins in different species are essential positive mediators of Hedgehog/Smoothened signal transduction in

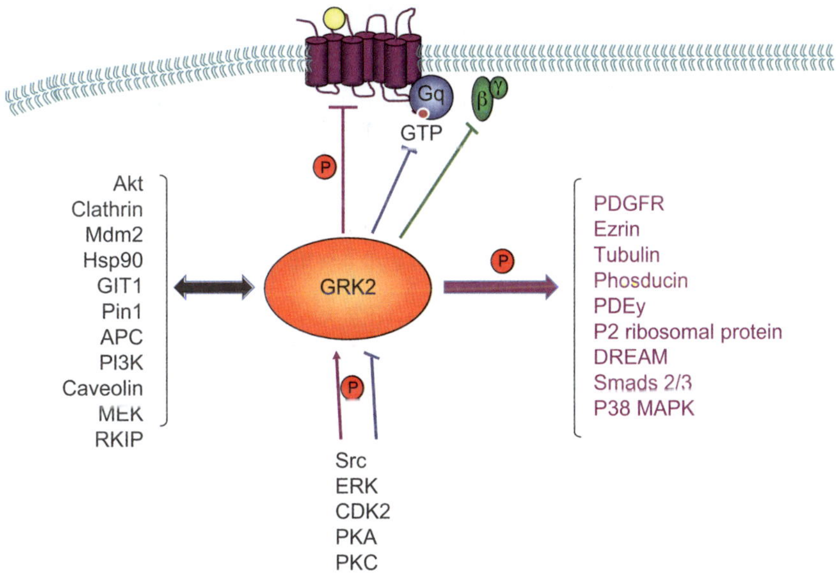

Figure 16.1 GRK2 displays a complex network of functional interactions. In addition to its 'classical' role in triggering agonist-mediated GPCR phosphorylation, GRK2 can modulate cell signalling by interacting with Gαq and Gβγ subunits. Moreover, GRK2 can positively or negatively modulate diverse signalling pathways by phosphorylating a variety of non-GPCR substrates or by interacting with other proteins involved in signal transduction. The activity of GRK2 towards its substrates and/or its association with many of these proteins can be dynamically regulated by the phosphorylation of GRK2 itself by the indicated kinases. See text for additional details and description of abbreviations. The P symbol denotes a phosphorylation-mediated modulation.

Drosophila, zebrafish and mice.[26–28] Overall, these data suggest that GRK2 may act as an 'effector', participating in the regulation of diverse cellular phenomena through the phosphorylation of substrates that are very varied functionally.

GRK2 might also control cellular responses in a phosphorylation-independent manner (Figure 16.1). GRK2 has been reported to associate with Gαq, Gβγ, PI3K, clathrin, GIT, caveolin, MEK, Akt, RKIP, RalA GTPase and APC protein in different experimental settings.[10] Such interactions appear to be promoted in a stimulus, cell-type and context-specific way and display a variety of functional consequences. For instance, GRK2/MEK association impairs chemokine induction of MAPK activation,[29] whereas dynamic GRK2 association with GIT1—a scaffold protein involved in multiple cellular processes such as cytoskeletal rearrangements, membrane trafficking and cell adhesion—underlies a positive effect of GRK2 in fibronectin or sphingosine-1-phosphate (S1P) stimulated epithelial cell migration.[30]

Other functional interactions have been shown to be involved in the regulation of GRK2 itself.[1–3] Phosphorylation by different kinases can either increase [protein kinase A (PKA), protein kinase C (PKC), Src kinases) or decrease (ERK) membrane targeting and/or the catalytic activity of GRK2.[31–33] Phosphorylation of GRK2 at given tyrosine or serine residues is also a key mechanism to dynamically modulate its interaction with cellular partners. Tyrosine phosphorylation by c-Src appears to enhance the interaction of GRK2 with Gαq[34] and with the GIT1 scaffold protein.[12] On the contrary, phosphorylation by ERK1/2 on S670 impairs GRK2/Gβγ interaction and inhibits kinase translocation and catalytic activity toward receptor membrane substrates, while also modulating GRK2 interaction with GIT1 and the prolyl-isomerase Pin1.[30,32] Apart from phosphorylation, *S*-nitrosylation has also been suggested as a novel mechanism to inhibit GRK2 activity.[35]

On the other hand, the proteasome pathway rapidly degrades GRK2. Its ubiquitination and turnover is enhanced by GPCR activation as a result of phosphorylation of GRK2 by c-Src and MAPK in a β-arrestin-dependent manner.[36,37] Interestingly, Mdm2, an E3-ubiquitin ligase involved in the control of cell growth and apoptosis, plays a key role in GRK2 degradation.[38] Mdm2 and GRK2 association and subsequent proteolysis are facilitated by the β-arrestin scaffold function upon β_2-adrenergic receptor stimulation. Interestingly, activation of the PI3K/Akt pathway by agonists such as IGF-1 alters Mdm2 phosphorylation and triggers its nuclear localization, thus hampering Mdm2-mediated GRK2 degradation and leading to enhanced GRK2 stability and increased kinase levels.[38] Since deregulated activity of the PI3K/Akt pathway takes place in pathophysiological contexts characterized by increased cell proliferation and survival, this might explain GRK2 up-regulation in some human tumour malignancies.[39]

In summary, the complex GRK2 'interactome' puts forward this kinase at the crossroads of multiple signalling pathways. However, how such multiple potential interactions are modulated by extracellular signals and the identification of the precise GRK2 interactome that underlies its participation in key biological functions in given cell types and/or in specific physiological and pathological contexts remain to be established. These questions are of particular interest when considering physiological and pathological situations characterized by changes in GRK2 expression levels and/or functionality. Some of these scenarios are discussed below.

16.5 GRK2 in Cardiovascular Cells: Implications in Heart Failure and Hypertension

Increased left ventricular GRK2 mRNA and activity have been reported in patients with ischemic or idiopathic dilated cardiomyopathy, cardiac ischemia, volume overload and left ventricular hypertrophy, which parallels a reduction in β1-adrenoceptor function.[2,11,40] Interestingly, in some animal models, the development of overt heart failure is preceded by an elevation of GRK2 levels. The levels and/or activity of GRK2 are also elevated in lymphocytes of patients with cardiac failure and a correlation can be drawn between GRK2 protein in the myocardium of these subjects and of those found in circulating mononuclear cells.[41] This has lead to the possibility that lymphocytic GRK2 could be used as surrogate readout of myocardial levels of GRK2. In fact, this very same 'mirror' effect can be detected in blood cells from hypertensive subjects and also in patients with ventricular overload.[42] Interestingly, not only did lymphocytic GRK2 levels increase with the progression of the disease, but they also decrease when heart functionality is improved. Two small clinical studies show that, after pharmacological therapy with ACE inhibitors in class II heart failure, or upon implantation of mechanical left ventricular support in end-stage heart failure patients, the amount of GRK2 in lymphocytes is diminished.[43,44] The presence of beta-agonists has been reported to up-regulate GRK2 mRNA, whereas ischemia might promote GRK2 degradation by the proteasome in some experimental models (reviewed in ref. 11). However, there is limited knowledge of the mechanisms modulating GRK2 expression in cardiovascular cells in pathological settings.

The importance of GRK2 in the ethiology and/or development of cardiovascular diseases and cardiac dysfunction has been established by a vast amount of experimental data obtained mainly from animal models of heart failure and has been extensively reviewed elsewhere.[11,40] In brief, hemizygous GRK2 mice expressing 50% less protein than control littermates are hyperresponsive to catecholamines and present increased cardiac contractility and function, whereas the opposite happens with transgenic mice overexpressing different levels of this kinase in which the adrenergic cardiac response is impaired. Genetic inhibition of GRK2 (by transgenic overexpression of a *C*-terminal peptide of this kinase that prevents its activation by Gβγ subunits) rescues cardiac dysfunction and improves survival in many different mouse models of heart failure.[45] This genetic inhibition acts in a synergic manner with established beta-blocker treatment, suggesting alternative mechanisms of action exist for both approaches.[46] It should be noted that overexpression of the GRK2 *C*-terminal peptide could also titrate Gβγ dimers from different signalling mediators and thus down-regulate other signalling routes.

The beneficial effects of inhibiting GRK2 in heart failure have been mainly related to its role in increasing adrenergic responsiveness.[45] However, chronic adrenergic activation appears to be more detrimental than beneficial for heart disease:[46,47]

- Increased adrenergic activation is a direct cause of cardiomyopathy.
- Chronic elevations in adrenergic signals achieved by transgenic over-expression of β1-receptors, Gαs proteins or adenylyl cyclase, or by means of chronic agonist treatment in mice, evolve to cardiac dysfunction.
- Genetic polymorphisms that enhance sympathetic tone are independent risk factors for heart failure.
- Most importantly, beta-blockers represent a successful standard treatment for this disease.

In addition, GRK2-deficient mice are embryonically lethal with cardiac hypoplasia,[13] which can be regarded as confirmation that GRK2 is required for the proper development and maintenance of heart structure. However, mice with specific ablation of GRK2 in cardiomyocytic cells are viable.[48] This result unmasks the fact that it is not myocardial GRK2 protein, but rather an extra-cardiac effect of GRK2 deficiency which accounts for hampered cardiac development. The fact that these germline cardiomyocyte-deficient GRK2 mice are also more prone to cardiac damage induced by adrenergic stimulation when adults speaks against a beneficial role of decreased GRK2 in cardiac function. However, it has been shown, by developing mice with an inducible GRK2-deficiency in cardiomyocytes, that deleting GRK2 either before or ten days after myocardial infarction impedes heart failure and improves survival.[49] This report clearly demonstrates a causal role for GRK2 in cardiac remodelling and dysfunction, and identifies this kinase as a clear target for therapeutic intervention. Moreover, the same group further established that long-term silencing of GRK2 by means of adeno-associated viruses injected in mouse myocardium increases contractility, prevents remodelling and, more importantly, halts the neurohumoral vicious circle of increased catecholamines and aldosterone characteristic of heart failure.[50]

The beneficial role of GRK2 inhibition in the treatment of heart failure may also arise from mechanisms different from β1-adrenergic receptor modulation. Alternative cardioprotective signalling pathways that may help improve cardiovascular function and survival outcomes have been identified.[51] These pathways have been shown to be more dependent on the arrestin-mediated branch downstream of GPCRs. In this line, lowering GRK2 has been shown to paradoxically increase arrestin-dependent ERK activation by GPCRs in a very specific manner.[52] Therefore, inhibiting GRK2 could lead to a 'signalling switch' that would selectively increase signalling to certain cardioprotective routes downstream of GPCRs, which could compensate for the cardiotoxic effects of the concomitant over-activation of second messenger-dependent pathways. The beneficial effects of the GRK2ct expression might also be related to its ability to disrupt the interaction of GRK2 with some of its additional potential partners or regulators present in cardiac cells such as RKIP, PI3K/Akt, tubulin, ezrin and GIT1 (see Section 16.4)—although this issue remains to be addressed in detail.

It should also be noted that adrenal GRK2 is also up-regulated in heart failure. Since adrenal α2-adrenergic receptors negatively regulate the release of catecholamines by the adrenal gland, such GRK2 alteration would trigger a catecholamine overdrive, which *per se* is detrimental for cardiac function in the long-term. In this regard, it has been demonstrated that inhibiting GRK2 in rat adrenal glands helps to restore cardiac function,[53] supporting GRK2 inhibition as an integrated approach to therapies that target catecholaminergic pathways in heart failure.[45]

As mentioned above, in addition to heart disease, increased GRK2 expression is also associated to several clinical manifestations of hypertension. These changes are observed in parallel with altered β-adrenergic receptor responsiveness, relevant to vasodilatation. Elevated GRK2 expression in the vasculature is not only a mere marker of the hypertensive state but rather a precipitating factor of hypertension. Transgenic mice with vascular smooth muscle-targeted over-expression of GRK2 display attenuated vascular β-adrenergic receptor signalling and vasodilatation, while α-adrenergic mediated vasoconstriction remains unaltered, leading to elevation of resting blood pressure and vascular remodelling.[42,54] Inhibition of vascular smooth muscle GRK2 by cell-specific gene ablation also enhances α1D-adrenergic receptor constriction.[55] In addition, both endothelin-1 receptor signalling and endothelial nitric oxide synthetase (eNOS) production (key controllers of portal tension) have been shown to be impaired by elevated GRK2 levels.[56] This new role of endothelial GRK2 in vascular homeostasis could be relevant in human primary hypertension, but whether GRK2 up-regulation in this condition is confined to vascular smooth muscle cells or also extended to the endothelium remains to be established. Overall, these studies suggest that enhanced vascular GRK2 might be an important contributor to the pathogenesis/maintenance of human essential hypertension.

16.6 GRK2 Interactome in Immune Cell Migration: Physiopathological Implications in Inflammation and Sepsis

In addition to modulate a variety of GPCR related to migration processes, GRK2 has been shown to interact with a variety of proteins involved in migration such as MEK, Akt, ezrin, PI3Kγ and GIT.[57] Consistently, recent evidence suggests that GRK2 plays an important, complex role in immune and epithelial cell migration (Figure 16.2).

GRK2 is highly expressed in different cellular types of the immune system and is an important regulator of cell responses during inflammation.[58] GRK2 phosphorylates a variety of chemokine receptors and chemotactic receptors for substance P, S1P or formyl peptide, responsible for leukocyte trafficking to the inflammatory foci, T cell egression from lymphoid organs, leukocyte activation

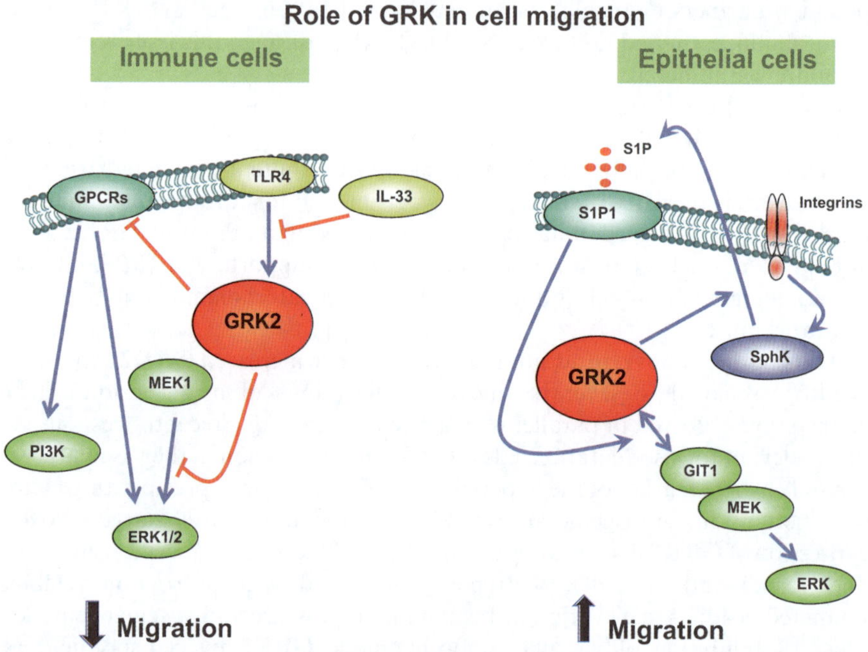

Figure 16.2 Specific GRK2 interactomes underlie its differential modulation of immune and epithelial cell migration. The diagram provides a schematic representation of the signalling molecules/pathways relevant for the effect of GRK2 in either immune or epithelial cell migration. The overall effect of GRK2 levels on cell migration is dependent on the cell type and the signalling context, which promote different interactions of GRK2 with its potential partners. S1P1R = sphingosine-1-phosphate receptor 1; Sphk = sphingosine kinase. See text for details and description of other abbreviations.

or proliferation.[58] According to the classical, 'negative' role of GRK2 in GPCR modulation, splenocytes and T lymphocytes isolated from GRK2+/− mice display increased agonist-induced activation of ERK and PI3K/Akt pathways[59] and increased migration compared with wild-type littermates in response to certain chemokines (CCL5, CXCL12). Consistently, it has been reported that the GRK2 and GRK5 transcriptional down-regulation caused by activation of the Toll-like receptor 4 (TLR4) pathway lowers chemokine receptor desensitization and enhances the migratory response of polymorphonuclear leukocytes.[58,60] The modulation of chemokine-mediated induction of ERK activity by altered GRK2 levels seems to involve both kinase-dependent and independent functions, the latter related to the ability of GRK2 to interfere with the MEK/ERK interface.[29] Whether other components of the GRK2 interactome may also modify immune cell migration remains to be investigated.[57]

These observations suggest that changes in GRK2 levels in cells of the immune system could be involved in the progression of inflammatory diseases. A decrease of GRK2 protein expression and kinase activity (~55%) was found in peripheral blood mononuclear cells of patients with rheumatoid arthritis compared with healthy subjects, and in splenocytes and mesenteric lymph node cells in animal models of experimental arthritis (reviewed in ref. 58). Transient down-regulation of GRKs in immune cells during inflammation may represent an initially adaptive mechanism to facilitate cell response, whereas chronic GRK2 down-regulation would lead to an aberrant inflammatory outcome. Interestingly, expression of GRK2 in leukocytes from patients with multiple sclerosis (MS) was also decreased by 40% compared with that of healthy individuals.[61] GRK2 levels appear to have a direct impact in the clinical course of experimental MS, since the onset of the relapsing–remitting experimental autoimmune encephalomyelitis in hemizygous GRK2+/− mice was accelerated, in parallel with a higher initial infiltration of T cells into the brain. Curiously, these animals display lower inflammatory infiltrates in the long term and do not develop relapses in the disease compared to wild-type animals.

Thus, the effects of altered GRK2 expression are far from being well-defined since many different cell types that are responsible for the progress of the disease, not only from the immune system but also from the inflamed tissue, may respond differently to GRK2 changes. For instance, whereas in T cells and monocytes decreased GRK2 levels lead to enhanced migration (consistent with its classical role in GPCR desensitization[57,58]), preliminary data suggest that in other cell types and in response to other stimuli (granulocytes stimulated with LTB4 or IL-8), decreased GRK2 levels do not affect chemotactic responses,[62] suggesting that the effect of GRK2 on cell migration would be dependent on the precise stimuli or cell type. In this regard, it is worth noting that a recent report has indicated that sepsis attenuation by interleukin-33 (IL-33) may involve modulation of GRK2 expression in neutrophils.[60] The amount of GRK2 is enhanced in neutrophils from individuals with sepsis, in parallel with a reduction of chemotaxis toward IL-8, by mechanisms probably involving the activation of the LPS–TLR4 signalling pathway. Such enhanced GRK2 levels would trigger CXCR2 receptor internalization and decreased response of neutrophils to CXCL2 or CXCL8 chemokines. IL-33 would reverse such TLR4-induced effects *via* inhibition of GRK2 expression, overriding the diminished neutrophil chemotaxis in these conditions (Figure 16.2).

In summary, dynamic modulation of GRK2 levels in specific cell types of the immune system may lead to relevant changes in the ability of these cells to respond to key messengers involved in inflammatory processes.

16.7 GRK2 Interactome in Epithelial Cell Migration

Our laboratory found that GRK2 promotes increased migration towards fibronectin in different epithelial cell lines models and in fibroblasts. These effects were independent of GRK2 kinase activity, since they were also observed upon expression of a catalytically inactive mutant. Therefore, and contrary to that described in immune cells, increased GRK2 expression facilitates migration towards fibronectin and GRK2 down-regulation impairs migration.[30] Fibronectin, acting through integrin receptors, induces sphingosine kinase activity and the subsequent release of S1P, which in turn triggers migration *via* Gi-coupled S1P1-receptors. Interestingly, GRK2 appears to affect this process at least at two levels (Figure 16.2). First, transmodulation of S1P1 receptors by fibronectin is potentiated by GRK2 expression, since enhanced GRK2 levels lead to increased S1P production, by mechanisms to be identified. In addition, GRK2 facilitates the activation of the ERK pathway and migration in response to fibronectin and S1P.[30] Therefore, contrary to the classical paradigm, GRK2 appears to be playing an overall 'positive' role in cell signalling in epithelial cells and fibroblasts.

Such effects involve the dynamic association of GRK2 with GIT1, a multidomain protein, with GTPase (GAP) activity towards the small G protein ARF able to interact with Rac modulators, paxillin and MEK1, first described as a GRK-interacting protein.[30] GIT1 can promote migration by increasing focal adhesion turnover, delivering active PAK (p21-activating kinase) to the leading edge and also by acting as a scaffold for ERK activation in focal adhesions. We found that S1P stimulation dynamically modulates the GRK2/GIT1 association by mechanisms involving sequential GRK2 phosphorylation by Src-like kinases and MAPK.[30] Notably, decreased GRK2 levels in hemizygous mice result in delayed wound healing rate *in vivo*, consistent with a physiological role for GRK2 as a regulator of coordinated integrin and GPCR-directed epithelial cell migration.[30]

These data put forward the interesting notion that altered GRK2 expression levels might alter migratory responses in pathological conditions. Aberrant epithelial cell motility plays a key role in cancer progression and metastasis. S1P and integrin signalling, as well as other GPCRs such as chemokine receptors or protease-activated receptors are involved in these processes.[63] Interestingly, certain signalling pathways instrumental in many cancers cause the up-regulation of GRK2 protein levels in malignant cell lines.[25,38] In addition, initial data indicate that GRK2 protein levels can be either up-regulated in tissue samples of patients with granulosa cell tumours and with differentiated thyroid carcinoma,[64] or down-regulated in a subgroup of prostate tumours.[65] Altogether these results suggest that altered GRK2 expression in specific tumour cells may affect migration in response to particular stimuli and play a role in carcinogenesis.

This hypothesis is further supported by the observed cooperation of GRK2 with known oncogenes in '*in vitro*' transformation assays[66] and by the emerging role of GRK2 in cell cycle progression (see below). A detailed characterization of GRK2 expression levels in different types of tumours and further insight on the effects of altered GRK2 expression in tumour progression are needed to further define its role in this process.

16.8 GRK2 and Cell Cycle Progression

GRK2 knockout mice are embryonic lethal[13] at day 9–12 and display marked cardiac abnormalities as a result of extracardiac GRK2 functions.[48] Germline

GRK2 ablation promotes generalized embryo growth retardation and additional alterations from normal development, indicating that this protein plays a critical role in basic cellular functions such as cell proliferation, differentiation and migration during development. GRK2 expression has been reported to have distinct impacts on cell proliferation and mitogenic signalling, depending on both the cell type and the mitogenic stimuli analysed. GRK2 inhibits TGF-mediated cell growth arrest and apoptosis in human hepatocarcinoma cells.[25] On the other hand, GRK2 attenuates serum- or PDGF-induced proliferation of thyroid cancer cell lines[64] and smooth muscle cells,[67] respectively, whereas its expression increases MAPK signalling in response to EGF in HEK-293 cells[68] and GRK2 kinase activity is required for insulin growth factor 1 (IGF-1) triggered proliferation and mitogenic signalling in osteoblasts.[69] We and other groups have found that GRK2 and orthologs potentiate Smoothened receptor signalling,[27,66] and knock-down of a GRK2 ortholog has been reported to cause growth arrest in zebrafish accompanied by abnormalities in somitogenesis, the hematopoietic system, and in patterning of the eyes and neural tube.[70] Such developmental arrest can be partially rescued by expression of a GRK2 kinase-inactive mutant, indicating that this phenotype relies both on kinase dependent and independent processes. Moreover, such phenotype resembles deficiencies in cell-cycle progression and cell differentiation caused by impaired Smoothened signalling.

In this regard, recent data point at a role for GRK2 as both an extrinsic and intrinsic cell-cycle regulator. It is known that Patched, a membrane protein that acts as a receptor for the Hedgehog ligand (and thus regulates Smoothened signalling), also inhibits cell growth by interacting with cyclin B1 and decreasing its nuclear accumulation. GRK2 has been recently reported to interact with Patched, which relieves the Patched-induced cytosolic retention of cyclin B1. This kinase activity-independent effect would explain why the absence of GRK2 causes defects in developing brain and eyes in the early zebrafish embryo, while the phosphorylation-mediated regulation of Smoothened by GRK2 could be responsible for the role of this kinase in somite patterning.[70] Such GRK2 regulatory roles depend on extrinsic cues promoting cell division, since the GRK2-mediated phosphorylation of Smoothened and the interaction of GRK2 with the Patched/cyclin B pathway are both promoted by the Hedgehog ligand.

On the other hand, a complex GRK2 interactome that is critical for timely G2/M transition has been recently reported during cell cycle progression.[71] GRK2 protein levels are transiently down-regulated during the G2/M transition by a mechanism involving CDK2-mediated phosphorylation of GRK2, which drives binding to the prolyl-isomerase Pin1 and subsequent degradation. Preventing GRK2 phosphorylation at a specific residue impedes normal GRK2 down-regulation and markedly delays cell-cycle progression.[71] Interestingly, the 'default' GRK2 protein decay in G2 is prevented in the presence of DNA damaging agents that trigger cell cycle arrest such as doxorubicin. Moreover, in cells with higher steady-state levels of the kinase, increased stabilized GRK2 levels in such conditions inversely correlate with the p53 response and the induction of apoptosis strongly suggests that GRK2 contributes to orchestrate G2/M checkpoint mechanisms, helping to restrict the apoptotic fate of arrested cells.[71] Since it has been reported that GRK2 is up-regulated in the context of oncogenic signalling (see above), it is tempting to suggest that inhibition of GRK2 expression might sensitize cells to drug-induced DNA damage.

16.9 GRK2 and Modulation of Insulin Signalling: Implications in Type 2 Diabetes

GRK2 can act as an inhibitor of insulin-mediated glucose transport stimulation in 3T3L1 adipocytes by interacting with Gαq/11 function independently of its kinase activity.[72] GRK2 also inhibits basal and insulin-stimulated glycogen synthesis in mouse liver FL83B cells.[73]

We have recently shown[74] that GRK2 is a key modulator of insulin sensitivity *in vivo*. In cultured adipocytes and myoblasts, increased GRK2 or GRK2–K220R levels inhibit insulin-stimulated glucose uptake and signalling in a kinase-activity independent manner, by mechanisms involving the formation of dynamic GRK2/IRS1 (Insulin Receptor Substrate 1) complexes.

It was previously shown that enhanced GRK2 levels favour insulin resistance induced by endothelin-1 in 3T3L1 cells[75] or by chronic β-adrenergic receptor stimulation in HEK-293 cells.[76] GPCR agonists promote GRK2 interaction with Gαq and also with IRS1, resulting in decreased insulin-stimulated glucose transport, IRS–serine phosphorylation and IRS1 degradation.[75] Our data in cultured cells, as well as in adipose tissue *in vivo*, show that IRS1 levels as well as the amount of basal GRK2–IRS1 complexes depend on GRK2 expression, and that insulin stimulation rapidly disrupts basal IRS1/GRK2 complexes, which is hampered in the presence of elevated levels of GRK2. Interestingly, GRK2 expression is enhanced by approximately two-fold in insulin-resistant human adipocytes, in blood mononuclear cells from insulin-resistant patients and in adipose and muscle tissues in either tumour necrosis factor alpha (TNFα), aging or high fat-diet-induced insulin resistance models.[74] Importantly, GRK2+/− mice maintain glucose tolerance and insulin signalling in the major insulin-responsive tissues under such experimental conditions, suggesting that enhanced GRK2 expression above a certain threshold markedly impairs insulin sensitivity *in vivo*. These results suggest a key role for GRK2 in the modulation of insulin sensitivity in physiological and pathological conditions.

16.10 GRK2 and Pain Modulation

Chronic pain associated with inflammation is a major clinical problem. Recent data unveil completely novel cell-specific roles of GRK2 in regulating acute and chronic inflammatory hyperalgesia. GRK2+/− mice have been shown to develop increased and markedly prolonged thermal hyperalgesia and mechanical allodynia after carrageenan-induced paw inflammation or after intraplantar injection of the GPCR-binding chemokine CCL3.[77] Interestingly,

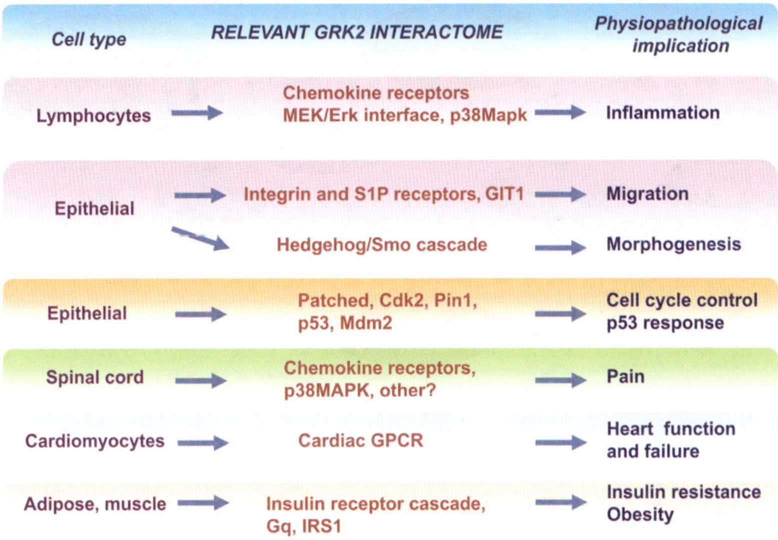

Figure 16.3 The distinct GRK2 interactome in different cells types has different physiological and pathological implications. As detailed in the text, ongoing research is helping to characterize the precise functional interactions that GRK2 establishes in different cell types, cellular contexts or in response to specific stimulus. This will help to understand why changes in GRK2 expression levels and functionality underlie different physiological and pathological processes.

Carrageenan- or CCL3-induced hyperalgesia was increased but not prolonged in mice with decreased GRK2 only in a specific set of neuronal nociceptors. On the other hand, reduced GRK2 complement specifically in microglia/monocytes was required and sufficient to transform acute carrageenan- or CCL3-induced hyperalgesia into chronic hyperalgesia. Chronic hyperalgesia in GRK2+/− mice was associated with ongoing microglial activation and increased phospho-p38 and TNFα in the spinal cord. Inhibition of spinal cord microglial, p38 or TNFα activity by intrathecal administration of specific inhibitors reversed ongoing hyperalgesia in GRK2+/− mice. Microglia/macrophage GRK2 expression was reduced in the lumbar ipsilateral spinal cord during neuropathic pain, underlining the pathophysiological relevance of microglial GRK2.

The same group has reported[78] that spinal microglia/macrophage GRK2 is reduced during chronic inflammation-induced hyperalgesia and that GRK2+/− mice develop prolonged hyperalgesia following a single intraplantar injection of the pro-inflammatory cytokine IL-1β. These authors applied CRE-Lox technology to create mice with low GRK2 in microglia/macrophages/granulocytes (LysM-GRK2(f/+)), or sensory neurons or astrocytes. Only mice deficient in microglial/macrophage/granulocyte GRK2 display prolonged IL-1β-induced hyperalgesia that lasts up to eight days. These data establish that chronic inflammatory hyperalgesia is associated with reduced GRK2 in

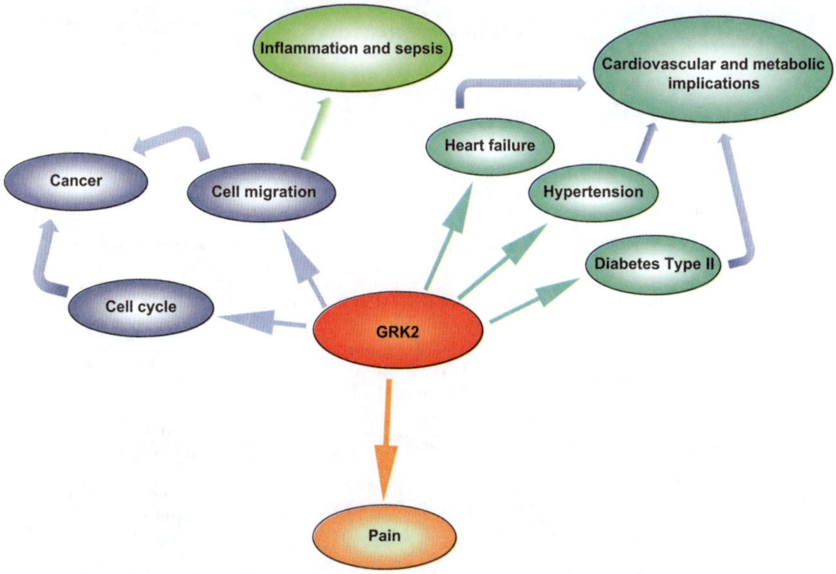

Figure 16.4 GRK2 is a relevant signalling node of the cellular transduction network, involved in many physiological and pathological situations. Alterations in GRK2 levels and/or activity have been reported in several relevant cardiovascular, inflammatory or cancer pathologies, suggesting that its changes may contribute to the onset or development of different diseases.

microglia/macrophages. Low GRK2 favours hyperalgesia by promoting enhanced spinal microglial/macrophage activity, fractalkine signalling, p38 activation and IL-1 signalling. Interestingly, previous data have indicated that GRK2 can act as a negative modulate of p38 MAPK.[24] It is tempting to propose that chronic inflammation decreases spinal microglial/macrophage GRK2, which prevents silencing of microglia/macrophage activity and thereby contributes to prolonged hyperalgesia.

16.11 Conclusions

The increasingly complex GRK2 'interactome' unveils that this kinase is a relevant signalling node of the cellular transduction network, beyond its initial role in GPCR desensitization. Further research is needed to better understand how the potential GRK2 'interactomes' are established depending on the specific stimulus, cell type and physiological context (Figure 16.3). The modulation of such potential GRK2 interactions by GPCR and other extracellular signals, and its spatial and temporal integration remains to be established. Moreover, the overall physiological and pathological implications arising from this emerging GRK2 interaction map are far from being elucidated (Figure 16.4). This is of particular relevance given that alterations in GRK2 levels and/or activity have been reported in a number of relevant cardiovascular, inflammatory and cancer pathologies.

In this regard, we need to better understand the mechanisms and signals governing the expression and activity of GRK2 (and other GRKs) in specific cell types during the progression of different diseases. Since it is likely that altered GRK2 expression would differentially affect the function of its interaction partners and impair homeostasis in distinct ways, depending on the cell type involved, it will also be important to assess the impact of such alterations in the complex integrated network of GRK2 cellular functions. In addition, the combined use of cellular and animal models with altered GRK2 complement/functionality in specific cell types or situations will be crucial in unveiling key cellular and physiological processes controlled by this protein.

Acknowledgements

Our laboratory is funded by grants from Ministerio de Educación y Ciencia (SAF2008-00552), Fundación Ramón Areces, the Cardiovascular Network (RECAVA) of Ministerio Sanidad y Consumo-Instituto Carlos III (RD06-0014/0037), Comunidad de Madrid (S-SAL-0159-2006) to F.M. and Comunidad de Madrid and Universidad Autónoma de Madrid (CCG08-UAM/BIO-4452) to P.P., and Instituto de Salud Carlos III (PI080461 and PS09/0128) to C.R. and C.M., respectively.

References

1. P. Penela, C. Ribas and F. Mayor, Jr., *Cell Signal.*, 2003, **15**, 973.
2. R. T. Premont and R. R. Gainetdinov, *Annu. Rev. Physiol.*, 2007, **69**, 511.
3. E. Reiter and R. J. Lefkowitz, *Trends Endocrinol. Metab.*, 2006, **17**, 159.
4. C. A. Moore, S. K. Milano and J. L. Benovic, *Annu. Rev. Physiol.*, 2007, **69**, 451.
5. S. M. DeWire, S. Ahn, R. J. Lefkowitz and S. K. Shenoy, *Annu. Rev. Physiol.*, 2007, **69**, 483.
6. J. J. Kovacs, M. R. Hara, C. L. Davenport, J. Kim and R. J. Lefkowitz, *Dev. Cell*, 2009, **17**, 443.
7. L. M. Luttrell and D. Gesty-Palmer, *Pharmacol. Rev.*, 2010, **62**, 305.
8. C. J. Hupfeld and J. M. Olefsky, *Annu. Rev. Physiol.*, 2007, **69**, 561.
9. M. Jurado-Pueyo, P. M. Campos, F. Mayor and C. Murga, *J. Recept. Signal Transduct. Res.*, 2008, **28**, 59.
10. P. Penela, C. Murga, C. Ribas, V. Lafarga and F. Mayor, Jr., *Br. J. Pharmacol.*, 2010, **160**, 821.
11. P. Penela, C. Murga, C. Ribas, A. S. Tutor, S. Peregrin and F. Mayor, Jr., *Cardiovasc. Res.*, 2006, **69**, 46.
12. C. Ribas, P. Penela, C. Murga, A. Salcedo, C. Garcia-Hoz, M. Jurado-Pueyo, I. Aymerich and F. Mayor, Jr., *Biochim. Biophys. Acta*, 2007, **1768**, 913.
13. M. Jaber, W. J. Koch, H. Rockman, B. Smith, R. A. Bond, K. K. Sulik, J. Ross, Jr., R. J. Lefkowitz, M. G. Caron and B. Giros, *Proc. Natl. Acad. Sci. U. S. A.*, 1996, **93**, 12974.

14. D. T. Lodowski, J. F. Barnhill, J. A. Pitcher, W. D. Capel, R. J. Lefkowitz and J. J. Tesmer, *Acta Crystallogr., Sect. D: Biol. Crystallogr.*, 2003, **59**, 936.
15. V. M. Tesmer, T. Kawano, A. Shankaranarayanan, T. Kozasa and J. J. Tesmer, *Science*, 2005, **310**, 1686.
16. J. J. Tesmer, V. M. Tesmer, D. T. Lodowski, H. Steinhagen and J. Huber, *J. Med. Chem.*, 2010, **53**, 1867.
17. C. A. Boguth, P. Singh, C. C. Huang and J. J. Tesmer, *EMBO J.*, 2010.
18. A. B. Tobin, A. J. Butcher and K. C. Kong, *Trends Pharmacol. Sci.*, 2008, **29**, 413.
19. J. M. Busillo, S. Armando, R. Sengupta, O. Meucci, M. Bouvier and J. L. Benovic, *J. Biol. Chem.*, 2010, **285**, 7805.
20. D. A. Zidar, J. D. Violin, E. J. Whalen and R. J. Lefkowitz, *Proc. Natl. Acad. Sci. U. S. A.*, 2009, **106**, 9649.
21. K. L. Hildreth, J. H. Wu, L. S. Barak, S. T. Exum, L. K. Kim, K. Peppel and N. J. Freedman, *J. Biol. Chem.*, 2004, **279**, 41775.
22. Y. Chen, H. Long, Z. Wu, X. Jiang and L. Ma, *Mol. Biol. Cell*, 2008, **19**, 2973.
23. S. Patial, J. Luo, K. J. Porter, J. L. Benovic and N. Parameswaran, *Biochem. J.*, 2010, **425**, 169.
24. S. Peregrin, M. Jurado-Pueyo, P. M. Campos, V. Sanz-Moreno, A. Ruiz-Gomez, P. Crespo, F. Mayor, Jr. and C. Murga, *Curr. Biol.*, 2006, **16**, 2042.
25. J. Ho, E. Cocolakis, V. M. Dumas, B. I. Posner, S. A. Laporte and J. J. Lebrun, *EMBO J.*, 2005, **24**, 3247.
26. W. Chen, X. R. Ren, C. D. Nelson, L. S. Barak, J. K. Chen, P. A. Beachy, F. de Sauvage and R. J. Lefkowitz, *Science*, 2004, **306**, 2257.
27. C. Molnar, H. Holguin, F. Mayor, Jr., A. Ruiz-Gomez and J. F. de Celis, *Proc. Natl. Acad. Sci. U. S. A.*, 2007, **104**, 7963.
28. M. Philipp, G. B. Fralish, A. R. Meloni, W. Chen, A. W. MacInnes, L. S. Barak and M. G. Caron, *Mol. Biol. Cell*, 2008, **19**, 5478.
29. M. C. Jimenez-Sainz, C. Murga, A. Kavelaars, M. Jurado-Pueyo, B. F. Krakstad, C. J. Heijnen, F. Mayor, Jr. and A. M. Aragay, *Mol. Biol. Cell.*, 2006, **17**, 25.
30. P. Penela, C. Ribas, I. Aymerich, N. Eijkelkamp, O. Barreiro, C. J. Heijnen, A. Kavelaars, F. Sanchez-Madrid and F. Mayor, Jr., *EMBO J.*, 2008, **27**, 1206.
31. A. Elorza, S. Sarnago and F. Mayor, Jr., *Mol. Pharmacol.*, 2000, **57**, 778.
32. J. A. Pitcher, J. J. Tesmer, J. L. Freeman, W. D. Capel, W. C. Stone and R. J. Lefkowitz, *J. Biol. Chem.*, 1999, **274**, 34531.
33. S. Sarnago, A. Elorza and F. Mayor, Jr., *J. Biol. Chem.*, 1999, **274**, 34411.
34. S. Mariggio, C. Garcia-Hoz, S. Sarnago, A. De Blasi, F. Mayor, Jr. and C. Ribas, *Cell Signal.*, 2006, **18**, 2004.
35. E. J. Whalen, M. W. Foster, A. Matsumoto, K. Ozawa, J. D. Violin, L. G. Que, C. D. Nelson, M. Benhar, J. R. Keys, H. A. Rockman, W. J. Koch, Y. Daaka, R. J. Lefkowitz and J. S. Stamler, *Cell*, 2007, **129**, 511.
36. P. Penela, A. Ruiz-Gomez, J. G. Castano and F. Mayor, Jr., *J. Biol. Chem.*, 1998, **273**, 35238.

37. A. Elorza, P. Penela, S. Sarnago and F. Mayor, Jr., *J. Biol. Chem.*, 2003, **278**, 29164.
38. A. Salcedo, F. Mayor, Jr. and P. Penela, *EMBO J.*, 2006, **25**, 4752.
39. T. Metaye, H. Gibelin, R. Perdrisot and J. L. Kraimps, *Cell Signal.*, 2005, **17**, 917.
40. G. Rengo, A. Lymperopoulos, D. Leosco and W. J. Koch, *J. Mol. Cell. Cardiol.*, 2011, **50**, 785.
41. G. Iaccarino, E. Barbato, E. Cipolletta, V. De Amicis, K. B. Margulies, D. Leosco, B. Trimarco and W. J. Koch, *Eur. Heart J.*, 2005, **26**, 1752.
42. D. M. Harris, H. I. Cohn, S. Pesant and A. D. Eckhart, *Clin. Sci. (Lond.)*, 2008, **115**, 79.
43. J. A. Hata, M. L. Williams, J. N. Schroder, B. Lima, J. R. Keys, B. C. Blaxall, J. A. Petrofski, A. Jakoi, C. A. Milano and W. J. Koch, *J. Card. Fail.*, 2006, **12**, 360.
44. N. Oyama, K. Urasawa, S. Kaneta, H. Sakai, T. Saito, C. Takagi, I. Yoshida, A. Kitabatake and H. Tsutsui, *Circ. J.*, 2006, **70**, 362.
45. H. Brinks and W. J. Koch, *J. Cardiovasc. Transl. Res.*, 2010, **3**, 499.
46. V. B. Harding, L. R. Jones, R. J. Lefkowitz, W. J. Koch and H. A. Rockman, *Proc. Natl. Acad. Sci. U. S. A.*, 2001, **98**, 5809.
47. A. Lymperopoulos, G. Rengo and W. J. Koch, *Trends Mol. Med.*, 2007, **13**, 503.
48. S. J. Matkovich, A. Diwan, J. L. Klanke, D. J. Hammer, Y. Marreez, A. M. Odley, E. W. Brunskill, W. J. Koch, R. J. Schwartz and G. W. Dorn, 2nd, *Circ. Res.*, 2006, **99**, 996.
49. P. W. Raake, L. E. Vinge, E. Gao, M. Boucher, G. Rengo, X. Chen, B. R. DeGeorge, Jr., S. Matkovich, S. R. Houser, P. Most, A. D. Eckhart, G. W. Dorn, 2nd and W. J. Koch, *Circ. Res.*, 2008, **103**, 413.
50. G. Rengo, A. Lymperopoulos, C. Zincarelli, M. Donniacuo, S. Soltys, J. E. Rabinowitz and W. J. Koch, *Circulation*, 2009, **119**, 89.
51. T. Noma, A. Lemaire, S. V. Naga Prasad, L. Barki-Harrington, D. G. Tilley, J. Chen, P. Le Corvoisier, J. D. Violin, H. Wei, R. J. Lefkowitz and H. A. Rockman, *J. Clin. Invest.*, 2007, **117**, 2445.
52. J. Kim, S. Ahn, X. R. Ren, E. J. Whalen, E. Reiter, H. Wei and R. J. Lefkowitz, *Proc. Natl. Acad. Sci. U. S. A.*, 2005, **102**, 1442.
53. G. W. Dorn, 2nd, *Nat. Rev. Cardiol.*, 2009, **6**, 283.
54. A. D. Eckhart, T. Ozaki, H. Tevaearai, H. A. Rockman and W. J. Koch, *Mol. Pharmacol.*, 2002, **61**, 749.
55. H. I. Cohn, D. M. Harris, S. Pesant, M. Pfeiffer, R. H. Zhou, W. J. Koch, G. W. Dorn, 2nd and A. D. Eckhart, *Am. J. Physiol. Heart Circ. Physiol.*, 2008, **295**, H1695.
56. S. Liu, R. T. Premont, C. D. Kontos, S. Zhu and D. C. Rockey, *Nat. Med.*, 2005, **11**, 952.
57. P. Penela, C. Ribas, I. Aymerich and F. Mayor, Jr., *Cell Adh. Migr.*, 2009, **3**, 19.
58. A. Vroon, C. J. Heijnen and A. Kavelaars, *J. Leukoc. Biol.*, 2006, **80**, 1214.
59. A. Vroon, C. J. Heijnen, M. S. Lombardi, P. M. Cobelens, F. Mayor, Jr., M. G. Caron and A. Kavelaars, *J. Leukoc. Biol.*, 2004, **75**, 901.

60. J. C. Alves-Filho, F. Sonego, F. O. Souto, A. Freitas, W. A. Verri, Jr., M. Auxiliadora-Martins, A. Basile-Filho, A. N. McKenzie, D. Xu, F. Q. Cunha and F. Y. Liew, *Nat. Med.*, 2010, **16**, 708.
61. A. Vroon, A. Kavelaars, V. Limmroth, M. S. Lombardi, M. U. Goebel, A. M. Van Dam, M. G. Caron, M. Schedlowski and C. J. Heijnen, *J. Immunol.*, 2005, **174**, 4400.
62. T. K. Tarrant, R. R. Rampersad, D. Esserman, L. R. Rothlein, P. Liu, R. T. Premont, R. J. Lefkowitz, D. M. Lee and D. D. Patel, *Clin. Immunol.*, 2008, **129**, 115.
63. R. T. Dorsam and J. S. Gutkind, *Nat. Rev. Cancer*, 2007, **7**, 79.
64. T. Metaye, P. Levillain, J. L. Kraimps and R. Perdrisot, *J. Endocrinol.*, 2008, **198**, 101.
65. I. Prowatke, F. Devens, A. Benner, E. F. Grone, D. Mertens, H. J. Grone, P. Lichter and S. Joos, *Br. J. Cancer*, 2007, **96**, 82.
66. A. R. Meloni, G. B. Fralish, P. Kelly, A. Salahpour, J. K. Chen, R. J. Wechsler-Reya, R. J. Lefkowitz and M. G. Caron, *Mol. Cell Biol.*, 2006, **26**, 7550.
67. K. Peppel, A. Jacobson, X. Huang, J. P. Murray, M. Oppermann and N. J. Freedman, *Circulation*, 2000, **102**, 793.
68. K. F. Wan, B. S. Sambi, R. Tate and C. Watersv N. J. Pyne, *J. Biol. Chem.*, 2003, **278**, 18658.
69. M. Bliziotes, M. Gunness, X. Zhang, R. Nissenson and K. Wiren, *Bone*, 2000, **27**, 367.
70. X. Jiang and P. Yangv L. Ma, *Proc. Natl. Acad. Sci. U. S. A.*, 2009, **106**, 10183.
71. P. Penela, V. Rivas, A. Salcedo and F. Mayor, Jr., *Proc. Natl. Acad. Sci. U. S. A.*, 2010, **107**, 1118.
72. I. Usui, T. Imamura, H. Satoh, J. Huang, J. L. Babendure, C. J. Hupfeld and J. M. Olefsky, *EMBO J.*, 2004, **23**, 2821.
73. G. Shahid and T. Hussain, *J. Biol. Chem.*, 2007, **282**, 20612.
74. L. Garcia-Guerra, I. Nieto-Vazquez, R. Vila-Bedmar, M. Jurado-Pueyo, G. Zalba, J. Diez, C. Murga, S. Fernandez-Veledo, F. Mayor, Jr. and M. Lorenzo, *Diabetes*, 2010, **59**, 2407.
75. I. Usui, T. Imamura, J. L. Babendure, H. Satoh, J. C. Lu, C. J. Hupfeld and J. M. Olefsky, *Mol. Endocrinol.*, 2005, **19**, 2760.
76. E. Cipolletta, A. Campanile, G. Santulli, E. Sanzari, D. Leosco, P. Campiglia, B. Trimarco and G. Iaccarino, *Cardiovasc. Res.*, 2009, **84**, 407.
77. N. Eijkelkamp, C. J. Heijnen, H. L. Willemen, R. Deumens, E. A. Joosten, W. Kleibeuker, I. J. den Hartog, C. T. van Velthoven, C. Nijboer, M. A. Nassar, G. W. Dorn, 2nd, J. N. Wood and A. Kavelaars, *J. Neurosci.*, 2010, **30**, 2138.
78. H. L. Willemen, N. Eijkelkamp, H. Wang, R. Dantzer, G. W. Dorn, 2nd, K. W. Kelley, C. J. Heijnen and A. Kavelaars, *Pain*, 2010, **150**, 550.

CHAPTER 17

The Mechanics of Arrestin–Receptor Interaction: How GPCRs and Arrestins Talk to Each Other

VSEVOLOD V. GUREVICH* AND
EUGENIA V. GUREVICH

Vanderbilt University, Nashville, TN 37232, USA

17.1 Arrestins: A Small Family of Proteins with Many Functions

Vertebrate species express 800–3500 different G protein-coupled receptors (GPCRs) (SEVENS database, http://sevens.cbrc.jp/),[1] but only four arrestins.[2] Note that in this chapter we use the systematic names of the arrestin proteins:

- arrestin-1 (historic names S-antigen, 48 kDa protein, visual or rod arrestin);
- arrestin-2 (β-arrestin or β-arrestin1);
- arrestin-3 (β-arrestin2);
- arrestin-4 (cone or X-arrestin; for unclear reasons its gene is called '*arrestin 3*' in the HUGO database).

Two of the four arrestins are visual: arrestin-1 is present at millimolar concentrations in both rod[3,4] and cone[5] photoreceptors, whereas arrestin-4 is

RSC Drug Discovery Series No. 8
G Protein-Coupled Receptors: From Structure to Function
Edited by Jesús Giraldo and Jean-Philippe Pin

Published by the Royal Society of Chemistry, www.rsc.org

specifically expressed in cones, although it constitutes only 2% of their total arrestin complement.[5] Non-visual arrestin-2 (first described as β-arrestin)[6] and arrestin-3[7] (first described as β-arrestin2)[8] are expressed in virtually every cell in the body, with the highest levels in the brain.[9] However, even in neurons their concentrations are sub-micromolar, with arrestin-2 outnumbering arrestin-3 by ~10–20 : 1 in most cells.[9,10]

Protochordate, *Ciona intestinalis*, has only one versatile arrestin doing its job in larval photoreceptors and in all other cells.[11] Insects have two sensory and one non-sensory arrestin (*kurtz* in *Drosophila*),[12] whereas *Caenorhabditis elegans* has only one for all purposes.[13] While the knockout of both non-visual arrestins in mice[14] and of *kurtz* in *Drosophila*[12] is embryonic lethal, *C. elegans* lacking arrestin is not only viable,[13] but even lives longer.[15]

Today it is hard to believe that only 15 years ago the binding to active phosphorylated GPCRs to block further G protein activation was considered the sole function of arrestin proteins. The first non-receptor binding partner of arrestins, clathrin, was described in 1996.[16] Now arrestins are recognized as versatile regulators of cell signalling, with the key function of assembling multi-protein signalling complexes and localizing them to particular cellular compartments, one of which is receptor-rich membranes.[17,18] Even though arrestins were reported to bind dozens or even hundreds of other proteins,[19] the sheer number of distinct GPCR subtypes in the animal kingdom makes receptors the most numerous type of arrestin binding partners.

17.2 Receptor Elements Engaged by Arrestins

Although it appears very likely that the majority of mammalian GPCRs are regulated by G-protein receptor kinase (GRK) phosphorylation followed by arrestin binding, receptor elements involved were experimentally identified in very few subtypes (reviewed in ref. 17). These typically include parts of the receptor that also engage G proteins, such as the third cytoplasmic loop,[20–27] E/DRY motif, as well as elements in the cavity between transmembrane helices that opens upon receptor activation.[28–30] This ensures direct competition between arrestins and G proteins, making their binding to the receptor mutually exclusive.[31,32] Interestingly, this competition can be rather complex, particularly with receptors coupling to several G proteins: in case of parathyroid hormone receptor (PTHR) both non-visual arrestins inhibit $G_{q/11}$-mediated IP3 accumulation much more robustly than G_s-mediated cAMP synthesis.[33] A recent report that arrestin-2 specifically binds Gβ1γ2 dimer, enhancing Gβ1γ2-mediated signalling,[34] adds another unexpected twist to the issue of competition. Receptor elements engaged by arrestins are largely those implicated in binding Gα subunits, suggesting the possibility that arrestin complex with Gβγ could bind the receptor in place of Gαβγ heterotrimer. The finding that *N*-formyl peptide receptor fused with Gαi2, which is constitutively associated with Gβγ, interacted with arrestins normally in living cells and *in vitro*,[35] also suggests that the competition between arrestins and G proteins may not be as simple as experiments with rhodopsin and β2-adrenergic receptor suggest.[6,8,32,36]

The additional set of receptor elements specifically mediating arrestin binding carries a significant negative charge. In most cases, it includes two or more phosphates,[32,37–39] usually in a cluster, that could be localized in the receptor *C*-terminus or in any of the cytoplasmic loops. The phosphates are attached to the receptor by G protein-coupled receptor kinases upon receptor activation. In several cases the critical negative charges are supplied by acidic residues present in the *C*-terminus[40,41] or the third loop.[21] Most receptors carry many more potential phosphorylation sites than the two or three phosphates that arrestins require for tight binding.[42] This provides an interesting opportunity for different GRKs to target distinct sets of serines and threonines on the cytoplasmic receptor surface. Receptors phosphorylated at different levels[39] or in different places appear to form functionally distinct complexes with their cognate arrestins.[43–45] The localization of phosphorylated sites likely determines the orientation of the arrestin molecule in the complex.[17] Thus, GRKs targeting specific serine/threonine clusters in the receptor can predetermine the functional outcome of subsequent arrestin binding.

17.3 Receptor-binding Arrestin Elements and GPCR Specificity

Arrestins are elongated molecules with long and short axes of ~75 and 40 Å, respectively, consisting of two domains, each with a β-strand sandwich at the core.[46–51] All known receptor-binding elements are localized on the concave sides of the two domains[37,38,52–57] and fall into two categories that appear to play opposite roles in receptor preference. Positively charged residues interact with the common element of arrestin targets, *i.e.* receptor-attached phosphates.[32,39,58–62] These are largely responsible for the ability of just two non-visual arrestins in vertebrates and a single one in *Drosophila* to interact with hundreds of GPCR subtypes.[63] The mechanism of phosphate sensing employed by arrestins contributes to this promiscuity: target receptor needs to activate arrestin by disrupting the key Arg–Asp salt bridge in the polar core that keeps arrestin in the basal state.[47,62,64–66] This only requires spatially concentrated negative charge, making arrestins largely insensitive to the sequence context of phosphorylated serines and threonines (reviewed in Refs. 2, 17). In contrast, a set of subtype-specific arrestin residues in the β-strands V–VI and XV–XVI and adjacent loops interacts with non-phosphorylated receptor elements and determines the receptor preference of arrestin proteins.[56] Arrestin-1 (formerly known as visual or rod arrestin), which is the most selective towards active phosphorylated form of its cognate receptor, rhodopsin (P-Rh*), has an additional phosphate-binding residue absent in other subtypes, which apparently contributes to its strict requirement for receptor phosphorylation.[50] Arrestin-1 binds only with high affinity to rhodopsin carrying three or more phosphates,[39] whereas non-visual arrestins often require no more then two,[38] and even bind certain unphosphorylated receptors, although in some cases negative charges of acidic receptor residues substitute for the phosphates.[21,40,41]

Both non-visual arrestins are designed to be promiscuous and serve hundreds of different GPCRs. Arrestin-3 binds with high affinity more GPCR subtypes than arrestin-2,[67] and demonstrates lower selectivity for phosphorylated active form of the receptors it binds.[38,66] Interestingly, the recent crystal structure of arrestin-3[51] revealed a unique feature; in contrast to the other three vertebrate arrestins, part of the β-strand XVI in the 'receptor discriminator' region is distorted and does not form a contiguous β-sheet with the adjacent β-strand. Replacement of this arrestin-3 element with highly homologous part of arrestin-2 significantly increases the specificity of the chimera for the active phosphorylated form of β2-adrenergic receptor,[51] indicating that this structural peculiarity contributes to lower selectivity of arrestin-3. The specificity of arrestin-4 appears to be largely determined by its exclusive expression in cone photoreceptors rather than by its inherent selectivity; *in vitro* it binds other GPCRs comparably to non-visual arrestins.[50] Arrestin-1 is the only subtype with inherently high receptor specificity; its binding to P-Rh* is much higher than to any other receptor tested,[38,56] although *in vivo* it can effectively quench signalling by cone opsins[68,69] and can bind certain non-visual GPCRs when overexpressed in cells.[67]

The remarkable preference of arrestin-1 for P-Rh* over active phosphorylated form of M2 muscarinic cholinergic receptor, and the opposing preference of arrestin-2, was successfully used to identify the elements responsible for receptor specificity.[56] It turned out that, within a rather extensive receptor-binding surface of arrestin proteins, only two elements with seven and 16 non-conservative substitutions in the *N*- and *C*-domain, respectively, fully define receptor preference, as evidenced by its complete reversal upon swapping of the two between arrestin-1 and -2.[56] This finding, along with very limited variability among the key receptor discriminator residues in arrestins,[2,70] creates an opportunity for artificial targeting of non-visual arrestins to particular GPCRs by judicious manipulation of very few residues on the receptor-binding surface. Inherently more selective arrestin-2 is a more suitable base protein for this endeavour.

17.4 Stoichiometry of the Arrestin–Receptor Complex

The original model of the complex, based on extensive structure–function studies showing that each arrestin domain has receptor-binding elements with unique functional characteristics,[37,38] posits that arrestin binds a single receptor molecule.[63] Many GPCRs form dimers at least at some stages of their functional cycle.[71,72] The provocative geometry, with the long axis of arrestin being almost twice as big as the diameter of the cytoplasmic tip of smaller GPCRs such as rhodopsin[73,74] or β2-adrenoreceptor,[75] leads to an alternative model that envisions a single arrestin simultaneously interacting with two receptors in a dimer.[71]

Recently this issue progressed from pure speculation to experimental testing. In rod photoreceptors, massive arrestin-1 translocation in bright light to the rhodopsin-containing outer segment is driven by arrestin-1 binding to P-Rh*.[76] As wild-type mice express arrestin-1 at a ~0.8 : 1 ratio to rhodopsin,[3,4] this phenomenon can be exploited to determine the stoichiometry of the interaction *in vivo*. The analysis of genetically modified mice expressing arrestin-1 at a wide range of levels relative to rhodopsin (from 0.4 : 1 to 2.4 : 1) showed that rhodopsin can recruit roughly equimolar amount of arrestin-1 to the outer segment.[4] *In vitro* binding experiments with two purified proteins also demonstrated that arrestin-1 saturates P-Rh* at a ratio of ~1 : 1.[4] However, arrestin-1 robustly self-associates, forming dimers and tetramers at physiological concentrations,[77–79] which opens up the possibility that the binding of the arrestin-1 dimer to the rhodopsin dimer could account for the ~1 : 1 binding ratio. Direct demonstration that only monomeric arrestin-1 can bind P-Rh*,[79] supported by the finding that in the solution tetramer receptor-binding surfaces of every arrestin molecule are shielded by sister subunits,[80] excluded this possibility.

Collectively, these data clearly demonstrated that each rhodopsin molecule binds its own arrestin. Strictly speaking, this does not exclude a possible role for rhodopsin dimers in arrestin binding; rhodopsin was in native disc membranes in all these experiments and could form any oligomers containing equal number of arrestin-1 and rhodopsin molecules. The first report that monomeric P-Rh* in nanodiscs binds arrestin-1[81] did not quite settle this issue. While the affinity of arrestin-1 for P-Rh* in native disc membranes and for monomeric P-Rh* reported in this work was roughly the same, the absolute values of estimated K_D were surprisingly high (*i.e.* >1 μM).[81] This was in sharp contrast to the K_D of 20–50 nM previously measured using the same 'extra-meta II' assay at the same low temperature,[82,83] as well as to even lower ~1 nM estimate obtained at a more physiological temperature.[84] Another recent study[85] tested monomeric P-Rh* in nanodiscs employing two different assays, direct binding of radiolabelled arrestin-1[37,86] and fluorescence quenching of purified arrestin-1 labelled with Texas red and a quencher QXL-610 attached to the membrane scaffold protein of the nanodisc. These measurements at near-physiological temperatures yielded K_D of 3–4 nM, in good agreement with previous work. Moreover, the stoichiometry of the arrestin-1 binding to monomeric P-Rh* was also shown to be 1:1,[85] the same as in native disc membranes.[4] Thus, monomeric P-Rh* binds arrestin-1 with physiological nanomolar affinity and 1:1 stoichiometry.[85] This work unambiguously settled the issue for arrestin-1-P-Rh* complex in vertebrates.

Interestingly, the same 1:1 stoichiometry of arrestin binding to rhodopsin was recently reported in *Drosophila*.[87] Since GRK1 (also known as rhodopsin kinase) 'prepares' rhodopsin for arrestin-1 binding, it was important to test the efficiency of the phosphorylation of monomeric rhodopsin in nanodiscs by purified GRK1. It was recently shown that GRK1 phosphorylates monomeric rhodopsin in nanodiscs at least as effectively as rhodopsin in native disc membranes.[85] Thus, all three proteins that preferentially bind light-activated rhodopsin, G protein transducin,[88–90] GRK1[85] and arrestin-1,[81,85] require only monomeric receptor. This makes biological sense in view of the single photon sensitivity of rod photoresponse.[91]

However, considering the impressive structural variability in the GPCR superfamily, one should not fall into a trap of assuming that what is true for one receptor must be true for all.[92,93] Direct equally definitive experiments with other class A (rhodopsin-like) and B (secretin receptor-like) GPCRs, as well as with constitutively dimeric class C receptors and non-visual GRKs and arrestins must be performed.

To summarize, at the moment we have acceptable proof that one monomeric class A receptor, rhodopsin, forms a 1 : 1 complex with arrestin-1 in mammals and insects, and no direct data on other GPCRs and arrestins. There is also no experimental evidence supporting any alternative model of the complex.

17.5 Arrestin Effects on the Receptor

Arrestin was first discovered in the visual system as a protein that binds light-activated phosphorylated rhodopsin.[94] It took several years to establish that arrestin binding reduces G protein activation.[36] Gradually it became clear that arrestin and G protein interactions with rhodopsin are mutually exclusive,[31,32] and that these two proteins simply compete for the receptor, with arrestin gaining an advantage upon rhodopsin phosphorylation. Thus, the earliest described arrestin effect on the receptor was the suppression of G protein activation by rhodopsin. This was soon shown to apply to other GPCRs and subsequently to cloned non-visual arrestins.[6–8]

The second arrestin function, the stabilization of active receptor conformation, was also discovered in the visual system. The spectrally distinct active form of rhodopsin, metarhodopsin II, was shown to be stabilized by the binding of the protein partners that prefer active rhodopsin, cognate G protein transducin[95] and arrestin-1.[82] This phenomenon, in the form of an 'extra meta II assay', which is essentially an easy spectroscopic measurement of the amount of meta II in excess of what is observed with free rhodopsin, was used to quantify transducin and arrestin-1 binding to rhodopsin.[82,83,95]

Other GPCRs demonstrate the same phenomenon, which manifests itself as high agonist affinity of the receptor associated with G protein[96] or non-visual arrestin.[97] The proportion of high and low affinity states in the receptor population can be determined from the agonist competition curve.[96] Interestingly, no receptor was reported to be fully converted into high agonist affinity state by a G protein in native membranes[96] or in purified reconstituted form,[98] although the effects of different G proteins on this process were never compared.

Arrestins also do not increase the agonist affinity of the whole receptor population, but individual arrestins (wild-type and mutant) have different ability to push the receptor into high agonist affinity state. Arrestin-3 was shown to be more effective in this regard than arrestin-2, whereas some conformationally loose pre-activated arrestin mutants are even stronger.[97] Unexpectedly, it was found that under certain circumstances, arrestins can bind the receptor without affecting agonist affinity: arrestin binding to *N*-formyl peptide receptor phosphorylated at one cluster of serines and threonines increases agonist affinity, whereas receptor phosphorylation at an alternative cluster promotes arrestin binding that does not affect agonist affinity.[43] Thus, different arrestins can form functionally (and likely structurally) different complexes with the same receptor, and even the same arrestin can form different complexes with receptor phosphorylated at different sites[44,45] or levels.[39] These results provide a plausible mechanistic explanation for intriguing observations that arrestin binding to differentially phosphorylated receptors has distinct functional consequences in the cell; phosphorylation with GRK-2 and -3 promotes internalization with minimal extracellular signal-regulated kinase (ERK) activation, whereas the action of GRK-5 and -6 has the opposite effect.[99,100]

It is becoming increasingly obvious that GPCRs, like all proteins, can assume multiple conformations.[101] The existence of biased receptor agonists that preferentially promote receptor interactions with certain subtypes of G proteins, or with GRKs and arrestins,[102] suggests that active receptor conformations stabilized by these partners are unlikely to be identical, although in case of rhodopsin they are spectroscopically indistinguishable.[82,95] This issue needs to be investigated experimentally using biophysical approaches that document conformational differences between active and inactive rhodopsin[28] and β2-adrenergic receptor occupied by different full and partial agonists.[103,104]

17.6 Receptor Binding-induced Conformational Changes in Arrestin

The fact that rhodopsin binding is accompanied by a global conformational change in arrestin-1 was established in 1989, based on an unusually high Arrhenius activation energy.[82] It is easy to envision how this rearrangement, by virtue of creating a high threshold for binding, would ensure remarkable specificity of arrestin for active phosphorylated receptor.[37] This model was proposed when a new binding assay with femtomolar sensitivity revealed specific low affinity binding of arrestin-1 to active unphosphorylated (Rh*) and inactive phosphorylated (P-Rh) rhodopsin, suggesting that arrestin-1 has independent elements which bind receptor-attached phosphates and parts of the receptor that change conformation upon activation.[37] However, the binding to P-Rh* is many times greater than to Rh* or inactive P-Rh.

As a simple cooperative two-site interaction cannot adequately explain the magnitude of this binding differential, a more complex mechanism was proposed.[37] The model suggests that separate arrestin elements recognize phosphates and the active state of the receptor acts as sensors controlling arrestin conformation. Simultaneous engagement of both sensors, which can only be achieved by active phosphorylated receptor, allows arrestin transition into active receptor-binding state, which was envisioned as a major conformational rearrangement bringing additional elements in contact with the receptor to increase binding energy, and hence the affinity of the interaction (reviewed in ref. 63). Initial mutagenesis of arrestin-1 suggested that the last 30–40 residues serve as a 'brake'. Their deletion does not appreciably reduce the binding to P-Rh*, while greatly increasing the binding to forms that can engage only one sensor, unphosphorylated Rh* and inactive P-Rh.[37,86] Increased susceptibility to proteolysis of the *C*-terminus of the receptor-bound arrestin-1[105] identified the movement of this C-tail as part of binding-induced conformational change. The fact that short splice variant of arrestin-1 lacking the C-tail retains half of the activation energy[83] suggests that other rearrangements in the arrestin molecule must account for the remaining energy barrier.

Two studies of the effects of phosphorylated peptide derived from the *C*-terminus of the vasopressin type II receptor on limited proteolysis of the two non-visual arrestins provided an interesting glimpse into these rearrangements.[106,107] In addition to the release of the C-tail, significant changes in the exposure of the inter-domain hinge and several elements in both domains were detected.[106,107] More intriguingly, the 'active' conformations of phosphopeptide-liganded arrestin-2 and -3 were found to be different.[107] Highly negatively charged polysulfated glycosaminoglycan heparin, believed to mimic phosphorylated receptor *C*-terminus, made the arrestin-1 C-tail more accessible for trypsin similarly to P-Rh*.[105] Heparin was shown to induce an increase of the mobility of the C-tail, with the effect growing from proximal to more distant residues, consistent with the complete release of the C-tail, allowing it to 'flop around'.[108] While heparin-induced release of the C-tail was also detected in arrestin-3, in other parts of the molecule the effects of heparin and receptor phosphopeptide differed.[106]

Receptor-binding elements, identified by several labs using a variety of methods,[37,38,52–62,109] populate a large surface on the concave sides of both arrestin domains. Comparison of solved structures of all four vertebrate arrestins,[46–51] inactive[73,74] and active[110,111] rhodopsin, as well as inactive β2-,[75] β1-adrenergic[112] and A2A adenosine[113] receptors suggests that a movement of the two arrestin domains relative to each other is necessary to bring all receptor-binding arrestin residues in contact with the cytoplasmic tip of the receptor. Indeed, an extended inter-domain hinge allowing such a movement was found to be necessary for receptor binding by visual and non-visual arrestins.[114,115] However, the actual extent of binding-induced movement of the two domains and other arrestin elements remains to be elucidated.

Differential engagement of several arrestin residues by distinct functional forms of the receptor[54,57] indicates that there are likely to be multiple binding-induced rearrangements, such as the movement of flexible loops. In particular, 'finger loop' (residues 69–78, 65–74 and 66–75 in bovine arrestin-1, -2, and -3, respectively) appears to change its position.[54,116] It was shown to be engaged upon 'pre-docking' to inactive P-Rh, and several residues in this loop are further immobilized upon tight binding to P-Rh*.[54] Residue 139 on the neighbouring loop in the central 'crest' on the receptor-binding surface demonstrated even more intriguing behaviour. It is immobilized by inactive P-Rh, but its mobility returns to that of the free arrestin upon rhodopsin activation to P-Rh*.[54] These data suggest that part of this loop participates in low-affinity P-Rh binding but not in the high affinity interaction with P-Rh*. This would obviously require the movement of this loop upon arrestin binding to P-Rh, P-Rh*, or both. Several loops in arrestins appear to be very flexible, assuming variable conformations in different monomers of arrestin-1 crystal tetramer[47] and in different crystal forms of arrestin-2[48,49] and -3.[51] Physical engagement of numerous residues in flexible loops by rhodopsin[54,57,116] suggests that binding-induced structural changes are likely to involve these elements as well.

The active phosphorylated receptor disrupts all intra-molecular interactions that make basal arrestin conformation rigid (reviewed in ref. 17). This makes 'activated' arrestin quite flexible, suggesting that the ultimate receptor-bound conformation must be determined by the receptor it interacts with. Indeed, a recent study using intra-molecular bioluminescence resonance energy transfer (BRET) in arrestin-3 tagged with *Renilla* luciferase and the yellow fluorescent protein at the *N*- and the *C*-termini, respectively, showed that the same arrestin can assume distinct conformations when bound to angiotensin 1a and parathyroid hormone receptors activated by natural and arrestin-biased agonists, as well as to wild-type β2-adrenergic receptor and its mutant that does not couple to G_s.[117] These data suggest that arrestin-3 bound to the same receptor in different functional states assumes distinct active conformations that are likely to determine the functional outcome of the binding.[117] While there is little doubt that receptor-bound arrestin is quite different from the free form structurally and functionally, the exact conformations it assumes remain to be elucidated by X-ray crystallography of the complex and/or intra- and inter-molecular distance measurements using site-directed spin labelling.

17.7 What Does the Receptor–Arrestin Complex Do that the Components Don't?

As the name suggests, the first member of the family, arrestin-1, was discovered as a protein that 'arrests' receptor coupling to the G protein.[36] Thus, active receptor activates G proteins, whereas the receptor–arrestin complex does not. The comparison of known arrestin and GPCR structures suggests that arrestin binding must shield much of the cytoplasmic tip of the receptor, simply crowding out G proteins. This notion is supported by experimental evidence of direct competition between arrestin-1 and transducin for rhodopsin,[31,32] and by the fact that arrestin and G protein engage some of the same receptor elements (reviewed in ref. 17), which makes their binding mutually exclusive.

Recent findings show that the formation of the arrestin–receptor complex also regulates receptor trafficking and receptor's ultimate fate, and initiates multiple G protein-independent signalling events (reviewed in Refs. 18,118). Clathrin, the organizing component of the internalization machinery of the coated pit, was the first discovered non-receptor binding partner of arrestin proteins,[16] with clathrin adaptor AP2 being a close second.[119] The clathrin and AP2 binding sites are localized on the arrestin C-tail,[16,119,120] and C-tail detachment upon arrestin binding to the receptor makes these sites more accessible for clathrin and AP2.[106,107,118] Conversely, tight anchoring of the C-tail to the body of the arrestin molecule prevents the competition of free arrestin with the arrestin–receptor complex for these partners. Thus, arrestin binding adds a strong trafficking signal directing receptor to the coated pit. However, some GPCRs that desensitize *via* arrestin binding are internalized in an arrestin-independent manner,[121] indicating that intrinsic trafficking signals in the receptor sequence can override those added by bound arrestin.

Arrestin binding also affects GPCR trafficking indirectly, *via* ubiquitination.[122] Bound arrestins were shown to recruit several ubiquitin ligases to the complex, including Mdm2,[122] AIP4[123] and parkin.[124] The interaction with Nedd4 that ubiquitinates β2-adrenoreceptor remains controversial. One study suggested that arrestin recruits this ligase to the receptor,[125] whereas another presented evidence that arrestin domain-containing protein 3 (rather than arrestin) promotes β2-adrenoreceptor ubiquitination by recruiting Nedd4.[126] AIP4[123] and Nedd4[126] ubiquitinate receptors and Mdm2 ubiquitinates arrestin,[122] whereas parkin enhances Mdm2 binding to arrestins but suppresses their ubiquitination.[124] The analysis of Mdm2 and parkin interaction illustrates the complexity of the interplay of the processes involved.

The discovery that arrestin ubiquitination by Mdm2 is stimulated by receptor activation led to the idea that receptor-bound arrestin preferentially binds Mdm2.[122] However, direct analysis of the effect of arrestin conformation on Mdm2 interaction showed that the mutants 'frozen' in the basal state by hinge deletions, and therefore incapable of proper receptor binding,[114,115] interact with Mdm2 more readily than corresponding wild-type arrestins and conformationally loose pre-activated mutants with enhanced receptor affinity.[127–129] Moreover, parkin enhances Mdm2 binding by stabilizing basal arrestin conformation. It increases Mdm2 recruitment by wild-type arrestin and pre-activated forms to a higher level observed with hinge deletion mutants, but does not further enhance Mdm2 binding to the mutants frozen in the basal state.[124] Collectively, these data suggest that receptor binding makes arrestin a better substrate for Mdm2-medited ubiquitination, but reduces the interaction affinity. This model predicts that arrestin comes to the receptor pre-loaded with Mdm2 (Figure 17.1A). Upon receptor binding, Mdm2 has very limited time to ubiquitinate arrestin before it is released due to reduced affinity. This scenario readily explains recent finding that different subsets of lysines in the same arrestin are ubiquitinated upon its binding to different receptors.[130] Arrestin-bound Mdm2 would only have time to hit very few most accessible targets before dissociating from receptor-bound arrestin.

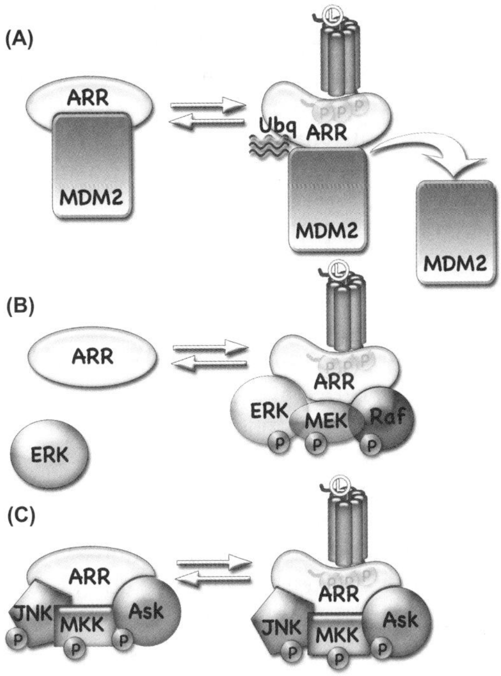

Figure 17.1 Conformational preferences of arrestin partners determine the sequence of events. (**A**) Ubiquitin ligase Mdm2 preferentially binds arrestins in the basal conformation. Thus, the most likely scenario is that free arrestins associate with Mdm2 and bring it to the receptor. Lower affinity of Mdm2 for the receptor-bound arrestin promotes its rapid dissociation, which limits the extent of arrestin ubiquitination in the complex. (**B**) ERK2 preferentially binds arrestin in complex with the receptor. Thus, it is likely recruited by the arrestin–receptor complex rather than by free arrestin, which explains why receptor activation followed by the arrestin binding is necessary for ERK activation on the arrestin scaffold. (**C**) JNK3 does not show strong preference for free or receptor-bound arrestins. Thus, JNK3 can be recruited by either. This explains how JNK3 phosphorylation is enhanced in response to the receptor activation and by increased levels of free arrestin-3.

When arrestin-dependent scaffolding of mitogen-activated protein (MAP) kinase cascades was first described, its dependence on receptor activation suggested that this is an exclusive function of the arrestin–receptor complex.[131,132] Indeed, the conformational preference of ERK2 is the opposite of that of Mdm2; its binding to free arrestins is relatively weak,[129] while arrestin recruitment to the receptor significantly enhances it. Thus, in this case receptor binding of arrestin likely precedes ERK2 recruitment (Figure 17.1B). Subsequent studies showed that free arrestin-3 also scaffolds the ASK1–MKK4–JNK3 cascade[129,133] and that even receptor binding-deficient hinge mutant of arrestin-3 frozen in the basal conformation promotes JNK3 activation.[129]

Thus, JNK3 appears to bind arrestin-3 in the basal and 'active' receptor-bound conformation comparably. However, arrestin-3 recruited to the active phosphoreceptor promotes JNK3 activation much more efficiently; the effects of free arrestin are only detectable upon ASK1 overexpression.[133] This suggests that free and receptor-associated arrestin can both recruit JNK3, although the probability of its activation is regulated by GPCRs (Figure 17.1C). In contrast, in case of arrestin-mediated recruitment of PDE4 to the receptor, which facilitates the degradation of cAMP near the plasma membrane,[134] the effect appears to be mediated by a simple re-localization of PDE4, which increases its local concentration in the vicinity of the receptor, as free non-visual arrestins bind PDE4 very well.[135]

These examples illustrate two new elements generated by the formation of the arrestin-receptor complex: (1) arrestin conformation significantly changes upon receptor binding; and (2) arrestins and any proteins bound to them become highly concentrated near receptor-rich membranes. The contribution of these two distinct factors in different arrestin-mediated signalling events triggered by GPCR activation remains to be elucidated. The most logical way of sorting this out is the use of pre-activated arrestin mutants with greatly enhanced conformational flexibility[54,136] (even though they do not exactly mimic receptor-bound state of arrestin) and arrestins tethered to the membrane *via* added sites for lipid modifications or fused transmembrane helices, as was successfully done with the G protein α-subunit.[137]

It should be noted that arrestin-dependent recruitment of signalling proteins to a particular cellular compartment is not limited to the receptor-rich membranes. Arrestins bind microtubules[76,115,138] and polymerized γ-tubulin in the centrosome,[139] largely *via* the same surface mediating receptor binding,[115,138] and therefore can recruit the same interaction partners to the cytoskeleton. Arrestin-dependent mobilization of Mdm2 to the microtubules was shown to dramatically increase the ubiquitination of microtubule-associated proteins,[115] but the consequences of cytoskeletal mobilization of the majority of arrestin-binding proteins remains unexplored.

17.8 What Do We Need to Know About Arrestins?

At first glance it may appear that we know more about arrestins and their interactions with various binding partners than about almost any other protein family. However, a lot of questions remain unanswered.

The most obvious gap in our knowledge is the structure of the receptor-bound arrestin. Several lines of evidence suggest that there must be more than one such structure, as arrestin binding to differentially phosphorylated receptors leads to distinct functional outcomes.[39,43–45,99] Like any protein, arrestin cannot be rock-solid—the molecule likely 'breathes', exploring a large conformational space. Crystal structures of arrestin-1,[46,47] -2,[48,49] -3[51] and -4[50] look remarkably similar. Judging by the conformation of arrestin-1 in solution deduced from site-directed spin labelling/electron paramagnetic resonance

(EPR) studies,[54,57] what we see in crystals closely resembles the basal arrestin conformation in solution. Apparently, part of the arrestin conformational space only becomes accessible when the energy barriers are lowered by unfastening of the clasps that hold arrestin in the basal state (*e.g.* the polar core and three-element interaction) by its interaction with active phosphorylated receptors (reviewed in ref. 63). It seems highly improbable that the receptor-bound state can be represented by a single conformation, especially considering that the disruption of stabilizing interactions by receptor-attached phosphates makes the molecule even more flexible[136] (reviewed in ref. 17). It is worth noting that the conformation of microtubule-bound arrestin is different from both basal and receptor-bound,[115,138] and it also needs to be elucidated.

We can be reasonably sure that arrestins bind GPCRs directly, as these interactions were demonstrated in several cases with purified proteins—rhodopsin and arrestin-1,[4,32,54,57,79,105] β2-adrenergic and M2 muscarinic receptors and arrestin-2 and -3.[38,97] However, direct arrestin interactions with very few other partners were confirmed with pure proteins. A disappointingly short list of these includes clathrin,[16] *N*-ethylmaleimide-sensitive fusion protein,[140] microtubules,[115,138] PDE4D,[135] Ca^{2+}-liganded calmodulin[141] and ubiquitin ligase parkin.[124] All other interactions were demonstrated by co-immunoprecipitation,[131,132] sometimes followed by proteomics analysis[19] and cell-based trafficking assays.[127,129] These assays show that the two proteins are in the same macro-molecular complex, but cannot exclude the interaction *via* an intermediary or prove direct binding.

The concentration of any arrestin has been determined with reasonable precision in surprisingly few cell types. Arrestin-1 in rod photoreceptors is expressed at a ~0.8 : 1 ratio to rhodopsin,[3,4] which allowed calculation of its intracellular concentration.[142] This number along with its self-association constants[79,80] yields a reasonable estimate of the concentration of the monomer, which is the only rhodopsin-binding species.[79] The concentrations of arrestin-1 and constitutively monomeric arrestin-4[143] in cones were also established.[5] The expression levels of arrestin-2 and -3 in the striatum have been determined,[9,10] but even there we do not know the intracellular concentrations of these proteins in different cell types (*e.g.* in neurons and glia). This issue came to the fore with the discovery that non-visual arrestins also oligomerize.[144,145] Although many interesting ideas regarding the functional differences of monomers, homo- and hetero-dimers of arrestin-2 and -3 were advanced, given their recently measured self-association constants[143] it is not entirely clear to what extent their self-association is possible at physiologically relevant endogenous expression levels, as opposed to artificial overexpression in cultured cells.

Quantitative aspects of the binding of arrestin to the majority of its partners are another large virtually unexplored area. Arrestin-1 affinity for P-Rh* has been measured by several groups, with K_D estimates ranging from ~1 to 50 nM.[81–85] Out of hundreds GPCRs that are regulated by arrestin-2 and -3, the affinity has been measured only for β2-adrenergic,[38,97] M2 muscarinic[38,97,109] and *N*-formyl peptide receptors.[43] The resulting K_Ds range from 0.1 to 60 nM, depending on the method used more than on the receptor–arrestin pair. Precise quantitative information about arrestin binding to non-receptor partners is also largely absent. The affinity for clathrin was only estimated for free arrestin-2 and -3,[16] whereas the biologically relevant affinity of the arrestin–receptor complex remains unknown. The only other numbers available are arrestin-2 affinity for microtubules ($K_D \sim 30\ \mu M$)[115] and for calmodulin ($K_D \sim 7\ \mu M$).[141]

Few would argue against the notion that biology needs to move from qualitative descriptions of different phenomena to quantitative understanding of the holistic behaviour of cells and organisms. Mathematical modelling, which allows us to translate available biochemical information into testable predictions of the behaviour of complex systems, requires a lot of precise numbers. As far as signalling pathways are concerned, with the exception of the light response of rod photoreceptors,[146] quantitative information is mostly lacking. The paucity of necessary data on the concentrations of key players, the affinity of their interactions with each other, as well as the activities of enzymes and channels in different functional states, severely hampers the attempts to model other types of signalling. Thus, we need to learn a lot more about arrestins and their interactions with GPCRs and other signalling proteins to understand different aspects of their function in the cell.

Acknowledgements

This work was supported in part by National Institutes of Health (NIH) grants EY011500, GM077561, GM081756 (VVG) and NS065868 (EVG), and the Vanderbilt University Medical School.

References

1. K. Kristiansen, *Pharmacol. Ther.*, 2004, **103**, 21–80.
2. E. V. Gurevich and V. V. Gurevich, *Genome Biol.*, 2006, **7**, 236.
3. K. J. Strissel, M. Sokolov, L. H. Trieu and V. Y. Arshavsky, *J. Neurosci.*, 2006, **26**, 1146–1153.
4. S. M. Hanson, E. V. Gurevich, S. A. Vishnivetskiy, M. R. Ahmed, X. Song and V. V. Gurevich, *Proc. Natl. Acad. Sci. U. S. A.*, 2007, **104**, 3125–3128.
5. S. S. Nikonov, B. M. Brown, J. A. Davis, F. I. Zuniga, A. Bragin, E. N. Pugh, Jr. and C. M. Craft, *Neuron*, 2008, **59**, 462–474.
6. M. J. Lohse, J. L. Benovic, J. Codina, M. G. Caron and R. J. Lefkowitz, *Science*, 1990, **248**, 1547–1550.
7. R. Sterne-Marr, V. V. Gurevich, P. Goldsmith, R. C. Bodine, C. Sanders, L. A. Donoso and J. L. Benovic, *J. Biol. Chem.*, 1993, **268**, 15640–15648.
8. H. Attramadal, J. L. Arriza, C. Aoki, T. M. Dawson, J. Codina, M. M. Kwatra, S. H. Snyder, M. G. Caron and R. J. Lefkowitz, *J. Biol. Chem.*, 1992, **267**, 17882–17890.

9. E. V. Gurevich, J. L. Benovic and V. V. Gurevich, *J. Neurochem.*, 2004, **91**, 1404–1416.
10. E. V. Gurevich, J. L. Benovic and V. V. Gurevich, *Neuroscience*, 2002, **109**, 421–436.
11. M. Nakagawa, H. Orii, N. Yoshida, E. Jojima, T. Horie, R. Yoshida, T. Haga and M. Tsuda, *Eur. J. Biochem.*, 2002, **269**, 5112–5118.
12. G. Roman, J. He and R. L. Davis, *Genetics*, 2000, **155**, 1281–1295.
13. A. Palmitessa, H. A. Hess, I. A. Bany, Y. M. Kim, M. R. Koelle and J. L. Benovic, *J. Biol. Chem.*, 2005, **280**, 24649–24662.
14. T. A. Kohout, F. S. Lin, S. J. Perry, D. A. Conner and R. J. Lefkowitz, *Proc. Natl. Acad. Sci. U. S. A.*, 2001, **98**, 1601–1606.
15. A. Palmitessa and J. L. Benovic, *J. Biol. Chem.*, 2010, **285**, 15187–15200.
16. O. B. Goodman, Jr., J. G. Krupnick, F. Santini, V. V. Gurevich, R. B. Penn, A. W. Gagnon, J. H. Keen and J. L. Benovic, *Nature*, 1996, **383**, 447–450.
17. V. V. Gurevich and E. V. Gurevich, *Pharmacol. Ther.*, 2006, **110**, 465–502.
18. S. M. DeWire, S. Ahn, R. J. Lefkowitz and S. K. Shenoy, *Annu. Rev. Physiol.*, 2007, **69**, 483–510.
19. K. Xiao, D. B. McClatchy, A. K. Shukla, Y. Zhao, M. Chen, S. K. Shenoy, J. R. Yates and R. J. Lefkowitz, *Proc. Natl. Acad. Sci. U. S. A.*, 2007, **104**, 12011–12016.
20. D. Raman, S. Osawa, V. V. Gurevich and E. R. Weiss, *J. Neurochem.*, 2003, **84**, 1040–1050.
21. S. Mukherjee, V. V. Gurevich, A. Preninger, H. E. Hamm, M.-F. Bader, A. T. Fazleabas, L. Birnbaumer and M. Hunzicker-Dunn, *J. Biol. Chem.*, 2002, **277**, 17916–17927.
22. R. S. Bhaskaran, L. Min, H. Krishnamurthy and M. Ascoli, *Biochemistry*, 2003, **42**, 13950–13959.
23. D. Raman, S. Osawa and E. R. Weiss, *Biochemistry*, 1999, **38**, 5117–5123.
24. E. I. Gelber, W. K. Kroeze, D. L. Willins, J. A. Gray, C. A. Sinar, E. G. Hyde, V. V. Gurevich, J. L. Benovic and B. L. Roth, *J. Neurochem.*, 1999, **72**, 2206–2214.
25. J. L. DeGraff, V. V. Gurevich and J. L. Benovic, *J. Biol. Chem.*, 2002, **277**, 43247–43252.
26. F. Schmidlin, D. Roosterman and N. W. Bunnett, *Am. J. Physiol. Cell Physiol.*, 2003, **285**, 945–958.
27. S. Mukherjee, K. Palczewski, V. V. Gurevich and M. Hunzicker-Dunn, *J. Biol. Chem.*, 1999, **274**, 12984–12989.
28. W. L. Hubbell, C. Altenbach, C. M. Hubbell and H. G. Khorana, *Adv. Protein Chem.*, 2003, **63**, 243–290.
29. C. Altenbach, J. Klein-Seetharaman, K. Cai, H. G. Khorana and W. L. Hubbell, *Biochemistry*, 2001, **40**, 15493–15500.
30. Z. T. Farahbakhsh, K. D. Ridge, H. G. Khorana and W. L. Hubbell, *Biochemistry*, 1995, **34**, 8812–8819.
31. J. G. Krupnick, V. V. Gurevich and J. L. Benovic, *J. Biol. Chem.*, 1997, **272**, 18125–18131.

32. U. Wilden, *Biochemistry*, 1995, **34**, 1446–1454.
33. M. Castro, F. Dicker, J. P. Vilardaga, C. Krasel, M. Bernhardt and M. J. Lohse, *Endocrinology*, 2002, **143**, 3854–3865.
34. M. Yang, R. L. He, J. L. Benovic and R. D. Ye, *Biochem. J.*, 2009, **417**, 287–296.
35. M. Shi, T. A. Bennett, D. F. Cimino, D. C. Maestas, T. D. Foutz, V. V. Gurevich, L. A. Sklar and E. R. Prossnitz, *Biochemistry*, 2003, **42**, 7283–7293.
36. U. Wilden, S. W. Hall and H. Kühn, *Proc. Natl. Acad. Sci. U. S. A.*, 1986, **83**, 1174–1178.
37. V. V. Gurevich and J. L. Benovic, *J. Biol. Chem.*, 1993, **268**, 11628–11638.
38. V. V. Gurevich, S. B. Dion, J. J. Onorato, J. Ptasienski, C. M. Kim, R. Sterne-Marr, M. M. Hosey and J. L. Benovic, *J. Biol. Chem.*, 1995, **270**, 720–731.
39. S. A. Vishnivetskiy, D. Raman, J. Wei, M. J. Kennedy, J. B. Hurley and V. V. Gurevich, *J. Biol. Chem.*, 2007, **282**, 32075–32083.
40. V. R. Jala, W. H. Shao and B. Haribabu, *J. Biol. Chem.*, 2005, **280**, 4880–4887.
41. E. Galliera, V. R. Jala, J. O. Trent, R. Bonecchi, P. Signorelli, R. J. Lefkowitz, A. Mantovani, M. Locati and B. Haribabu, *J. Biol. Chem.*, 2004, **279**, 25590–25597.
42. A. B. Tobin, A. J. Butcher and K. C. Kong, *Trends Pharmacol. Sci.*, 2008, **29**, 413–420.
43. T. A. Key, T. D. Foutz, V. V. Gurevich, L. A. Sklar and E. R. Prossnitz, *J. Biol. Chem.*, 2003, **278**, 4041–4047.
44. X. R. Ren, E. Reiter, S. Ahn, J. Kim, W. Chen and R. J. Lefkowitz, *Proc. Natl. Acad. Sci. U. S. A.*, 2005, **102**, 1448–1453.
45. J. Kim, S. Ahn, X. R. Ren, E. J. Whalen, E. Reiter, H. Wei and R. J. Lefkowitz, *Proc. Natl. Acad. Sci. U. S. A.*, 2005, **102**, 1442–1447.
46. J. Granzin, U. Wilden, H. W. Choe, J. Labahn, B. Krafft and G. Buldt, *Nature*, 1998, **391**, 918–921.
47. J. A. Hirsch, C. Schubert, V. V. Gurevich and P. B. Sigler, *Cell*, 1999, **97**, 257–269.
48. M. Han, V. V. Gurevich, S. A. Vishnivetskiy, P. B. Sigler and C. Schubert, *Structure*, 2001, **9**, 869–880.
49. S. K. Milano, H. C. Pace, Y. M. Kim, C. Brenner and J. L. Benovic, *Biochemistry*, 2002, **41**, 3321–3328.
50. R. B. Sutton, S. A. Vishnivetskiy, J. Robert, S. M. Hanson, D. Raman, B. E. Knox, M. Kono, J. Navarro and V. V. Gurevich, *J. Mol. Biol.*, 2005, **354**, 1069–1080.
51. X. Zhan, L. E. Gimenez, V. V. Gurevich and B. W. Spiller, *J. Mol. Biol.*, 2011, **406**, 467–478.
52. H. Ohguro, K. Palczewski, K. A. Walsh and R. S. Johnson, *Protein Sci.*, 1994, **3**, 2428–2434.
53. A. Pulvermuller, K. Schroder, T. Fischer and K. P. Hofmann, *J. Biol. Chem.*, 2000, **275**, 37679–37685.

54. S. M. Hanson, D. J. Francis, S. A. Vishnivetskiy, E. A. Kolobova, W. L. Hubbell, C. S. Klug and V. V. Gurevich, *Proc. Natl. Acad. Sci. U. S. A.*, 2006, **103**, 4900–4905.
55. S. M. Hanson and V. V. Gurevich, *J. Biol. Chem.*, 2006, **281**, 3458–3462.
56. S. A. Vishnivetskiy, M. M. Hosey, J. L. Benovic and V. V. Gurevich, *J. Biol. Chem.*, 2004, **279**, 1262–1268.
57. S. A. Vishnivetskiy, D. J. Francis, N. Van Eps, M. Kim, S. M. Hanson, C. S. Klug, W. L. Hubbell and V. V. Gurevich, *J. Mol. Biol.*, 2010, **395**, 42–54.
58. V. V. Gurevich and J. L. Benovic, *J. Biol. Chem.*, 1995, **270**, 6010–6016.
59. V. V. Gurevich, C.-Y. Chen, C. M. Kim and J. L. Benovic, *J. Biol. Chem.*, 1994, **269**, 8721–8727.
60. V. V. Gurevich and J. L. Benovic, *Mol. Pharmacol.*, 1997, **51**, 161–169.
61. S. A. Vishnivetskiy, C. Schubert, G. C. Climaco, Y. V. Gurevich, M.-G. Velez and V. V. Gurevich, *J. Biol. Chem.*, 2000, **275**, 41049–41057.
62. S. A. Vishnivetskiy, C. L. Paz, C. Schubert, J. A. Hirsch, P. B. Sigler and V. V. Gurevich, *J. Biol. Chem.*, 1999, **274**, 11451–11454.
63. V. V. Gurevich and E. V. Gurevich, *Trends Pharmacol. Sci.*, 2004, **25**, 59–112.
64. V. V. Gurevich, *J. Biol. Chem.*, 1998, **273**, 15501–15506.
65. A. Kovoor, J. Celver, R. I. Abdryashitov, C. Chavkin and V. V. Gurevich, *J. Biol. Chem.*, 1999, **274**, 6831–6834.
66. J. Celver, S. A. Vishnivetskiy, C. Chavkin and V. V. Gurevich, *J. Biol. Chem.*, 2002, **277**, 9043–9048.
67. R. H. Oakley, S. A. Laporte, J. A. Holt, M. G. Caron and L. S. Barak, *J. Biol. Chem.*, 2000, **275**, 17201–17210.
68. S. Chan, W. W. Rubin, A. Mendez, X. Liu, X. Song, S. M. Hanson, C. M. Craft, V. V. Gurevich, M. E. Burns and J. Chen, *Invest. Ophthalmol. Vis. Sci.*, 2007, **48**, 1968–1075.
69. G. Shi, K. W. Yau, J. Chen and V. J. Kefalov, *J. Neurosci.*, 2007, **27**, 10084–10093.
70. V. V. Gurevich and E. V. Gurevich, *Expert Rev. Mol. Med.*, 2010, **12**, e13.
71. D. Fotiadis, B. Jastrzebska, A. Philippsen, D. J. Muller, K. Palczewski and A. Engel, *Curr. Opin. Struct. Biol.*, 2006, **16**, 252–259.
72. G. Milligan, *Curr. Opin. Pharmacol.*, 2010, **10**, 23–29.
73. K. Palczewski, T. Kumasaka, T. Hori, C. A. Behnke, H. Motoshima, B. A. Fox, I. LeTrong, D. C. Teller, T. Okada, R. E. Stenkamp, M. Yamamoto and M. Miyano, *Science*, 2000, **289**, 739–745.
74. J. Li, P. C. Edwards, M. Burghammer, C. Villa and G. F. Schertler, *J. Mol. Biol.*, 2004, **343**, 1409–1438.
75. S. G. Rasmussen, H. J. Choi, D. M. Rosenbaum, T. S. Kobilka, F. S. Thian, P. C. Edwards, M. Burghammer, V. R. Ratnala, R. Sanishvili, R. F. Fischetti, G. F. Schertler, W. I. Weis and B. K. Kobilka, *Nature*, 2007, **450**, 383–387.
76. K. S. Nair, S. M. Hanson, A. Mendez, E. V. Gurevich, M. J. Kennedy, V. I. Shestopalov, S. A. Vishnivetskiy, J. Chen, J. B. Hurley, V. V. Gurevich and V. Z. Slepak, *Neuron*, 2005, **46**, 555–567.

77. C. Schubert, J. A. Hirsch, V. V. Gurevich, D. M. Engelman, P. B. Sigler and K. G. Fleming, *J. Biol. Chem.*, 1999, **274**, 21186–21190.
78. Y. Imamoto, C. Tamura, H. Kamikubo and M. Kataoka, *Biophys. J.*, 2003, **85**, 1186–1195.
79. S. M. Hanson, N. Van Eps, D. J. Francis, C. Altenbach, S. A. Vishnivetskiy, V. Y. Arshavsky, C. S. Klug, W. L. Hubbell and V. V. Gurevich, *EMBO J.*, 2007, **26**, 1726–1736.
80. S. M. Hanson, E. S. Dawson, D. J. Francis, N. Van Eps, C. S. Klug, W. L. Hubbell, J. Meiler and V. V. Gurevich, *Structure*, 2008, **16**, 924–934.
81. H. Tsukamoto, A. Sinha, M. Dewitt and D. L. Farrens, *J. Mol. Biol.*, 2010, **399**, 501–511.
82. A. Schleicher, H. Kuhn and K. P. Hofmann, *Biochemistry*, 1989, **28**, 1770–1775.
83. A. Pulvermuller, D. Maretzki, M. Rudnicka-Nawrot, W. C. Smith, K. Palczewski and K. P. Hofmann, *Biochemistry*, 1997, **36**, 9253–9260.
84. L. Zhang, C. D. Sports, S. Osawa and E. R. Weiss, *J. Biol. Chem.*, 1997, **272**, 14762–14768.
85. T. H. Bayburt, S. A. Vishnivetskiy, M. McLean, T. Morizumi, C.-C. Huang, J. J. Tesmer, O. P. Ernst, S. G. Sligar and V. V. Gurevich, *J. Biol. Chem.*, 2011, **285**, 1420–1428.
86. V. V. Gurevich and J. L. Benovic, *J. Biol. Chem.*, 1992, **267**, 21919–21923.
87. A. K. Satoh, H. Xia, L. Yan, C. H. Liu, R. C. Hardie and D. F. Ready, *Neuron*, 2010, **67**, 997–1008.
88. S. Banerjee, T. Huber and T. P. Sakmar, *J. Mol. Biol.*, 2008, **377**, 1067–1081.
89. T. H. Bayburt, A. J. Leitz, G. Xie, D. D. Oprian and S. G. Sligar, *J. Biol. Chem.*, 2007, **282**, 14875–14881.
90. M. R. Whorton, B. Jastrzebska, P. S. H. Park, D. Fotiadis, A. Engel, K. Palczewski and R. K. Sunahara, *J. Biol. Chem.*, 2008, **283**, 4387–4394.
91. D. A. Baylor, T. D. Lamb and K. W. Yau, *J. Physiol.*, 1979, **288**, 613–634.
92. V. V. Gurevich and E. V. Gurevich, *Trends. Neurosci.*, 2008, **31**, 74–81.
93. V. V. Gurevich and E. V. Gurevich, *Mol. Pharmacol.*, 2008, **74**, 312–316.
94. H. Kuhn, S. W. Hall and U. Wilden, *FEBS Lett.*, 1984, **176**, 473–478.
95. K. P. Hofmann, *Biochim. Biophys. Acta*, 1985, **810**, 278–281.
96. J. M. Stadel, A. DeLean and R. J. Lefkowitz, *J. Biol. Chem.*, 1980, **255**, 1436–1441.
97. V. V. Gurevich, R. Pals-Rylaarsdam, J. L. Benovic, M. M. Hosey and J. J. Onorato, *J. Biol. Chem.*, 1997, **272**, 28849–28852.
98. M. R. Whorton, M. P. Bokoch, S. G. Rasmussen, B. Huang, R. N. Zare, B. Kobilka and R. K. Sunahara, *Proc. Natl. Acad. Sci. U. S. A.*, 2007, **104**, 7682–7687.
99. J. Chen, G. Shi, F. A. Concepcion, G. Xie, D. Oprian and J. Chen, *J. Neurosci.*, 2006, **26**, 11929–11937.

100. L. Pan, E. V. Gurevich and V. V. Gurevich, *J. Biol. Chem.*, 2003, **278**, 11623–11632.
101. B. K. Kobilka and X. Deupi, *Trends Pharmacol. Sci.*, 2007, **28**, 397–406.
102. J. D. Violin and R. J. Lefkowitz, *Trends Pharmacol. Sci.*, 2007, **28**, 416–422.
103. X. J. Yao, G. Vélez Ruiz, M. R. Whorton, S. G. Rasmussen, B. T. DeVree, X. Deupi, R. K. Sunahara and B. Kobilka, *Proc. Natl. Acad. Sci. U. S. A.*, 2009, **106**, 9501–9506.
104. M. P. Bokoch, Y. Zou, S. G. Rasmussen, C. W. Liu, R. Nygaard, D. M. Rosenbaum, J. J. Fung, H. J. Choi, F. S. Thian, T. S. Kobilka, J. D. Puglisi, W. I. Weis, L. Pardo, R. S. Prosser, L. Mueller and B. K. Kobilka, *Nature*, 2010, **463**, 108–112.
105. K. Palczewski, A. Pulvermuller, J. Buczylko and K. P. Hofmann, *J. Biol. Chem.*, 1991, **266**, 18649–18654.
106. K. Xiao, S. K. Shenoy, K. Nobles and R. J. Lefkowitz, *J. Biol. Chem.*, 2004, **279**, 55744–55753.
107. K. N. Nobles, Z. Guan, K. Xiao, T. G. Oas and R. J. Lefkowitz, *J. Biol. Chem.*, 2007, **282**, 21370–21381.
108. T. Zhuang, S. A. Vishnivetskiy, V. V. Gurevich and C. R. Sanders, *Biochemistry*, 2010, **49**, 10473–10485.
109. V. V. Gurevich, R. M. Richardson, C. M. Kim, M. M. Hosey and J. L. Benovic, *J. Biol. Chem.*, 1993, **268**, 16879–16882.
110. P. Scheerer, J. H. Park, P. W. Hildebrand, Y. J. Kim, N. Krauss, H. W. Choe, K. P. Hofmann and O. P. Ernst, *Nature*, 2008, **455**, 497–502.
111. J. H. Park, P. Scheerer, K. P. Hofmann, H. W. Choe and O. P. Ernst, *Nature*, 2008, **454**, 183–187.
112. T. Warne, M. J. Serrano-Vega, J. G. Baker, R. Moukhametzianov, P. C. Edwards, R. Henderson, A. G. Leslie, C. G. Tate and G. F. Schertler, *Nature*, 2008, **454**, 486–491.
113. V. P. Jaakola, M. T. Griffith, M. A. Hanson, V. Cherezov, E. Y. Chien, J. R. Lane, A. P. Ijzerman and R. C. Stevens, *Science*, 2008, **322**, 1211–1217.
114. S. A. Vishnivetskiy, J. A. Hirsch, M.-G. Velez, Y. V. Gurevich and V. V. Gurevich, *J. Biol. Chem.*, 2002, **277**, 43961–43968.
115. S. M. Hanson, W. M. Cleghorn, D. J. Francis, S. A. Vishnivetskiy, D. Raman, S. Song, K. S. Nair, V. Z. Slepak, C. S. Klug and V. V. Gurevich, *J. Mol. Biol.*, 2007, **368**, 375–387.
116. M. E. Sommer, D. L. Farrens, J. H. McDowell, L. A. Weber and W. C. Smith, *J. Biol. Chem.*, 2007, **282**, 25560–25568.
117. A. K. Shukla, J. D. Violin, E. J. Whalen, D. Gesty-Palmer, S. K. Shenoy and R. J. Lefkowitz, *Proc. Natl. Acad. Sci. U. S. A.*, 2008, **105**, 9988–9993.
118. V. V. Gurevich and E. V. Gurevich, *Structure*, 2003, **11**, 1037–1042.
119. S. A. Laporte, R. H. Oakley, J. Zhang, J. A. Holt, s. S. G. Ferguson, M. G. Caron and L. S. Barak, *Proc. Natl. Acad. Sci. U. S. A.*, 1999, **96**, 3712–3717.

120. Y. M. Kim and J. L. Benovic, *J. Biol. Chem.*, 2002, **277**, 30760–30768.
121. R. Pals-Rylaarsdam, V. V. Gurevich, K. B. Lee, J. Ptasienski, J. L. Benovic and M. M. Hosey, *J. Biol. Chem.*, 1997, **272**, 23682–23689.
122. S. K. Shenoy, P. H. McDonald, T. A. Kohout and R. J. Lefkowitz, *Science*, 2001, **294**, 1307–1313.
123. D. Bhandari, J. Trejo, J. L. Benovic and A. Marchese, *J. Biol. Chem.*, 2007, **282**, 36971–36979.
124. M. R. Ahmed, X. Zhan, X. Song, V. V. Gurevich and E. V. Gurevich, *Biochemistry*, 2011, **50**, 3749–3763.
125. S. K. Shenoy, K. Xiao, V. Venkataramanan, P. M. Snyder, N. J. Freedman and A. M. Weissman, *J. Biol. Chem.*, 2008, **283**, 22166–22176.
126. J. F. Nabhan, H. Pan and Q. Lu, *EMBO Rep.*, 2010, **11**, 605–611.
127. X. Song, D. Raman, E. V. Gurevich, S. A. Vishnivetskiy and V. V. Gurevich, *J. Biol. Chem.*, 2006, **281**, 21491–21499.
128. X. Song, E. V. Gurevich and V. V. Gurevich, *J. Neurochem.*, 2007, **103**, 1053–1062.
129. X. Song, S. Coffa, H. Fu and V. V. Gurevich, *J. Biol. Chem.*, 2009, **284**, 685–695.
130. S. K. Shenoy and R. J. Lefkowitz, *J. Biol. Chem.*, 2005, **280**, 15315–15324.
131. P. H. McDonald, C. W. Chow, W. E. Miller, S. A. Laporte, M. E. Field, F. T. Lin, R. J. Davis and R. J. Lefkowitz, *Science*, 2000, **290**, 1515–1518.
132. L. M. Luttrell, F. L. Roudabush, E. W. Choy, W. E. Miller, M. E. Field, K. L. Pierce and R. J. Lefkowitz, *Proc. Natl. Acad. Sci. U. S. A.*, 2001, **98**, 2449–2454.
133. W. E. Miller, P. H. McDonald, S. F. Cai, M. E. Field, R. J. Davis and R. J. Lefkowitz, *J. Biol. Chem.*, 2001, **276**, 27770–27777.
134. S. J. Perry, G. S. Baillie, T. A. Kohout, I. McPhee, M. M. Magiera, K. L. Ang, W. E. Miller, A. J. McLean, M. Conti, M. D. Houslay and R. J. Lefkowitz, *Science*, 2002, **298**, 834–836.
135. G. S. Baillie, D. R. Adams, N. Bhari, T. M. Houslay, S. Vadrevu, D. Meng, X. Li, A. Dunlop, G. Milligan, G. B. Bolger, E. Klussmann and M. D. Houslay, *Biochem. J.*, 2007, **404**, 71–80.
136. J. M. Carter, V. V. Gurevich, E. R. Prossnitz and J. R. Engen, *J. Mol. Biol.*, 2005, **351**, 865–878.
137. T. W. Lee, R. Seifert, X. Guan and B. K. Kobilka, *Biochemistry*, 1999, **38**, 13801–13809.
138. S. M. Hanson, D. J. Francis, S. A. Vishnivetskiy, C. S. Klug and V. V. Gurevich, *J. Biol. Chem.*, 2006, **281**, 9765–9772.
139. H. Shankar, A. Michal, R. C. Kern, D. S. Kang, V. V. Gurevich and J. L. Benovic, *J. Biol. Chem.*, 2010, **285**, 8316–8329.
140. P. H. McDonald, N. L. Cote, F. T. Lin, R. T. Premont, J. A. Pitcher and R. J. Lefkowitz, *J. Biol. Chem.*, 1999, **274**, 10677–10680.

141. N. Wu, S. M. Hanson, D. J. Francis, S. A. Vishnivetskiy, M. Thibonnier, C. S. Klug, M. Shoham and V. V. Gurevich, *J. Mol. Biol.*, 2006, **364**, 955–963.
142. V. V. Gurevich, S. M. Hanson, E. V. Gurevich and S. A. Vishnivetskiy, in *Signal Transduction in the Retina*, ed. S. J. Fliesler and O. Kisselev, (ed.), CRC Press, Boca, Raton, FL, 2007, pp. 55–88.
143. S. M. Hanson, S. A. Vishnivetskiy, W. L. Hubbell and V. V. Gurevich, *Biochemistry*, 2008, **47**, 1070–1075.
144. S. K. Milano, Y. M. Kim, F. P. Stefano, J. L. Benovic and C. Brenner, *J. Biol. Chem.*, 2006, **281**, 9812–9823.
145. H. Storez, M. G. Scott, H. Issafras, A. Burtey, A. Benmerah, O. Muntaner, T. Piolot, M. Tramier, M. Coppey-Moisan, M. Bouvier, C. Labbé-Jullié and S. Marullo, *J. Biol. Chem.*, 2005, **280**, 40210–40215.
146. L. Shen, G. Caruso, P. Bisegna, D. Andreucci, V. V. Gurevich, H. E. Hamm and E. DiBenedetto, *IET Syst. Biol.*, 2010, **4**, 12–32.

Section III
Modelling G Protein-coupled Receptor Structure and Function

CHAPTER 18

Structure-based Virtual Screening for Ligands of G Protein-coupled Receptors

STEFANO COSTANZI

Laboratory of Biological Modeling, National Institute of Diabetes and Digestive and Kidney Diseases, National Institutes of Health, DHHS, Bethesda, MD 20892, USA

18.1 Introduction

Knowledge of the three-dimensional structure of proteins of pharmaceutical interest is of utmost importance for drug discovery, as it can provide a solid basis for the discovery of ligands capable of modulating their activity by means of computational, or *in silico*, techniques. The first successes of this rational approach, which goes under the name of structure-based drug discovery (SBDD), can be traced back to the early 1990s, which for example, saw the rational design of the anti-influenza drug zanamivir (Relenza®) on the basis of the crystal structure of the influenza virus sialidase.[1] Thanks to the constant augmentation of computer power and the sophistication of the computational algorithms, SBDD is becoming increasingly a more practical and adopted means to rationally drive drug discovery campaigns, with conspicuous cost savings for the pharmaceutical industry. As a result, as recently reviewed by Congreve and Marshall, 25% of the drugs approved in 2006 were discovered through SBDD.[2]

RSC Drug Discovery Series No. 8
G Protein-Coupled Receptors: From Structure to Function
Edited by Jesús Giraldo and Jean-Philippe Pin

Published by the Royal Society of Chemistry, www.rsc.org

Among the various SBDD approaches, one of the most widespread is structure-based virtual screening (SBVS), which consists of the computational analysis of the interaction of a large collection of compounds—hundreds of thousands or even millions—with a three-dimensional (3-D) representation of the structure of the target protein. This is typically done through molecular docking, a computational technique that consists of the exploration of the possible binding modes—or poses—of a ligand inside the binding cavity of a target protein, followed by the calculation of a scores that reflect the predicted binding energies of each of the generated poses. Upon screening a large collection of compounds, a possible binding mode is typically found for most of the molecules subjected to the docking procedures, including those that, in fact, cannot bind to the receptor. However, ideally true ligands will receive better docking scores than non-binders. In reality, a complete prioritization of ligands *versus* non-binders is never achieved, and the practical aim of a docking-based virtual screening is the concentration of most ligands within the top scoring portion of the database. The top scoring few hundred molecules, or sometimes a subset of these selected through a careful visual inspection based on biochemical intuition and knowledge of the molecular system under investigation, are then acquired and subjected to experimental testing to hopefully yield a certain number of experimentally confirmed hits that can be further developed into lead compounds and finally drug candidates to be introduced into preclinical and clinical studies.

In conjunction with docking, or sometimes as an alternative to it, SBVS can also be based on structure-based pharmacophore searches, which filter the screened molecules on the basis of their ability to match certain chemical features—such as hydrogen bond donors or acceptors, aromatic groups, *etc.*—that are supposed to be required for ligand binding on the basis of the physicochemical characteristics of the protein binding site. Docking-based and pharmacophore-based virtual screening can also be conveniently combined, with the pharmacophore searches acting as filters to select, among all the docked molecules, only those that establish certain crucial interactions with the receptor.

Given the numerous physiological and pathological implications of their signalling, G protein-coupled receptors (GPCRs) have a very high pharmaceutical appeal and constitute the target of a large share of the drugs on the market.[3] Moreover, many more GPCRs are regarded as very promising targets for pharmaceutical applications and are the object of intense drug discovery efforts. However, up until recently, the structure-based discovery of GPCR ligands has been confined to rhodopsin-based homology models. In fact, despite the large size of the superfamily, which in humans includes about 800 members,[4] for years rhodopsin has been the only GPCR with experimentally elucidated 3-D structures.[5]

However, recent breakthroughs in GPCR crystallography led to the recent solution of the structures of a small number of additional GPCRs, namely the β_1- and β_2-adrenergic receptors in 2007 and the adenosine A_{2A} receptor in 2008, all solved in complex with competitive antagonists or inverse agonists.[6]

The new pace with which the field is progressing suggests that the structures of more receptors will be solved in the near future—solution of the crystal structure of the dopamine D3 and of chemokine CXCR4 receptors has been announced at the time of this writing. In this context, the field of structure-based discovery of GPCR ligands has gained considerable interest and is likely to undergo a rapid expansion both in academia and industry.

18.2 The Interhelical Binding Cavity as a Target for Virtual Screening

The helical bundle of GPCRs encloses an internal cavity located towards the extracellular opening and partially lined by the second extracellular loop (EL2).[7] This cavity hosts the binding site of the natural ligand for all the receptors with solved three-dimensional structures; Figure 18.1 shows 11-*cis*-retinal, carazolol and ZM241385, bound to rhodopsin, the β_2-adrenergic receptor and the adenosine A_{2A} receptor, respectively. Besides these receptors, this cavity is considered to have the same role for all the receptors belonging to the rhodopsin family (or class A) that are naturally activated by small molecules.[7] Other GPCRs are supposed to utilize only marginally or not utilize at all the interhelical cavity for the binding of their natural ligands. For example, glycoprotein hormone receptors (GPHRs) and receptors belonging to the glutamate family (or class C) bind their ligands to their large extracellular

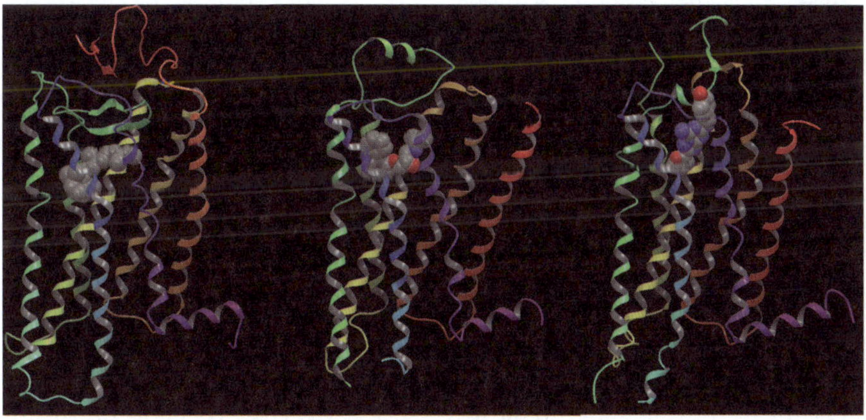

Figure 18.1 The transmembrane helical bundle of G protein-coupled receptor encloses a cavity that is thought to host the orthosteric binding site for all the receptors belonging to the rhodopsin family (class A) naturally activated by small molecules. Here we see three crystal structures in which the co-crystallized ligands are bound to this cavity: rhodopsin with the covalently bound inverse agonist 11-*cis*-retinal, on the left, the β_2-adrenergic receptor in complex with the competitive inverse agonist carazolol in the centre, and the adenosine A_{2A} receptor in complex with the competitive neutral antagonist ZM241385, on the right.

N-termini. However, a number of studies, such as those that we have conducted at the Ca^{2+} sensing receptor,[8] a class C receptor, and the thyroid-stimulating hormone (TSH) receptor,[9–11] a GPHR, have demonstrated that the activity of these receptors can also be modulated by compounds that bind to the interhelical binding cavity in an allosteric manner. Thus, the interhelical binding cavity can be considered an excellent general target for the discovery of compounds capable of modulating the activity of GPCRs. The following sections illustrate the effectiveness of its use in virtual screening campaigns.

18.3 The Use of GPCR Crystal Structures for Virtual Screening Purposes

18.3.1 Virtual Screening Campaigns

Crystal structures provide atomic-level pictures of proteins and protein ligand complexes and thus offer an ideal platform for structure-based virtual screening. Until recently, such an experimental structure directly applicable to SBDD was not available for any of the members of the GPCR superfamily. In fact, rhodopsin, which as mentioned has been for a long time the only GPCR with experimentally elucidated structures, is a peculiar receptor that is not activated by a diffusible ligand but features an inverse agonist, 11-*cis*-retinal, covalently bound to its interhelical binding cavity. By the action of light photon, this is isomerized to the all-*trans* form, which behaves like an agonist and triggers the activation of the receptor.[5] However, in 2007, the solution of the crystal structure of the β_2-adrenergic receptor finally offered an experimental GPCR model amenable to virtual screening.[12,13]

On the basis of this structure, Sabio and co-workers[14] and Schoichet, Kobilka and co-workers[15] independently embarked on docking-based virtual screening campaigns for ligands of the β_2-adrenergic receptor and probed, for the first time, the applicability of the crystal structure of a GPCR to drug discovery. In particular, in the campaign described by Shoichet and co-workers, after docking about one million commercially available 'lead-like' compounds, the authors selected 25 molecules from the top 500 ranking ones and subjected these to experimental tests. This procedure led to the identification of six actual ligands, with a notable hit rate—defined as the number of the identified active compounds divided by the number of the tested compounds—of 24%.

One of the principal aims of structure-based virtual screening is the identification of ligands endowed with novel chemical scaffolds, of the kind that one could not design simply on the basis of the structure of known binders. Notably, two out of the six newly identified ligands are substantially different from previously seen β_2-adrenercigic receptor ligands. Thus, despite their μM affinity that makes them considerably weaker than the best known compounds, they could constitute an excellent starting point for the design of novel series of ligands based on two different chemotypes. This study provided a substantial proof of principle that GPCR crystal structures, besides providing invaluable insights into the three-dimensional architecture and the structure–function relationships of the receptor, can indeed be applied to drug discovery.

The results obtained for the β_2-adrenergic receptor were amply confirmed and corroborated by two virtual screens subsequently conducted on the basis of the crystal structure of the adenosine A_{2A} receptor, solved in 2008.[16] These virtual screening campaigns were conducted independently by Abagyan, Stevens, Ijzerman and co-workers[17] and Jacobson, Shoichet and co-workers[18] in a fashion similar to that described previously for the β_2-adrenergic receptors and yielded impressive hit rates of 41% and 35%, respectively.

Notably, the virtual screening approach taken by Shoichet's group is also based on a careful visual inspection of the top ranking compounds prior to the selection of the molecules to be subjected to experimental testing.[15,18] Thus, the high-throughput screening capabilities ensured by the calculation power of the computers and the sophistication of the docking algorithms is supplemented by the biochemical intuition, knowledge and experience of the researchers.

18.3.2 Controlled Virtual Screening Experiments

Besides the actual virtual screening campaigns described above, a number of controlled computational experiments have been conducted to quantitatively assess the effectiveness of virtual screening based on crystal structures of GPCRs. In these studies, known ligands have been collected from the literature, sketched, and dispersed within a substantially larger number of presumably inactive molecules with similar physicochemical properties used as decoys. The resulting molecular databases have been subjected to molecular docking at the crystal structure of the receptor, finally resulting in the ranking of all the compounds on the basis of their docking scores.

These controlled screens have been performed by various research groups, including mine, for the β_2-adrenergic receptor[19–22] and the adenosine A_{2A} receptors.[23] They revealed a remarkable prioritization of ligands *versus* decoys, with a consequent substantial concentration of the ligands within the top ranking position of the database. For example, our controlled screening based on the crystal structure of the β_2-adrenergic receptor, in which we dispersed 60 known ligands within about 60 000 decoys, resulted in the recovery of 24 ligands within the 50 top scoring compounds, with a hit rate of 48%, when the most effective combination of docking algorithm and scoring function was applied. A plot of the hit rates obtained in this controlled virtual screening experiment with various combinations of docking algorithms and scoring functions is shown in Figure 18.2.[22,24]

We also demonstrated that, since crystal structures are static representations of molecules that actually have a flexible nature, the screening performances could be further improved by taking into account multiple configurations of the receptor.[24] In particular, we applied an approach—known as receptor–ensemble docking—based on the generation of alternative conformations of the interhelical cavity through *in silico* conformational searches and the subsequent parallel docking of the compounds at each of the generated structures in

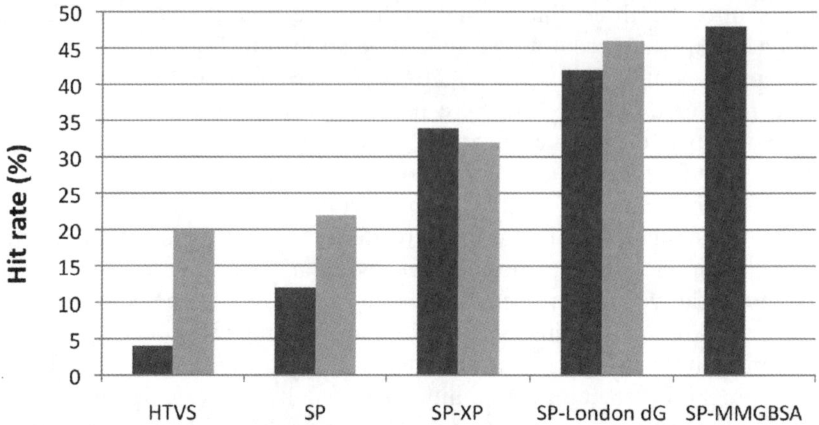

Figure 18.2 We subjected a database composed of 60 known ligands of the β_2-adrenergic receptor and ~60 000 decoys to docking-based virtual screening employing different docking algorithms and different scoring functions.[22,24] Some of them are very fast but less accurate, and are primarily intended for the initial screening of large databases of millions compounds. Others are more sophisticated and accurate, but also inevitably slower, and are primarily intended to be used in the subsequent stages of the screening. In particular, the two docking algorithms that we used are the high-throughput virtual screening (HTVS, 13 hours for ~60 000 molecules) and standard precision (SP, 98 hours for ~60 000 molecules) docking modes implemented in Glide.[28] We also rescored the poses obtained with SP docking mode with the extra precision (XP, 18 hours to rescore ~60 000 molecules), the London dG (London dG, 4 hours to rescore ~60 000 molecules), and the MM-GBSA scoring functions implemented in Glide,[28] the Molecular Operating Environment (MOE)[39] and Prime,[40] respectively. This figure reports the hit rates found within the top 50 compounds for the screening conducted at the crystal structure, in dark grey, and those conducted combining the crystal structure with three alternative conformations of the receptor according to the receptor ensemble docking technique, in light grey; at the time of this writing, we have not applied MM-GBSA rescoring to the screening based on receptor ensemble docking. Notably, the hit rate resulting from a random selection of the compounds would be ~0.1%.

addition to the crystal structure. The final ranking of the compounds is then compiled using, for each molecule, the highest docking score obtained across the parallel dockings, thus allowing the molecules to select their preferred conformation of the receptor (Figure 18.3).

The increase in the performance of the screening was most evident when using less accurate but faster scoring functions, which are intended to be employed in the initial stages of the screening campaigns to rapidly evaluate large databases of millions of compounds and narrow them down to a smaller subset that can be subjected to screening with more accurate, but also more time-consuming, docking algorithms and scoring functions. For example, with the fastest of the docking algorithms that we applied, we retrieved two ligands

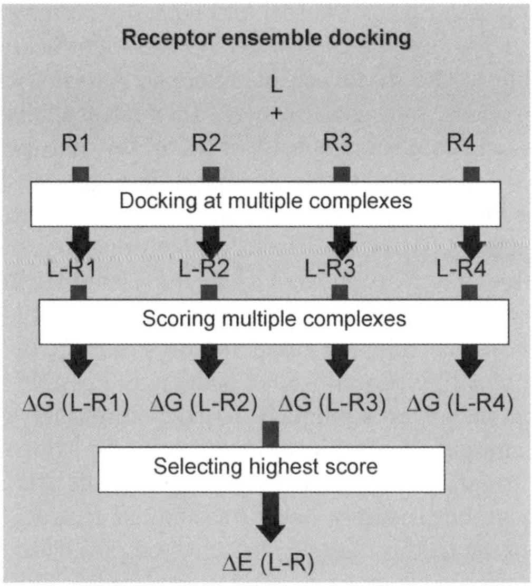

Figure 18.3 Schematic representation of the receptor–ensemble docking strategy. Rather than docking the compounds only at the crystal structure of the receptor, each compound is docked in parallel at the crystal structure and several alternative *in silico* generated conformations of the receptor. For each compound, the best obtained score is selected and used for ranking purposes. A comparison of the performances of virtual screening based on the crystal structure alone and on receptor–ensemble docking is provided in Figure 18.2.

within the top 50 scoring molecules, with a hit rate of 4%, when using the crystal structure alone. When using the crystal structure in combination with three *in silico* generated alternative conformations, we retrieved ten ligands within the top 50 scoring molecules, with a hit rate of 20%. A complete comparison of the performances of the virtual screening based on the crystal structure alone *versus* those based on receptor–ensemble docking is given in Figure 18.2.

18.3.3 Identification of GPCR Agonists and Blockers Through Virtual Screening

None of the crystal structures of GPCRs with diffusible ligands that have been published at the time of writing have been solved in complex with agonists. Instead, so far they have all been obtained in complex with either inverse agonists or neutral antagonists.[12,13,16,25–27] [Note: After this chapter was written, crystal structures of GPCRs in complex with agonists have been reported.]

Hence, a natural question is whether the virtual screening campaigns based on these structures are suitable exclusively for the identification of blockers or are also applicable to the discovery of agonists. As GPCRs signal not only through the G proteins, but also through other intracellular proteins (*e.g.* β-arrestin), a compound that is an agonist or a blocker of a specific pathway will not necessarily have the same behaviour on a different pathway. Here, the term blocker is used to indicate collectively neutral antagonists and inverse agonists of the G protein related pathway; similarly, the term agonist is used to indicate compounds capable of activating the G protein related signalling pathway.

The virtual screening campaigns illustrated in Section 18.3.1 yielded exclusively blockers. However, this fact alone does not indicate that agonists cannot be identified through this procedure. In fact, it is possible that the screened databases were biased towards blockers, *i.e.* they might have contained a substantially higher amount of compounds capable of blocking the activity of the receptor than compounds capable of stimulating it. Thus, the lack of retrieval of agonists might have been simply due to a significant lower concentration of compounds belonging to this class in the screened databases, possibly because agonists may have more stringent structural requirements than blockers.

The controlled virtual screening experiments performed with the β_2-adrenergic receptor offered a more direct way to assess the possibility of retrieving agonists, since the known ligands utilized in these exercises comprised not only blockers but also agonists.[19,20,24] Surprisingly, perhaps, the results of these controlled experiments revealed a substantial prioritization not only of the blockers, but also of the agonists over the decoys. For example, in our study, when using the standard precision (SP) docking algorithm of the Glide[28] docking program followed by rescoring with the extra-precision (XP) scoring function, we retrieved 16 blockers and eight agonists within the top 100 compounds out of the 31 blockers and 29 agonists that were dispersed within the decoys database, indicating that although obtained in complex with a blocker, the structure of the β_2-adrenergic receptor can recognize agonists too. Nevertheless, the screening showed a substantial tendency to privilege blockers over agonists, with 52% of the total number of blockers and 28% of the total number of agonists contained in the screened database retrieved within the top 100 compounds.

It is worth noting that the docking procedures at the basis of these virtual screening experiments consider the flexibility of the ligands but not that of the receptors. Thus, the docking complexes obtained as a result of the screening do not reflect the conformational changes that agonists are thought to induce to the receptor. Instead, they probably reflect only the initial stage of the agonist binding, prior the occurrence of the conformational changes that will lead to the formation of optimal receptor-agonist complexes and to the triggering of the signalling cascade that initiates with G protein activation. Consequently, the preference of our controlled virtual screening for blockers is not surprising, since agonists are scored in complex with a state of the receptor for which they have low affinity. However, the same virtual screening experiments have also demonstrated the possibility to obviate the disadvantage shown by the agonists and bias the screening towards the retrieval of compounds belonging to this class.[19,20,24]

Several strategies have been successfully applied to this purpose, all implying the use of a modified structure of the receptor. In particular, Rognan and co-workers succeeded in reverting the tendency of the screening by changing the rotameric state of specific residues in the binding pocket and utilizing a topological scoring function based on molecular interaction fingerprints,[20] while Abagyan and co-workers[19] and Vaidehi and co-workers achieved similar results by means of rotations and/or translations of specific transmembrane domains.[29]

In our study, we adopted a different approach that is not based on the application to the receptors of conformational changes known to be associated with the activation process. Instead, we used a strategy based on the Induced Fit docking procedure, as implemented in the Schrödinger package,[30] which incorporates the flexibility of the receptors into the docking process by sampling the degrees of freedom of the side chains in the dihedral space while also granting flexibility to the backbone. Although too slow to be practically applied to the screening of large databases of compounds, the Induced Fit procedure can be used to dock one or a few known agonists at the receptor, thus generating an agonist-adapted conformation of the binding pocket. Notably, this approach has the advantage of not requiring prior knowledge of the conformational changes involved in the receptor activation process and necessitates the availability of only one known agonist, such as the natural ligand of the receptor.

In particular, through the Induced Fit procedure, we docked the full agonist isoproterenol at β_2-adrenergic receptor and used the resulting isoproterenol-adapted structure to conduct a controlled virtual screening according to the same modalities described above for the screening based on the crystal structure. Using this procedure we succeeded in obtaining a substantial bias of the screening towards the identification of agonists, with 16 agonists and one blocker retrieved within the top 100 compounds when using the standard precision (SP) docking algorithm of Glide followed by a rescoring with the extra-precision scoring function (XP). Thus, within the top 100 compounds we retrieved 55% of the total number of agonists contained in the screened database, as opposed to the 28% retrieved with the screening based on the crystal structure. At the same time, the retrieval of blockers within the top 100 compounds decreased from the 52% obtained with the crystal structure to a mere 3%.

18.4 Homology Models of GPCRs and their Use for Virtual Screening Purposes

18.4.1 Useful Homology Models can be Constructed

As mentioned, GPCRs are a large superfamily, which in humans includes about 800 members.[4] According to the GRAFS classification scheme, they can be divided into five main families:[4]

- glutamate (G; also class C or family III);
- rhodopsin (R; also class A or family I);

- adhesion (A; also class B or family 2, together with the secretin family);
- frizzled/taste2 (F);
- secretin (S, also class B or family 2, together with the adhesion family).

Among these, the rhodopsin family is by far the largest, accounting for 84% of the entire superfamily. However, notwithstanding its conspicuous dimensions and the vast pharmaceutical interests revolving around it, the GPCR superfamily has historically been characterized by a paucity of structural information. For these reasons, homology modelling has been amply utilized to construct three-dimensional structures of the receptors of interest on the basis of the experimentally solved structure of a known receptor used as a template.

For years, rhodopsin itself has been the only available template, first with the two-dimensional electron crystallography structures of Schertler and co-workers in the 1990s, and then with the various X-ray structures solved in the first decade of the new millennium, beginning with the one published by Palczewski and co-workers in 2000.[5,31,32] These rhodopsin-based models have been widely employed to:

- design site-directed mutagenesis experiments intended to explore the structure–function relationships of the receptors;
- characterize the mechanisms of ligand recognition; and
- identify compounds capable of modulating the activity of the receptors.

Of note is the neoceptor/neoligand approach taken by Jacobson and co-workers which combined site-directed mutagenesis with complementary modifications made to the ligands, providing a particularly powerful tool to investigate receptor/ligand contacts and to generate experimentally supported molecular models.[33]

The breakthroughs in GPCR crystallography, which as mentioned in the introduction, brought the recent solutions of several receptors belonging to the rhodopsin family (besides producing structural information directly applicable to drug discovery), also afforded for the first time a means of ultimately validating the accuracy of homology models through a direct comparison between the newly solved crystal structures with the computer-generated models of the same receptors. Of note is my comparison between molecular models of the β_2-adrenergic receptor in complex with the inverse agonist carazolol, obtained through rhodopsin-based homology modelling followed by fully flexible molecular docking of the ligand, with the crystal structure of the same receptor.[34] This study, which I published soon after the solution of the crystal structure of the receptor, offered the first direct evaluation of the accuracy of a GPCR homology model in light of its experimentally elucidated structure and revealed that not only the structure of the receptor, but also the binding mode of the ligand and the receptor ligand interactions could be approximated reasonably well by the models.

More recently, a wider evaluation of the state of the art was provided by the first 'community-wide assessment of GPCR structure modelling and ligand docking', organized in coordination with the solution of the structure of the adenosine A_{2A} receptor.[35] This time, models of the A_{2A} receptor in complex with the neutral antagonist ZM241385 were submitted to the organizers of the assessment by molecular modellers prior to the unveiling of the experimentally derived coordinates of the receptor–ligand complex. In line with what I had found for the β_2-adrenergic receptor, this blind test revealed that the seven transmembrane domains of the receptor could be built with a good level of accuracy while the modelling of the loops that connect these domains, especially the long ones, was confirmed to be problematic. A wide distribution was found for the accuracy of the prediction of the pose of the ligand within the binding cavity, suggesting that the docking of the ligand and the prediction of the receptor–ligand interactions are more challenging than the construction of the model of the receptor. However, the organizers found that the top three scoring models (submitted by Costanzi, Abagyan/Katrich and Abagyan/Lam) predicted correctly over 40% of the total number of the receptor-ligand contacts.

The current pace of the field promises that the structure of a considerable number of GPCRs, possibly in complex with several ligands, will be solved in the near future. As mentioned, the solution of the structure of a few GPCRs has been announced at the time of writing. In coordination with this, a novel community-assessment of GPCR modelling and docking has been organized (http://cmpd.scripps.edu/GPCRDock2010/index.html). All these efforts promise a considerable furthering of the field in the years to come, with substantial benefits for structure-based drug discovery campaigns.

18.4.2 Controlled Virtual Screening Experiments

In parallel with the controlled virtual screening based on the crystal structure of the β_2-adrenergic receptor illustrated in the Section 18.3.2, we have subjected three rhodopsin based models to the same experiments.[24] Just as for the crystal structure, we docked a database containing 60 ligands and 60 000 decoys at the homology models to then assess how well the ligands could be prioritized over the decoys. The three models on which we based the screening are those we had described previously in an article[34] featuring the comparison of *in silico* and X-ray derived structures of the receptor in complex with carazolol.

The three models are endowed with three different levels of structural accuracy when compared to the crystal structure:

- Model 1 features six residues in the binding pocket in the wrong rotameric state, an RMSD of the binding pocket residues of 3.3 Å and an RMSD of the ligand of 3.7 Å.
- Model 2 features one residues in the binding pocket in the wrong rotameric state, an RMSD of the binding pocket residues of 2.8 Å and an RMSD of the ligand of 2.9 Å.

- Model 3 features all residues in the binding pocket in the correct rotameric state, an RMSD of the binding pocket residues of 2.7 Å and an RMSD of the ligand of 1.7 Å.

For further details on the differences between the three models and the way they were constructed see ref. 34. As illustrated in Figure 18.4, the homology models performed according to their levels of accuracy, with model 3 yielding the highest hit rates. However all the models yielded a substantial prioritization of the ligands *versus* the decoys and, with the most effective combinations of docking algorithms and scoring functions, resulted in hit rates of 4%, 10%, and 22% within the top scoring 50 compounds for model 1, model 2, and model 3, respectively. Notably, these hit rates, although lower than those registered with the crystal structure, are enormously higher than what would result from a random selection, which would produce a hit rate of about 0.1%.

Similarly, Abagyan and co-workers have subjected to a controlled virtual screening experiment the crystal structure of the adenosine A_{2A} receptor in parallel with the top scoring homology models that resulted from the blind community-wide assessment described above.[23] In particular, the authors subjected to the screening 14,000 molecules, including 348 adenosine A_{2A} antagonists, extracted from the GPCR-Ligand Database (GLIDA, http://pharminfo.pharm.kyoto-u.ac.jp/services/glida/index.php)[36] with the aim of

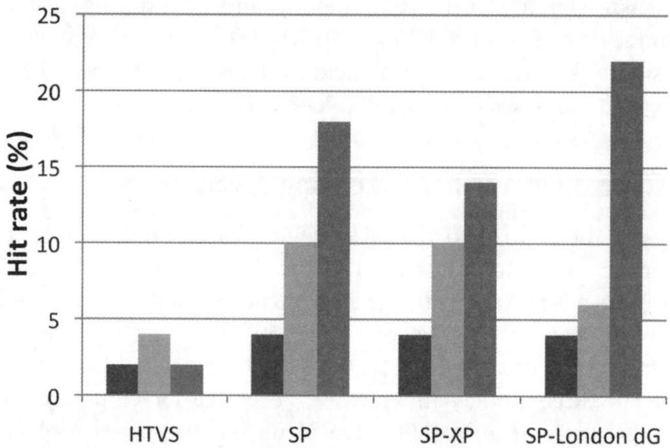

Figure 18.4 Not only crystal structures, but also homology models of G protein-coupled receptors can be applied to docking-based virtual screening. This figure reports the hit rates found within the top 50 compounds for the screening conducted on the basis of three different homology models endowed with three different levels of accuracy (see text). The dark grey, light grey and medium grey bars refer to the screening experiments conducted on the basis of model 1, model 2, and model 3, respectively. Notably, the hit rate resulting from a random selection of the compounds would be ~0.1%. Further details on the docking algorithms and the scoring functions are given in Figure 18.2; at the time of this writing, we have not applied MM-GBSA rescoring to the screening experiments based on the homology models.

assessing how well these compounds would be prioritized over the rest. The virtual screening conducted at the crystal structure resulted in an area under the receiver operating characteristic (ROC) curve of 0.88, while the top three models yielded areas under the ROC curve of 0.76 (Costanzi), 0.75 (Katritch/Abagyan), and 0.74 (Lam/Abagyan)—the area under the ROC curve is a parameter often used to monitor virtual screening results; a perfect screening that prioritized all the true ligands *versus* the decoys would yield an area under the ROC curve of 1 and a random screening that could not distinguish the ligands from the decoys would yield an area under the ROC curve of 0.5. The less accurate models, instead, gave results closer to a random selection. Thus, in line with what we have shown for the β_2-adrenergic receptor, this study also suggests that, in the absence of crystal structures, good homology models are valuable drug discovery tools.

18.4.3 Examples of Virtual Screening Campaigns based on GPCR Homology Models

The applicability of GPCR homology models to virtual screening suggested by the controlled experiments illustrated in Section 18.4.2 is confirmed by the success of numerous virtual screening campaigns. As it is impossible to describe all the cases reported in the literature, I will just mention, by way of example, three successful virtual screening campaigns with which I was personally involved.

In one of these campaigns, with an approach that combined molecular docking with a screening based on a pharmacophoric representation of the binding cavity (both based on a homology model), we identified a number of antagonists of the thyrotropin-releasing hormone receptor (TRH-R)—one of which resulted in the most potent known to date as well as the most selective for the TRH-R1 *versus* the TRH-R2 subtype.[37] In another study, by means of a virtual screening based on molecular docking at a homology model of the receptor optimized in complex with a known agonist, we identified several novel ligands of the free fatty acid receptor 1 (FFA1), which is considered a promising target for the treatment of type 2 diabetes.[38] Notably, most of those ligands turned out to be agonists of the receptor, thus confirming—besides the applicability of GPCR homology models to virtual screening—the possibility of biasing the screening towards the retrieval of agonists discussed in Section 18.3.2.

More recently, on the basis of a homology model, we conducted a virtual screening campaign for antagonists of the $P2Y_1$ receptor, which are highly sought after as antithrombotic agents. This effort resulted in the identification of numerous novel antagonists based on nine diverse scaffolds (data not published at the time of writing). While most of the known $P2Y_1$ antagonists are analogs of the nucleotide adenosine-3′,5′-bisphophate (A3P5P), the newly identified compounds are all non-charged or minimally charged drug-like molecules and thus provide excellent scaffolds for the rational design of novel potential drugs.

18.5 Conclusions

Due to the vast implications of their signalling, G protein-coupled receptors play a very important role in the pharmaceutical sector. A relatively small number of them already constitute the target of a large share of the drugs currently on the market, while many more members of the large superfamily are under study for the development of agents for the treatment of a wide range of diseases. Thanks to the progress made recently in the field of structural determination, the structure-based discovery of GPCR ligands is becoming increasingly more widely applied, as it can focus tremendously the drug discovery process with conspicuous savings of time and money.

Controlled experiments and actual screening campaigns have clearly shown that the crystal structures of the receptors are indeed applicable to structure-based virtual screening. With the testing of a only few dozens of molecules, this rational approach can lead to the identification of several novel and diverse hit compounds amenable to further optimization through medicinal chemistry. Notably, controlled experiments and actual campaigns have also demonstrated the possibility of biasing the virtual screening towards the selective retrieval of agonists or blockers, which is not a trivial thing, since all the X-ray structures of GPCRs solved at the time of writing have been obtained in complex with either inverse agonists or neutral antagonists. [Note: After this chapter was written, crystal structures of GPCRs in complex with agonists have been reported.] Besides directly offering platforms for the virtual screening, experimental GPCR structures also provide templates for the constructions of homology models of other members of the superfamily. Controlled experiments and actual campaigns have demonstrated that such models can indeed be successfully applied to structure-based drug discovery and structure-based virtual screening.

Given the pace at which the field is proceeding, it is certain that the structure of a considerable number of receptors will be solved in the near future—in complex not only with blockers but with agonists too. These structures will also provide additional templates and will undoubtedly afford the construction of more accurate homology models.

Thus, the expected expansion of the experimentally derived structural knowledge, the continuous improvement of the computational algorithms, and the incessant increase of the computer power suggest that the field of computer-assisted structure-based discovery of GPCR ligands will keep growing stronger, with great benefit for the pharmaceutical sciences and for the healthcare sector in general.

Acknowledgements

I thank Dr Santiago Vilar for the stimulating discussions and help with the collection of the data for the construction of Figures 18.2 and 18.4. This work was supported by the intramural research program of the National Institute of Diabetes and Digestive and Kidney Diseases of the National Institutes of Health.

References

1. M. von Itzstein, W. Wu, G. Kok, M. Pegg, J. Dyason, B. Jin, T. Van Phan, M. Smythe, H. White and S. Oliver, *Nature*, 1993, **363**, 418–423.
2. M. Congreve and F. Marshall, *Br. J. Pharmacol.*, 2010, **159**, 986–996.
3. K. Pierce, R. Premont and R. Lefkowitz, *Nat. Rev. Mol. Cell. Biol.*, 2002, **3**, 639–650.
4. D. Gloriam, R. Fredriksson and H. Schiöth, *BMC Genomics*, 2007, **8**, 338.
5. S. Costanzi, J. Siegel, I. Tikhonova and K. Jacobson, *Curr. Pharm. Des.*, 2009, **15**, 3994–4002.
6. M. Hanson and R. Stevens, *Structure*, 2009, **17**, 8–14.
7. J. Surgand, J. Rodrigo, E. Kellenberger and D. Rognan, *Proteins*, 2006, **62**, 509–538.
8. J. Hu, J. Jiang, S. Costanzi, C. Thomas, W. Yang, J. Feyen, K. Jacobson and A. Spiegel, *J. Biol. Chem.*, 2006, **281**, 21558–21565.
9. H. Jäschke, S. Neumann, S. Moore, C. Thomas, A. Colson, S. Costanzi, G. Kleinau, J. Jiang, R. Paschke, B. Raaka, G. Krause and M. Gershengorn, *J. Biol. Chem.*, 2006, **281**, 9841–9844.
10. S. Moore, H. Jaeschke, G. Kleinau, S. Neumann, S. Costanzi, J. Jiang, J. Childress, B. Raaka, A. Colson, R. Paschke, G. Krause, C. Thomas and M. Gershengorn, *J. Med. Chem.*, 2006, **49**, 3888–3896.
11. S. Neumann, G. Kleinau, S. Costanzi, S. Moore, J. Jiang, B. Raaka, C. Thomas, G. Krause and M. Gershengorn, *Endocrinology*, 2008, **149**, 5945–5950.
12. D. Rosenbaum, V. Cherezov, M. Hanson, S. Rasmussen, F. Thian, T. Kobilka, H. Choi, X. Yao, W. Weis, R. Stevens and B. Kobilka, *Science*, 2007, **318**, 1266–1273.
13. V. Cherezov, D. Rosenbaum, M. Hanson, S. Rasmussen, F. Thian, T. Kobilka, H. Choi, P. Kuhn, W. Weis, B. Kobilka and R. Stevens, *Science*, 2007, **318**, 1258–1265.
14. M. Sabio, K. Jones and S. Topiol, *Bioorg. Med. Chem. Lett.*, 2008, **18**, 5391–5395.
15. P. Kolb, D. Rosenbaum, J. Irwin, J. Fung, B. Kobilka and B. Shoichet, *Proc. Natl. Acad. Sci. U. S. A.*, 2009, **106**, 6843–6848.
16. V. Jaakola, M. Griffith, M. Hanson, V. Cherezov, E. Chien, J. Lane, A. Ijzerman and R. Stevens, *Science*, 2008, **322**, 1211–1217.
17. V. Katritch, V. Jaakola, J. Lane, J. Lin, A. Ijzerman, M. Yeager, I. Kufareva, R. Stevens and R. Abagyan, *J. Med. Chem.*, 2010, **53**, 1799–1809.
18. J. Carlsson, L. Yoo, Z. Gao, J. Irwin, B. Shoichet and K. Jacobson, *J. Med. Chem.*, 2010, **53**, 3748–3755.
19. K. Reynolds, V. Katritch and R. Abagyan, *J. Comput. Aided. Mol. Des.*, 2009, **23**, 273–288.
20. C. de Graaf and D. Rognan, *J. Med. Chem.*, 2008, **51**, 4978–4985.
21. S. Vilar, J. Karpiak and S. Costanzi, *J. Comput. Chem.*, 2010, **31**, 707–720.
22. S. Vilar, G. Ferino and S. Costanzi, presented at the 236th American Chemical Society National Meeting, Washington, DC, 2009.

23. V. Katritch, M. Rueda, P. Lam, M. Yeager and R. Abagyan, *Proteins*, 2010, **78**, 197–211.
24. S. Vilar, G. Ferino, S. S. Phatak, B. Berk, C. N. Cavasotto and S. Costanzi, *J. Mol. Graph. Model.*, 2011, **29**, 614–623.
25. T. Warne, M. Serrano-Vega, J. Baker, R. Moukhametzianov, P. Edwards, R. Henderson, A. Leslie, C. Tate and G. Schertler, *Nature*, 2008, **454**, 486–491.
26. S. Rasmussen, H. Choi, D. Rosenbaum, T. Kobilka, F. Thian, P. Edwards, M. Burghammer, V. Ratnala, R. Sanishvili, R. Fischetti, G. Schertler, W. Weis and B. Kobilka, *Nature*, 2007, **450**, 383–387.
27. M. Hanson, V. Cherezov, M. Griffith, C. Roth, V. Jaakola, E. Chien, J. Velasquez, P. Kuhn and R. Stevens, *Structure*, 2008, **16**, 897–905.
28. Glide, version 5.0, Schrodinger, LLC, New York, www.schrodinger.com.
29. S. Bhattacharya and N. Vaidehi, *J. Am. Chem. Soc.*, 2010, **132**, 5205–5214.
30. W. Sherman, T. Day, M. Jacobson, R. Friesner and R. Farid, *J. Med. Chem.*, 2006, **49**, 534–553.
31. G. Schertler, C. Villa and R. Henderson, *Nature*, 1993, **362**, 770–772.
32. K. Palczewski, T. Kumasaka, T. Hori, C. Behnke, H. Motoshima, B. Fox, I. Le Trong, D. Teller, T. Okada, R. Stenkamp, M. Yamamoto and M. Miyano, *Science*, 2000, **289**, 739–745.
33. K. Jacobson, Z. Gao and B. Liang, *Trends Pharmacol. Sci.*, 2007, **28**, 111–116.
34. S. Costanzi, *J. Med. Chem.*, 2008, **51**, 2907–2914.
35. M. Michino, E. Abola, G. 2008 Participants, C. r. Brooks, J. Dixon, J. Moult and R. Stevens, *Nat. Rev. Drug Discov.*, 2009.
36. Y. Okuno, A. Tamon, H. Yabuuchi, S. Niijima, Y. Minowa, K. Tonomura, R. Kunimoto and C. Feng, *Nucleic Acids Res.*, 2008, **36**, D907–912.
37. S. Engel, A. Skoumbourdis, J. Childress, S. Neumann, J. Deschamps, C. Thomas, A. Colson, S. Costanzi and M. Gershengorn, *J. Am. Chem. Soc.*, 2008, **130**, 5115–5123.
38. I. Tikhonova, C. Sum, S. Neumann, S. Engel, B. Raaka, S. Costanzi and M. Gershengorn, *J. Med. Chem.*, 2008, **51**, 625–633.
39. The Molecular Operating Environment (MOE), version 2008.10, Chemical Computing Group, Inc., Montreal, Canada, www.chemcomp.com.
40. Prime, version 2.0, Schrodinger, LLC, New York, USA, www.schrodinger.com.

CHAPTER 19

Probing the Activation Mechanism of Heptahelical Receptors: Experimental Validation of Molecular Dynamics Simulations

P. MUKHOPADHYAY, T. HUBER AND T.P. SAKMAR*

Laboratory of Molecular Biology & Biochemistry, Rockefeller University, 1230 York Avenue, New York, NY 10065, USA

19.1 Introduction

Heptahelical G protein-coupled receptors (GPCRs) located in the cell's plasma membrane are responsible for transmitting chemical signals across the lipid bilayer. GPCRs comprise a large family of related receptors that have evolved to bind a wide range of extracellular ligands from biogenic amines and neuromodulatory peptides to peptide hormones and proteins, and to lipids and fatty acids—to name but a few. GPCRs are also probably the most important single class of pharmaceutical drug targets in the human genome. According to Overington and co-workers, of the 266 human targets for approved drugs, a

RSC Drug Discovery Series No. 8
G Protein-Coupled Receptors: From Structure to Function
Edited by Jesús Giraldo and Jean-Philippe Pin

Published by the Royal Society of Chemistry, www.rsc.org

remarkable 27% correspond to rhodopsin-like or type A GPCRs.[1] Despite recent dramatic advances in the therapeutic targeting of kinases, the continued success of monoclonal antibody-based therapeutics and the advent of new drug entities like siRNAs, GPCRs remain the pre-eminent class of drug targets.

With perhaps 100 receptors still categorized as orphan (*i.e.* tissue-specific receptor expression can be documented but no endogenous ligand has been identified) out of a total of about 726 GPCR genes in the human genome, it is likely that GPCRs will continue to be robust drug targets in the foreseeable future. The discovery and exploitation of small-molecule allosteric GPCR ligands, so-called biased ligands and 'pepducins' also might allow targeting of 'refractory' GPCRs that resist conventional drug development strategies. Receptor oligomerization, or in some cases hetero-oligomerization, is thought to modulate receptor cell surface expression, ligand-binding affinity and downstream signalling specificity for at least some classes of GPCRs, and targeting receptor oligomers might also prove possible.

19.2 Results and Discussion

19.2.1 General Strategies

Recent reports of the crystal structures of native and engineered GPCRs such as rhodopsin, opsin, A2A adenosine, β2- and β1-adrenergic receptors (ARs) provide insights into the molecular mechanism of GPCR activation.[2,3] With the exception of the opsin crystal structures, the rhodopsin, β1AR, β2AR and A2A structures represent the inactive conformation, which are stabilized by inverse agonist or antagonist binding. In contrast, the structure of opsin (the apoprotein form of rhodopsin lacking its inverse agonist ligand 11-*cis*-retinal, which is the archetypical GPCR of visual phototransduction) bound to an 11-residue-long peptide representing a variant of the *C*-terminus of the alpha subunit of transducin (the cognate G protein of rhodopsin) may provide clues to a putative active-state receptor.

However, a wealth of pharmacological and biochemical data suggest that probably for most GPCRs, a fully-active state of the receptor might only exist when it is in a ternary complex with an agonist ligand and a heterotrimeric G protein.[4] The key question can be formulated as follows: 'How does the binding of an agonist ligand, generally originating on the outside surface of the cell membrane receptor, cause the release of bound guanosine diphosphate (GDP) on the alpha subunit of a heterotrimeric G protein located on the inside of the cell at a distance of perhaps 7 to 10 nm?

GPCRs can be thought of as allosteric signalling conduits that direct and relay information in a complex interplay between extracellular ligands and intracellular protein adapter and transducer molecules. In the absence of a high-resolution structure of at least a representative of such a ligand-receptor-G-protein ternary complex, the mechanism of receptor-mediated G-protein activation remains formally unresolved. However, a number of technologies are beginning to provide a new set of tools to address with chemical precision how ligands activate receptors and how receptors talk to cellular G proteins. The end result of interdisciplinary studies on the GPCR signalosome will create a systems biology description of a complex chemical signalling network.

We are interested in uncovering the principles that underlie ligand recognition in GPCRs and understanding with chemical precision how receptors change conformation in the membrane bilayer and activate G proteins when ligands bind. We also want to know how ligand binding specificity and receptor activation is modulated allosterically by other proteins, other receptors, small molecules and the lipid environment of the bilayer itself.

Initially, using vertebrate vision as a model system, we have developed an interdisciplinary approach that employs a number of new converging technologies:

(i) all atom and coarse grain molecular dynamics (MD) computer simulations of GPCRs in membrane bilayers in concert with experimental validation;[5–7]
(ii) unnatural amino acid mutagenesis of GPCRs using amber codon suppression technology;
(iii) advanced Fourier transform infrared spectroscopy (FTIR) and solid-state NMR methods to interrogate receptor dynamics;[8,9] and
(iv) nanoscale apolipoprotein bound bilayers (NABBs) as membrane mimic support structures for GPCRs.[10]

19.2.2 Computational Approaches

Comparative or homology protein structure modelling is now a well-established method for building three-dimensional (3-D) model of proteins of unknown structures (target) based on one or more proteins of known structure (template).[11] The quality of the model increases with the sequence similarity and availability of a correct alignment between template and target. Both of these conditions are generally satisfied in building a 3-D model of a protein complex using components with known crystal structures. Thus, comparative modelling techniques can be adapted to predict a 3-D structure of a protein complex using structures of the components that constitute the macromolecular assembly. For example, a putative functional receptor–G protein complex can be modelled using existing high-resolution structures of the constituent receptor and G protein.

In 2008, Scheerer *et al.* reported the crystal structure of the complex of opsin and a peptide variant of the *C*-terminus of alpha-subunit of transducin.[3] The bound peptide provided some clues as to how the entire G protein might interact with the receptor. Superposition of the transducin ($Gt\alpha\beta\gamma \cdot GDP$) and opsin–peptide crystal structures, however, resulted in a severe steric clash between $Gt\alpha\beta\gamma \cdot GDP$ and the lipid bilayer. The authors suggested that the opsin–peptide structure might represent the opsin–transducin complex in which

GDP has just left the binding pocket. Based on this assumption, they proposed a conceptual model of signal transmission from the active-state receptor to the G protein through a massive change in alpha5 helix packing in the alpha-subunit of transducin by a 40° tilt of the helix away from the remainder of alpha–beta–gamma subunits of transducin.

We used a slightly different approach to model the complex of an active-state opsin with the heterotrimeric G protein, transducin (Gtαβγ · GDP). We modelled the entire opsin–Gtαβγ · GDP complex by building it up from bits and pieces using available high-resolution structures. Using the NMR structure of a 11-residue peptide from the *C*-terminus of the alpha-subunit of transducin (GtαCT; PDB access code: 1AQG) and the structures of opsin in its G protein interacting conformation (PDB access code: 3DQB), rhodopsin in an inactive state (PDB access code: 1U19) and the heterotrimeric G protein, transducin (Gtαβγ · GDP; PDB access code: 1GOT), we built a plausible model and then performed an all-atom (~150 000) molecular dynamics (MD) computer simulation of the opsin–Gtαβγ · GDP complex in a palmitoyl oleoyl phosphatidylcholine (POPC) bilayer in a salt solution with physiological ionic strength to equilibrate the system. We also included all post-translational modifications of both the receptor and G protein.

Preliminary analysis shows that our new alternate model of the opsin–Gtαβγ · GDP complex does not require massive changes in alpha5 helix packing for receptor-mediated G protein activation. The opsin–GtαCT complex model also differs from the crystal structure of opsin–GtαCTK341L complex in other ways. For example, whereas the GtαCTK341L peptide in the receptor-binding pocket has no side chain–side chain interactions with the receptor and is oriented towards helix 5 and helix 6 in opsin, in our molecular model of the opsin–GtαCT complex, GtαCT orients towards the solvent exposed cleft between helix 6 and helix 8 in opsin.

Figure 19.1 shows a snapshot derived from the MD simulations of the opsin–Gtαβγ · GDP complex in the POPC bilayer. Opsin is oriented in the bilayer with helix 4 perpendicular to the plane of the bilayer. The palmitoyl modifications on Cys322 and Cy323, in helix 8 (H8) in opsin, anchor the receptor to the lipid bilayer with H8 positioned inside the lipid bilayer interface. In contrast, the *N*-terminus of alpha-subunit (GtαNT) and the *C*-terminus of gamma-subunits (GtγCT) of transducin are positioned outside the lipid bilayer interface with myristoyl and farnesyl groups on the alpha- and gamma-subunits of transducin inside the bilayer. GtαNT lies parallel to the plane of the bilayer interface with the positively charged surface of the amphipathic helix interacting with the negatively charged bilayer interface.

Our preliminary molecular model of the opsin–Gtαβγ · GDP complex provides structural insights into the role of the receptor's cytoplasmic loops in G protein recognition and activation. The model of the complex is also amenable to long timescale MD simulations that we hope will provide the structural basis for understanding receptor-mediated G protein activation.

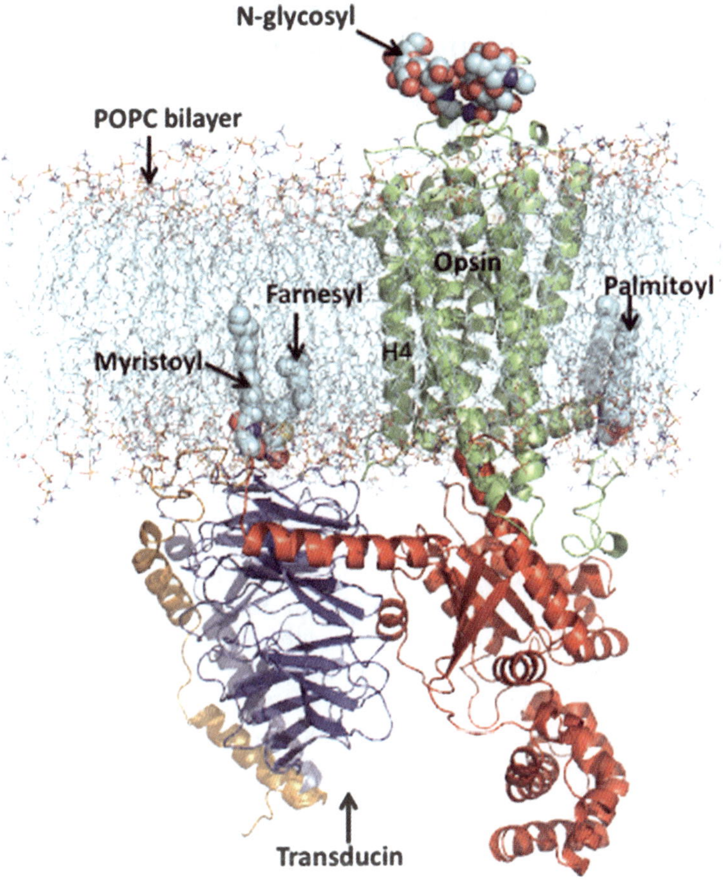

Figure 19.1 MD simulation model of the opsin–transducin complex in a POPC lipid bilayer. Opsin and transducin are drawn in ribbon format and post-translational modification on opsin and transducin is represented in space-fill style. Opsin is oriented in the bilayer such that helix 4 (H4) is perpendicular to the plane of the bilayer. The extracellular surface is oriented toward the top of the figure.

For example, the opsin–Gt$\alpha\beta\gamma \cdot$GDP complex model reveals several contact sites between the opsin and the alpha-subunit (Gtα) of transducin. These types of interactions, which are predicted by the model, can be used to design crosslinking experiments for biochemical validation using the site-directed unnatural amino acid mutagenesis technology developed in parallel. For example, opsin and Gtα can be co-expressed in mammalian cells in culture and various combinations of amber codon mutations can be introduced into one or the other construct. In preliminary photocrosslinking work as described below, we are interested in targeting the specific interactions between transducin and opsin H8 and to determine precisely how H8 modulates transducin activation.[12]

19.2.3 Probing Receptor Activation Using Genetically Encoded Non-Natural Amino Acids

Unnatural amino acid mutagenesis is based on the principle of amber codon suppression. If an amber codon (UAG) is suppressed, read-through occurs during mRNA translation. Protein synthesis will be terminated at the next incidental nonsense codon to create a fusion protein with the native *N*-terminal portion and a fusion at its *C*-terminal end that represents a translation of 3′-mRNA sequence between the original amber codon and the next functional nonsense codon. Amber codon suppression requires a so-called suppressor tRNA and a corresponding amino acyl-tRNA synthetase, which charges the tRNA with an activated amino acid.

Because the amber codon (UAG) is used rarely in most cells, engineering on-demand amber codon suppression is a reasonable strategy. If the amber codon were used at high frequency, then any strategy to suppress amber codon termination would result in large numbers of read-through fusion proteins that might prevent proper function and cause cell toxicity. When a suppressor tRNA/amino acyl-tRNA synthetase pair is orthogonal, they form an exclusive and specific pair and will always introduce a discrete amino acid at the amber codon. The efficiency of amber codon suppression depends upon a number of factors, but can be easily measured. Now suppose that the amino acyl-tRNA synthetase can be engineered to charge its orthogonal suppressor tRNA, not with one of the 20 naturally occurring amino acids, but with an unnatural amino acid with an interesting chemical property. If the engineered orthogonal suppressor tRNA/amino acyl tRNA synthetase pair was introduced in the presence of the proper unnatural amino acid, then that amino acid would be incorporated into the growing peptide chain during protein synthesis at an amber codon. Furthermore, if a particular gene of interest were mutated to contain an amber codon, then the specific unnatural amino acid corresponding to the orthogonal tRNA/synthetase pair would be introduced at the site of the mutation.

A particular unnatural amino acid might contain a reactive chemical group or a spectroscopic probe. Every full-length expressed protein should contain the unnatural amino acid. Amber codon suppression might then be used to label a heterologously expressed protein of interest at a specific location. For example, *p*-azido-L-phenylalanine is a particularly useful unnatural amino acid. The azido group is a useful chemical tag, its vibrational signature can be detected by infrared spectroscopy, and it can be used in photochemical cross-linking studies. Schultz and others have refined and exploited the amber codon suppression technology for use in *E. coli* protein expression systems.[13] However, GPCRs do not generally express to yield a functional receptor in *E. coli*. The main technical challenge to using the technology in mammalian cells is to find and express the proper suppressor tRNA.

We recently developed a highly efficient amber suppression system for use in mammalian cells in culture.[14–16] We used a luciferase reporter system to optimize successively the expression of an effective suppressor tRNA and then a high-fidelity orthogonal amino acyl-tRNA synthetase. The amino acyl-tRNA synthetase genes were engineered to recognize particular unnatural amino acids that could be added to the cell culture media. We created several specific orthogonal pairs that incorporate *p*-benzoyl-L-Phe, *p*-acetyl-L-Phe, or *p*-azido-L-Phe. The system was efficient enough to be able to incorporate these unnatural amino acids into heterologously expressed GPCRs in transiently transfected HEK cells.

The idea of site-directed unnatural amino acid mutagenesis is to introduce a reactive group or probe without loss of function. The ideal site in a GPCR is at a location that does not affect expression level, membrane targeting or function, but is close enough to some site of interest to be useful as a probe of function. We recently introduced *p*-azido-L-Phe into rhodopsin at various locations and expressed the protein in a mammalian cell system. We then carried out FTIR difference spectroscopy on purified mutant receptors. We showed that the vibrational signatures of the azido groups change after rhodopsin is activated by light and undergoes a series of progressive conformational changes. Since the vibrational frequency of the azido group is sensitive to the electrostatic environment, we could correlate changes in azido FTIR spectra with protein conformation. This method is particularly informative since the azido stretching frequency is far removed from protein vibrations, which can be recorded simultaneously.

FTIR spectroscopy may not be practical as a general method to study GPCRs. However, we have adapted a form of SEIRA (surface-enhanced infrared absorption) spectroscopy to study rhodopsin immobilized on a gold film surface, which raises the possibility that similarly tethered recombinant expressed GPCRs can be studied in the future.[17] Along these lines, it is currently possible to encode genetically azido-Phe into GPCRs other than rhodopsin. We have recently focused our attention on chemokine receptors, for example. We are using photo-crosslinking approaches to study receptor heterodimerization as outlined in Figure 19.2. In addition, unnatural amino acids can be tagged with fluorophores or other useful probes using emerging bioorthogonal chemical ligation strategies. Experiments along these lines are under way.

19.2.4 Reconstitution of Expressed Receptors in Membrane Nanoparticles

As part of our aim to study reconstituted receptors in membrane bilayer systems as much as possible, we have developed a nanoscale apolipoprotein-bound bilayer (NABB) system.[10] Based on naturally occurring high density lipoprotein particles, NABBs are essentially self-assembling membrane scaffold particles. NABBs can be prepared so that they contain one, or perhaps two, functional GPCRs in each particle. The NABBs are small enough that they remain in solution and create a long-term stable environment for GPCRs. One nice advantage of NABB particles is that both topological surfaces of the receptors are accessible simultaneously to the aqueous phase.

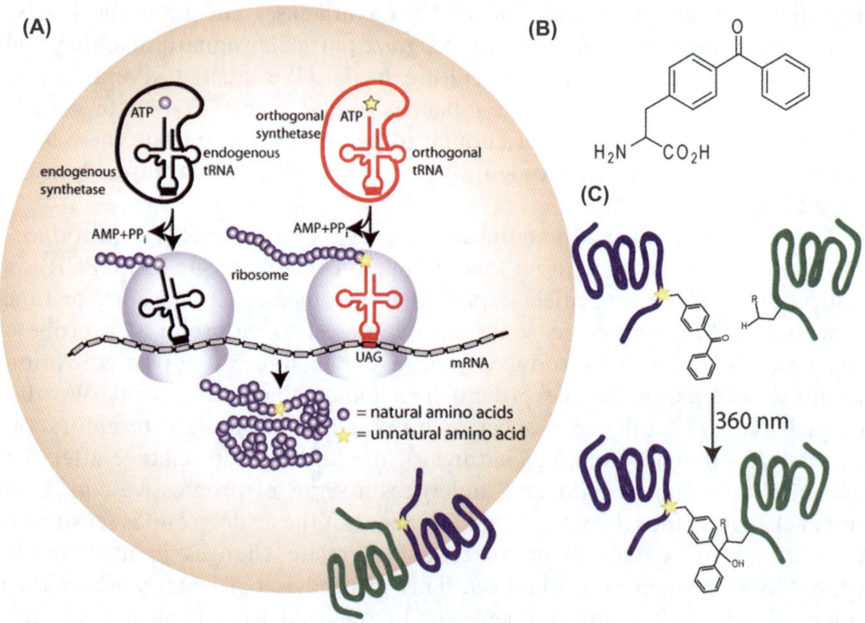

Figure 19.2 Scheme to encode genetically *p*-benzoyl-L-Phe into expressed GPCRs. (**A**) At the amber codon UAG, an engineered suppressor tRNA and orthogonal amino-acyl tRNA synthetase can incorporate an unnatural amino acid. (**B**) Structure of *p*-benzoyl-L-phenylalanine (*p*Bpa). (**C**) In a photo-crosslinking approach site-directed unnatural amino acid mutagenesis is used to genetically encode *p*Bpa into a particular GPCR (blue), which is hypothesized to interact with another GPCR (green). Upon irradiation with 360 nm light, *p*Bpa will form a covalent bond to a side chain of the green receptor.

19.3 Conclusions

GPCR signalling complexes are allosteric machines. Agonist receptor ligands outside of the cell induce guanine-nucleotide exchange on a heterotrimeric guanine-nucleotide binding regulatory protein (G protein) inside the cell where the ligand-binding site on the receptor and the nucleotide-binding site and the G protein are of the order of 7–10 nm or more apart. Additional non-canonical signalling pathways facilitate crosstalk between linear GPCR-signalling pathways and receptor tyrosine kinase (RTK) mediated signalling pathways. Receptor phosphorylation by receptor-specific kinases and the binding of various cellular adaptor proteins also regulate receptor desensitization, internalization, sequestration and recycling. Receptor olgomerization, or in some cases hetero-oligomerization, is thought to modulate receptor cell surface expression, ligand-binding affinity and downstream signalling specificity for at least some classes of GPCRs.

To understand how GPCRs really work will require the application of new technologies and ongoing interdisciplinary approaches. Biomolecular simulations are useful to predict and compare the stabilities of closely related structures in highly dynamic allosteric systems like GPCR signalling complexes. However, MD simulations require significant pre-existing structural information, which happens to be the case in particular for rhodopsin and transducin, but not for many other interesting receptor targets. Simultaneously, we need to develop new experimental methods to probe signalosome dynamics in native membrane environments, or at least in reasonable reconstituted model membrane systems. We expect that experimental validation of computational models will eventually lead to the identification of key conformational states of individual proteins and corresponding molecular assemblies in the G protein activation reaction pathway.

References

1. J. P. Overington, B. Al-Lazikani and A. L. Hopkins, *Nat. Rev. Drug Discov.*, 2006, **5**, 993.
2. D. M. Rosenbaum, S. G. Rasmussen and B. K. Kobilka, *Nature*, 2009, **459**, 356.
3. P. Scheerer, J. H. Park, P. W. Hildebrand, Y. J. Kim, N. Krauss, H. W. Choe, K. P. Hofmann and O. P. Ernst, *Nature*, 2008, **455**, 497.
4. D. Mustafi and K. Palczewski, *Mol. Pharmacol.*, 2009, **75**, 1.
5. X. Periole, T. Huber, S. J. Marrink and T. P. Sakmar, *J. Am. Chem. Soc.*, 2007, **129**, 10126.
6. T. Huber, S. Menon and T. P. Sakmar, *Biochemistry*, 2008, **47**, 11013.
7. A. V. Botelho, T. Huber, F. C. Peterson, T. P. Sakmar and M. F. Brown, *Biophys. J.*, 2006, **91**, 4464.
8. S. Ahuja, V. Hornak, E. C.-Y. Yan, N. Syrett, J. A. Goncalves, A. Hirshfeld, M. Ziliox, T. P. Sakmar, M. Sheves, P. J. Reeves, S. O. Smith and M. Eilers, *Nat. Struct. Mol. Biol.*, 2009, **16**, 168.
9. R. Vogel, M. Mahalingam, S. Lüdeke, T. Huber, F. Siebert and T. P. Sakmar, *J. Mol. Biol.*, 2008, **380**, 648.
10. S. Banerjee, T. Huber and T. P. Sakmar, *J. Mol. Biol.*, 2008, **377**, 1067.
11. N. Eswar, B. John, N. Mirkovic, A. Fiser, V. A. Ilyin, U. Pieper, A. C. Stuart, M. A. Marti-Renom, M. S. Madhusudhan, B. Yerkovich and A. Sali, *Nucleic Acids Res.*, 2003, **31**, 3375.
12. O. Ernst, K. P. Hofmann and T. P. Sakmar, *J. Biol. Chem.*, 2000, **275**, 1937.
13. L. Wang, A. Brock, B. Herberich and P. G. Schultz, *Science*, 2001, **292**, 498.
14. S. Ye, C. Köhrer, T. Huber, M. Kazmi, P. Sachdev, E. C. Yan, A. Bhagat, U. L. Rajbhandary and T. P. Sakmar, *J. Biol. Chem.*, 2008, **283**, 1525.
15. S. Ye, T. Huber, R. Vogel and T. P. Sakmar, *Nat. Chem. Biol.*, 2009, **5**, 397.
16. S. Ye, E. Zaitseva, G. Caltabiano, G. F. X. Schertler, T. P. Sakmar, X. Deupi and R. Vogel, *Nature*, 2010, **5**, 397.
17. E. Zaitseva, M. Saavedra, S. Banerjee, T. P. Sakmar and R. Vogel, *Biophys. J.*, 2010, **99**, 2327.

CHAPTER 20

Probing the Conformational Dynamics of GPCRs with Molecular Dynamics Simulation

RON O. DROR,*[a] ALBERT C. PAN,[a] DANIEL H. ARLOW[a] AND DAVID E. SHAW*[ab]

[a] D. E. Shaw Research, New York, NY 10036, USA; [b] Center for Computational Biology and Bioinformatics, Columbia University, New York, NY 10032, USA

20.1 Introduction

The function of a G protein–coupled receptor (GPCR) is intimately linked to its dynamics. In order to transmit a signal from an extracellular ligand to an intracellular G protein, a GPCR must undergo a conformational change. Moreover, an expanding body of pharmacological evidence indicates that a given GPCR can signal through multiple distinct intracellular pathways and that different ligands can selectively regulate different pathways. This implies that a GPCR must be capable of adopting multiple conformations with distinct signalling profiles, as opposed to just one active and one inactive conformation. In addition, because GPCRs possess intrinsic flexibility, each ligand induces not a single, static receptor conformation but rather an ensemble of interconverting conformations.

Despite recent breakthroughs in GPCR crystallography, the identities of most of these conformations remain a mystery, as do the mechanisms by which they lead to distinct signalling profiles and by which ligands select among them. Crystallography typically identifies only the lowest energy conformations,

RSC Drug Discovery Series No. 8
G Protein-Coupled Receptors: From Structure to Function
Edited by Jesús Giraldo and Jean-Philippe Pin

Published by the Royal Society of Chemistry, www.rsc.org

as opposed to the many conformations that appear to play a role in signalling pathways, and crystal structures provide little information on the dynamics of each conformation or on the pathways and intermediates connecting them. Experimental techniques such as fluorescence quenching and NMR have provided evidence for distinct receptor conformations at equilibrium and for differences in conformation and dynamics in the presence of different ligands,[1,2] but these techniques do not provide a direct high-resolution picture of the structures and motions involved.

Molecular dynamics (MD) simulation offers a potential solution to this problem. Using a physics-based approach, such simulations can capture the motion of every atom in a simulated system, including not only a receptor but also a lipid bilayer, the surrounding water molecules, and perhaps ligands and other proteins. Such simulations may be used to predict conformational changes, elucidate the effects of mutations or of a receptor's environment on its behaviour, and capture the interactions of GPCRs with ligands and intracellular signalling proteins. They provide very high resolution in both space (sub-angstrom) and time (a few femtoseconds).

Two primary challenges have historically limited the applicability of MD simulations, particularly to GPCRs. First, simulations on timescales of many microseconds or longer, on which major conformational changes of GPCRs typically take place, have been beyond the reach of traditional computers. Second, an accurate initial protein structure is generally required to generate reliable simulation results and determination of high-resolution GPCR structures has historically proven difficult.

Recent developments—particularly dramatic increases in achievable simulation length and an explosion in available GPCR crystal structures—are now beginning to remove these obstacles, promising to make MD a much more powerful tool for understanding the conformational dynamics of GPCRs. In this chapter, we summarize MD simulation methodology, discuss the major factors determining its applicability, provide a brief history of the use of MD to study GPCRs, and discuss future prospects for MD in characterizing GPCR conformational dynamics.

20.2 Molecular Dynamics Simulation: Training a Computational Microscope on GPCR Function

An atomistic MD simulation typically comprises thousands to millions of individual atoms; in the case of a GPCR, the simulation usually represents not only the receptor but also the lipid bilayer and water that surround it (Figure 20.1). The simulation progresses in a series of short, discrete time steps. At each time step, the force acting on each atom due to all the other atoms is computed, and the position and velocity of each atom are then updated according to Newton's laws of motion. Forces are evaluated according to a model known as a molecular mechanics force field (or simply force field), which is fitted to a combination of experimental data and quantum mechanical measurements,

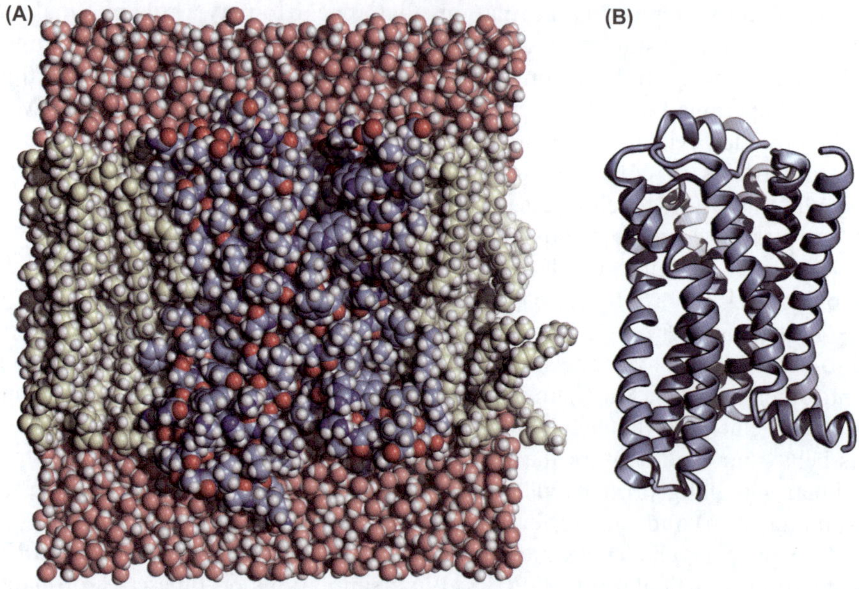

Figure 20.1 Two renderings of a snapshot from an all-atom MD simulation of a β_2-adrenergic receptor in a hydrated phospholipid bilayer. (**A**) Hydrogen atoms are in white, phospholipid non-hydrogen atoms are in yellow, water oxygen atoms are in pink, protein oxygen atoms are in red, protein carbon are in light blue, and nitrogen atoms are in dark blue. Phospholipids and water molecules located in front of the protein are not shown. (**B**) The protein is shown in a cartoon representation, while the phospholipids and water molecules are not shown. Molecular rendering was performed using OpenStructure (ref. 4).

typically under the assumption that covalent bonds are not broken or formed. A sufficiently long and accurate MD simulation can serve as a 'computational microscope', allowing the observation of biomolecular processes at otherwise inaccessible spatial and temporal scales.[3]

Historically, MD simulations of GPCRs have faced two principal challenges: the availability of high-resolution structures from which to initiate a simulation and the timescales accessible to simulation. The last several years have seen dramatic advances on both fronts.

First, the current explosion in GPCR crystallography provides a set of starting points for simulation. The utility of an MD simulation is often dictated by the quality of the initial structure. While a simulation may be initiated from a rough structural model, it is difficult to determine whether the resulting behaviour accurately characterizes the dynamics of the real protein. This concern is exacerbated by the fact that accurate computational prediction of GPCR structure has proven highly challenging.[5,6]

Indeed, the appearance in 2000 of the three-dimensional crystal structure of bovine rhodopsin[7]—the first high-resolution structure of a GPCR—engendered the first major wave of GPCR MD simulations (see Section 20.3).

Most of these focused on rhodopsin itself, with some using models of other GPCRs based on homology to rhodopsin. Bovine rhodopsin remained the only GPCR for which crystal structures were available until 2007, when the structure of the β_2-adrenergic receptor appeared,[8,9] followed in 2008 by the structures of the β_1-adrenergic receptor,[10] the A_{2A} adenosine receptor,[11] squid rhodopsin[12] and opsin.[13,14] These structures reflect the development of a number of new crystallographic techniques, which are expected to yield an increasing number of GPCR structures in the next few years, paving the way for broader application of MD simulation.

Second, recently developed computer hardware and software accelerate MD simulations by orders of magnitude, allowing them to capture events that take place over substantially longer periods of time. In experimental work, phenomena that occur very quickly are typically harder to characterize than those that occur over long timescales. In MD simulations, the opposite is true—the longer the physical time to be simulated, the more time is required to execute the simulation. The highest atomic vibrational frequencies limit each time step to a few femtoseconds, such that simulating even a microsecond requires nearly a billion sequential time steps, with evaluation of the forces on every atom required at each step. Because the force on each atom depends on the positions of every other atom (through long-range electrostatic effects), parallelizing the simulation by splitting it across multiple processors requires substantial inter-processor communication, limiting the speedup achievable through parallelization. Until recently, atomistic simulations of most biomolecular systems on timescales above a microsecond were not feasible, and even simulation timescales above one hundred nanoseconds were rare. On the other hand, many of the events most critical to the function of a GPCR—from binding of a diffusible ligand to activation—take place on timescales of microseconds or longer[15] (Figure 20.2).

Over the past several years, advances in both parallel algorithms and computer hardware have enabled substantially longer simulations. We and others have designed novel algorithms to address the communication bottleneck in highly parallel MD simulation; in some cases, these algorithms asymptotically reduce the communication burden compared with traditional methods.[22] By exploiting these algorithms, together with improved numerical techniques and software infrastructure, our Desmond software achieved nearly an order of magnitude speedup at high levels of parallelism over any previously available MD code on the same hardware.[23] The parallel scalability of several other MD codes including NAMD,[24] GROMACS[25] and AMBER[26] has also seen substantial improvement over the past several years. Current trends in computer architecture, including the design of highly parallel architectures such as IBM's Blue Gene machines,[27] have led to additional speedups. Recently, we completed a special-purpose parallel machine for MD simulations, called Anton, which enabled the first millisecond-scale all-atom MD simulation of a protein—more than 100 times longer than any reported previously.[16,28]

Longer-timescale simulations such as those enabled by Anton are just now being brought to bear on GPCRs, and the biggest payoffs remain to be seen, but MD simulations of GPCRs have already advanced significantly

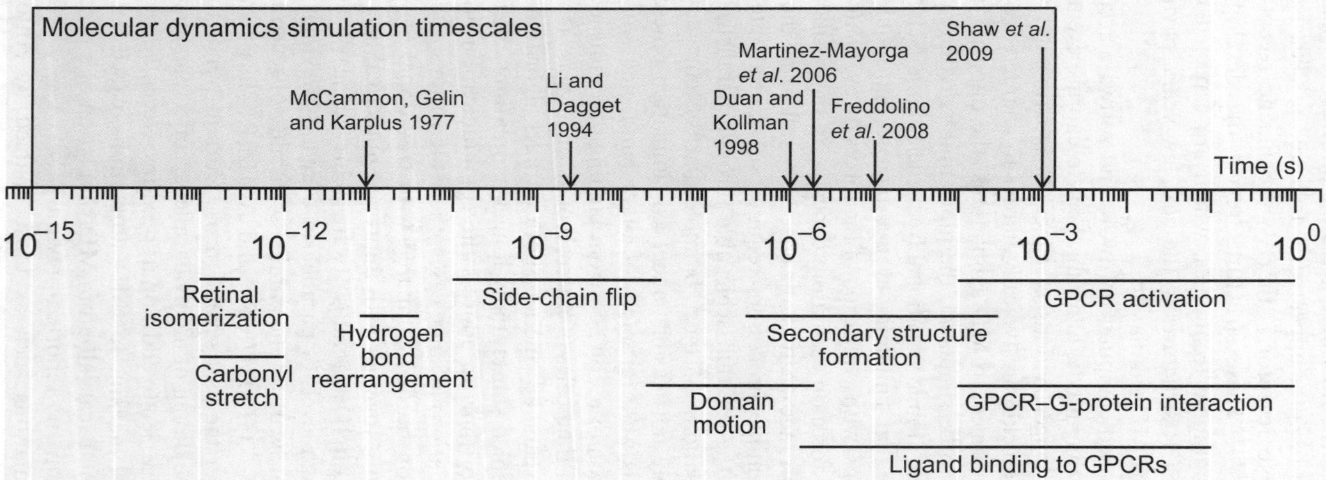

Figure 20.2 A comparison of timescales reachable by all-atom molecular dynamics simulation (grey bar above the time axis) with those on which various molecular processes relevant to GPCR function take place (bottom). The arrows and corresponding references (refs. 16–21) inside the grey bar indicate the longest published all-atom MD simulations of a protein in explicitly represented water at various points over the past several decades. The solid lines below the time axis indicate the range of timescales on which various events occur. Longer-timescale simulations, such as those enabled by Anton (ref. 16), are now being brought to bear on GPCRs.

in the past few years. In the following sections, we discuss several examples illustrating successful applications of MD to GPCRs to date, and then speculate on what the future has in store.

20.3 A Brief History of Molecular Dynamics Simulation of GPCRs

Computational modelling has been used for decades to explain the conformations and dynamics of molecular constituents of GPCRs such as retinal, the covalently bound chromophore of rhodopsin.[29–31] The first MD simulations of whole GPCRs, however, followed the appearance of the high-resolution crystal structure of rhodopsin in 2000,[7] with the goal of explaining how conformational changes in retinal coupled to larger-scale conformational changes in the receptor. The earliest MD simulations of rhodopsin, limited by accessible timescales, focused on the initial stages of activation after retinal isomerization[32–35] and the characterization of inactive rhodopsin in a realistic membrane environment.[36,37] Saam *et al.*,[33] for example, performed an atomistic MD simulation of rhodopsin embedded in a palmitoyl oleoyl phosphatidyl choline (POPC) lipid bilayer. To study the initial steps of activation, the 11-*cis* retinal was 'photoisomerized' to an all-*trans* conformation by transiently adding a strong biasing potential to the C11=C12 dihedral for 200 fs. The relaxation of the system in the subsequent 10 ns, after the bias was removed, involved a loosening of the contact between the β-ionone ring of the retinal and the conserved residue Trp$^{6.48}$. [Superscripts refer to Ballesteros–Weinstein residue numbering,[38] in which x.50 denotes the most-conserved residue in helix x across the class A (rhodopsin family) GPCRs and other residues in that helix are numbered relative to this most-conserved residue.] These studies, which were among the first atomistic simulations of membrane proteins, set the stage for MD simulations of GPCRs over the next decade.

More recent simulations of rhodopsin have reached timescales up to a few microseconds. These simulations investigated the interactions of lipids and cholesterol molecules with rhodopsin,[39,40] the coupling of retinal dynamics to changes in protein structure during the early stages of activation,[17,34,41–43] the hydration of the protein core after retinal photoisomerization[44] and the implications of receptor dimerization and oligomerization.[45–47] In order to establish the molecular basis for observed effects of lipid composition on receptor function, for example, Grossfield *et al.*[40] performed 26 independent 100-ns simulations of dark-adapted rhodopsin in a realistic membrane environment including cholesterol and lipids containing both saturated fatty acids and docosahexaenoic acid (DHA), an ω-3 polyunsaturated fatty acid. They discovered that, unlike cholesterol and saturated lipid chains, DHA associates frequently with rhodopsin at several specific sites, weakening the receptor's interhelical packing in the vicinity of those sites. These results offer an explanation for the observation that ω-3 polyunsaturated fatty acids such as DHA destabilize rhodopsin and accelerate its activation and deactivation processes.

Other studies have used MD simulation to interpret and integrate experimental results. Smith and co-workers[48,49] modelled the active state of rhodopsin using NMR-derived distance restraints to bias an MD simulation. A set of 64 restraints from ^{13}C dipolar recoupling measurements between atoms in the retinal binding pocket of activated rhodopsin was used to guide molecular dynamics simulations of rhodopsin from the inactive state into a modelled active conformation.

The absence of non-rhodopsin GPCR structures before late 2007 motivated the use of computational models based in part on the rhodopsin template, especially of drug targets such as the β_2-adrenergic receptor (β_2AR),[50,51] the D-prostanoid receptor,[52] the cholecystokinin-2 receptor,[53] and the cannabinoid receptor.[54] Javitch and co-workers,[51] for example, used simulations of such a model, together with experiments, to argue for the involvement of a 'rotamer toggle switch'[55] in β_2AR activation. Earlier mutagenesis experiments had revealed that the cluster of aromatic residues in helix 6 surrounding the highly conserved Pro$^{6.50}$ played a role in ligand binding and activation in the biogenic amine receptors,[56] and that the neighbouring Cys$^{6.47}$ became accessible to thiol-labelling reagents in constitutively active mutants.[57] Javitch and co-workers[51] used Monte Carlo simulations (closely related to MD simulations) of a homology model of helix 6 in β_2AR to provide a mechanistic interpretation of the increase in reactivity of Cys$^{6.47}$ in the active conformation. Based on this interpretation, they went on to propose an activation mechanism whereby agonist binding causes a rotameric transition in Phe$^{6.52}$ and Trp$^{6.48}$ (the 'rotamer toggle switch'), which permits Cys$^{6.47}$ to rotate into the binding pocket and causes helix 6 to straighten around the Pro$^{6.50}$ kink, ultimately leading to an outward movement of its intracellular end.

Beginning in 2007, a series of crystallographic breakthroughs have produced structures of several additional GPCRs, including β_2AR with various ligands bound,[8,9,58–60] the β_1-adrenergic receptor (β_1AR),[10] the A_{2A} adenosine receptor (A_{2A}AR),[11] squid rhodopsin,[12] and opsin with and without a G protein fragment bound.[13,14] Simulation studies have begun to exploit these new structures to make a variety of contributions, including proposing a role for cholesterol in stabilizing A_{2A}AR,[61] elucidating ligand binding in β_2AR,[62,63] describing the dynamics of the internal water molecules in squid rhodopsin[64] and probing dimerization mechanisms[65] and sodium ion dynamics[66] using homology models based on the β_2AR structure.

Several of these recent studies have focused on elucidating the conformations of the conserved ionic lock region of the adrenergic and adenosine receptors[67–71] (Figure 20.3). In dark-adapted rhodopsin, the ionic lock involves a salt bridge between Glu$^{6.30}$ on helix 6 and Arg$^{3.50}$ on helix 3, which appears to stabilize the inactive state, breaking upon activation when helices 3 and 6 move apart. Despite their general similarity to the structure of dark-adapted rhodopsin, the recently determined antagonist-bound adrenergic and adenosine receptor crystal structures all exhibited a broken ionic lock, a puzzling feature that raised questions about their signalling mechanism. Biochemical data had previously suggested that a salt bridge between the corresponding residues of many other GPCRs—and β_2AR in particular—also helped stabilize the inactive state. Although the adrenergic and adenosine receptors were crystallized in a presumably inactive state, the ionic lock was disrupted in the crystal structures, provoking uncertainty about the true conformation(s) of the inactive state as well as the role of the ionic lock in receptor activation and signaling.[8,72]

To address these questions, we performed MD simulations of β_2AR in multiple wild-type and mutant forms[67] (Figure 20.3). In simulations of wild-type β_2AR, the ionic lock formed reproducibly within a few hundred nanoseconds, with the salt bridge network matching that suggested by the earlier biochemical studies.[74] In microsecond-timescale simulations, we observed that the lock remained formed most of the time but occasionally broke for tens or hundreds of nanoseconds. Mutations associated with increased activity of the unliganded receptor increased the fraction of time the lock was broken in our simulations. To obtain the high-resolution β_2AR structure, a flexible loop of the receptor had been replaced by a small, rigid protein, T4 lysozyme (T4L).[9] We hypothesized that this replacement altered the ionic lock conformational equilibrium. Indeed, when we simulated the crystallographic β_2AR–T4L fusion protein, the lock-broken conformation predominated. Our results suggest that inactive β_2AR exists in a context-sensitive equilibrium and that the lock-formed conformation may predominate in the inactive wild-type receptor, despite the broken lock in the crystal structures. Subsequent studies by other groups demonstrated that the ionic lock forms in simulations of wild-type β_1AR and A_{2A}AR as well.[61,68,71]

We also observed the formation of several secondary structural elements in both intracellular and extracellular loops in our simulations of inactive β_2AR. In certain simulations, for example, intracellular loop 2, which exhibited a random coil conformation in the β_2AR crystal structures, formed a stable α-helix.[67] Intriguingly, this helix was essentially identical to that observed in the crystal structure of inactive β_1AR by Warne *et al.*,[10] who noted that such a helix could not be accommodated in existing β_2AR structures due to crystal lattice contacts. Our simulations suggest that β_2AR may well form such a helix in its native environment.

In more recent work[75] using Anton, we were able to capture the transition of a GPCR from an active conformation to an inactive conformation, addressing a puzzle posed by two newly solved agonist-bound structures of β_2AR. One of these structures, which has a G protein–mimetic nanobody bound to its intracellular surface, appears to represent an active conformation.[76] The other, which lacks a cytoplasmic binding partner but has a covalently bound agonist, is almost identical to the antagonist-bound structure and thus appears to represent an inactive conformation.[75] Although the crystallographic construct differs in several regards from a wild-type protein bound to a diffusible agonist, this latter structure raises the somewhat counterintuitive possibility that the lowest energy conformational state of an agonist-bound β_2AR in the absence of a G protein or alternative cytoplasmic binding partner may in fact have an inactive-like conformation.

To test this hypothesis, we performed a 30-microsecond simulation of β_2AR starting from the nanobody-bound active structure, leaving the agonist bound but removing the nanobody. In this simulation, the receptor spontaneously transitioned to a conformation that closely matched the antagonist-bound inactive crystal structure and did not revert to the original active conformation. An additional simulation with certain receptor residues in different protonation states yielded similar results. These simulations suggest that, although agonist binding makes the active receptor conformation less energetically unfavourable

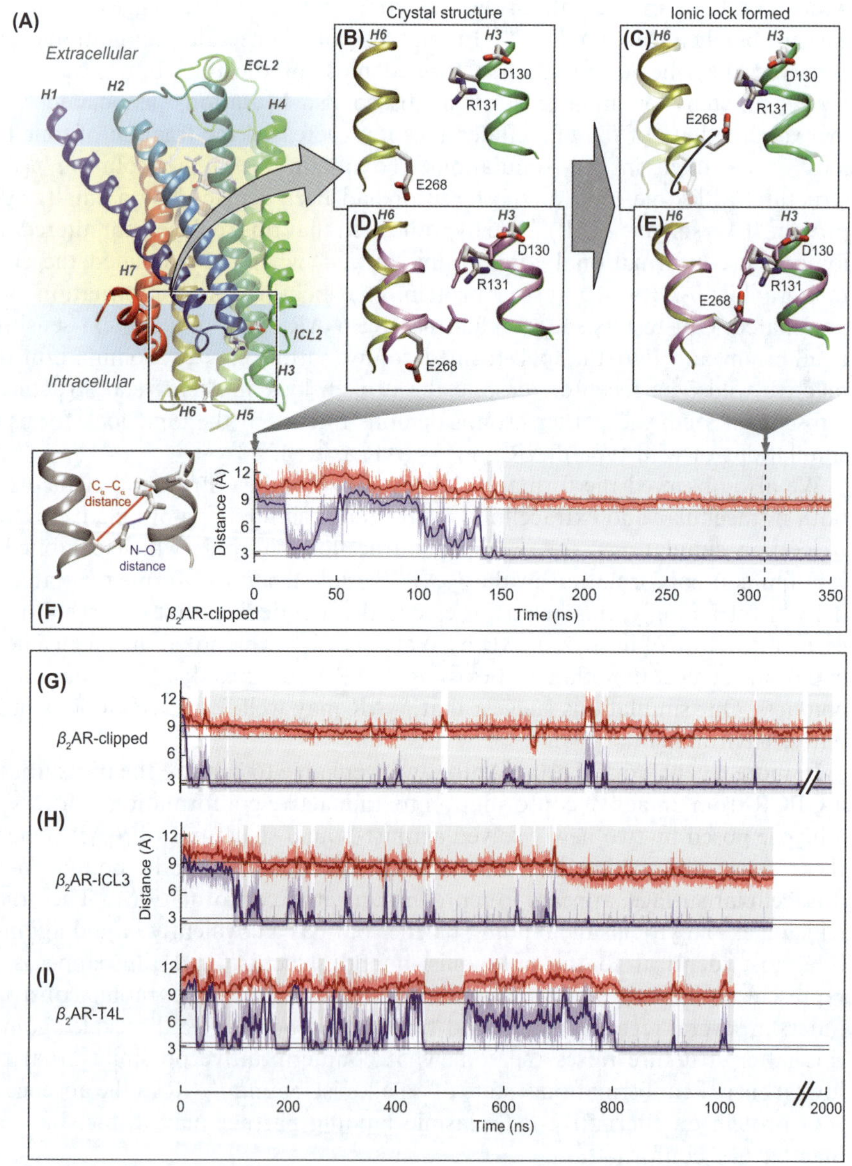

and thus increases its population, the majority of the receptor population remains in an inactive-like conformation in the absence of a cytoplasmic binding partner such as a G protein or nanobody. Notably, these also represent the first reported simulations in which a GPCR has transitioned spontaneously between two crystallographically observed conformations representing substantially different signalling states.

20.4 Future Prospects

As longer timescales become accessible and more crystal structures become available, MD simulations promise to provide an ever more effective microscope onto GPCR conformational dynamics. In addition to describing the conformational ensemble of a receptor in a given state, such simulations may be able to capture the process of ligand binding, the distinct receptor conformations stabilized by different ligands and the mechanisms by which receptors interconvert between these conformations. Moreover, MD may elucidate the interactions of GPCRs with one another and with intracellular signalling proteins, including G proteins and arrestins.

Understanding the functional dynamics of GPCRs is important not only from a scientific perspective, but also pharmaceutically. Differences in drug selectivity among receptors, for example, can be difficult to explain based on static structures alone, as illustrated by the striking similarity of the binding pockets of the β_1- and β_2-adrenergic receptor crystal structures.[10] By taking dynamics into account, simulations may be able to explain such differences,

Figure 20.3 Simulations of β_2AR show an equilibrium between conformations with the ionic lock open and closed (ref. 67). (**A**) Cartoon rendering of β_2AR crystal structure (ref. 9; PDB ID 2RH1). (**B**) Close-up of the intracellular ends of helices 3 and 6 in the crystal structure showing the broken ionic lock, and (**C**) a representative conformation with the ionic lock formed from a simulation of carazolol-bound β_2AR with the T4 lysozyme (T4L) removed. (**D**,**E**) Same conformations as shown in (**B**) and (**C**) with the corresponding residues of inactive rhodopsin (ref. 73; 1GZM; purple) superimposed. (**F**) Time series of distances between the C_α atoms of Arg131$^{3.50}$ and Glu268$^{6.30}$ (C_α–C_α distance; light red) and between the closest nitrogen and oxygen atoms of their corresponding side chains (N–O distance; light blue) for the same simulation, with smoothed versions in dark red and blue; grey shading indicates when the ionic lock is closed. The upper pair of grey horizontal lines indicates C_α–C_α distances of two inactive rhodopsin structures, and the lower pair indicates the corresponding N–O distances. (**G–I**) Corresponding time series for several additional simulations of carazolol-bound β_2AR, including (**G**) a longer simulation with T4L removed (ionic lock formed 90% of time), (**H**) a simulation in which T4L has been replaced by a model of intracellular loop 3 (ionic lock formed 92% of time), and (**I**) a simulation of the crystallized β_2AR-T4L construct (ionic lock formed 41% of time). Based on figures published previously by Dror *et al.* (ref. 67).

which underlie the function of widely used drugs, including β_1-selective beta-blockers. In addition, MD may facilitate the identification of allosteric binding sites and the design of drugs that bind to those sites; such drugs may offer improved target selectivity, due to the increased divergence between receptors outside the orthosteric binding site.[77]

MD may also enable the design of drugs that up-regulate desirable signalling pathways while down-regulating undesirable ones controlled through the same receptor. β-adrenergic receptors, for example, signal along both G protein mediated pathways and β-arrestin mediated pathways. Beta-blockers, which are widely used to treat heart failure, inhibit G protein mediated signalling. Most also inhibit β-arrestin mediated signalling, but some have recently been found to stimulate β-arrestin mediated signalling, which appears to have a cardioprotective effect. Rationally designing such 'functionally selective' ligands[78]—or even determining to what extent such fine-grained control is possible—requires understanding which conformations a receptor can adopt, how ligands stabilize these conformations and how they affect coupling to cytoplasmic binding partners. MD provides a promising route to tackle such problems.

Additionally, MD may be used as part of the structure prediction process, along with crystallography and NMR. *Ab initio* folding of proteins from an extended conformation by MD has recently become possible, but so far only for small globular proteins.[28] MD has been used to determine the structural effects of receptor modifications necessary for crystallization, and to synthesize NMR data into structural predictions (see Section 20.3). In the future, it may also prove useful for refining GPCR homology models.[79]

When applying MD simulations to any of these problems, however, a number of caveats should be kept in mind. First, although the most recent generation of biomolecular force fields has been shown to reproduce a variety of experimental data,[80–82] a number of shortcomings remain, and more will likely be revealed through longer-timescale simulations. Force field parameters must be selected with care; parameters validated on specific molecules (*e.g.* proteins or lipids) are typically much more reliable than parameters generated automatically. Force field errors tend to impact certain results more than others. The molecular conformations predicted by MD simulations are typically more robust than the relative populations of those different conformations or the rates of interconversion between them, because populations are sensitive to minor errors in relative energies and rates depend strongly on the heights of energy barriers.

Secondly, classical MD simulations treat covalent bonds as unchanging. In particular, protonation states of titratable residues remain fixed during a simulation. Care must be taken to choose these protonation states appropriately, as incorrect protonation can lead to a variety of artefacts, including the loss of atomic interactions that stabilize a certain protein conformation or a substantial change in hydration of the protein interior. Experimental techniques such as FTIR can sometimes determine protonation states, but in many cases, they remain uncertain, sometimes requiring multiple simulations with appropriate residues alternately protonated and deprotonated.[32,67,75] Methods have been developed to allow protonation states to evolve dynamically, generally at the cost of a substantial increase in computational requirements.[83] More general methods are available to simulate chemical reactions in which covalent bonds are formed or broken; in quantum mechanics/molecular mechanics (QM/MM) simulations, for example, the bulk of the system is simulated as in classical MD, but a small part is evaluated using more computationally intensive quantum mechanical approaches.[84]

Thirdly, even when a simulation captures an event of interest, limitations on the length and number of simulations performed can lead to errors. One must ask whether results are robust to minor changes in the initial conditions, whether observations represent initial transients or steady-state behaviour, and whether the simulations have reached all the relevant conformational states. Repeated simulations—typically starting from the same structure, but with different initial velocities—help ensure robustness, but do not always substitute for longer simulations, as repeated simulations may experience the same initial transients. Our recent experiences with long simulations have indicated that, in some cases, transient behaviour can last for microseconds. When possible, a simulation long enough to capture repeated transitions between states is desirable,[67] as this provides evidence for an equilibrium on the simulated timescale.

A number of researchers have been developing enhanced sampling methods to allow MD to capture conformational changes on timescales exceeding those of individual atomistic simulations. Some of these methods have been devised to guide a simulation over energy barriers by applying biasing forces. Others employ coarse-grained simulations, in which each simulated particle represents multiple atoms, to substantially reduce computational requirements. Such methods have been used to study activation, allostery and aggregation in GPCRs.[65,66,85,86] While these approaches are promising, especially when combined with the ability to perform longer simulations, they must be used carefully as they can sometimes introduce unrealistic behaviour.

Despite these caveats, molecular dynamics simulations, when used properly, allow investigation of dynamic phenomena at spatial and temporal scales that are difficult to access through experimental techniques.[3] As the connections between GPCR pharmacology and structure have begun to emerge, the importance of receptor flexibility and dynamics in explaining and exploiting these relationships has become increasingly clear.[6] MD has become an effective method to probe the conformational dynamics of GPCRs and the molecules with which they interact, thanks to recent advances in both simulation technology and crystallography. Coupled with experimental investigation, simulations may lead not only to a better picture of the mechanisms of GPCR function, but also to the discovery of new ways to regulate these mechanisms pharmaceutically.

Acknowledgements

We thank Ansgar Philippsen for performing molecular renderings, David Borhani and Thomas Mildorf for helpful comments, and Rebecca Kastleman and Amy Perry for editorial assistance.

References

1. B. K. Kobilka and X. Deupi, *Trends Pharmacol. Sci.*, 2007, **28**(8), 397.
2. M. P. Bokoch, Y. Zou, S. G. F. Rasmussen, C. W. Liu, R. Nygaard, D. M. Rosenbaum, J. J. Fung, H.-J. Choi, F. S. Thian, T. S. Kobilka, J. D. Puglisi, W. I. Weis, L. Pardo, R. S. Prosser, L. Mueller and B. K. Kobilka, *Nature*, 2010, **463**(7277), 108.
3. R. O. Dror, M. Ø. Jensen, D. W. Borhani and D. E. Shaw, *J. Gen. Physiol.*, 2010, **135**(6), 555.
4. M. Biasini, V. Mariani, J. Haas, S. Scheuber, A. D. Schenk, T. Schwede and A. Philippsen, *Bioinformatics*, 2010, **26**(20), 2626.
5. M. Michino, E. Abola, C. L. Brooks III, J. S. Dixon, J. Moult and R. C. Stevens, *Nat. Rev. Drug Discovery*, 2009, **8**, 455.
6. D. M. Rosenbaum, S. G. F. Rasmussen and B. K. Kobilka, *Nature*, 2009, **459**(7245), 356.
7. K. Palczewski, T. Kumasaka, T. Hori, C. A. Behnke, H. Motoshima, B. A. Fox, I. Le Trong, D. C. Teller, T. Okada, R. E. Stenkamp, M. Yamamoto and M. Miyano, *Science*, 2000, **289**(5480), 739.
8. S. G. F. Rasmussen, H.-J. Choi, D. M. Rosenbaum, T. S. Kobilka, F. S. Thian, P. C. Edwards, M. Burghammer, V. R. P. Ratnala, R. Sanishvili, R. F. Fischetti, G. F. X. Schertler, W. I. Weis and B. K. Kobilka, *Nature*, 2007, **450**(7168), 383.
9. V. Cherezov, D. M. Rosenbaum, M. A. Hanson, S. G. F. Rasmussen, F. S. Thian, T. S. Kobilka, H. J. Choi, P. Kuhn, W. I. Weis, B. K. Kobilka and R. C. Stevens, *Science*, 2007, **318**(5854), 1258.
10. T. Warne, M. J. Serrano-Vega, J. G. Baker, R. Moukhametzianov, P. C. Edwards, R. Henderson, A. G. W. Leslie, C. G. Tate and G. F. X. Schertler, *Nature*, 2008, **454**(7203), 486.
11. V.-P. Jaakola, M. T. Griffith, M. A. Hanson, V. Cherezov, E. Y. T. Chien, J. R. Lane, A. P. Ijzerman and R. C. Stevens, *Science*, 2008, **322**(5905), 1211.
12. M. Murakami and T. Kouyama, *Nature*, 2008, **453**(7193), 363.
13. J. H. Park, P. Scheerer, K. P. Hofmann, H.-W. Choe and O. P. Ernst, *Nature*, 2008, **454**(7201), 183.
14. P. Scheerer, J. H. Park, P. W. Hildebrand, Y. J. Kim, N. Krauß, H.-W. Choe, K. P. Hofmann and O. P. Ernst, *Nature*, 2008, **455**(7212), 497.
15. M. J. Lohse, P. Hein, C. Hoffmann, V. O. Nikolaev, J.-P. Vilardaga and M. Bünemann, *Br. J. Pharmacol.*, 2008, **153**(S1), S125.
16. D. E. Shaw, R. O. Dror, J. K. Salmon, J. P. Grossman, K. M. Mackenzie, J. A. Bank, C. Young, M. M. Deneroff, B. Batson, K. J. Bowers, E. Chow, M. P. Eastwood, D. J. Ierardi, J. L. Klepeis, J. S. Kuskin, R. H. Larson, K. Lindorff-Larsen, P. Maragakis, M. A. Moraes, S. Piana, Y. Shan and B. Towles, in *Proc. Conf. High Performance Computing, Networking, Storage and Analysis (SC09)*, ACM Press, New York, 2009.
17. K. Martinez-Mayorga, M. C. Pitman, A. Grossfield, S. E. Feller and M. F. Brown, *J. Am. Chem. Soc.*, 2006, **128**(51), 16502.

18. J. A. McCammon, B. R. Gelin and M. Karplus, *Nature*, 1977, **267**(5612), 585.
19. A. Li and V. Daggett, *Proc. Natl. Acad. Sci. U.S.A.*, 1994, **91**(22), 10430.
20. Y. Duan and P. A. Kollman, *Science*, 1998, **282**(5389), 740.
21. P. L. Freddolino, F. Liu, M. Gruebele and K. Schulten, *Biophys. J.*, 2008, **94**(10), L75.
22. K. J. Bowers, R. O. Dror and D. E. Shaw, *J. Comput. Phys.*, 2007, **221**(1), 303.
23. K. J. Bowers, E. Chow, H. Xu, R. O. Dror, M. P. Eastwood, B. A. Gregersen, J. L. Klepeis, I. Kolossváry, M. A. Moraes, F. D. Sacerdoti, J. K. Salmon, Y. Shan and D. E. Shaw, in *Proc. ACM/IEEE Conf. on Supercomputing (SC06)*, ACM Press, New York, 2006.
24. A. Bhatelé, S. Kumar, C. Mei, J. C. Phillips, G. Zheng and L. V. Kalé, in *Proc. IEEE Int. Parallel and Distributed Processing Symp. (IPDPS)*, IEEE Press, New York, 2008.
25. B. Hess, C. Kutzner, D. van der Spoel and E. Lindahl, *J. Chem. Theory Comput.*, 2008, **4**(3), 435.
26. D. A. Case, T. E. Cheatham III, T. Darden, H. Gohlke, R. Luo, K. M. Merz Jr., A. Onufriev, C. Simmerling, B. Wang and R. Woods, *J. Comput. Chem.*, 2005, **26**(16), 1668.
27. B. G. Fitch, A. Rayshubskiy, M. Eleftheriou, T. J. C. Ward, M. Giampapa and M. C. Pitman, in *Proc. ACM/IEEE Conf. on Supercomputing (SC06)*, ACM Press, New York, 2006.
28. D. E. Shaw, P. Maragakis, K. Lindorff-Larsen, S. Piana, R. O. Dror, M. P. Eastwood, J. A. Bank, J. M. Jumper, J. K. Salmon, Y. Shan and W. Wriggers, *Science*, 2010, **330**(6002), 341.
29. B. Honig and M. Karplus, *Nature*, 1971, **229**(5286), 558.
30. A. Warshel, *Nature*, 1976, **260**(5553), 679.
31. D. Polli, P. Altoè, O. Weingart, K. M. Spillane, C. Manzoni, D. Brida, G. Tomasello, G. Orlandi, P. Kukura, R. A. Mathies, M. Garavelli and G. Cerullo, *Nature*, 2010, **467**(7314), 440.
32. U. F. Röhrig, L. Guidoni and U. Rothlisberger, *Biochemistry*, 2002, **41**(35), 10799.
33. J. Saam, E. Tajkhorshid, S. Hayashi and K. Schulten, *Biophys. J.*, 2002, **83**(6), 3097.
34. P. S. Crozier, M. J. Stevens, L. R. Forrest and T. B. Woolf, *J. Mol. Biol.*, 2003, **333**(3), 493.
35. U. F. Röhrig, L. Guidoni, A. Laio, I. Frank and U. Rothlisberger, *J. Am. Chem. Soc.*, 2004, **126**(47), 15328.
36. T. Huber, A. V. Botelho, K. Beyer and M. F. Brown, *Biophys. J.*, 2004, **86**(4), 2078.
37. S. E. Feller, K. Gawrisch and T. B. Woolf, *J. Am. Chem. Soc.*, 2003, **125**(15), 4434.
38. J. A. Ballesteros and H. Weinstein, in *Methods Neurosci.*, eds. P. M. Conn and S. C. Sealfon, Academic Press, San Diego, 1995, vol. 25, pp. 366–428.

39. M. C. Pitman, A. Grossfield, F. Suits and S. E. Feller, *J. Am. Chem. Soc.*, 2005, **127**(13), 4576.
40. A. Grossfield, S. E. Feller and M. C. Pitman, *Proc. Natl. Acad. Sci. U.S.A.*, 2006, **103**(13), 4888.
41. P.-W. Lau, A. Grossfield, S. E. Feller, M. C. Pitman and M. F. Brown, *J. Mol. Biol.*, 2007, **372**(4), 906.
42. V. Lemaître, P. Yeagle and A. Watts, *Biochemistry*, 2005, **44**(38), 12667.
43. Y. Kong and M. Karplus, *Structure*, 2007, **15**(5), 611.
44. A. Grossfield, M. C. Pitman, S. E. Feller, O. Soubias and K. Gawrisch, *J. Mol. Biol.*, 2008, **381**(2), 478.
45. M. Filizola, S. X. Wang and H. Weinstein, *J. Comput.-Aided Mol. Des.*, 2006, **20**(7–8), 405.
46. A. Cordomí and J. J. Perez, *J. Biomol. Struct. Dyn.*, 2009, **27**(2), 127.
47. M. Neri, S. Vanni, I. Tavernelli and U. Rothlisberger, *Biochemistry*, 2010, **49**(23), 4827.
48. S. Ahuja, V. Hornak, E. C. Y. Yan, N. Syrett, J. A. Goncalves, A. Hirshfeld, M. Ziliox, T. P. Sakmar, M. Sheves, P. J. Reeves, S. O. Smith and M. Eilers, *Nat. Struct. Mol. Biol.*, 2009, **16**(2), 168.
49. V. Hornak, S. Ahuja, M. Eilers, J. A. Goncalves, M. Sheves, P. J. Reeves and S. O. Smith, *J. Mol. Biol.*, 2010, **396**(3), 510.
50. P. Spijker, N. Vaidehi, P. L. Freddolino, P. A. J. Hilbers and W. A. Goddard III, *Proc. Natl. Acad. Sci. U.S.A.*, 2006, **103**(13), 4882.
51. L. Shi, G. Liapakis, R. Xu, F. Guarnieri, J. A. Ballesteros and J. A. Javitch, *J. Biol. Chem.*, 2002, **277**(43), 40989.
52. Y. Li, F. Zhu, N. Vaidehi, W. A. Goddard III, F. Sheinerman, S. Reiling, I. Morize, L. Mu, K. Harris, A. Ardati and A. Laoui, *J. Am. Chem. Soc.*, 2007, **129**(35), 10720.
53. E. Marco, M. Foucaud, I. Langer, C. Escrieut, I. G. Tikhonova and D. Fourmy, *J. Biol. Chem.*, 2007, **282**(39), 28779.
54. D. P. Hurst, A. Grossfield, D. L. Lynch, S. Feller, T. D. Romo, K. Gawrisch, M. C. Pitman and P. H. Reggio, *J. Biol. Chem.*, 2010, **285**(23), 17954.
55. I. Visiers, J. A. Ballesteros and H. Weinstein, *Methods Enzymol.*, 2002, **343**, 329.
56. L. Shi and J. A. Javitch, *Annu. Rev. Pharmacol. Toxicol.*, 2002, **42**, 437.
57. J. A. Javitch, D. Fu, G. Liapakis and J. Chen, *J. Biol. Chem.*, 1997, **272**(30), 18546.
58. D. M. Rosenbaum, V. Cherezov, M. A. Hanson, S. G. F. Rasmussen, F. S. Thian, T. S. Kobilka, H.-J. Choi, X.-J. Yao, W. I. Weis, R. C. Stevens and B. K. Kobilka, *Science*, 2007, **318**(5854), 1266.
59. M. A. Hanson, V. Cherezov, M. T. Griffith, C. B. Roth, V.-P. Jaakola, E. Y. T. Chien, J. Velasquez, P. Kuhn and R. C. Stevens, *Structure*, 2008, **16**(6), 897.
60. D. Wacker, G. Fenalti, M. A. Brown, V. Katritch, R. Abagyan, V. Cherezov and R. C. Stevens, *J. Am. Chem. Soc.*, 2010, **132**(33), 11443.

61. E. Lyman, C. Higgs, B. Kim, D. Lupyan, J. C. Shelley, R. Farid and G. A. Voth, *Structure*, 2009, **17**(12), 1660.
62. T. Huber, S. Menon and T. P. Sakmar, *Biochemistry*, 2008, **47**(42), 11013.
63. T. Wang and Y. Duan, *J. Mol. Biol*, 2009, **392**(4), 1102.
64. E. Jardón-Valadez, A.-N. Bondar and D. J. Tobias, *Biophys. J.*, 2009, **96**(7), 2572.
65. D. Provasi, J. M. Johnston and M. Filizola, *Biochemistry*, 2010, **49**(31), 6771.
66. J. Selent, F. Sanz, M. Pastor and G. De Fabritiis, *PLoS Comput. Biol.*, 2010, **6**(8), e100884.
67. R. O. Dror, D. H. Arlow, D. W. Borhani, M. Ø. Jensen, S. Piana and D. E. Shaw, *Proc. Natl. Acad. Sci. U.S.A.*, 2009, **106**(12), 4689.
68. S. Vanni, M. Neri, I. Tavernelli and U. Rothlisberger, *Biochemistry*, 2009, **48**(22), 4789.
69. S. Vanni, M. Neri, I. Tavernelli and U. Rothlisberger, *J. Mol. Biol.*, 2010, **397**(5), 1339.
70. T. D. Romo, A. Grossfield and M. C. Pitman, *Biophys. J.*, 2010, **98**(1), 76.
71. B. Jójárt, R. Kiss, B. Viskolcz, I. Kolossváry and G. M. Keserű, *J. Phys. Chem. Lett.*, 2010, **1**(6), 1008.
72. R. J. Lefkowitz, J.-P. Sun and A. K. Shukla, *Nat. Biotechnol.*, 2008, **26**(2), 189.
73. J. Li, P. C. Edwards, M. Burghammer, C. Villa and G. F. X. Schertler, *J. Mol. Biol.*, 2004, **343**(5), 1409.
74. J. A. Ballesteros, A. D. Jensen, G. Liapakis, S. G. F. Rasmussen, L. Shi, U. Gether and J. A. Javitch, *J. Biol. Chem.*, 2001, **276**(31), 29171.
75. D. R. Rosenbaum, C. Zhang, J. Lyons, R. Holl, D. Aragao, D. H. Arlow, S. G. F. Rasmussen, H.-J. Choi, B. T. DeVree, R. K. Sunahara, P. S. Chae, S. H. Gellman, R. O. Dror, D. E. Shaw, W. I. Weis, M. Caffrey, P. Gmeiner and B. K. Kobilka, *Nature*, 2011, **469**(7329), 236.
76. S. G. F. Rasmussen, H.-J. Choi, J.-J. Fung, E. Pardon, P. Casarosa, P. S. Chae, B. T. DeVree, D. M. Rosenbaum, F. S. Thian, T. S. Kobilka, A. Schnapp, I. Konetzki, R. K. Sunahara, S. H. Gellman, A. Pautsch, J. Steyaert, W. I. Weis and B. K. Kobilka, *Nature*, 2011, **469**(7329), 175.
77. Z.-G. Gao and K. A. Jacobson, *Drug Discovery Today*, 2006, **11**(5/6), 191.
78. J. D. Urban, W. P. Clarke, M. von Zastrow, D. E. Nichols, B. Kobilka, H. Weinstein, J. A. Javitch, B. L. Roth, A. Christopoulos, P. M. Sexton, K. J. Miller, M. Spedding and R. B. Mailman, *J. Pharmacol. Exp. Ther.*, 2007, **320**(1), 1.
79. T. Yarnitzky, A. Levit and M. Y. Niv, *Curr. Opin. Drug Discovery Dev.*, 2010, **13**(3), 317.
80. R. B. Best and G. Hummer, *J. Phys. Chem. B.*, 2009, **113**(26), 9004.
81. K. Lindorff-Larsen, S. Piana, K. Palmo, P. Maragakis, J. L. Klepeis, R. O. Dror and D. E. Shaw, *Proteins: Struct., Funct., Bioinf.*, 2010, **78**(8), 1950.

82. J. B. Klauda, R. M. Venable, J. A. Freites, J. W. O'Connor, D. J. Tobias, C. Mondragon-Ramirez, I. Vorobyov, A. D. MacKerell Jr. and R. W. Pastor, *J. Phys. Chem. B*, 2010, **114**(23), 7830.
83. J. Mongan and D. A. Case, *Curr. Opin. Struct. Biol.*, 2005, **15**(2), 157.
84. H. M. Senn and W. Thiel, *Angew. Chem., Int. Ed. Engl.*, 2009, **48**(7), 1198.
85. X. Periole, T. Huber, S.-J. Marrink and T. P. Sakmar, *J. Am. Chem. Soc.*, 2007, **129**(33), 10126.
86. B. Isin, K. Schulten, E. Tajkhorshid and I. Bahar, *Biophys. J.*, 2008, **95**(2), 789.

CHAPTER 21

Investigating Mechanisms of Ligand Recognition, Activation and Oligomerization in GPCRs Using Enhanced Molecular Dynamics Methods

JENNIFER M. JOHNSTON AND MARTA FILIZOLA*

Mount Sinai School of Medicine, Department of Structural and Chemical Biology, One Gustave L. Levy Place, New York, NY, 10029, USA

21.1 Introduction

Despite the current availability of multiprocessor computers combined with efficiently parallelized dynamics codes, the timescales accessible to standard molecular dynamics (MD) simulations of G protein-coupled receptors (GPCRs) are still of the order of a few microseconds, and therefore much shorter than average experimental timescales of seconds. As an alternative to multiple long-time MD simulations, enhanced MD-based methods and/or simplified physical models have recently proven useful in the study of the rugged energy landscape of GPCRs by providing mechanistic details of important events in the life-cycle of these receptors such as ligand recognition, activation and oligomerization.

In the following sections, we focus on the simulation approaches that have helped smooth the energy landscape of GPCRs by (1) reducing the number of particles in the system and/or (2) reducing the dimensionality of the GPCR

RSC Drug Discovery Series No. 8
G Protein-Coupled Receptors: From Structure to Function
Edited by Jesús Giraldo and Jean-Philippe Pin

Published by the Royal Society of Chemistry, www.rsc.org

energy surface. Since the number of calculations required to simulate a GPCR system scales with the square of the number of particles, reducing the system size increases considerably the speed with which simulations are performed. This size reduction can be achieved by eliminating the explicit representation of a component of the system (*e.g.* the solvent or the membrane, as in Generalized Born models), or by grouping individual atoms into interaction sites (*e.g.* coarse graining bead models, elastic network models). Nevertheless, the reduced representation of a GPCR system may not be sufficient to bridge the gap between the timescales accessible to standard MD simulations and average experimental timescales. With the exception of rather exclusive, extremely long, all-atom MD simulations conducted on special-purpose machines such as IBM's Blue Gene[1] and D. E. Shaw Research's Anton,[2] the timescale accessible to most research groups using standard MD simulations is too short to observe so-called 'rare events' such as ligand binding, long-range conformational changes leading to activation, or protein–protein association.

We report here on how MD can be manipulated using different biasing techniques to improve the efficacy with which it explores the conformational phase space of GPCRs, thus allowing the study of complex processes such as molecular recognition, activation and oligomerization for these receptors. Transmembrane (TM) residues of GPCRs throughout this chapter are referenced according to both the sequence number and the corresponding Ballesteros–Weinstein generic numbering scheme,[3] where the first number (*e.g.* 1 in 1.48) indicates the helix and the second number (*e.g.* 48 in 1.48) represents the residue position in that helix relative to the most conserved residue in the helix (numbered 50 by definition).

21.2 Ligand Recognition in GPCRs

A thorough understanding of the mechanisms by which GPCRs recognize their ligands is fundamental to successful drug discovery. While standard MD simulations can only provide limited insights into molecular recognition by a GPCR, biased MD techniques have been successfully employed to reveal details of ligand-binding pathways for the β2-adrenergic receptor, rhodopsin and the δ-opioid receptor. A survey of these studies is provided below.

21.2.1 Insights from Random Acceleration Molecular Dynamics

The probability of spontaneous ligand exit from a binding cavity on the timescale accessible to MD simulations can be enhanced by addition of an artificial, randomly directed force imposed upon the ligand. Thus, from knowledge of the pathways by which a ligand can exit a receptor, one can infer specific details of the binding mechanism. This is the premise of Random Acceleration Molecular Dynamics (RAMD),[4,5] a technique first introduced by Wade and colleagues under the name of Random Expulsion Molecular Dynamics (REMD).[6]

According to the formulation, the direction of the force applied to the centre of mass (COM) of the ligand is randomly chosen and is maintained for a number of steps, N. During these steps, the COM of the ligand is expected to move a minimum distance r_{min}, *i.e.* the average velocity, $\langle v \rangle$, of the ligand will maintain a threshold value of at least:

$$\langle v \rangle = r_{\min}/N\Delta t, \tag{21.1}$$

where Δt is the timestep of the MD simulation.

When the ligand encounters a rigid/unyielding portion of the binding site and its velocity falls below the predetermined threshold velocity, the trajectory is considered to be complete, and a new random direction is assigned. Each time, the new direction is maintained as long as the velocity of the ligand exceeds the threshold value, or until N steps of MD have been completed. In this way, multiple trajectories will permit the ligand to thoroughly probe the binding pocket until it finds an exit pathway or pathways, or no egress at all. The key feature of this technique, which makes it particularly effective for the purpose of exploring ligand binding pathways compared with for example steered MD, is that no knowledge of an egress pathway is required *a priori*. This enables a thorough and unbiased exploration of the ligand-binding pocket.

Wang and Duan have recently applied the RAMD technique to both the β_2-adrenergic receptor (B2AR)[4] and rhodopsin.[5] A total of 100 RAMD trajectories were performed on the B2AR crystal structure. In this crystal structure (PDB ID: 2RH1),[7] carazolol (an inverse agonist) is located in a binding pocket within the TM region forming strong interactions with polar residues in TM3, TM5 and TM7. The second extracellular loop (EL2) forms a short helix and is extended outward, rendering the binding pocket slightly open to the extracellular side. However, two bulky aromatic residues (F193 on EL2 and Y308$^{7.35}$ on TM7) and a salt bridge between EL2 and TM7 (formed by D192–K305$^{7.32}$), restrict egress from the binding site.

The RAMD trajectories suggested that the dominant exit pathway of carazolol from the B2AR crystallographic binding pocket (PDB: 2RH1) was *via* the opening at the top of the binding pocket, towards the extracellular side. The salt bridge between EL2 and TM7 (D192–K305$^{7.32}$) was broken during exit along this pathway. Although this was the dominant pathway (termed pathway A), five additional exit pathways (termed pathways B–F) were observed through interhelical clefts in around 30% of the trajectories. Of these, pathway B was the most statistically significant, offering a ligand exit through transient breakage of the interhelical interactions between TM4 and TM5. All proposed RAMD exit pathways are shown in Figure 21.1. The predominant barrier to ligand egress was presented by the interactions between the ligand and the polar groups within the binding pocket.

Furthermore, by simulating the receptor in the absence of the ligand, using standard MD over 120 ns, the authors observed that:

- the D192–K305$^{7.32}$ salt bridge was dynamic in nature; and
- that a conformational change in F193 caused it to rotate outward toward TM7, enabling its phenyl ring to pack face-to-face against the

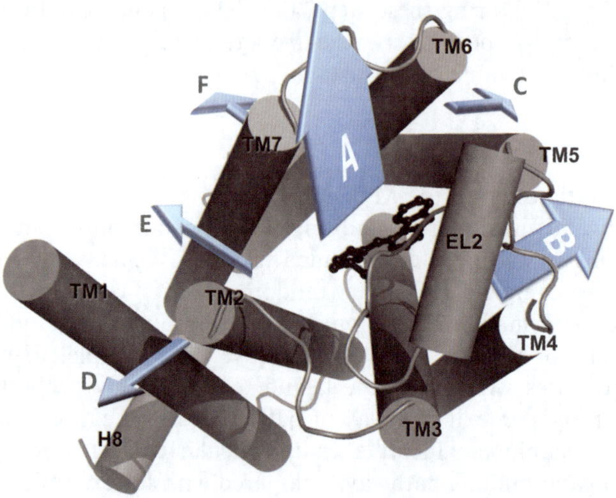

Figure 21.1 Exit pathways (**A–F**) of carazolol (black, ball and stick representation) from the binding site of B2AR (PDB ID: 2RH1) (grey, cartoon representation) as determined by random acceleration molecular dynamics (RAMD) and viewed from the extracellular side. The largest arrows indicate the most commonly followed exit pathways.

D192–K305$^{7.32}$ salt bridge and its backbone oxygen atom to hydrogen bond to K305$^{7.32}$.

This constituted a 'hydrophobic cluster' between EL2, TM7 and incorporating the neighbouring F194, Y308$^{7.35}$ and I309$^{7.36}$ residues. This cluster presented a possible barrier to ligand exit through pathway A. This new conformation was proposed to represent a putative 'ligand-free' conformation for the receptor. The ligand was then re-inserted into the binding pocket of this new conformation and a further 100 RAMD trajectories were performed. Three new exit pathways from this conformation were found, and the exit pathways observed in the simulations of the crystal structure, A–F, were followed with differing frequency. In some cases the aforementioned 'hydrophobic cluster' was disrupted and in others it was not. Overall, the average ligand egress time was slightly longer from the putative 'ligand-free' conformation than from the crystal structure. This increase was attributed to the barrier formed by the clustering of the F193 and the salt bridge connecting TM7 and EL2.

Wang and Duan used these results to propose a binding pathway for carazolol. First, the ligand ring head was found to enter *via* the cleft between TM2, TM3 and TM7 at the extracellular opening. Subsequently, the ligand interacted with F193 as it passed through the TM7–EL2 junction on the way to the bottom of the binding pocket, where it was oriented and stabilized by the polar interactions with TM3, TM5 and TM7. Such a pathway is in agreement with the conformation of the crystal structure, but may not be expected to be transferable to other receptor subtypes of the β-adrenoreceptor family, since the salt bridge linking EL2 and TM7 (residues D192–K305$^{7.32}$) is absent in these receptors, removing one of the major steric obstacles to ligand entrance/exit that are present in B2AR.

The same authors conducted a similar, albeit smaller scale, investigation on the rhodopsin crystal structure, (PDB ID. 1U19[8]), embedded in a palmitoyl oleoyl phosphatidyl choline (POPC) bilayer and solvated with explicit water. The binding site of endogenous ligand 11-*cis*-retinal is deep within the TM domain of rhodopsin but, unlike carazolol in the B2AR, the ligand is covalently bonded to K296$^{7.43}$ through a protonated Schiff base. Moreover, in contrast to the B2AR, in which the conformation of EL2 offers a putative egress, EL2 in rhodopsin forms a beta-hairpin fold that completely blocks access to the binding pocket from the extracellular side. The system was prepared for the RAMD by a *cis*-to-*trans* isomerisation, followed by manual hydrolysis for the retinal, and 2 ns of standard MD simulation of all-*trans* retinal in the binding pocket. The protein structure was not altered and remained in the dark state. A total of 38 RAMD simulations were performed on the rhodopsin system. The predominant exit pathways were observed to be towards the extracellular side, as observed for the B2AR, and in particular, through interhelical clefts, either between TM4 and TM5, or between TM5 and TM6. These exit pathways once again involved transient breakage of the interhelical interactions, which reformed immediately upon complete exit of the ligand from the binding site. A further ten trajectories were performed with the disulfide bridge that connects EL2 to TM3 (C110$^{3.25}$–C187) deliberately removed—in an attempt to encourage the EL2 to move away from the ligand binding site—but the main ligand exit pathways remained *via* the interhelical clefts, not by perturbation of the beta-hairpin of the EL2.

21.2.2 Binding Pathways as Assessed by Well-Tempered Metadynamics

Well-tempered metadynamics[9] is an enhanced sampling algorithm put forward by the Parrinello lab that works within the framework of classical MD and has been shown to be a powerful tool for the efficient exploration of the multidimensional free energy surfaces of a wide range of biological systems.[10,11] The metadynamics method requires a careful *a priori* choice of a set of collective variables (CVs) to provide a satisfactory description of the process of interest. The dynamics in the space of the chosen CVs is driven by the free energy of the system and is biased by a history-dependent (non-Markovian) potential, constructed as a sum of Gaussians localized along the trajectory. The accumulation of this biasing potential allows a system to overcome high energy barriers and thereby explore efficiently its free energy surface as defined by the CVs. Unlike standard metadynamics,[10] in the well-tempered variant of the method[9] the height of the added Gaussians depends on the underlying bias, decreasing to zero when a given energy threshold is reached. Thus, not only the convergence of the algorithm to the correct free energy profile can be proven rigorously, but this method also ensures exploration of physically relevant regions of the conformational space for complex systems where it is difficult to select *a priori* a small number of relevant CVs.

We recently applied[12] well-tempered metadynamics to perform flexible ligand/flexible protein docking of the non-selective antagonist naloxone (NLX) from the water environment into the well-accepted alkaloid-binding pocket of δ-opioid receptor (DOR), spanning residues in TM3, TM5, TM6 and TM7.[12] To efficiently exploit parallel cluster computing resources, we used the multiple walker approach,[13] with several simulations running simultaneously and contributing to the same history-dependent bias potential. Thus, a B2AR-based homology model of DOR was simulated for half a microsecond in a hydrated dipalmitoyl phosphatidyl choline (DPPC) cholesterol lipid bilayer, using ten walkers whose collective bias potential was updated frequently. All simulations were performed using GROMACS 4.0.5[14] with PLUMED.[15]

Two CVs were chosen to obtain an overall free energy profile of the flexible docking of NLX to DOR. Specifically, CV_1 referred to the distance between the COM of the heavy atoms of NLX and the COM of the heavy atoms of the alkaloid-binding pocket of opioid receptors, which is composed of selected conserved residues in TM3, TM5, TM6 and TM7 of DOR that are known to affect alkaloid binding (see ref. 12 for details). The other collective variable, CV_2, described the distance between the COM of the DOR alkaloid binding pocket and the COM of the heavy atoms of the middle residues of the EL2. This was selected to enable enhanced conformational sampling of the EL2 region of DOR. Using these two CVs, we were able to reconstruct the free energy surface of the NLX binding event and to determine that the non-selective antagonist NLX exhibits a first molecular recognition site on the DOR surface at a cleft formed by EL2 and EL3, and ends in a preferred orientation into the receptor alkaloid binding pocket, after visiting a few less stable states. In agreement with experimental data from mutagenesis and competition binding assays,[16–22] the most stable NLX-bound state of DOR corresponded to a conformation in which NLX anchored itself to $D128^{3.32}$ *via* a salt bridge through its ammonium group and a polar interaction with its alcohol moiety. The ligand was stabilized into a specific orientation through strong interactions with a number of aromatic residues within the binding pocket including $H278^{6.52}$, $W274^{6.48}$, $Y308^{7.43}$, $F218^{5.43}$ and $M132^{3.36}$. Before assuming this binding mode within the alkaloid binding pocket, however, NLX visited two metastable states characterized by a different opening of EL2 with respect to this binding pocket. In the state characterized by a more closed configuration of EL2, NLX was found to interact with the EL2/EL3 cleft through a salt bridge with EL2 D290, van der Waals contacts with $W284^{6.58}$, hydrophobic interactions with $L300^{7.35}$, EL3 P205 and EL3 F202, and a water-mediated interaction with $Y208^{5.33}$. In contrast, in the state characterized by a more 'open' conformation of EL2, the ligand positioned itself further down in the helix bundle interacting with L300$^{7.35}$, V297$^{7.32}$ and L110$^{2.65}$. From this location, NLX moved through different less stable alternative pockets within the TM region before accessing the final most stable binding mode.

From the results of the metadynamics simulations described above, corrected to improve NLX sampling in the bulk region using a method put forward by the Roux lab,[23,24] we were able to calculate equilibrium constant values for the final bound state of NLX very close to experimental results. Specifically, we allowed NLX to only move in a conical region centred at the COM of the binding pocket, and containing the EL2/EL3 cleft, using a steep repulsive potential (see Figure 21.2). We then used metadynamics with collective variables CV_1 and CV_2 to obtain a restrained free-energy profile, following the protocol and equations described in ref. 12. Given that the calculated equilibrium constant, of $K_{eq} = 80 \pm 13$ nM, for the final bound state of NLX at DOR was remarkably close to the majority of reported experimental values (see, for example, ref. 25), the free energy calculations described above hold great potential for possible application to other GPCRs.

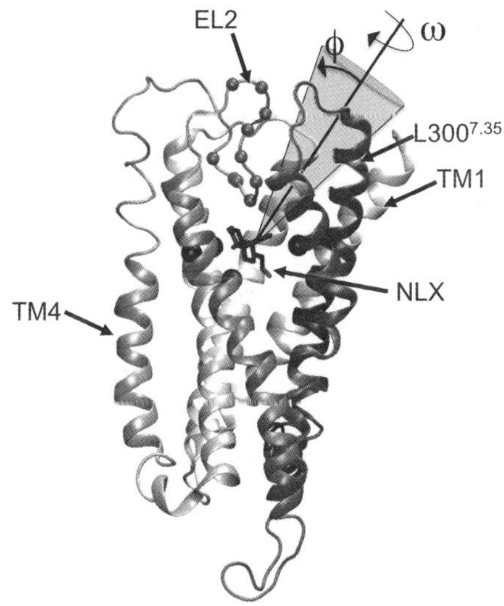

Figure 21.2 Flexible docking of Naloxone (NLX) (black, licorice representation) into the binding site (represented by black spheres) of the δ-opioid receptor. The collective variables used to describe the docking were: CV_1—the distance of the centre of mass (COM) of NLX from the COM of the binding pocket; and CV_2—the distance of the COM of the heavy atoms of residues C198-P209 (Cα represented by grey spheres) of EL2 from the COM of the binding pocket. Steep repulsive potentials were used to restrict the exploration of NLX to a conical region, where $\phi = 5$ degrees, and is the angle between the COM of NLX, the COM of the binding pocket and residue L300 (7.35) (white sphere). The protein is coloured by residue number from white (TM1) to black (H8).

21.3 Activation Mechanisms in GPCRs

The activation process of GPCRs is known to occur after seconds or even minutes in nature,[26] and thus it is inaccessible to simulations using standard MD. Here we describe the enhanced MD techniques and reduced system representations that have been used recently to study the prototypic GPCR rhodopsin in a solvated membrane environment.

21.3.1 Exploration of Global Conformational Motions of GPCRs with Elastic Network Models and Normal Mode Analysis

Elastic network model (ENM) or Gaussian network model (GNM) representations of molecules have their origins in polymer chemistry and the work of, among others, Flory[27] and Taketomi and Go.[28] Tirion[29] was first to propose an atomistic elastic network representation for application to proteins in 1996, successfully demonstrating that a single parameter model might be used to represent the vibrational properties of a complex biomolecule. This original model was later modified to a residue level representation by Bahar and co-workers,[30,31] and subsequently applied to GPCRs using a few variants of the method, including the anisotropic network model (ANM).[32,33]

To describe full atomic protein fluctuations, ENMs rely on the presence of nodes (*e.g.* *C*-alpha atoms) joined to their neighbouring nodes within a pre-defined cut-off distance, by means of a harmonic restraining potential, somewhat akin to an elastic spring:

$$E(\vec{r}_a, \vec{r}_b) = \begin{cases} \frac{k}{2}\left(\|\vec{r}_a - \vec{r}_b\| - \|\vec{r}_a{}^0 - \vec{r}_b{}^0\|\right)^2 & \text{for } \|\vec{r}_a{}^0 - \vec{r}_b{}^0\| \leq R_C \\ 0 & \text{for } \|\vec{r}_a{}^0 - \vec{r}_b{}^0\| > R_C \end{cases} \tag{21.2}$$

where $\vec{r}_a - \vec{r}_b$ denotes the vector joining nodes a and b, the zero superscript indicates the initial configuration of the nodes, R_C is the spatial cut-off for interconnections between nodes and k is the force constant for connecting 'springs' (see Figure 21.3).

Connections between nodes may represent both the covalent and non-covalent interactions in the real protein structure that stabilize the native conformation. No distinction is made between types of atoms or amino acids in an ENM, and the force constant of the harmonic restraint is the same for all interacting pairs of residues. Furthermore, when the ENM is used in conjunction with normal mode analysis (NMA)[34] to investigate dynamic processes, the global motions of the protein are decomposed into vibrational modes. The normal modes are the eigenvectors obtained by diagonalization of the Hessian matrix resulting from the potential in eqn (21.2). The mode shapes are insensitive to the absolute value of k. The cross correlation between the fluctuations of nodes a and b

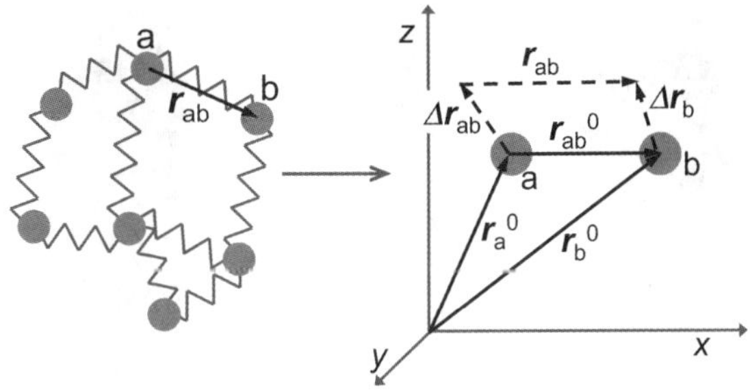

Figure 21.3 Schematic of an elastic network model. Nodes *a* and *b* (grey circles) are connected by a harmonic potential that behaves like a spring (grey zigzag line). The vector interpretation is given on the right.

[eqn (21.3)] can be expressed as a sum over the contributions of all modes (m) using the eigenvectors ($\mathbf{u}_m$) and eigenvalues (λ_m) of the Hessian matrix, $\mathbf{H}$, evaluated from the second derivatives of the potential equation (21.2).

$$\langle \Delta\vec{\mathbf{r}}_a \cdot \Delta\vec{\mathbf{r}}_b \rangle = \sum_m (3k_bT/K)\left[\lambda_m^{-1}\mathbf{u}_m\mathbf{u}_m^T\right] \tag{21.3}$$

where T is the absolute temperature, k_B is the Boltzmann constant and a total of $1 \le m \le 3N-6$ non-zero modes contribute to the cross correlation. $\mathbf{U}_m$ describes the displacements of residues induced by mode m. $\lambda_m{}^{1/2}$ scales with its frequency.

By eliminating the fast high-frequency vibrations between individual atoms, global motions larger than those that can be observed by standard MD simulation methods may be characterized.

In a first application of ENM to rhodopsin (see an example of an ENM representation in Figure 21.4), the Bahar lab[35] monitored the fast modes of the receptor and confirmed that the most rigid regions of rhodopsin corresponded to the naturally occurring disulfide bond between C110$^{3.25}$ and C187, and the retinal ligand binding pocket. They further analysed the ENM slow modes of rhodopsin in the dark state[36] in an attempt to identify molecular mechanisms underlying receptor activation. The results of these studies led to the proposal of an active meta II state model of rhodopsin, which was validated based on available experimental data. Notably, these authors confirmed the presence of a hinge site near the retinal binding pocket, which appeared to be connected by concerted motions to the opening of the helical bundle at the cytoplasmic G protein binding site.

In collaboration with the Weinstein lab, we also conducted a NMA of an ENM representation of the dark state of rhodopsin in order to identify the few lowest frequency normal modes capable of describing the transition between the inactive structure of the receptor and putative activated structures obtained using simulated annealing MD simulations and different sets of structural constraints extracted from the literature. These constraints comprised both

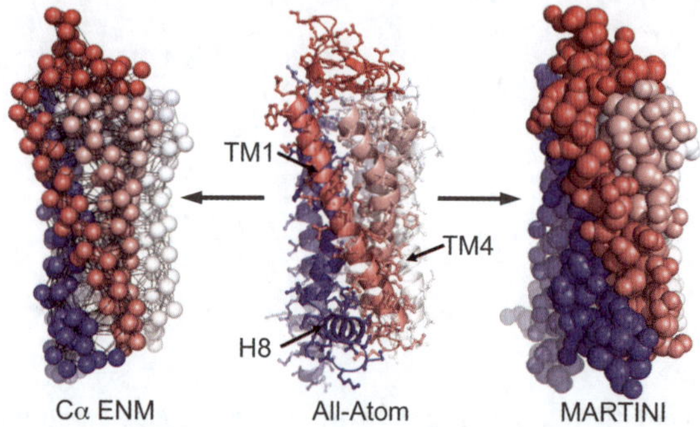

Figure 21.4 Physical representations of rhodopsin. In the centre is the atomic level representation of rhodopsin (PDB ID: 1GZM[36]) showing the helical backbone (cartoon representation) with the heavy atoms of the side chains in ball and stick representation. The left-hand image shows an elastic network model of the Cα of 1GZM ($R_C = 10$ Å). The right-hand image is a coarse grained representation of 1GZM according to the MARTINI forcefield. In all cases, the proteins are coloured from red (TM1) to white to blue (H8).

protein/protein and protein/retinal constraints and were applied, using the NOE functionality in CHARMM,[37] to chain A of the 1GZM[38] crystal structure of the dark state rhodopsin. This structure, with six combinations of all, or subsets, of the constraints applied, was subjected to simulated annealing/energy minimization iterations and the lowest energy structure for each combination of the restraints was selected.

Analysis of the six resulting activated models revealed previously undocumented structural changes that were not directly imposed by the constraints and could possibly be attributed to activation; specifically, changes in the distances between TM2/5, TM6/H8 and TM3/4. Furthermore, comparison of the activated models showed variations in the bending of TM5 and TM6, as well as previously noted orientational fluctuations for well-known structural motifs—the 'ionic lock' and the $W^{6.48}$ 'toggle-switch'—indicating the relevance of TM3 and TM5 in activation, alongside the well-documented TM6 rearrangement. The results of our ENM–NMA experiment excluded the existence of individual or pairs of lowest frequency normal modes that dominated the conformational rearrangement from the dark state of rhodopsin to any of the six pre-determined putative active models. This observation confirmed that activation of rhodopsin is not a spontaneous process, but a more complicated event that must depend strongly on the absorption of a photon.

21.3.2 Signal Transmission Mechanisms Assessed By MD Simulations Restrained by Normal Modes

To explore functionally relevant conformational changes produced in rhodopsin by the *cis–trans* photoisomerization of retinal, Isin and colleagues proposed a computational strategy that utilizes normal modes derived from an ANM[32,33] of the receptor as harmonic restraints to bias short MD simulations towards possible target structures. A detailed description of the ANM can be found in ref. 32, but simply, the ANM is a directional variation of the GNM. The underlying assumption is that the ANM-derived restraints will drive the excursion of the molecule along a direction toward a target conformation, which would eventually have been chosen naturally during a standard MD simulation but in a much longer time.

A flow diagram of the process is given in Figure 21.5, loosely separated into four phases. In phase I, a pool of ANM modes was generated and a subset of the lowest energy modes was chosen. In phase II, for each mode, (m), two target conformations were defined, lying in opposite directions along the mode, and referred to as the 'plus' and 'minus' displacements. The conformations represented by the 3N-dimensional position vectors, r_m^+ and r_m^-, are derived from:

$$r_m^{\pm} = r^0 \pm s\lambda_m^{-1/2} u_m^{ANM}, \tag{21.4}$$

where r^0 is the conformation before the application of restraints, s is a scaling parameter and $u_m^{ANM} = [\Delta\vec{r}_1(m)\Delta\vec{r}_2(m)\Delta\vec{r}_3(m) \ldots \Delta\vec{r}_N(m)]^T$ is the eigenvector corresponding to the mode m. λ and m are as defined in eqn (21.3). Short MD

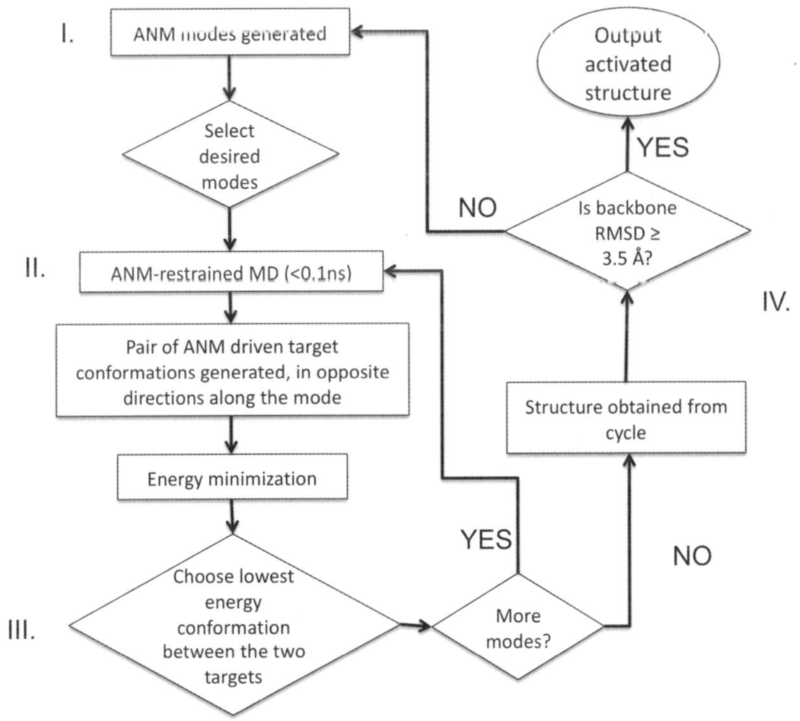

Figure 21.5 Flowchart for the ANM-biased MD adapted from Isin and colleagues.[89]

simulations (of 20 ps in length) were conducted in the presence of the ANM restraints to favour the $r^{\pm}$ derived target structures. In phase III, the resulting lowest energy conformation of the two targets was chosen, against which the ANM modes were updated. In the final stage, IV, this procedure was iterated (see Figure 21.5) until the final structure deviated from the initial structure by a backbone RMSD of 3.5 Å.

By adopting this ANM-restrained MD protocol, the authors were able to use steered MD to drive an explicit representation of the receptor, lipid membrane and water environment towards an atomic resolution model of a putative activated state. They were able to identify two hinge sites that can propagate communications through the body of the protein. The first site is near the retinal binding pocket, comprising residues on TM3 and TM6 that participate in activation. These residues are either located in the binding pocket of retinal or form new interactions to stabilize all-*trans* retinal in the activated structures. The second site involves residues at the intracellular ends of TM1, TM2 and TM7, along with EL2. These same helices are involved in an extensive hydrogen-bonding network involving highly conserved residues $N302^{7.49}$ and $Y306^{7.53}$ in the NPXXY motif of TM7 and conserved water molecules spanning the TM bundle. By comparing the initial and final structures from their simulations, the authors noted that the intracellular ends of TM3, TM4, TM5 and TM6, as well as the loops IL2 and IL3, were shown to be highly mobile; this is congruent with their role in binding the G protein.

Compared with the authors' previous work,[36] the activated model from this study has the advantage of atomistic resolution for the residue side chains. Whilst the hinge regions in this model are not identical to those from the previous study, the following residues are common to both:

- $A124^{3.39}$ and $I125^{3.40}$ from TM3;
- $W161^{4.50}$ from TM4;
- $F212^{5.47}$ from TM5 and P180; and
- C187 from the EL2.

The atomistic resolution also enabled them to determine a close contact between retinal and $C167^{4.56}$, previously considered to be a steric clash.[36] Inclusion of explicit water molecules was able to more precisely pinpoint the positions of the hinge regions, as well as identifying water-mediated hydrogen bonding networks which span the whole protein from the intracellular to the extracellular regions.

21.3.3 Insights from Biased MD Using Mass-Weighted RMSD Restraints and an Implicit Membrane Mimetic

Biophysical data were also combined with existing crystallographic data by the Costanzi lab to generate an all atom model of the meta II state of rhodopsin.[39] These authors sequentially applied two sets of experimentally derived distance restraints to the LUMI crystal structure (PDB ID: 2HPY[40,41]) and performed biased MD with mass-weighted RMSD restraints to obtain putative activated models. Specifically, they obtained first an all-atom model by imposing distance restraints that provided a comprehensive description of possible movements within the binding pocket during activation. A second stage model was then constructed using an ENM (C-alpha nodes) of the rhodopsin structure resulting from the MD simulations and a second set of experimentally determined distance restraints describing the global rearrangements of the TMs. The global motions owing to the restraints were revealed by the smoothened energy surface of the low-resolution model.

The authors were able to compare the resulting models with the other existing crystallographic structures, *i.e.* the ground-, LUMI-,[40,41] and photo-activated deprotonated intermediate[42] states. Interhelical movements produced significant changes in the prevalent interhelical hydrogen bonding networks involving TM7, TM3 and TM1, alongside breakage of the salt bridge between R135$^{3.50}$ and E247$^{6.30}$ in the highly conserved D(E)RY motif, caused by outward movements of TM6 away from TM3. Hydrogen bonds were broken between TM1 and H8, IL1 and H5, TM3 and TM5, and in IL3, TM1 and TM7. Additional hydrogen bond networks were disrupted by movement in TM3, TM6, TM7 and H8, but new compensating hydrogen bond networks were formed involving TM2, TM3, TM4, TM7, H8 and EL2. The authors suggested that the change in flexibility associated with the disruption of the complicated hydrogen bonding networks may make the receptor more amenable to interaction with the G protein. Similar fluctuations to those observed by Niv and colleagues[43] were also noted for the orientation of the Trp 'toggle switch' at position W$^{6.48}$.

To investigate the dynamic aspect of the activation models generated, the authors used biased MD simulations to drive the activation transition with mass-weighted RMSD restraints. The simulations were carried out in CHARMM (version c33b2)[37] employing the CHARMM27 forcefield.[44,45] To expedite the calculations, an implicit solvent model was used to represent the membrane. This involved using a dielectric constant to represent the electrostatic influence of the solvent, whilst explicitly representing the protein of interest.

Specifically, the authors used an adaptation of the Generalized Born (GB) formulation referred to as GBSW.[46,47] Consistent with Poisson–Boltzmann (PB) continuum electrostatics, in this model the influence of the membrane is included as a solvent-inaccessible, infinite planar slab of low dielectric constant. A simple smoothing function is included to approximate the dielectric boundary between the 'bulk water' and the 'membrane'. Since a discontinuity at the boundary between the solvent and the membrane dielectrics may induce instability into the calculation of solvation energies using the finite difference methods, a continuous, smooth dielectric boundary is employed in this model to avoid this issue. The detailed derivation of the method can be found in ref. 46.

The progress of the transition throughout the simulations was followed by monitoring two reaction coordinates; first the rotation of the W$^{6.48}$ toggle switch (*i.e.* the value of the dihedral χ_1 angle), and secondly the length of the salt bridge that forms the 'ionic lock' (*i.e.* the distance between R135$^{3.50}$ and E247$^{6.30}$) since these are considered to be the definitive differences between the inactive and active states in rhodopsin. The activation transition was treated as an equilibrium process, so both activation (forward) and deactivation (reverse) transitions were calculated. Five trajectories were calculated for wild-type (WT) rhodopsin, and in general, upon activation, rotation of the 'toggle switch' was observed prior to breaking of the salt bridge.

The results of the calculations on the WT rhodopsin were compared with identical calculations for several mutants, and these mutants were then classified as possessing the ability to hamper or promote activation. Specifically, T94I$^{2.61}$, G121V$^{3.36}$ and M257Y$^{6.40}$ were seen to favour activation, meaning that their forward transitions were fast and their backward transitions were slower, ultimately shifting the equilibrium towards the meta II state. Notably, clear correlations with phenotypes of these mutations were found as follows:

- T94I is responsible for night-blindness and shows an extremely long-lived meta II state compared with WT rhodopsin.[48]
- G121 mutated to larger side chains causes a progressive blue shift in the λ_{max} value of the pigment.[49]
- Despite rhodopsin having a uniquely low basal receptor activity, M257 mutations induce increased constitutive activity.[50]

21.3.4 Activation Pathways from Adiabatic Biased Molecular Dynamics Combined with Metadynamics

We recently developed, and tested on prototypic GPCR rhodopsin, a computational strategy that combines Adiabatic Biased Molecular Dynamics (ABMD) with path CV metadynamics to characterize possible metastable active states of GPCRs.[51] Considering that the available crystal structures of ligand-free bovine opsin contain features of active GPCR structures, we steered the crystal structure of a photoactivated deprotonated intermediate of rhodopsin (PDB ID: 2I37)[42] towards the low pH crystal structure of opsin (PDB ID: 3CAP)[52] in the presence of 11-*trans*-retinal and an explicit POPC membrane bilayer, using ten statistically independent ABMD simulations.[53] Like targeted molecular dynamics (TMD),[54] ABMD forces a reduction of the RMSD between initial and target endpoints to produce a conformational transition in a single trajectory. Unlike TMD, the ABMD algorithm[53] ensures exploration of low energy pathways by keeping the total potential energy of the system continuous during the MD run through application of a time-dependent biasing potential.

To obtain the least perturbed paths in a reasonable time frame, we carried out ten different 3 ns long ABMD simulations with an elastic constant of 0.01 kJ/nm^2, and independently drawn Maxwellian initial velocities, for a total of

30 ns. The square of the RMSD between the backbone atoms of the TM domains of 2I37 and 3CAP, as well as the side chain rotamer of W265$^{6.48}$, were chosen as reaction coordinates. The latter was imposed to promote the 'toggle switch' expected upon activation, similar to the choice made by Tikhonova and colleagues[39] in the previous section.

All ABMD trajectories were then pooled, and the sampled conformations clustered based on the RMSD of the TM domain, to define path collective variables for further metadynamics simulations. Clustering of the conformations sampled by the ten different ABMD simulations revealed two distinct groups of qualitatively different conformational transitions, characterized by a different temporal sequence of cluster occupancy, but the same definite clusters of conformations.

To obtain information about the relative stability of the states populated by rhodopsin during transition from 2I37 to 3CAP, we performed metadynamics simulations using as CVs the position (s) along and the distance (z) from the ABMD-predetermined trajectories. Thus, for smoother descriptions of the ABMD conformational transitions, we linearly interpolated their structures, obtaining $k = 10$ uniformly spaced reference conformations $R^{(i)}$ with $1 \leq i \leq k$. These reference frames, k, were used to define as CVs the position of the progression ($0 \leq s \leq 1$) of the system during the pre-determined transition, and its distance (z) from it, according to eqn (21.5):

$$s(R(t)) = Z^{-1} \sum_{i=1}^{k} \frac{i-1}{k-1} e^{-\lambda d_{\mathrm{TM}}\left(R(t), R^{(i)}\right)} \tag{21.5}$$

where $\delta = \left\langle d_{\mathrm{TM}}\left(R^{(i)}, R^{(i \pm 1)}\right)\right\rangle$ is the average distance between two adjacent frames in the transition discretization, the constant exponent, λ, was chosen in such a way that $\lambda \times \delta \cong 1$, and Z is a normalization factor defined by:

$$Z = \sum_{i=1}^{k} e^{-\lambda d_{\mathrm{TM}}\left(R(t), R^{(i)}\right)} \tag{21.6}$$

while

$$z(R(t)) = -\lambda^{-1} \log Z \tag{21.7}$$

The free-energy of rhodopsin as a function of s and z was then reconstructed from 80 ns well-tempered metadynamics simulations[9] (see ref. 51 for details of the simulation protocol). Simulations were performed for wild-type rhodopsin with either a charged or an uncharged residue E134$^{3.49}$ within the E(D)RY motif (to simulate proton uptake from the bulk occurring in the late stage of rhodopsin activation). Convergence of the reconstructed free energy was assessed by ensuring (1) small drift in the free energy difference between the minima revolving around the experimental endpoints during time and (2) frequent re-crossing of the values of the collective variables during time.

Reconstruction of the rhodopsin free energy landscape pointed to three common metastable states between 2I37 and 3CAP along the ABMD pre-calculated transition paths. Intramolecular distance analysis of these states, and comparison with results from double electron–electron resonance spectroscopy,[55] revealed that two of the identified minima may correspond to active intermediates of bovine rhodopsin, characterized by a different amplitude of the outward movement of TM6. The predicted largest separation between TM3 and TM6 was in line with data obtained for meta IIb from spectroscopy[56] and for opsin from crystallography.[52] Further support to these predictions was offered by additional simulations carried out on mutant K231A$^{5.66}$ rhodopsin—a mutant with engineered stability relative to wild-type rhodopsin. Specifically, residue K231$^{5.66}$ was identified as the residue whose contacts differed the most between the two proposed active states of rhodopsin exhibiting a significant separation of TM6 apart from TM3.

An interaction between this residue and E247$^{6.30}$ appeared to stabilize the predicted active conformation with the largest separation between TM3 and TM6, as also seen in the opsin crystal structure[52] (see Figure 21.6). We hypothesized that removal of a polar interaction between K231$^{5.66}$ and E247$^{6.30}$ by replacing K231$^{5.66}$ with an alanine would switch the equilibrium toward the predicted conformation with a smaller separation between TM3 and TM6. Thus, we carried out metadynamics simulations of the K231A$^{5.66}$ rhodopsin with either a charged or an uncharged residue E134$^{3.49}$ within the E(D)RY motif, and according to prediction, the putative active conformation of rhodopsin with the largest separation between TM3 and TM6 was destabilized in the presence of a neutral E134$^{3.49}$, mimicking the activation-dependent proton uptake from the bulk. Based on inferences from these simulations, we proposed this mutant as worthy of experimental testing, since this mutation would contribute more significantly to the stabilization of one state over the other, and help in assessing whether a smaller separation between TM3 and TM6 than that seen in the opsin conformation might be sufficient for G protein coupling and functional modulation.

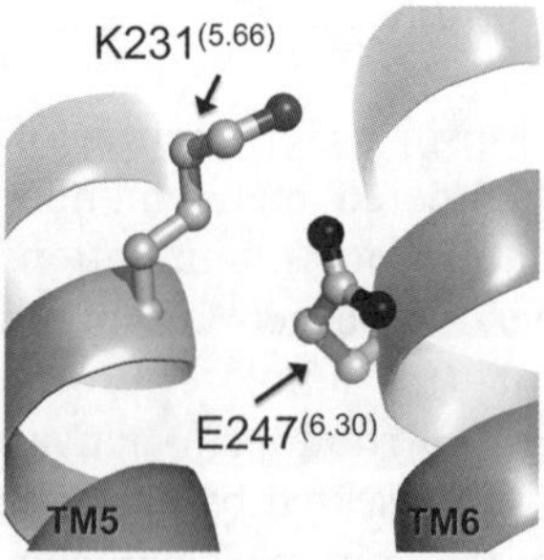

Figure 21.6 Interaction between K231$^{5.66}$ and E247$^{6.30}$ as seen in the opsin crystal structure PDB ID: 3CAP.[50]

21.4 Oligomerization Mechanisms in GPCRs

The ability of GPCRs to assemble into stable, heteromeric or homomeric, multi-protomer arrangements remains a subject of much debate.[57,58] From a computational perspective, understanding the thermodynamics and kinetics of GPCR oligomerization is not a trivial problem owing to (1) the large size of the complex system and (2) the stochastic nature of the process. A complete understanding of such a mechanism would require multiple, long timescale MD simulations of fully atomistic representations of GPCR complexes embedded into an explicit lipid–water environment, which are currently unapproachable even on modern multiprocessor computers combined with efficiently parallelized dynamics codes. In this section we survey the enhanced MD-based methods that have recently been used in combination with reduced representations of the GPCR systems to study the dynamic behaviour of dimeric/oligomeric arrangements of these receptors.

21.4.1 Coarse-Grained Representations of GPCR Complexes

Undeniably, the most accurate representation of a GPCR is when all atoms are considered. In this way, no interactions or their influences on the conformational change, or biophysical process of interest, will be neglected or approximated beyond the classical limit. However, as previously stated, the number of calculations required by a biological molecule scales as the square of the number of particles included in its representation and, consequently, a lengthy simulation of a large, explicitly represented GPCR—particularly in a dimeric or oligomeric state—very quickly becomes intractable. This issue has prompted several recent efforts to focus on reducing the number of particles included in a GPCR and a number of strategies have been proposed.

Several 'bead models' have been devised to coarsely represent biomolecules (see refs. 59 and 60 for recent reviews), ranging in degree of complexity from one bead representing 2–6 heavy atoms. The MARTINI forcefield,[61–63] originally developed for lipids, offers coarse-grained (CG) representations of proteins and cholesterol,[63] as well as carbohydrates.[64] A MARTINI CG model for prototypic GPCR rhodopsin (PDB: 1GZM) is shown in Figure 21.4 in comparison with all atom and ENM representations of the protein. Developed primarily by Marrink and co-workers, the MARTINI forcefield offers residue-level detail, stemming from an extensive calibration of a large library of chemical building blocks against thermodynamic data (notably, oil–water partition coefficients). This ensures close comparability to both experimental and atomistic MD approaches, and the flexibility to be transferable to a large range of biomolecules without the need to re-parameterize the model each time. The MARTINI forcefield is compatible with GROMACS[14] and its CG model consists (on average) of a four-to-one mapping of heavy atoms (including water molecules) to one bead. An exception is made in the case of molecules containing rings where a mapping of up to 2 : 1 is used.

There are four main types of beads: P = polar; N = non-polar; C = apolar; and charged = Q. Within the main types, there are subtypes, depending on hydrogen bonding capacity (a = acceptor, d = donor, da = both, 0 = neither) or by polarity (1 = low polarity through 5 = highly polar). For reasons of efficiency, each of the CG beads is set to have a mass of 72 amu (the mass of four water molecules), except for ring structure beads, which have a mass of 45 amu. This enables a simulation timestep of 20–40 fs, approximately ten times that possible with an all-atom GROMACS simulation, which corresponds to an effective time of 80–160 fs. Based on comparison of diffusion rates for atomistic and MARTINI CG models, the CG model affords a scaling factor of four for the effective time sampled.

To date, one of the most extensive applications of the MARTINI CG model to GPCRs has been the self-aggregation of rhodopsin monomers in four different explicit, CG, lipid membrane environments. Systems with 16 CG rhodopsin monomers (based on the 1L9H structure[65]), at a protein : lipid ratio of 1 : 100, were simulated for up to 8 μs. The results of these simulations indicate a localized membrane adaptation to the presence of the receptor, supposedly driven by the motivation to overcome the hydrophobic mismatch between the length of the hydrophobic part of the monomeric receptor and the equilibrium hydrophobic bilayer thickness. Hydrophobic mismatch has been shown to promote self-assembly of rhodopsin reconstituted in membrane bilayers.[66,67] Such phenomena could not be modelled in a simulation using an implicit representation of a membrane and an atomistic simulation of this scale would be prohibitively expensive, even on the most effective computer architectures.

Rhodopsin was simulated in different phosphocholine (PC) lipid environments and a clear dependence on chain length was observed. Bilayer adaptation was reproducibly most pronounced as local thickening near helices TM2, TM4 and TM7 in $(C12:0)_2$PC, $(C16:1)_2$PC, and $(C20:1)_2$PC bilayers and as local thinning near helices TM1/H8 and TM5/TM6 in $(C20:1)_2$PC and $(C20:0)_2$PC. The persistent extent of these thickness deformations was 1–2 nm, consistent with predictions from continuum models and atomistic simulations.

The evolution of protein contacts from the initially dispersed proteins in the membrane was analysed in terms of the buried accessible surface area (ASA) per protein, a_b, *i.e.* the surface area of the proteins that is involved in protein–protein contacts. The results indicated a higher propensity for protein–protein contact interactions in $(C16:1)_2$PC. During the simulations, clear reorganization and increase of the rhodopsin dimer interfaces was observed, indicating a search for shape complementarity that maximized a_b. The time-dependent a_b, averaged over all proteins, was correlated to the bilayer thickness.

The authors concluded that hydrophobic mismatch drives self-assembly of rhodopsin into liquid-like structures with short-range order. In order to evaluate these aggregates, the Cα density of TM helical regions surrounding the protomers was analysed in the plane of the membrane for each of the different lipid types. The high and low Cα density regions in the first shell surrounding the central reference protein density exhibited some similarities in the different lipid environments, which suggested the existence of preferential protein–protein interaction sites in rhodopsin. The number of contact interfaces was clearly higher in $(C12{:}0)_2PC$ and $(C16{:}1)_2PC$, where strong forces were introduced by the hydrophobic mismatch, than in $(C20{:}1)_2PC$ and $(C20{:}0)_2PC$, where the forces were more balanced. Three contact zones were clearly visible on the 6–8-μs time scale in $(C20{:}1)_2PC$; these included previously suggested homo- and heterodimerization interfaces in rhodopsin and other GPCRs, involving the exposed surfaces of the helices TM1/TM2/H8, TM4/TM5 and TM6/TM7. The results suggested that local distortion of the bilayer is likely to influence protein interactions and that the interactions may not be specific to individual residues in specific helices, although the latter needs to be tested both computationally and experimentally.

21.4.2 Estimates of Dimerization Constants from Umbrella Sampling Methods

Umbrella sampling was introduced in 1977 by Torrie and Valleau,[68] and has been considered the prototypic biased MD technique for improving sampling of a potential of mean force (PMF) as defined by the Kirkwood equation[69] [eqn (21.8)] from the average distribution, $\langle \rho(\zeta) \rangle$, along a predefined reaction coordinate, ζ:

$$W(\zeta) = W(\zeta^*) - k_B T \ln\left[\frac{\langle \rho(\zeta) \rangle}{\langle \rho(\zeta^*) \rangle}\right] \tag{21.8}$$

where k_BT is the Boltzmann factor, and ζ^* and $W(\zeta^*)$ are arbitrary constants.

In this method, the reaction coordinate is divided into sections (windows), and a biasing potential $[w(\zeta)]$ is introduced to tether the system to the centre of each window. The biasing potential acts to restrict the variations of the measured variable in order to enhance configurational sampling within each window and thereby along the whole of the reaction coordinate when the windows are combined. The biased simulations—one for each window (i) along the reaction coordinate—are generated using the potential energy $[U(\mathbf{R}) + w_i(\zeta)]$, (where $\mathbf{R}$ is the system coordinates) and commonly, $w_i(\zeta)$ is a harmonic potential of the form:

$$w_i(\zeta) = {}^1\!/_2 K(\zeta - \zeta_i)^2 \tag{21.9}$$

centred on the successive values of ζ_i. K represents the force constant of the harmonic potential.

One major advantage of umbrella sampling from a computational perspective is that the relative independence of the windows from one another allows the biased simulations along the length of ζ to be conducted in parallel. To obtain the potential of mean force (PMF), $[W(\zeta)]$ for the range of ζ that is of interest, it is necessary to unbias and combine the simulations from each of the windows. It is of utmost importance that these last two stages are done correctly in order to obtain an accurate PMF. Thus, it is crucial that the variations in the coordinate obtained from each of the windows offer sufficient overlap such that the whole of the coordinate space is covered.

Roux offers a thorough description of the derivation in ref. 70, but briefly, the biased distribution function, $\langle \rho(\zeta) \rangle_{(i)}$ as obtained from the *i*th biased ensemble, is:

$$\langle \rho(\zeta) \rangle_{(i)} = e^{-w_i(\zeta)/k_B T} \langle \rho(\zeta) \rangle \left\langle e^{-w_i(\zeta)/k_B T} \right\rangle^{-1}, \tag{21.10}$$

where $k_B T$ is the Boltzmann factor and the unbiased PMF from the *i*th biased ensemble is:

$$W_i(\zeta) = W(\zeta^*) - k_B T \ln \left[\frac{\langle \rho(\zeta) \rangle_{(i)}}{\langle \rho(\zeta^*) \rangle} \right] - w_i(\zeta) + F_i. \tag{21.11}$$

ζ^* and $W(\zeta^*)$ are arbitrary constants. F_i is an undetermined constant representing the free energy associated with introducing the biasing potential:

$$e^{-F_i/k_B T} = \left\langle e^{-w_i(\zeta)/k_B T} \right\rangle. \tag{21.12}$$

Several methods have been used to unbias PMFs obtained from umbrella sampling, but the most commonly used is the Weighted Histogram Analysis Method, or WHAM,[71,72] for which the Grossfield lab (University of Rochester, NY) has created a user-friendly implementation. WHAM focuses on optimizing the estimate of the coordinate-dependent unbiased distribution function as a weighted sum over all the data from the biased simulations and determining the functional form of the weight factors that minimize the statistical error. The derivation of the WHAM equations can be found in ref. 71 but the key expressions can be distilled down to:

$$\langle \rho(\zeta) \rangle = \frac{\sum_{i=1}^{N_w} n_i \langle \rho(\zeta) \rangle_i}{\sum_{j=1}^{N_w} n_j e^{-[w_j(\zeta) - F_j]/k_B T}} \tag{21.13}$$

$$F_j = -k_B T \ln \left(\int \langle \rho(\zeta) \rangle e^{-w_j(\zeta)/k_B T} d\zeta \right) \tag{21.14}$$

where N_w is the number of windows; n_i is the number of independent data points, used to construct the biased distribution function, in the *i*th window at specific value of ζ; $\langle \rho(\zeta) \rangle_i^b$ is the biased distribution function in the *i*th window; $w_j(\zeta)$ is the window potential of the *j*th window, at specific ζ; and F_j is the (unknown and to be determined) free energy constant.

The unbiased distribution function, $\langle\rho(\zeta)\rangle$ is dependent on $\{F_j\}$, and thus the simultaneous equations (21.13) and (21.14) must be solved iteratively until a consistent solution is obtained for both of them. This unbiased distribution function may then be substituted into the Kirkwood equation[69] (21.8), to obtain the PMF along the coordinate ζ.

We have recently applied umbrella sampling and WHAM to investigate the free energy of binding of a putative dimeric arrangement of the DOR.[73] Specifically, we performed umbrella sampling MD simulations of a CG representation of a homodimeric arrangement of mouse DOR in an explicitly CG represented, POPC : 10% cholesterol–water environment.

Molecular modelling was employed to generate all-atom structural models of DOR protomers from *Mus musculus*, using the same strategy we had reported in the literature for human DOR.[12] Briefly, we built the TM region of mouse DOR by homology modelling using the crystal structure of the B2AR at 2.4 Å resolution (PDB ID: 2RH1)[7] and a sequence alignment based on conserved residues and motifs that are present in family A GPCRs as inputs to Modeller 9v3.[74] The loop regions were built *ab initio* using Rosetta 2.2.[75] Protein *N*- (residues 1–44) and *C*- (residues 335–372) termini were not included in the models. A pair of the resultant DOR models was placed facing one another at a putative symmetrical interface, involving residue 4.58, inferred from cysteine cross-linking data on this and other GPCRs (see, for example, refs. 76, 77).

In an effort to improve the stability of the secondary structure of the CG representation of the receptor, we combined an ENM with the MARTINI CG representation, in a model recently referred to as ELNEDIN.[78] Bond *et al.*[79] and Periole *et al.*[80] have previously used versions of a classical elastic network with the aim of maintaining the secondary or tertiary structure of modelled biomolecules. ENMs are ideally suited to this pursuit since they are structure-based and therefore introduce an intrinsic bias toward the structure upon which they are established. In the case of ELNEDIN,[78] Periole and colleagues formalized this combination and concluded that there is a compensatory relationship between R_C (the cut-off distance between nodes within which the harmonic restraint is applied) and K_{SPRING} (the force constant of the harmonic restraint to maintain the overall structure of the protein), and that a region of (R_C/K_{SPRING}) space can be defined that provides a qualitative agreement with atomistic simulations for a particular protein.

In a novel extension to this method, we have applied a secondary-structure-dependent ELNEDIN construct to models of the DOR dimer.[73] The strength of K_{SPRING} was determined by the secondary structure of each of the residues. If the residue was determined to have a defined secondary structure (by DSSP[81]), *e.g.* α-helix as in the case of the TM regions of the DOR, then a force constant of $K_{SPRING} = 1000$ kJ/mol/nm^2 was applied. For a sequence of >2 residues with undefined secondary structure (*i.e.* coil, bend, hydrogen bonded turn or other undefined structure), a force constant of $K_{SPRING} = 250$ kJ/mol/nm^2 was applied. A comparison of the RMSF for the Cα beads for the CG representation with an equivalent atomistic simulation indicated that this

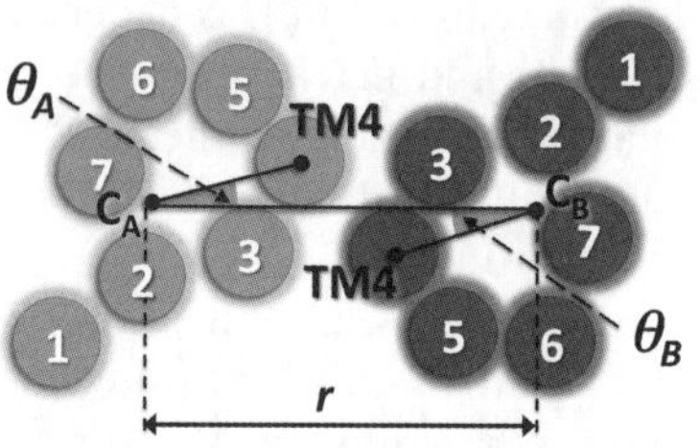

Figure 21.7 Schematic representation of the CVs (reaction coordinates) used in the umbrella sampling simulations of DOR δ-opioid receptor dimers. Specifically, the relative position of the two protomers was described by: (i) the distance *r* between centres of mass C_A and C_B of the two TM regions; (ii) the rotational angle θ_A, defined by the projection onto the plane of the membrane of the center of mass of $TM4_A$ (defined by residues $162^{4.39}$–$185^{4.62}$) C_A of the protomer bearing TM4, and C_B of the second protomer; and (iii) the equivalent rotational angle θ_B. The protomers are represented as a plan view of the approximate arrangement of their TM helices.

allowed the secondary structure of the helices to be maintained, without compromising the flexibility of the loop regions.[73]

We used the CVs illustrated in Figure 21.7 to describe the relative position of interacting DOR protomers *A* and *B* during simulation. CV_1 represented the distance *r* between the COM of protomers *A* (C_A) and *B* (C_B), CV_2 described the rotational angle θ_A, defined by the projection on to the plane of the membrane of the COM of the TM4 of *A*, the C_A, and the C_B, and CV_3 corresponded to the equivalent rotational angle θ_B.

To allow exploration of an experimentally supported TM4 interface of DOR homodimers involving position 4.58 in a reasonable timescale, we limited the sampling of the two rotational angles θ_A and θ_B to a $\sim 25°$ interval, using steep repulsive potentials. For the umbrella sampling simulation, 43 starting points with distances between $r_1 = 3.0$ nm and $r_{43} = 4.90$ nm were prepared and equilibrated for 50 ns, using harmonic restraints on the distance. The distribution $p(r)$ of the distance was harvested for 250 ns after the 'burn-in' phase and the resulting probability distributions were combined using WHAM to calculate the free energy as a function of the distance.

The free energy surface identified two different dimeric states of DOR (D_1 and D_2) involving the TM4 interface, which were separated from each other by a transition state at $r_{TS1} = 3.28$ nm, and from the monomeric state ($r \geq 4.90$ nm) by a transition state at $r_{TS2} = 3.75$ nm. The two dimeric states exhibited slightly different arrangement of the protomers relative to one another at the interface, and had similar free energy values (D_1 was more stable than D_2 by approximately 1 kcal/mol). In D_1, the structure indicated that TM4 from one protomer inserts into a groove on the opposite protomer formed by helices TM2, the *C*-terminal half of TM3, and TM4. The D_2, structure was similar to D_1 in the overall orientation of the protomers, but corresponded to a less

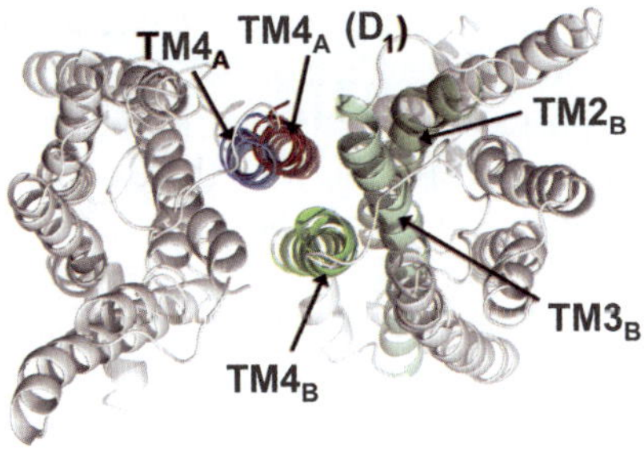

Figure 21.8 Representative δ-opioid receptor homodimer configurations from the energy basins identified by umbrella sampling (view of the extracellular side). To help comparison between the D1 and D2 dimer configurations, the position of TM4 of protomer A from D1 is reported (in red) after structural alignment of protomer B.

compact ($r = 3.40$ nm), slightly asymmetrical version of the interface. Figure 21.8 shows a comparison of the homodimeric arrangements in D_1 and D_2.

Several observables were calculated from these simulations using the derived free energy surface of DOR homodimers and the formalism described by Roux and co-workers in 23 and 24. Notably, we were able to calculate a dimerization constant, $K_D = 1.02\ \mu m^2$, for the identified lowest energy DOR homodimer with position 4.58 at the interface from the calculated free energy surface (see details of the derivation in ref. 73). In combination with a diffusion coefficient D_T value of 0.08 μm^2/s determined experimentally for the μ-opioid receptor[82] and in line with a value of 0.1 μm^2/s obtained for several other GPCRs,[83–86] this calculated dimerization constant led to an estimate of 4.4 s for the half-time of DOR dimers involving position 4.58 in the lipid bilayer. This calculated lifetime of DOR homodimers approximates the association and dissociation kinetics of M1 muscarinic receptors recently assessed by single-molecule studies.[87] Whether this suggested short lifetime of DOR homodimers and/or other GPCRs within the membrane has implications for the functional role of receptor complexes and/or the specificity of their interactions remains to be established.

21.4.3 Normal Mode Analysis of Oligomeric Assemblies

We recently explored[88] a number of different oligomeric arrangements for rhodopsin using NMA and ENMs based on the Tirion formulation.[29] By comparing the lowest frequency normal modes of the rhodopsin monomer with the equivalent for experimentally supported models of both inactive and active rhodopsin dimer and tetramer involving the TM4 and TM5 regions (see ref. 88 for details), we were able to determine the effect that oligomerization has on the rhodopsin structure.

Previous comparison of a monomer *vs.* a dimeric arrangement of rhodopsin indicated a dimerization-induced effect on the conformational preferences of interacting rhodopsin protomers, resulting in asymmetric structures after nanosecond-scale simulations.[89] In line with these observations, we found that there was significant perturbation of the normal modes of the rhodopsin monomer upon oligomerization, manifested at the interfacial regions. We also observed increased positive correlation between motions of the TM region, and between the TM region and the extracellular loops. This was compared to fluctuations of the intracellular regions, including H8, which were quenched upon tetramer formation. Covariance matrices of the 50 lowest normal modes of the rhodopsin monomer, and the equivalent protomer in both the dimer and the tetramer, indicated that the most pronounced positive correlations arose between sequential and structurally contacting domains, *e.g.* TM7 was correlated with TM1, TM6, TM2 and TM3. Inter-residue correlations differed significantly between monomer, dimer and tetramer. However, in general, positive correlations were enhanced at the interfaces between protomers, which was attributed to the need to minimize the conformational entropy lost upon complexation.

These studies also proposed the establishment of specific tetrameric arrangements upon activation. Specifically, we calculated the contribution of the lowest frequency normal modes calculated for a supposedly inactive rhodopsin tetramer (based on inferences from atomic force microscopy; the 1N3M model[90]) to each of three putative tetrameric arrangements of activated rhodopsin interfaces of three putative activated rhodopsin interfaces[76] (see ref. 88 for details). Based on the poor overlap between normal modes, we concluded that the hypothesized rigid body rotation of interacting TM4s along their own helical axes was an unlikely mechanism for the conformational rearrangement of a GPCR intradimeric interface observed upon activation.[76] The other two movements possibly occurring upon receptor activation (*i.e.* protomer displacement or sliding) were found to be equally feasible dynamic motions, based on normal mode analysis. Notably, these motional changes could be discriminated experimentally, and specific hypotheses of distance changes between specific residues (*e.g.* upon activation, distances between residues 1.57 and 1.59, or 5.62 and 5.63, would increase in one of the active interface arrangements but decrease by the same magnitude in the other) were put forward for experimental testing.

Acknowledgements

The authors are grateful to Dr Davide Provasi for comments on this chapter. The authors' work on GPCRs is supported by National Institutes of Health grants: DA020032, DA017976, and DA026434 from the National Institute on Drug Abuse, and MH091360 from the National Institute of Mental Health.

References

1. D. P. Hurst, A. Grossfield, D. L. Lynch, S. Feller, T. D. Romo, K. Gawrisch, M. C. Pitman and P. H. Reggio, *J. Biol. Chem.*, 2010, **285**, 17954–17964.

2. D. E. Shaw, P. Maragakis, K. Lindorff-Larsen, S. Piana, R. O. Dror, M. P. Eastwood, J. A. Bank, J. M. Jumper, J. K. Salmon, Y. B. Shan and W. Wriggers, *Science*, 2010, **330**, 341–346.
3. J. A. Ballesteros and H. Weinstein, in *Methods in Neurosciences*, ed. S. C. Sealfon and P. M. Conn, Academic Press, San Diego, 1995, vol. 25, pp. 366–428.
4. T. Wang and Y. Duan, *J. Am. Chem. Soc.*, 2007, **129**, 6970–6971.
5. T. Wang and Y. Duan, *J. Mol. Biol.*, 2009, **392**, 1102–1115.
6. S. K. Ludemann, V. Lounnas and R. C. Wade, *J. Mol. Biol.*, 2000, **303**, 797–811.
7. V. Cherezov, D. M. Rosenbaum, M. A. Hanson, S. G. F. Rasmussen, F. S. Thian, T. S. Kobilka, H. J. Choi, P. Kuhn, W. I. Weis, B. K. Kobilka and R. C. Stevens, *Science*, 2007, **318**, 1258–1265.
8. T. Okada, M. Sugihara, A. N. Bondar, M. Elstner, P. Entel and V. Buss, *J. Mol. Biol.*, 2004, **342**, 571–583.
9. A. Barducci, G. Bussi and M. Parrinello, *Phys. Rev. Lett.*, 2008, **100**, 020603.
10. A. Laio and M. Parrinello, *Proc. Natl. Acad. Sci. U. S. A.*, 2002, **99**, 12562–12566.
11. V. Leone, F. Marinelli, P. Carloni and M. Parrinello, *Curr. Opin. Struct. Biol.*, 2010, **20**, 148–154.
12. D. Provasi, A. Bortolato and M. Filizola, *Biochemistry*, 2009, **48**, 10020–10029.
13. P. Raiteri, A. Laio, F. L. Gervasio, C. Micheletti and M. Parrinello, *J. Phys. Chem. B*, 2006, **110**, 3533–3539.
14. D. Van der Spoel, E. Lindahl, B. Hess, G. Groenhof, A. E. Mark and H. J. C. Berendsen, *J. Comput. Chem.*, 2005, **26**, 1701–1718.
15. M. Bonomi, D. Branduardi, G. Bussi, C. Camilloni, D. Provasi, P. Raiteri, D. Donadio, F. Marinelli, F. Pietrucci, R. A. Broglia and M. Parrinello, *Comput. Phys. Commun.*, 2009, **180**, 1961–1972.
16. K. Befort, L. Tabbara, S. Bausch, C. Chavkin, C. Evans and B. Kieffer, *Mol. Pharmacol.*, 1996, **49**, 216–223.
17. K. Befort, L. Tabbara, D. Kling, B. Maigret and B. L. Kieffer, *J. Biol. Chem.*, 1996, **271**, 10161–10168.
18. G. Bot, A. D. Blake, S. X. Li and T. Reisine, *J. Neurochem.*, 1998, **70**, 358–365.
19. J. G. Li, C. G. Chen, J. L. Yin, K. Rice, Y. Zhang, D. Matecka, J. K. de Riel, R. L. DesJarlais and L. Y. Liu-Chen, *Life Sci.*, 1999, **65**, 175–185.
20. A. Mansour, L. P. Taylor, J. L. Fine, R. C. Thompson, M. T. Hoversten, H. I. Mosberg, S. J. Watson and H. Akil, *J. Neurochem.*, 1997, **68**, 344–353.
21. C. E. Spivak, C. L. Beglan, B. K. Seidleck, L. D. Hirshbein, C. J. Blaschak, G. R. Uhl and C. K. Surratt, *Mol. Pharmacol.*, 1997, **52**, 983–992.
22. C. K. Surratt, P. S. Johnson, A. Moriwaki, B. K. Seidleck, C. J. Blaschak, J. B. Wang and G. R. Uhl, *J. Biol. Chem.*, 1994, **269**, 20548–20553.
23. B. Roux, *Biophys. J.*, 1999, **77**, 139–153.
24. T. W. Allen, O. S. Andersen and B. Roux, *Proc. Natl. Acad. Sci. U. S. A.*, 2004, **101**, 117–122.

25. L. Toll, I. P. Berzetei-Gurske, W. E. Polgar, S. R. Brandt, I. D. Adapa, L. Rodriguez, R. W. Schwartz, D. Haggart, A. O'Brien, A. White, J. M. Kennedy, K. Craymer, L. Farrington and J. S. Auh, *NIDA Research Monographs*, 1998, **178**, 440–466.
26. K. Palczewski, *Annu. Rev. Biochem.*, 2006, **75**, 743–767.
27. P. J. Flory, M. Gordon and N. G. McCrum, *Proc. R. Soc. London, Ser. A*, 1976, **351**, 351–380.
28. H. Taketomi, Y. Ueda and N. Go, *Int. J. Pept. Protein Res.*, 1975, **7**, 445–459.
29. M. M. Tirion, *Phys. Rev. Lett.*, 1996, **77**, 1905–1908.
30. I. Bahar, A. R. Atilgan and B. Erman, *Folding Des.*, 1997, **2**, 173–181.
31. T. Haliloglu, I. Bahar and B. Erman, *Phys. Rev. Lett.*, 1997, **79**, 3090–3093.
32. A. R. Atilgan, S. R. Durell, R. L. Jernigan, M. C. Demirel, O. Keskin and I. Bahar, *Biophys. J.*, 2001, **80**, 505–515.
33. P. Doruker, A. R. Atilgan and I. Bahar, *Proteins: Struct., Funct., Genet.*, 2000, **40**, 512–524.
34. N. Go and H. A. Scheraga, *J. Chem. Phys.*, 1969, **51**, 4751–4768.
35. A. Rader, G. Anderson, B. Isin, H. Khorana, I. Bahar and J. Klein-Seetharaman, *Protein Sci.*, 2004, **13**, 358.
36. B. Isin, A. J. Rader, H. K. Dhiman, J. Klein-Seetharaman and I. Bahar, *Proteins: Struct., Funct., Bioinf.*, 2006, **65**, 970–983.
37. B. R. Brooks, R. E. Bruccoleri, B. D. Olafson, D. J. States, S. Swaminathan and M. Karplus, *J. Comput. Chem.*, 1983, **4**, 187–217.
38. J. Li, P. C. Edwards, M. Burghammer, C. Villa and G. F. X. Schertler, *J. Mol. Biol.*, 2004, **343**, 1409–1438.
39. I. G. Tikhonova, R. B. Best, S. Engel, M. C. Gershengorn, G. Hummer and S. Costanzi, *J. Am. Chem. Soc.*, 2008, **130**, 10141–10149.
40. H. Nakamichi and T. Okada, *Angew. Chem., Int. Ed.*, 2006, **45**, 4270–4273.
41. H. Nakamichi and T. Okada, *Proc. Natl. Acad. Sci. U. S. A.*, 2006, **103**, 12729–12734.
42. D. Salom, D. T. Lodowski, R. E. Stenkamp, I. Le Trong, M. Golczak, B. Jastrzebska, T. Harris, J. A. Ballesteros and K. Palczewski, *Proc. Natl. Acad. Sci. U. S. A.*, 2006, **103**, 16123–16128.
43. M. Y. Niv, L. Skrabanek, M. Filizola and H. Weinstein, *J. Comput. Aided Mol. Des.*, 2006, **20**, 437–448.
44. A. D. MacKerell, D. Bashford, M. Bellott, R. L. Dunbrack, J. D. Evanseck, M. J. Field, S. Fischer, J. Gao, H. Guo, S. Ha, D. Joseph-McCarthy, L. Kuchnir, K. Kuczera, F. T. K. Lau, C. Mattos, S. Michnick, T. Ngo, D. T. Nguyen, B. Prodhom, W. E. Reiher, B. Roux, M. Schlenkrich, J. C. Smith, R. Stote, J. Straub, M. Watanabe, J. Wiorkiewicz-Kuczera, D. Yin and M. Karplus, *J. Phys. Chem. B*, 1998, **102**, 3586–3616.
45. A. D. MacKerell, M. Feig and C. L. Brooks, *J. Am. Chem. Soc.*, 2004, **126**, 698–699.
46. W. P. Im, M. S. Lee and C. L. Brooks, *J. Comput. Chem.*, 2003, **24**, 1691–1702.
47. W. Im, M. Feig and C. L. Brooks, *Biophys. J.*, 2004, **86**, 3330–3330.

48. E. Ramon, L. J. del Valle and P. Garriga, *J. Biol. Chem.*, 2003, **278**, 6427–6432.
49. M. Han, S. W. Lin, M. Minkova, S. O. Smith and T. P. Sakmar, *J. Biol. Chem.*, 1996, **271**, 32337–32342.
50. M. Han, S. O. Smith and T. P. Sakmar, *Biochemistry*, 1998, **37**, 8253–8261.
51. D. Provasi and M. Filizola, *Biophys. J.*, 2010, **98**, 2347–2355.
52. J. H. Park, P. Scheerer, K. P. Hofmann, H. W. Choe and O. P. Ernst, *Nature*, 2008, **454**, 183–187.
53. M. Marchi and P. Ballone, *J. Chem. Phys.*, 1999, **110**, 3697–3702.
54. J. Schlitter, M. Engels and P. Kruger, *J. Mol. Graphics*, 1994, **12**, 84–89.
55. C. Altenbach, A. K. Kusnetzow, O. P. Ernst, K. P. Hofmann and W. L. Hubbell, *Proc. Natl. Acad. Sci. U. S. A.*, 2008, **105**, 7439–7444.
56. B. Knierim, K. P. Hofmann, O. P. Ernst and W. L. Hubbell, *Proc. Natl. Acad. Sci. U. S. A.*, 2007, **104**, 20290–20295.
57. N. A. Lambert, *Sci. Signal.*, 2010, **3**, pe12.
58. V. V. Gurevich and E. V. Gurevich, *Trends Pharmacol. Sci.*, 2008, **29**, 234–240.
59. V. Tozzini, *Curr. Opin. Struct. Biol.*, 2005, **15**, 144–150.
60. V. Tozzini, *Acc. Chem. Res.*, 2010, **43**, 220–230.
61. S. J. Marrink, A. H. de Vries and A. E. Mark, *J. Phys. Chem. B*, 2004, **108**, 750–760.
62. S. J. Marrink, H. J. Risselada, S. Yefimov, D. P. Tieleman and A. H. de Vries, *J. Phys. Chem. B*, 2007, **111**, 7812–7824.
63. L. Monticelli, S. K. Kandasamy, X. Periole, R. G. Larson, D. P. Tieleman and S. J. Marrink, *J. Chem. Theory Comput.*, 2008, **4**, 819–834.
64. C. A. Lopez, A. J. Rzepiela, A. H. de Vries, L. Dijkhuizen, P. H. Hunenberger and S. J. Marrink, *J. Chem. Theory Comput.*, 2009, **5**, 3195–3210.
65. T. Okada, Y. Fujiyoshi, M. Silow, J. Navarro, E. M. Landau and Y. Shichida, *Proc. Natl. Acad. Sci. U. S. A.*, 2002, **99**, 5982–5987.
66. A. Kusumi and J. S. Hyde, *Biochemistry*, 1982, **21**, 5978–5983.
67. N. J. P. Ryba and D. Marsh, *Biochemistry*, 1992, **31**, 7511–7518.
68. G. M. Torrie and J. P. Valleau, *Chem. Phys. Lett.*, 1974, **28**, 578–581.
69. J. G. Kirkwood, *J. Chem. Phys.*, 1935, **3**, 300–313.
70. B. Roux, *Comput. Phys. Commun.*, 1995, **91**, 275–282.
71. S. Kumar, D. Bouzida, R. H. Swendsen, P. A. Kollman and J. M. Rosenberg, *J. Comput. Chem.*, 1992, **13**, 1011–1021.
72. S. Kumar, J. M. Rosenberg, D. Bouzida, R. H. Swendsen and P. A. Kollman, *J. Comput. Chem.*, 1995, **16**, 1339–1350.
73. D. Provasi, J. M. Johnston and M. Filizola, *Biochemistry*, 2010, **49**, 6771–6776.
74. A. Sali and T. L. Blundell, *J. Mol. Biol.*, 1993, **234**, 779–815.
75. C. Wang, P. Bradley and D. Baker, *J. Mol. Biol.*, 2007, **373**, 503–519.
76. W. Guo, L. Shi, M. Filizola, H. Weinstein and J. A. Javitch, *Proc. Natl. Acad. Sci. U. S. A.*, 2005, **102**, 17495–17500.
77. W. Guo, E. Urizar, M. Kralikova, J. C. Mobarec, L. Shi, M. Filizola and J. A. Javitch, *EMBO J.*, 2008, **27**, 2293–2304.

78. X. Periole, M. Cavalli, S. J. Marrink and M. A. Ceruso, *J. Chem. Theory Comput.*, 2009, **5**, 2531–2543.
79. P. J. Bond, J. Holyoake, A. Ivetac, S. Khalid and M. S. P. Sansom, *J. Struct. Biol.*, 2007, **157**, 593–605.
80. X. Periole, T. Huber, S. J. Marrink and T. P. Sakmar, *J. Am. Chem. Soc.*, 2007, **129**, 10126–10132.
81. W. Kabsch and C. Sander, *Biopolymers*, 1983, **22**, 2577–2637.
82. A. N. Sauliere-Nzeh, C. Millot, M. Corbani, S. Mazeres, A. Lopez and L. Salome, *J. Biol. Chem.*, 2010, **285**, 14514–14520.
83. L. S. Barak, S. S. G. Ferguson, J. Zhang, C. Martenson, T. Meyer and M. G. Caron, *Mol. Pharmacol.*, 1997, **51**, 177–184.
84. O. Hegener, L. Prenner, F. Runkel, S. L. Baader, J. Kappler and H. Haberlein, *Biochemistry*, 2004, **43**, 6190–6199.
85. Y. I. Henis, M. Hekman, E. L. Elson and E. J. M. Helmreich, *Proc. Natl. Acad. Sci. U. S. A.*, 1982, **79**, 2907–2911.
86. M. M. Poo and R. A. Cone, *J. Supramol. Struct.*, 1973, **1**, 354.
87. J. A. Hern, A. H. Baig, G. I. Mashanov, B. Birdsall, J. E. T. Corrie, S. Lazareno, J. E. Molloy and N. J. M. Birdsall, *Proc. Natl. Acad. Sci. U. S. A.*, 2010, **107**, 2693–2698.
88. M. Y. Niv and M. Filizola, *Proteins: Struct., Funct., Bioinf.*, 2008, **71**, 575–586.
89. M. Filizola, S. X. Wang and H. Weinstein, *J. Comput. Aided Mol. Des.*, 2006, **20**, 405–416.
90. Y. Liang, D. Fotiadis, S. Filipek, D. A. Saperstein, K. Palczewski and A. Engel, *J. Biol. Chem.*, 2003, **278**, 21655–21662.

CHAPTER 22

Functional Selectivity of Drugs for Seven Transmembrane Receptors: Biased Agonism and Antagonism

TERRY KENAKIN

Molecular Discovery Research, GlaxoSmithKline Research and Development, 5 Moore Drive, Research Triangle Park, NC 27709, USA

22.1 Receptors as Allosteric Proteins

Seven transmembrane receptors (7TMRs) are nature's prototypic allosteric protein; they bind molecules to domains accessible from the extracellular space to change their shape to alter their interaction with proteins in the cell cytosol. Allosterism through proteins like 7TMRs is vectorial; that is, a modulator binds to a site on the protein (this need not be the endogenous agonist binding site) to stabilize a conformation of the receptor that alters other regions (which could be in the cytosol) to, in turn, alter the interaction of those regions with other proteins. The vector could direct energy from an extracellular modulator binding site toward another extracellular binding site (classic 'guest' allosterism) to change the affect of ligands that bind to that site, or along the plane of the membrane to affect the receptor interaction with other receptors or membrane proteins (receptor oligomerization) or towards cytosolic signalling proteins. This chapter concentrates on this latter effect.

RSC Drug Discovery Series No. 8
G Protein-Coupled Receptors: From Structure to Function
Edited by Jesús Giraldo and Jean-Philippe Pin

Published by the Royal Society of Chemistry, www.rsc.org

Agonists bind to 7TMRs and the resulting stabilized altered receptor conformation interacts with cytosolic signalling proteins to elicit cellular response. It is now known that different ligands can direct such activation to various signalling pathways differentially, *i.e.* they can bias stimulus towards some pathways and not others. It is important to consider the properties of 7TMRs with regard to this biased signalling.

Proteins can adopt a myriad of interchangeable conformations according to the thermal energy in the system to form a dynamic equilibrium. Evidence for such ensembles comes from theoretical modeling[1,2] and experimental studies (*i.e.* site-directed spin labelling, site-directed fluorescence quenching, sulfydryl accessibility, disulfide crosslinking methods) reviewed in refs. 3–7. A given ligand interacts with this population of conformations (referred to as an 'ensemble') to preferentially stabilize those for which it has the highest affinity. According to Le Chatelier's principle, the equilibrium of the system will shift towards the stabilized conformation(s) thereby producing a new ligand-associated ensemble. This is referred to as 'conformation selection'.[8,9] The interconversion of conformations in the ensemble can be viewed as the 7TMR rolling on what is referred to as an 'energy landscape.[10–20]

The ease with which a ligand can alter protein conformation is related to how disordered the protein is and molecular dynamics predicts that signalling proteins such as receptors have an unusual amount of intrinsic disorder thereby favouring allosteric modulation.[21,22] The 7TMR goes on thermodynamic excursions away from canonical conformations through ligand binding. These stabilized conformations are represented by energy wells on the landscape, *i.e.* the ligand moves the receptor onto another energy landscape;[3] see Figure 22.1A.

Various members of stabilized ensembles may or may not have distinct biological activity;[24,25] see Figure 22.1B. If they do and if the proportions of these conformational ensembles vary with the ligand type, then biased cellular signalling may result. The bottom of these ligand-stabilized energy wells may be thought of as being 'rugged', allowing for a range of nearly isoenergetic conformers;[26] in fact these may be the ligand states stabilized to produce biased signalling.

Allosterism is related to protein disorder and mutation studies suggest that disorder within 7TMRs may favour formation of an 'active' state (*i.e.* one that produces a physiological response) by virtue of the fact that a rigid conformation with residues locked away from cytosolic proteins may be made accessible by the conformational change. For example, a point mutation with 22 different amino acids at position 293 of α_{1B}-adrenoceptor leads to the production of 19 constitutively active states for inositol phosphate production.[27] In general, this suggests that there are a number of 'active' states for 7TMRs that are associated with disorder and relatively fewer inactive states.

Allosterically linked binding sites need not exist within close geographical proximity to each other, *i.e.* a modulator may stabilize a global conformation that has aberrations in conformation far removed from the allosteric binding site. Population dynamics relates allosteric effect to order/disorder transitions

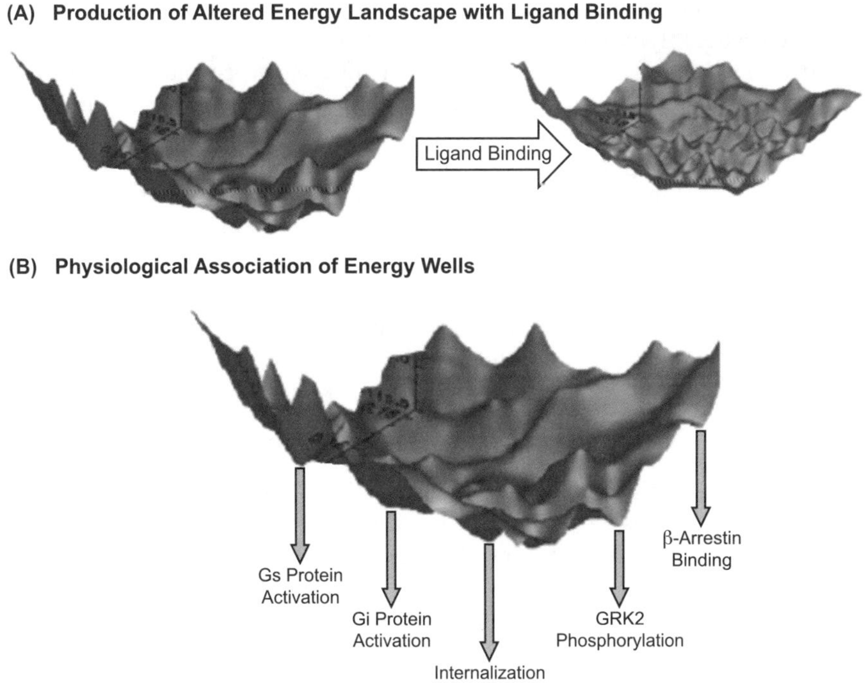

Figure 22.1 7TMR energy landscapes. The interchanging 7TMR conformations can be visualized as the protein rolling on an energy landscape and falling into energy wells of preferred low energy conformations. (**A**) The binding of a ligand to the 7TMR changes the receptor thermodynamically and causes it to traverse a new energy landscape. Whereas the unliganded receptor had a certain probability of forming various stabilized conformations, the binding of the ligand necessarily changes these probabilities to another set unique to the ligand-receptor pair. (**B**) Certain of the preferred stabilized 7TMR conformations may have cellular activity by virtue of the fact that they have high affinity for cytosolic signalling proteins. Under these circumstances, these conformations become so called 'active states' in that they produce physiologically relevant effects.

mediating long-range allosteric communication.[28] In these terms, the energy balance within the protein (*i.e.* what receptor states are most stable and which states bind ligand) controls the energetics of allosterism.[21] Thus, multiple sites may be affected by the modulator causing differential alteration of binding of more than one cytosolic signalling protein. For example, studies with C-C chemokine receptor type 5 (CCR5) and HIV-1 entry inhibitors indicate that portions of CCR5 that interact with endogenous chemokines differ from those that interact with HIV-1 gp120.[29–31] In addition, there is evidence to suggest that the chemokine CCL5 binds to regions of the receptor different from those that bind the allosteric modulator Sch-C.[32,33] However, the allosteric modulator aplaviroc, which blocks HIV-1 entry, does not block the binding of the chemokine CCL5.[34,35] Thus, binding at an allosteric site presumably stabilizes an ensemble of conformations the members of which may have regions of the protein considerably different from the native ensemble.

Another important aspect of biased signalling is the consideration of the thermodynamic importance of viewing allosteric effect on 7TMRs as changing both the affinity and efficacy of ligands binding to the receptor. Thus, the affinity of a cytosolic protein may be affected by the binding of a modulator or the efficacy of modulator (if it is an agonist) may be affected or both. The corollary to this idea is the fact that allosteric energy transfer is reciprocal in nature;[36] thus while a modulator affects the affinity of the 7TMR for a cytosolic binding protein, so too does the binding of cytosolic binding protein affect the affinity of the 7TMR for the modulator. This becomes important in terms of the observed cell phenotype dependence of biased agonism.

22.2 Functional Selectivity: Historical Perspective

The fact that agonists and antagonists can produce functional selectivity was only really appreciated once multiple measures of receptor function could be taken. Prior to this, receptors were classified and studied on the basis of whole cell readouts with null procedures, *i.e.* equiactive potency ratios were used as system-independent measures of agonist activity. The basic assumption for these procedures was that all agonists produced uniform activation of 7TMRs and thus the only difference in agonist efficacy was the strength of signal imparted to the receptor. However, this simple model was seen to be lacking in early quantitative studies of receptor ligand function (see refs. 37–45), where it was seen that agonists did not appear to produce uniform effects when activating the same receptors in different cells.

The first theoretical model to account for functional selectivity described agonists as stabilizing different active states to induce heterogeneous activation of cytosolic signalling proteins;[46] this was referred to as 'stimulus trafficking'. This model predicts that if the receptor is pleiotropic with respect to the number of signalling proteins with which it can interact, *i.e.* multiple G proteins[47–54] or β-arrestin (*vide infra*), then different agonist-stabilized receptor conformations would produce heterogeneous activation of different signalling pathways leading to functional selectivity in cells.

This model was specifically designed to explain striking data like that seen with pituitary adenylate cyclase-activating polypeptide (PACAP) agonists in LLC-PK1 cells transfected with PACAP receptors. In this system, the relative potencies of $PACAP_{1-27}$ and $PACAP_{1-38}$ were measured for increasing cellular cyclic AMP and inositol phosphate. Uniform activation of the PACAP receptor by these agonists predicts that the resulting stimulation would be constant for both pathways. In striking contrast, the two agonists *reversed* their relative order of potency for the two signalling pathways mediated by the same receptor. Specifically, whereas the relative potency for cyclic AMP was $PACAP_{1-27} > PACAP_{1-38}$, this order was reversed for inositol phosphate production ($PACAP_{1-27} < PACAP_{1-38}$).[55] The most simple hypothesis to explain this divergent behaviour is to postulate that the two agonists produce different receptor active states with varying affinity for the cyclic AMP and inositol signalling pathways.

Over the past 15 years there have been numerous reports of how different ligands bind to receptors to cause activation of only portions of the repertoire of pathways that those receptors mediate (see reviews in refs. 56–60). Thus, in contrast to a linear system of efficacy whereby receptor occupancy by an agonist leads to receptor activation (of all pathways linked to that receptor), phosphorylation, desensitization and internalization, there are now numerous reports of ligands producing only some of these results. One of the most striking is the active internalization of receptors by antagonists that otherwise do not produce signaling.[61,62]

As discussed previously, the mechanism proposed for this diversity is the production of ligand-selective receptor active states.[46] In fact, there is much independent evidence to support the notion that ligands can produce selective 7TMR conformations.[56,63–65] The techniques to show this are diverse and range from studying fluorescent probes on receptors, fluorescence resonance energy transfer (FRET), bioluminescence resonance energy transfer (BRET), plasmon-waveguide resonance spectroscopy, circular dichroism, X-ray crystallography, antibody binding site directed mutagenesis, molecular modelling and data from kinetic studies.[66–82] Recently, a direct link between biased signalling of the mitogen-activated protein kinase (MAPK) pathway through G protein and G protein-independent pathways and unique receptor conformations as measured by BRET has been described.[67]

Selective activation of pathways through agonism by different agonists has been given many names including 'stimulus trafficking',[46] 'functional dissociation',[83] 'biased inhibition',[84] 'biased agonism',[85] 'differential engagement',[86] 'discrete activation of transduction',[43] and 'functional selectivity'.[87,88] 'Functional selectivity', popularized by the Mailman group as early as 1994,[40] is a common terminology for this effect. Proposed as a universal thermodynamic mechanism for all 7TMRs,[46] there are now reported functionally selective ligands for a large number of 7TMRs.[54,89]

22.3 Biased Agonism

Consistent with the notion that stabilization of global conformations of 7TMRs can alter portions of the receptor in different ways is the fact that multiple cytosolic signalling partners for receptors sense this differential conformational change and functional selectivity in direct signalling results (biased agonism; see Figure 22.2). This can occur at the level of multiple G proteins but recent evidence also suggests that a major division in signalling may occur at the level of G proteins and β-arrestin. This latter protein had been known to mediate cessation of G protein signaling[90–92] and the internalization of 7TMRs,[93–97] but now evidence suggests that this cytosolic protein is involved in direct cellular signalling as well.[98–103] Thus, low levels of persistent cellular signal through activation of extracellular signal-regulated kinase (ERK)1/2, p38 MAPK, c-Jun *N*-terminal kinase (JNK), tyrosine kinase c-Src, PI-3K-AKT

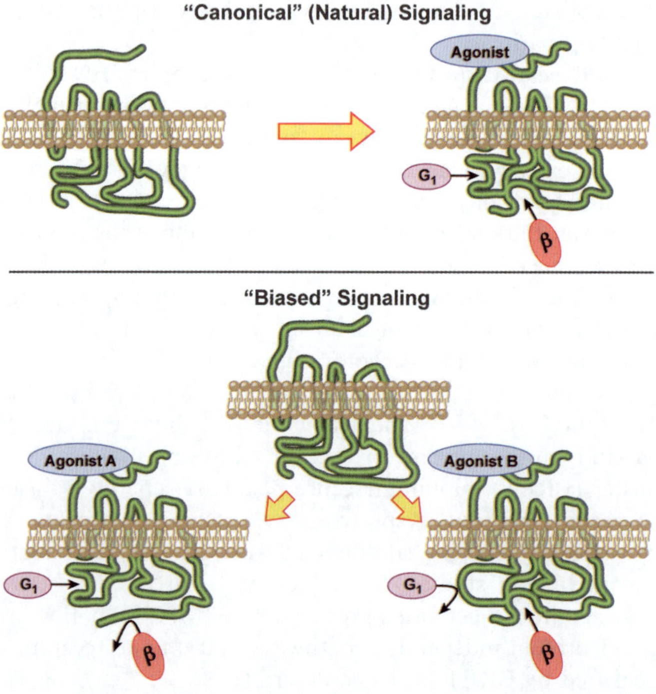

Figure 22.2 Agonists stabilize 7TMR active states that then can interact with various cytosolic signalling proteins (*i.e.* G proteins, β-arrestin). Natural ligands (*i.e.* neurotransmitters, hormones) probably have the ability to initiate all behaviours of the 7TMR to activate all pathways mediated by the receptor. Functionally selective ('biased') agonists may stabilize conformations that interact with only certain signalling proteins thus they will preferentially activate some cellular pathways over others.

and NF-κB pathways have been reported for β-arrestin activation through 7TMRs.[99,104–108] For example, binding of angiotensin to angiotensin 1A receptors produces activation of G proteins and β-arrestin. However, the angiotensin agonist analog SII (Sar^1, Ile^4, Ile^8-AngII)[109] produces activation of only the β-arrestin pathway through this receptor;[110,111] see Figure 22.3.

Similarly, parathyroid hormone (PTH) activates ERK through separate G protein-related and G protein-independent pathways. Analogs of PTH can produce separate stimulation of these pathways through the PTH receptor. For example, ERK1/2 is stimulated primarily *via* the G protein pathway by [Trp^1]PTHrp-(1–36) and mainly through the β-arrestin dependent (and G protein independent) pathway by PTH-1A {[D-Trp^{12},Tyr^{34}]PTH-(7–34)}.[112] In light of known interactions of 7TMRs with other cytosolic proteins,[113–117] there are no data at present to suppose that G proteins and β-arrestin are the only ways in which 7TMR ligand activation can affect cytosolic signalling.

In general, there are data to show that many agonists can selectively activate receptor signalling pathways, *i.e.* it is an established phenomenon. However,

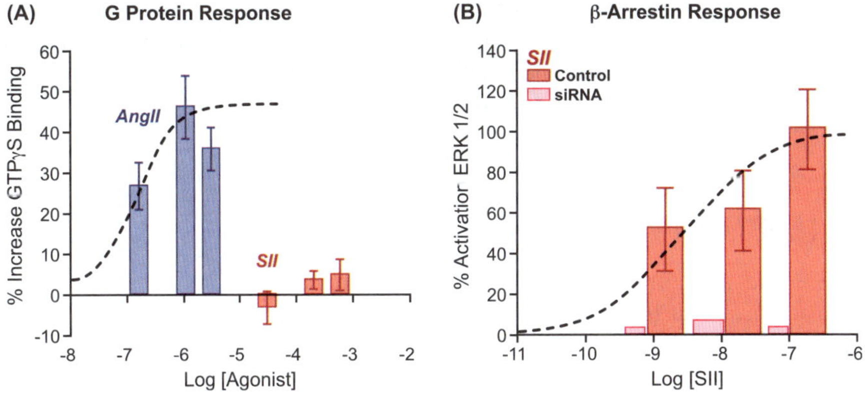

Figure 22.3 Signalling profile of the functionally selective angiotensin AT1 receptor agonist SII. (**A**) Whereas angiotensin produces activation of G proteins, SII is devoid of this activity. (**B**) In contrast, SII produces ERK1/2 activation through the signalling protein β-arrestin; this effect is eliminated by removal of β-arrestin through siRNA. Data redrawn from ref. 111.

what is not yet clear is the therapeutic relevance of this effect. In some cases, there are *a priori* reasons for believing a selective pathway activation may be therapeutically useful. For example, PTH regulates calcium homeostasis and bone metabolism therefore activation of the PTH receptor should be useful in the treatment of osteoporosis. In mice devoid of β-arrestin 2, PTH does not stimulate bone formation[118] suggesting that a beneficial effect of PTH-stimulation in osteoporosis could be obtained through selective activation of β-arrestin 2.[119] In fact, recent data show that biased analogues of PTH can produce selective stimulation of G protein and β-arrestin pathways.[112,120]

Another example of therapeutic superiority may be found in the selective activation of the angiotensin AT1 receptor. Activation of G_q protein through this receptor can lead to hypertension and other deleterious cardiovascular effects. However, selective activation of β-arrestin may be beneficial both through a reduction in G protein signalling by angiotensin II and a β-arrestin mediated anti-apoptotic effect.[121–123] This favourable profile is seen with Sar[1], Ile[4], Il[8]-Ang II (SII), a β-arrestin biased agonist of the AT_1 receptor which blocks G_q-activation by angiotensin II and causes AT_1-receptor association with β-arrestin and with no concomitant G_q protein effect;[109–111] see Figure 22.3.

In addition to selective active signalling, biased agonists could also be unique in terms of not causing receptor desensitization and/or receptor internalization. For instance, opioids activate and internalize opioid receptors, but while enkephalins avidly internalize receptors, internalization by morphine is cell type dependent and variable.[124,125] The functionally selective opioid agonist, herkinorin, produces little β-arrestin receptor interaction;[126,127] this profile may provide the key to opioid receptor-mediated analgesia with less concomitant desensitizing effect.[128] It has been shown that the analgesic effects of morphine are enhanced and prolonged in β-arrestin 2 knockout mice.[129,130] The negative effects of morphine, such as respiratory depression and constipation, are decreased in β-arrestin knockout mice. In addition to predicted advantages of biased agonists,[131] translational medicine will no doubt furnish data linking clinically identifiable phenotype effects of agonists with functional selectivity as new agonists are tested.

22.4 Biased Antagonism

Orthosteric antagonists induce blockade by preventing the binding of the agonist; this effect is pre-emptive,[132] *i.e.* there is never a receptor species with both the antagonist and agonist bound. This being the case, there is no mechanism by which the antagonist can modify the signalling of the agonist other than complete elimination of agonist response.

For one molecule to bias the activity of another on a receptor, both molecules must occupy the receptor simultaneously, *i.e.* the system must be permissive.[132] This can occur in all instances of 7TMR agonism (the two molecules being the agonist and the cytosolic signalling protein) but only occurs for antagonism when the antagonist is an allosteric modulator. In cases of allosteric antagonism, functional antagonism has been described—specifically in cases where allosteric modulators impose functional selectivity on previously unselective agonists; see Figure 22.4. For example, the natural NK2 receptor

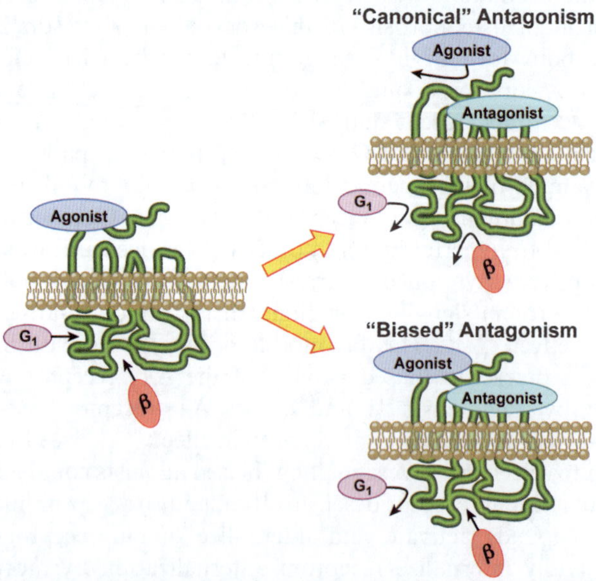

Figure 22.4 Antagonism of 7TMRs. Orthosteric antagonists (which prevent agonist binding to the receptor) necessarily block all signalling effects of the agonist. In contrast, an allosteric functionally selective antagonist binds to the receptor to alter its conformation so that the agonist can still bind and produce a limited (biased) signal.

agonist neurokinin A produces activation of both G_s and G_q proteins; this pattern is changed by the allosteric modulator LP1805 to one of enhanced G_q and blockade of G_s activation.[133] Similarly, the allosteric modulator for CRTH2 receptors *N*α-tosyltryptophan, when bound to the receptor, causes the natural agonist prostaglandin D_2 to change its signalling pattern from G_i and β-arrestin activation to solely G_i activation (with no concomitant β-arrestin interaction).[134] The peptide antagonist of the IL-1 receptor 101.10 blocks certain IL-1 signalling pathways but not others,[135] while the antagonist RB213 shows pathway-selective antagonism for type 2 CCK receptor-mediated inositol phosphate formation and arachidonic acid release.[136] There may be undiscovered antagonist functional selectivity in the current allosteric modulators known due to the fact that functional antagonist selectivity requires specific testing with multiple probe agonists to be detected.

22.5 The Impact of Functional Selectivity on New Drug Discovery

The fact that ligands can differentially affect 7TMR function has a profound impact on the way new drugs are discovered. As described by Stephenson,[137] agonist efficacy was linear in nature, *i.e.* an agonist activated a receptor to produce all the physiological actions that the receptor mediates (given sufficient strength of stimulus). This was based on the assumption that all agonists stabilize a single receptor active state. Within these ideas, a single functional assay (assuming it is very sensitive to detect weak stimulus) is sufficient to detect all agonism. With the advent of more sophisticated technology that allows the observation of multiple 7TMR mediated physiology (*i.e.* separate signalling pathway activation, phosphorylation, internalization) has come the realization that different agonists can mediate different activities. Thus, the lack of observed effect on one signalling pathway need not impose a mandatory similar behaviour on 7TMR interaction with another pathway. Under these circumstances, a single biochemical functional assay may be insufficient to detect all agonism. One approach around this limitation is to use more general assays that encompass multiple receptor-linked pathways (*i.e.* whole cell activity with label free assays[138–141] or ERK1/2 from phosphorylation in multiple pathways[142]).

Another conclusion to be drawn from the ability to observe multiple 7TMR behaviours is that there are 7TMR effects that may be therapeutically relevant that do not involve active cellular signalling. For example, receptor internalization is a receptor behaviour that could be important in the termination of agonist effect (*i.e.* opioid analgesia, see above) or removal of 7TMR target from the cell surface (*i.e.* removal of CCR5 to prevent HIV-1 infection). Therefore, the discovery process for these effects may require specialized assays in the screening process.

A possible shortcoming of using whole cell assays (that may be more inclusive in terms of detecting multiple signalling effects) is that, for functionally selective agonists, these effects may be cell type dependent. This is because of the mandatory reciprocal effects on cytosolic coupling proteins on the affinity of the receptor for agonists. Therefore, this raises the spectre that activity detected in a whole cell screening assay may not manifest itself in the therapeutic cell line. More to the point, the cell-based screen may not detect a therapeutically relevant cell type dependent effect.

In general, there are two possible strategies to deal with functional selectivity. One would be to work, if possible, in the therapeutically relevant primary cell line. A second would be to measure bias with a system-independent scale to guide structure–activity relationships. To this end, the Black/Leff operational model can be used to quantify the relative excitation of various signalling pathways mediated by the receptor. In this model, affinity (denoted as K_A, the equilibrium dissociation constant of the agonist–receptor complex) and efficacy (denoted as τ, a parameter encompassing both the efficacy of the agonist and the sensitivity of the system)[143,144] are quantified by fitting data to the Black/Leff operational equation. It is important that the index of activation includes affinity and efficacy as both are relevant in agonist 7TMR systems where the agonist is modulator, the receptor the conduit and the signalling protein the guest.[145] Under these circumstances, a measure of activation of a 7TMR for a given pathway can be quantified with a transduction coefficient, namely $Log(\tau/K_A)$; see Figure 22.5. This scale cancels system effects and the resulting bias measured will be strictly ligand dependent. Theoretically, such measures

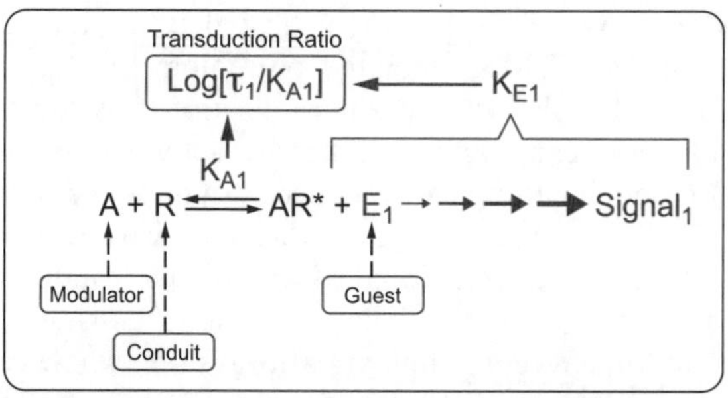

Figure 22.5 Cellular signalling as depicted by the Black/Leff operational model. In this scheme, the receptor is bound by the agonist with affinity K_A and the ligand-bound complex further binds to an effector coupling protein (*i.e.* G protein, β-arrestin) to elicit a cellular response. The affinity of the receptor for the agonist is determined by the nature of the coupling protein (E) according to allosteric theory and therefore the affinity of the agonist will be unique to the specific signalling pathway being activated. Similarly, the efficacy of the [AR] complex is determined by the chemical structure of the agonist; this is encompassed in the ligand-dependent portion of τ. The $\log(\tau/K_A)$ transducer coefficient thereby characterizes the power of any agonist to produce activation of any cellular pathway. Ratios of τ/K_A values $[\Delta Log(\tau/K_A)]$ then can be used to quantify ligand-dependent bias.

of ligand bias can be used by medicinal chemists to systematically explore functional selectivity in a chemical series.

22.6 Impact on Drug Taxonomy: Agonist *vs.* Antagonist

The fact that ligands can produce diverse signals from 7TMRs poses interesting challenges for drug nomenclature. When agonism was defined in terms of whole cell response, the presence or absence of activation and concomitant receptor occupancy clearly delineated agonists and antagonists. Now that pharmacologists have access to signalling on a subcellular level, it has been observed that some ligands are agonists for some pathways and antagonists for others, *i.e.* 'efficacy' now must be considered to be pluridimensional.[146] For example, propranolol, carvedilol and bucindolol block β_2-adrenoceptor-mediated elevations in cyclic AMP but activate ERK through β-arrestin.[67,147,148] Similarly, [D-Trp^{12},Tyr^{34}PTH(7-34)] is an PTH inverse agonist for PTH-receptor mediated $G_{\alpha s}$ protein activation[149] but a positive agonist *via* β-arrestin in the absence of $G_{\alpha s}$ or $G_{\alpha q}$ stimulation.[112] The cannabinoid CB_1 receptor agonist desacetyllevonantradol is a positive agonist for G_{i1} and G_{i2}, but an inverse agonist for G_{i3} while the CB_1 ligand methanandamide is an inverse agonist for G_{i1} and G_{i2} and a positive agonist for G_{i3}.[150] Also, as mentioned previously, some antagonists have been shown to actively internalize 7TMRs in the absence of any cellular signalling.

22.7 Conclusions

In general, the discovery of agonist and antagonist functional selectivity has changed the drug discovery landscape in many ways. On one hand, the discovery process has been made more complex in terms of discovering drugs in test systems with defined activities that are relevant in therapeutic systems. On the other hand, functional selectivity shows ligands to be potentially complex switches that can be used to unlock pre-programmed conformationally driven portals to cellular signalling that can be controlled through judicious choice of chemical structure. It will be most interesting to see how the new generation of functionally selective drugs affects the chemical control of pathophysiology in the clinic.

References

1. P. Spijker, N. Vaidehi, P. L. Freddolino, P. A. Hilbers and W. A. Goddard, *Proc. Natl. Acad. Sci. U. S. A.*, 2006, **103**, 4882.
2. P. S. Crozier, M. J. Stevens, L. R. Forrest and T. B. Woolf, *J. Mol. Biol.*, 2003, **333**, 493.
3. U. Gether, *Endocrinol. Rev.*, 2000, **21**, 90.

4. E. C. Meng and H. R. Bourne, *Trends Pharmacol. Sci.*, 2001, **22**, 587.
5. B. K. Kobilka, *J. Pept. Res.*, 2002, **60**, 317.
6. W. L. Hubbell, C. Altenbach, C. M. Hubbell and H. G. Khorana, *Adv. Protein Chem.*, 2003, **63**, 243.
7. P. S.-H. Park, D. T. Lodowski and K. Palczewski, *Annu. Rev. Pharmacol. Toxicol.*, 2008, **48**, 107.
8. A. S. V. Burgen, *Fed. Proc.*, 1981, **40**, 2723.
9. B. K. Kobilka and X. Deupi, *Trends Pharmacol. Sci.*, 2007, **28**, 397.
10. H. Fraunfelder, *Nature*, 1989, **338**, 623.
11. H. Fraunfelder, F. Parak and R. D. Young, *Annu. Rev. Biophys. Biophys. Chem.*, 1988, **17**, 451.
12. H. Fraunfelder, S. G. Sligar and P. G. Wolynes, *Science*, 1991, **254**, 1598.
13. J. Hilser and E. Freire, *Protein Struct. Funct. Genet.*, **27**, 171.
14. V. J. Hilser, D. Dowdy, T. G. Oas and E. Freire, *Proc. Natl. Acad. Sci. U. S. A.*, 1998, **95**, 9903.
15. V. J. Hilser, E. B. Garcia-Moreno, T. G. Oas, G. Kapp and S. T. Whitten, *Chem. Rev.*, 2006, **106**, 1545.
16. B. Ma, M. Shatsky, H. J. Wolfson and R. Nussinov, *Protein Sci.*, 2002, **11**, 184–197.
17. E. Freire, *Adv. Protein Chem.*, 1998, **51**, 255.
18. K. A. Dill and H. S. Chan, *Nat. Struct. Biol.*, 1997, **4**, 10.
19. D. W. Miller and K. A. Dill, *Protein Sci.*, 1997, **6**, 2166.
20. C. Woodward, *Trends Biochem. Sci.*, 1993, **18**, 359.
21. V. J. Hilser and E. B. Thompson, *Proc. Natl. Acad. Sci. U. S. A.*, 2007, **104**, 8311.
22. T. Liu, S. T. Whitten and V. J. Hilser, *Proteins*, 2006, **62**, 728.
23. G. Peleg, P. Ghanoun, B. K. Kobilka and R. N. Zare, *Proc. Natl. Acad. Sci. U. S. A.*, 2001, **98**, 8469.
24. H. O. Onaran and T. Costa, *Ann. N. Y. Acad. Sci.*, 1997, **812**, 98.
25. H. O. Onaran, A. Scheer, S. Cotecchia and T. Costa, in *The Pharmacology of Functional, Biochemical and Recombinant Systems*, ed. T. P. Kenakin and J. A. Angus, Handbook of Experimental Pharmacology, vol. 148, Springer, Heidelberg, 2002, pp. 217–260.
26. S. Kumar, B. Ma, C.-J. Tsai, N. Sinha and R. Nussinov, *Prot. Sci.*, 2000, **9**, 10.
27. M. A. Kjelsberg, S. Cotechhia, J. Ostrowski Caron and R. J. Lefkowitz, *J. Biol. Chem.*, 1992, **267**, 1430.
28. J. Liu, N. B. Perumal, C. J. Oldfield, E. W. Su, V. N. Uversky and A. K. Dunker, *Biochemistry*, 2006, **45**, 6873.
29. C. C. Blanpain, B. J. Doranz, J. Vaklil, J. Rucker, C. Govaerts, S. S. W. Baik, O. Lorthoir, I. Migeotte, F. Libert, F. Baleux, G. Vassart, R. W. Doms and M. Parmentier, *J. Biol. Chem.*, 1999, **274**, 34719.
30. C. C. Blanpain, B. Lee, J. Vaklil, B. J. Doranz, C. Govaerts, I. Migeotte, M. Sharron, V. Dupriez, G. Vassart, R. W. Doms and M. Parmentier, *J. Biol. Chem.*, 1999, **274**, 18902.

31. O. M. Z. Howard, A.-K. Shirakawa, J. A. Turpin, A. Maynard, G. J. Tobin, M. Carrington, J. J. Oppenheim and M. Dean, *J. Biol. Chem.*, 1999, **274**, 16228.
32. F. Tsamis, S. Gavrilov, F. Kajumo, C. Seibert, S. Kuhmann, T. Ketas, A. Trkola, A. Palani, J. W. Clader, J. R. Tagat, S. McCombie, B. Baroudy, J. P. Moore, T. P. Sakmar and T. Dragic, *J. Virol.*, 2003, **77**, 5201.
33. L. Wu, G. LaRosa, N. Kassam, C. J. Gordon, H. Heath, N. Ruffing, J. Chen, J. Humblias, M. Samson, M. Parmentier, J. P. Moore and C. R. MacKay, *J. Exp. Med.*, 1997, **186**, 1373.
34. K. Maeda, H. Nakata, Y. Koh, T. Miyakawa, H. Ogata, Y. Takaoka, S. Sbibayama, K. Sagawa, D. Fukushima, J. Moravek, Y. Kovanagi and H. Misuya, *J. Virol.*, 2004, **78**, 8654.
35. C. Watson, S. Jenkinson, W. Kazmierski and T. P. Kenakin, *Mol. Pharmacol.*, 2005, **67**, 1268.
36. C. Trankle, A. Weyand, A. Schroter and K. Mohr, *Mol. Pharmacol.*, 1999, **56**, 962.
37. B. L. Roth and D.-M. Chuang, *Life Sci.*, 1987, **41**, 1051.
38. R. B. Mailman, C. P. Lawler, M. M. Lewis and B. Blake, in *Dopamine Receptor Subtypes: From Basic Science to Clinical Application*, ed. P. Jenner and R. Demirdamar, IOS Press, Amsterdam and Oxford, 1998.
39. C. O. Lawler, V. J. Watts, R. G. Booth, S. B. Southerland and R. B. Mailman, *Abstr. Soc. Neurosci.*, 1994, **20**, 525.
40. C. P. Lawler, C. Prioleua, M. M. Lewis, C. Mak, D. Jiang, J. A. Schetz, A. M. Gonzalez, D. R. Sibley and R. B. Mailman, *Neuropsychopharmacology*, 1999, **20**, 612.
41. D. M. Mottola, L. L. Cook, S. R. Jones, R. G. Booth, D. E. Nichols and R. B. Mailman, *Abstr. Soc. Neurosci.*, 1991, **17**, 818.
42. A. Fisher, E. Heldman, D. Gurwitz, R. Haring, D. Barak, H. Meshulam, D. Marciano, R. Brandeis, Z. Pittel and M. Segal *et al.*, *Ann N. Y. Acad. Sci.*, 1993, **695**, 300.
43. D. Gurwitz, R. Haring, E. Heldman, C. M. Fraser, D. Manor and A. Fisher, *Eur. J. Pharmacol.*, 1994, **267**, 21.
44. J. S. Ward, L. Merrit, D. O. Calligaro, F. P. Bymaster, H. E. Shannon, B. D. Sawyer, C. H. Mitch, J. B. Deeter, S. C. Peters, M. J. Sheardown, P. H. Olesen, M. D. B. Swedberg and P. Sauerberg, *J. Med. Chem.*, 1995, **38**, 3469.
45. E. Heldman, J. Barg, A. Fisher, R. Levy, Z. Pittel, R. Zimlichman, M. Kushnir and Z. Vogel, *Eur. J. Pharmacol.*, 1996, **297**, 283.
46. T. P. Kenakin, *Trends Pharmacol. Sci.*, 1995, **16**, 232.
47. P. L. Prather, H. H. Loh and P. Y. Lee, *Mol. Pharmacol.*, 1994, **45**, 997.
48. P. L. Prather, N. A. Martin, C. S. Breivogel and S. R. Childers, *Mol. Pharmacol.*, 2000, **57**, 1000.
49. P. R. Albert and L. Robillard, *Cell Signal.*, 2002, **14**, 407.

50. N. T. Burford, D. X. Wang and W. Sadee, *Biochem. J.*, 2000, **348**, 531.
51. T. Gudermann, F. Kalbrenner and G. Schultz, *Annu. Rev. Pharmacol. Toxicol.*, 1996, **36**, 429.
52. S. Offermanns, T. Wieland, D. Homann, J. Sandmann, D. Bombien, K. Spicher, G. Schultz and K. Jakobs, *Mol. Pharmacol.*, 1994, **45**, 890.
53. A. Wise, M. Sheehan, S. Rees, M. Lee and G. Milligan, *Biochemistry*, 1999, **38**, 2272.
54. E. Hermans, *Pharmacol. Ther.*, 2003, **99**, 25.
55. D. Spengler, C. Waeber, C. Pantaloni, F. Holsboer, J. Bockaert, P. H. Seeburgt and L. Journot, *Nature*, 1993, **365**, 170.
56. T. P. Kenakin, *Nat. Rev. Drug Discov.*, 2002, **1**, 103.
57. T. P. Kenakin, *Expert Opin. Drug Discov.*, 2006, **1**, 635.
58. T. P. Kenakin, *Trends Pharmacol. Sci.*, 2007, **28**, 407.
59. D. M. Perez and S. S. Karnick, *Pharmacol. Rev.*, 2005, **57**, 147.
60. J. P. Kukkonen, *Recept. Channels*, 2004, **10**, 167.
61. B. F. Roettger, D. Ghanekar, R. Rao, C. Toledo, J. Yingling, D. Pinon and L. J. Miller, *Mol. Pharmacol.*, 1997, **51**, 357.
62. J. A. Gray and B. L. Roth, *Brain Res. Bull.*, 2001, **56**, 441.
63. T. Palanche, B. Ilien, S. Zoffmann, M.-P. Reck, B. Bucher, S. J. Edelstein and J.-L. Galzi, *J. Biol. Chem.*, 2001, **276**, 34853.
64. T. P. Kenakin, *FASEB J.*, 2001, **15**, 598.
65. T. P. Kenakin, *Trends Pharmacol. Sci.*, 2003, **24**, 346.
66. J.-L. Baneres, D. Mesnier, A. Martin, L. Joubert, A. Dumuis and J. Bockaret, *J. Biol. Chem.*, 2005, **280**, 20253.
67. S. Galandrin, G. Oligny-Longpre, H. Bonin, K. Ogawa, C. Gales and M. Bouvier, *Mol. Pharmacol.*, 2008, **74**, 162.
68. U. Gether, S. Lin and B. K. Kobilka, *J. Biol. Chem.*, 1995, **270**, 28268.
69. P. Ghanouni, Z. Gryczynski, J. J. Steenhuis, T. W. Lee, D. L. Farrens, J. R. Lakowicz and B. K. Kobilka, *J. Biol. Chem.*, 2001, **276**, 24433.
70. S. Granier, T. Kim, A. M. Shafer, V. R. Ratnala, J. J. Fung, R. N. Zare and B. K. Kobilka, *J. Bio. Chem.*, 2007, **282**, 13895.
71. T. Georgieva, S. Devanathan, D. Stropova, P. K. Park, Z. Salamon, G. Tollin, H. J. Hruby, W. R. Roeske, H. I. Yamamura and E. Varga, *Eur. J. Pharmacol.*, 2008, **581**, 19.
72. V. J. Hruby and G. Tollin, *Curr. Opin Pharmacol.*, 2007, **7**, 507.
73. B. K. Kobilka and U. Gether, *Methods Enzymol.*, 2002, **343**, 170.
74. M. J. Lohse, P. Hein, C. Hoffmann, V. O. Nikolaev, J.-P. Viladarga and M. Bunemann, *Br. J. Pharmacol.*, 2008, **153**, S125.
75. T. Okada and K. Palczewski, *Curr. Opin. Struc. Biol.*, 2001, **1**, 420.
76. L. P. Pellissier, J. Sallander, M. Campillo, F. Gaven, E. Queffeulou, M. Pillot, A. Dumuis, S. Claeysen, J. Bockaert and L. Pardo, *Mol. Pharmacol.*, 2009, **75**, 982.
77. G. Swaminath, Y. Xiang, T. W. Lee, J. Steenhuis, C. Parnot and B. K. Kobilka, *J. Biol. Chem.*, 2004, **279**, 686.
78. G. Swaminath, X. Deupi, T. W. Lee, W. Zhu, F. S. Thian, S. Kobilka and B. Kobilka, *J. Biol. Chem.*, 2005, **280**, 22165.

79. A. Tutor, P. Penela and F. Mayor, *Cardiocasc. Res.*, 2007, **76**, 51.
80. J. P. Vilardaga, M. Bunemann, C. Krasel, M. Castro and M. J. Lohse, *Nat. Biotechnol.*, 2003, **21**, 807.
81. J. P. Viladarga, R. Steinmeyer, G. S. Harms and M. J. Lohse, *Nat. Chem Biol.*, 2005, **1**, 25.
82. A. Zurn, U. Zabel, J.-P. Viladarga, H. Schindlein, M. J. Lohse and C. Hoffmann, *Mol. Pharmacol.*, 2009, **75**, 534.
83. J. L. Whistler, H. H. Chuang, P. Chu, L. Y. Jan and M. van Zastrow, *Neuron*, 1999, **23**, 737.
84. O. Kudlacek, M. Waldoer, M. U. Kassack, P. Nickel, J. I: Salmi, M. Freissmuth and C. Nanoff, *Naunyn Schmiedebergs Arch. Pharmacol.*, 2002, **365**, 8.
85. M. B. Jarpe, C. Knall, F. M. Mitchell, A. M. Buhl, E. Duzic and G. L. Johnson, *J. Biol. Chem.*, 1998, **273**, 3097–3104.
86. D. R. Manning, *Mol. Pharmacol.*, 2002, **62**, 451.
87. C. P. Lawler, C. Prioleua, M. M. Lewis, C. Mak, D. Jiang, J. A. Schetz, A. M. Gonzalez, D. R. Sibley and R. B. Mailman, *Neuropsychopharmacology*, 1999, **20**, 612.
88. J. D. Kilts, H. S. Connery, A. G. Arrington, M. M. Kewis, C. P. Lawler, G. S. Oxford, K. L. O'Malley, R. D. Todd, B. L. Blake and D. E. Nichols, *J. Pharmacol. Exp. Ther.*, 2002, **301**, 1179.
89. D. Gurwitz and R. Haring, *Drug Disc. Today*, 2003, **8**, 1108.
90. S. S. Ferguson, *Pharmacol. Rev.*, 2001, **53**, 1.
91. K. L. Pierce and R. J. Lefkowitz, *Nat. Rev. Neurosci.*, 2001, **2**, 727.
92. L. M. Luttrell and R. J. Lefkowitz, *J. Cell Sci.*, 2002, **115**, 455.
93. Y. M. Kim and J. L. Benovic, *J. Biol. Chem.*, 2002, **277**, 30760.
94. S. A. Laporte, R. H. Oakley, J. A. Holt, L. S. Barak and M. G. Caron, *J. Biol. Chem.*, 2000, **275**, 23120.
95. S. A. Laporte, W. E. Miller, K. M. Kim and M. G. Caron, *J. Biol. Chem.*, 2002, **277**, 9247.
96. M. A. Edeling, S. K. Mishra, P. A. Keyel, A. L. Steinhauser, B. M. Collins, R. Roth, J. E. Heuser, D. J. Owen and L. M. Traub, *Dev. Cell*, 2006, **10**, 329.
97. O. B. Goodman, J. G. Krupnik, F. Santini, V. V. Gurevich, R. B. Penn, A. W. Gagnon, J. H. Keen and J. L. Benovic, *Nature*, 1996, **383**, 447.
98. S. M. DeWire, S. Ahn, R. J. Lefkowitz and S. K. Shenoy, *Annu. Rev. Physiol.*, 2007, **69**, 483.
99. R. J. Lefkowitz and S. K. Shenoy, *Science*, 2005, **308**, 512.
100. S. Terrillon and M. Bouvier, *EMBO J.*, 2004, **23**, 3950.
101. R. J. Lefkowitz, K. Rajagopal and E. J. Whalen, *Mol. Cell*, 2006, **24**, 643.
102. L. M. Luttrell, *J. Mol. Neurosci.*, 2005, **26**, 253.
103. N. J. Smith and L. M. Luttrell, *Hyptertension*, 2006, **48**, 173.
104. T. Van Biesen and L. M. Luttrell, *Endocr. Rev.*, 1996, **17**, 698.
105. A. Claing, S. A. Laporte, M. G. Caron and R. J. Lefkowitz, *Prog. Neurobiol.*, 2002, **66**, 61.
106. R. J. Lefkowitz and E. J. Whalen, *Curr. Opin. Cell Biol.*, 2004, **16**, 162.

107. Z. Garorik and L. Hunyady, *Trends Endocrinol. Metab.*, 2004, **15**, 286.
108. A. Vroon, M. S. Lombardi, A. Kavelaars and C. J. Heijnen, *J. Neuroimmunol.*, 2003, **137**, 79.
109. A. C. Holloway, H. Qian, L. Pipolo, J. Ziogas, S. Miura, S. Karnik, B. R. Southwell, M. J. Lew and W. G. Thomas, *Mol. Pharmacol.*, 2002, **61**, 768.
110. S. Ahn, S. K. Shenoy, H. Wei and R. J. Lefkowitz, *J. Biol. Chem.*, 2004, **279**, 35518.
111. H. Wei, Ahn, S. K. Shenoy, S. S. Karnik, L. Hunyady, L. M. Luttrell and R. J. Lefkowitz, *Proc. Natl. Acad. Sci. U. S. A.*, 2003, **100**, 10782.
112. D. Gesty-Palmer, M. Chen, E. Reiter, S. Ahn, C. D. Nelson, S. Wang, A. E. Eckhardt, C. L. Cowan, R. F. Spurney, L. M. Luttrell and R. J. Lefkowitz, *J. Biol. Chem.*, 2006, **281**, 10856.
113. A. E. Brady and L. E. Limbird, *Cell. Signal.*, 2002, **14**, 297.
114. S. Gavarini, C. Becamel, C. Altier, C. P. Lory, P. J. Poncet, J. Wijnholds, J. Bockaert and P. Marin, *Mol. Biol. Cell*, 2006, **17**, 4619.
115. J. Bockaert and J. P. Pin, *EMBO J.*, 1999, **18**, 1723.
116. J. Bockaert, L. Fagni, A. Dumuis and P. Marin, *Pharmacol. Ther.*, 2004, **103**, 203.
117. Y. Sun, D. McGarrigle and X.-Y. Huang, *Mol. BioSyst.*, 2007, **3**, 849.
118. S. L. Ferrari, D. D. Pierroz, V. Glatt, D. S. Goddard, E. N. Bianchi, F. T. Lin, D. Manen and M. L. Bouxsein, *Endocrinology*, 2005, **146**, 1854.
119. C. L. Schmid, K. M. Raehal and L. M. Bohn, *Proc. Natl. Acad. Sci. U. S. A.*, 2008, **105**, 1079.
120. A. Bisselo, M. Chorev, M. Rosenblatt, L. Monticelli L, D. F. Mierke and D. F. Ferrari, *J. Biol. Chem.*, 2002, **277**, 38524.
121. J. D. Violin, S. M. DeWire, W. G. Barnes and R. J. Lefkowitz, *J. Biol. Chem.*, 2006, **281**, 36411.
122. K. Rajagopal, E. J. Whalen, J. D. Violin, J. A. Stiber, P. B. Rosenberg, R. T. Premont, T. M. Coffman, H. A. Rockman and R. J. Lefkowitz, *Proc. Natl. Acad. Sci. U. S. A.*, 2006, **103**, 16284.
123. C. M. Revenkar, C. M. Vines, D. Cimino and E. R. Prossnitz, *J. Biol. Chem.*, 2004, **279**, 24578.
124. D. E. Keith, B. Anton, S. R. Murray, P. A. Zaki, P. C. Chu, D. V. Lissin, G. Monteillet-Agius, P. L. Stewart, C. J. Evans and M. von Zastrow, *Mol. Pharmacol.*, 1998, **53**, 377.
125. J. Zhang, Z. J. Zhangm, S. S. G. Ferguson, L. S. Barak, S. R. Bodduluri, S. A. LaPorte, P.-Y. Law and M. G. Caron, *Proc. Natl. Acad. Sci. U. S. A.*, 1998, **95**, 7157.
126. C. E. Groer, K. Tidgewell, R. A. Moyer, W. W. Harding, R. B. Rothman, T. E. Prisinzano and L. M. Bohn, *Mol. Pharmacol.*, 2007, **71**, 549.
127. H. Xu, J. S. Partilla, X. Wang, J. M. Rutherford, K. Tidgewell, T. E. Prisinzano, L. M. Bohn and R. B. Rothman, *SYNAPSE*, 2007, **61**, 166.
128. E. V. Varga, E. Navratilova, D. Stropova, J. Jambrosic, W. R. Roeske and H. I. Yamamura, *Life Sci.*, 2003, **76**, 599.
129. L. Bohn, R. J. Lefkowitz, R. R. Gainetdinov, K. Peppel, M. G. Caron and L. T. Lin, *Science*, 1999, **286**, 2495.

130. K. M. Raehal, J. K. L. Walker and L. M. Bohn, *J. Pharmacol. Exp. Ther.*, 2005, **314**, 1195.
131. T. P. Kenakin, *J. Recept. Signal Trans.*, 2010.
132. T. P. Kenakin, *Nat. Rev. Drug Discov.*, 2005, **4**, 919.
133. E. L. Maillet, N. Pellegrini, C. Valant, B. Bucher, M. Hibert, J.-J. Bourguignon and J.-L. Galzi, *FASEB J.*, 2007, **21**, 2124.
134. J. M. Mathiesen, T. Ulven, L. Martini, L. O. Gerlach A. Heineman and E. Kostenis, *Mol. Pharmacol.*, 2005, **68**, 393.
135. C. Quiniou, P. Saphieha, I. Lahaie, X. Hou, S. Brault, M. Beauchamp, M. Leduc, L. Rihakova, J.-S. Joyal and S. Nadeau *et. al.*, *J. Immunol.*, 2008, **180**, 6977.
136. B. Pommier, S. Da Nascimento, S. Dumont, B. Bellier, E. Million, C. Garbay, B. P. Roques and F. Noble, *J. Neurochem.*, 1999, **73**, 281.
137. R. P. Stephenson, *Br. J. Pharmacol. Chemother.*, 1956, **11**, 379.
138. T. P. Kenakin, *Nature Rev. Drug Discov.*, 2009, **8**, 617.
139. Y. Fang, *ASSAY Drug Dev. Technol.*, 2006, **4**, 583.
140. Y. Fang, A. M. Ferrie, N. Fontaine, J. Mauro and J. Balikrishnan, *Biophys. J.*, 2006, **91**, 1925.
141. Y. Fang and A. M. Ferrie, *FEBS Lett.*, 2008, **582**, 558.
142. J. Kim, S. Ahn, K. Rajagopal and R. J. Lefkowitz, *J. Biol. Chem.*, 2009, **284**, 11953.
143. J. W. Black and P. Leff, *Proc. R. Soc. Lond. [Biol.]*, 1983, **220**, 141.
144. J. W. Black, P. Leff and N. P. Shankley (Appendix J. Wood), *Br. J. Pharmacol.*, 1985, **84**, 561.
145. T. P. Kenakin and L. Miller, *Pharmacol. Rev.*, 2010.
146. S. Galandrin and M. Bouvier, *Mol. Pharmacol.*, 2006, **70**, 1575.
147. M. Azzi, P. G. Charest, S. Anger, G. Rousseau, T. Kohout and M. Bouvier, *Proc. Natl. Acad. Sci. U. S. A.*, 2004, **100**, 11406.
148. J. W. Wisler, S. M. DeWire, E. J. Whalen, J. D. Violin, M. T. Drake, S. K. Shenoy and R. J. Lefkowitz, *Proc. Natl. Acad. Sci. U. S. A.*, 2007, **104**, 16657.
149. T. J. Gardella, M. D. Luck, E. Jensen, J. T. Schipani, J. T. Potts and H. Juppner, *Endocrinology*, 1996, **137**, 3936.
150. S. Mukhopadhyay and A. C. Howlett, *Mol. Pharmacol.*, 2005, **67**, 2016.

CHAPTER 23

Functional Selectivity of G Protein-coupled Receptors: Bridging the Gap Between Monomeric and Dimeric Receptors

X. ROVIRA AND J. GIRALDO*

Laboratory of Systems Pharmacology and Bioinformatics, Institut de Neurociències and Unitat de Bioestadística, Universitat Autònoma de Barcelona, 08193 Bellaterra, Spain

23.1 Introduction

G protein-coupled receptors (GPCRs) are the largest family of cell surface receptors and key components of the signal transduction machinery. The activation of these receptors, which is driven by a wide variety of extracellular entities, regulates the function of most cells in the body. Ligands acting on GPCRs are commonly used in drug therapy for numerous diseases and it is estimated that a 27% of the gene distribution of human targets of currently approved drugs corresponds to class A GPCRs.[1]

The process of GPCR activation has been studied by many experimental[2–8] and theoretical[9,10] approaches. A vast array of data has been collected identifying a number of receptor micro-switches along the complex functional network that interconnects the binding of the ligand to the receptor and the activation processes of both the receptor and the heterotrimeric G protein.[11,12] However,

RSC Drug Discovery Series No. 8
G Protein-Coupled Receptors: From Structure to Function
Edited by Jesús Giraldo and Jean-Philippe Pin

Published by the Royal Society of Chemistry, www.rsc.org

the issue of the number of receptor protomers involved in receptor activation remains elusive. Thus, it is known that a monomeric receptor is sufficient to activate G proteins.[13–17] However, there is increasing evidence indicating that GPCRs group into dimers or higher oligomers in live systems.[18–24]

The aggregation of receptors into oligomers seems unlikely to occur by mere consequence of an increase in their concentration level in the membrane. Indeed, there is evidence showing the involvement of receptor oligomerization in different processes, amongst them receptor maturation, the correct addressing of the receptor to the membrane and receptor internalization.[25] Importantly, valuable data providing insights on how receptor dimerization can be helpful for receptor functioning have been recently published. Using the mouse luteinizing hormone receptor (LHR) as a GPCR model, Rivero-Müller *et al.* showed that dimerization may supply a structural arrangement suitable for restoring function in defective receptors.[22] Working on the melatonin MT1 receptor, Maurice *et al.* characterized the binding of MT1 to both a Gi protein and the regulator of G protein signalling 20 (RGS20) and proposed a model in which one Gi and one RGS20 bind to separate protomers of MT1 dimers in a pre-associated ternary complex that rearranges upon agonist activation.[26] The authors generalized this result and provided a new functional justification for GPCR dimerization, which applies to both homo- and heterodimers, by suggesting that dimerization gives receptors new functional opportunities by the simultaneous and direct binding of two GPCR-interacting proteins (GIPs) to the two dimer protomers.[26] These complex interactions involving oligomeric GPCRs and multiple GIPs have also been suggested by others.[27]

From data described above, it is apparent that a functional role for receptor assembly in the early steps of ligand–receptor interactions is plausible. It is expected, therefore, that a mechanistic analysis of the activation process of GPCR dimers may provide new insights into the reasons why receptor oligomerization is present at the very beginning of the signalling process.[21] What is more, models for receptor oligomers may reveal singular mechanisms that, formalized in a mathematical framework, can be useful for a quantitative description of pharmacological phenomena. Hopefully, these models will be able to furnish agonist efficacy with the structural elements required for the currently observed functional complexity.[28] Yet, what is the current meaning of the classical concept of agonist efficacy?

With the aim of providing an answer to this question, this review attempts to summarize, in a mathematical modelling context, the evolution of the concept of efficacy from early monomeric, single pathway receptors to current oligomeric, multiple pathway receptors.

23.2 Originally Signal Transduction was Simple: A Monomeric Receptor and a Single Signalling Pathway

The concept of efficacy has been revisited many times since, forced by the need to explain the results obtained with drugs (partial agonists) showing activities lower than maximum, Ariens included intrinsic activity (α) in concentration-effect pharmacological equations.[29] However, the concept remains elusive.[30,31]

Stephenson introduced the notion of receptor stimulus in an attempt to separate ligand binding from the effect produced on the receptor.[32] This led to the definition of efficacy (e) as a proportionally term denoting the capacity of a drug to elicit a pharmacological response. Later, Furchgott defined intrinsic efficacy (ε) as a system-independent drug property by dividing e by the total number of receptors.[33]

Subsequently Monod, Wyman and Changeaux laid the foundations of one of the most widely accepted conceptions of receptor theory (the conformational selection-based MWC model) by establishing a model wherein the receptor is in equilibrium between inactive and active states whose relative populations depend on preferential ligand binding.[34] Concurrently with the MWC model, the Koshland–Nemethy–Filmer (KNF) model appeared in which the ligand induces the activation of the receptor, originally in an inactive state, as a consequence of the binding process.[35] Both the MWC and KNF models can be considered part of the so-called two-state receptor model.[36–38]

In the two-state receptor model (Figure 23.1A and Appendix A23.1), intrinsic efficacy ε is defined as either Z/T [ratio of the dissociation constants of the ligand for the inactive (R) and active (R*) states of the receptor, following the MWC hypothesis] or Y/X (ratio of the induction of receptor active state of the ligand-receptor complex relative to the free receptor, following the KNF model). It is worth noting that both definitions are formally equivalents and produce the same quantity for ε because these equilibrium constants form part of a thermodynamic cycle.

The main contribution of the two-state model has been its capacity to account for receptor basal response ($X>0$) and the quantitative distinction between agonists ($T<Z$ or $Y>X$), inverse agonists ($T>Z$ or $Y<X$) and neutral antagonists ($T=Z$ or $Y=X$). Concentration–effect theoretical curves are constructed by defining the effect as the fraction of active receptors (f_{R*}), the component of the signal–transduction cascade closer to the receptor action. The shape of f_{R*} provides quantitative information on the capacity of the agonist-receptor system for eliciting the physiological response. In particular, the two asymptotes of the curves ($[A]=0$, basal response, and $[A]\rightarrow\infty$, maximum response; see Appendix A23.1) will give us a measure of receptor constitutive activity and agonist efficacy, respectively. We see that f_{R*} basal response depends uniquely on receptor constitutive activity through equilibrium constant X whereas f_{R*} maximum (minimum in the case of inverse agonists) response depends on both receptor constitutive activity (X) and agonist intrinsic efficacy (Z/T).

The two-state receptor model has been used in different theoretical analyses, in particular for the examination of pharmacological response after receptor mutation,[39] the extension of the Schild and Cheng–Prusoff methods for antagonist affinity constant determination to the case of inverse agonists[40] and the discussion on whether neutral antagonists do really exist.[41] The two-state model can be considered a subset of the extended ternary complex model, in

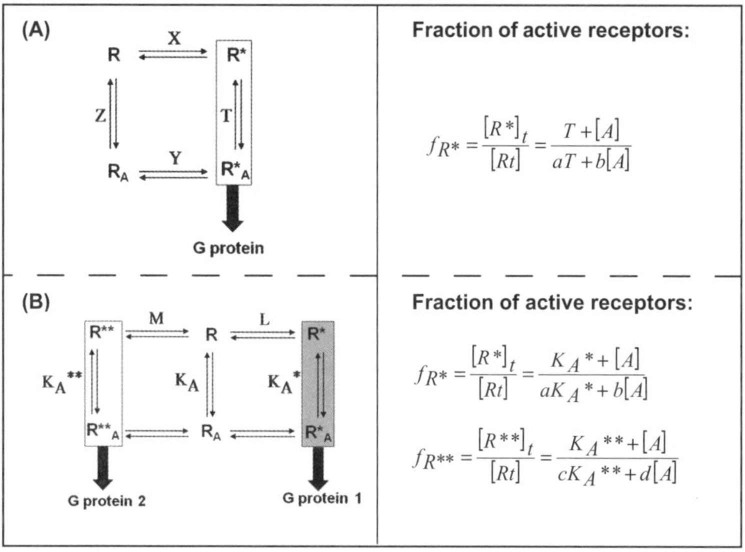

Figure 23.1 Monomeric models. (**A**) The two-state receptor model of agonism. The receptor is monomeric and there is only one signalling pathway, associated to the receptor active state R* (originally a G protein-dependent pathway but could also be G protein-independent). (**B**) The three-state receptor model of agonism: extension of the two state receptor model to account for functional selectivity. The receptor is monomeric and there are two signalling pathway arising from two interconvertible active conformations, R* and R** (originally both were G protein-dependent pathways but one or both of them could be a G protein-independent pathway). See Appendix A23.1 and A23.2, respectively, for definition of constants. The notations of the models **A** and **B** are as presented in previous studies (refs. 39 and 48, respectively). In the models, horizontal arrows correspond to interconversions between receptor states and vertical arrows to ligand binding to receptor states. Note the conceptual equivalence between constants X and both L and M in Figures 23.1A and 23.1B, respectively. It is worth noting that both models yield the same empirical equation—a Hill equation with a Hill coefficient of 1.

which the coupling of the G protein to both R* and AR* activated receptors was included.[42] The explicit inclusion of the G protein in mathematical models can be necessary in particular circumstances such as, for instance, to account for biphasic concentration–effect curves which cannot be explained by conformational interconversions of a monomeric receptor. However, for consistency with this review as a whole, the G protein component is not considered in the mathematical models presented and the discussion is limited to extensions of the two-state model to include the possibility of more than one signalling pathway both in monomeric and dimeric receptors.

The two-state model of agonism contains a single pathway for signal transduction. That is one receptor–one G protein. However, recently, many data have appeared indicating that a single receptor type may lead to more than one signalling pathway, which can be differentially ligand-regulated. This finding has received several names such as ligand-directed signalling, agonist-receptor trafficking, biased agonism and functional selectivity.[43]

23.3 Later Signal Transduction was Multiple

Because GPCRs are proteins, they are flexible molecules. Their inherent flexibility confers on these receptors a complexity which makes them not simple 'on–off' switches but ensembles of active and inactive conformations.[44] Thus, it is not unexpected that this structural plasticity has been exploited by evolution to associate specific receptor conformations with specific signalling pathways. Binding of a ligand to a receptor may alter the multiple active basal state of the free receptor, driving the equilibrium towards ligand-specific receptor conformations that can result in differential activation of signalling pathways (functional selectivity).[28] In this way, and for a given receptor, a particular ligand can act as an inverse agonist for one pathway and as a positive agonist for another. A typical example is that of propranolol on β_2-adrenoceptor, which acts as an inverse agonist and as a positive agonist for the $G_{\alpha s}$[45] and β-arrestin–extracellular signal-regulated kinase (ERK)[46] signalling pathways, respectively. Functional selectivity is currently accepted as a genuine feature of GPCRs after the many experimental findings that cannot be explained by a simple receptor–G protein binary interaction (for a review see refs. 30, 31 and 47).

The three-state receptor model of agonist action (Figure 23.1B and Appendix A23.2) was proposed as an extension of the two-state receptor model to account for the possibility of the coupling of the receptor to more than one G protein.[48] In this model an additional active receptor state (R**) was included and it was proposed that R* interacts with G protein G1 whereas R** interacts with G protein G2. This structural multiplicity provides functional versatility to the system: (i) one particular ligand can have different efficacy orders or even opposite (agonist–antagonist) effects in the same receptor type depending on the signalling pathway considered (biased ligand); and (ii) for a particular receptor (in the absence of ligand), we can consider both R* and R** conformations as competitive partners.

This leads to the concept of unbiased receptor if both pathways are equipotent and biased receptor if one pathway prevails over the other.[49] It is worth noting that, although the three-state receptor model was originally proposed to account for G protein–multiple pathways, it can also be applied to the case of signalling pathways including accessory proteins other than G proteins (as for instance β-arrestins) as it is currently known that GPCRs signal not exclusively through the G protein cascade. This knowledge can have important applications in human health. As it has been pointed out, the multiplicity of signalling involving G proteins, β-arrestins and other accessory proteins could constitute in the future a new source for drug development and therapeutics.[50]

23.4 Currently Signal Transduction is Complex: Oligomeric Receptors and Multiple Signalling Pathways

Though it has been shown that monomeric GPCRs are able to activate their cognate G proteins,[13–16] there is a large body of evidence indicating that GPCRs group into dimers and higher oligomers in live systems (see ref. 51 for review). A number of mechanistic approaches can be found in the literature[36,52–60] that include receptor oligomerization to account for the shape of binding and response curves (see ref. 61 for review). Curve shape can provide useful information concerning receptor oligomerization because Hill coefficients lower and higher than one are often taken as indicators of the cooperativity effects, negative and positive, respectively, that happen in oligomeric systems. It is worth noting that two- or multiple-states receptor models involving conformational interconversions in monomeric receptors yield monophasic curves in all cases with a Hill coefficient equal to one (see refs. 62 and 63 for a discussion on this issue).

The two-state dimer receptor model[64,65] (Figure 23.2A and Appendix A23.3) is the extension of the two-state receptor model of agonism to dimeric receptors. In the model the receptor can have two states—an inactive (RR) and an active (RR)*. The receptor can either be free (Figure 23.2A top row), or occupied by one (Figure 23.2A middle row) or two ligands (Figure 23.2A bottom row). Vertical arrows correspond to ligand–receptor dissociation constants (conformational selection) whereas horizontal arrows correspond to induction of receptor active state (conformational induction). The model is able to explain negative and positive cooperative effects and their effects on curve shapes: Hill coefficients lower and greater than one, respectively (see ref. 66 for a discussion of the Hill coefficient as a curve–slope parameter).

The three-state receptor model was proposed as an extension of the earlier two-state receptor model to account for functional selectivity in monomeric receptors (see above). For the same reasons and in an analogous way, a three-state dimer receptor model (Figure 23.2B and Appendix A23.4) containing one inactive (RR) and two active receptor states [(RR)* and (RR)**] was constructed to account for functional selectivity in a dimer receptor context.[67] The model was efficient in explaining the striking curve–shape patterns of some 5-HT$_{2A}$ receptor antagonists in two signalling pathways—one associated with inositol phosphate accumulation (monophasic) and the other with arachidonic acid release (biphasic).[67] Systematic data analyses following different receptor models suggested that interconversion between multiple active dimer receptor states was the simplest model to account for the observed curve profiles. Fluorescence resonance energy transfer and immunoprecipitation studies confirmed the hypothesis of 5-HT$_{2A}$ dimer receptors.[67]

In the above-mentioned dimer receptor models,[65,67] the receptor was considered as a global entity and the protomers within each dimer state could

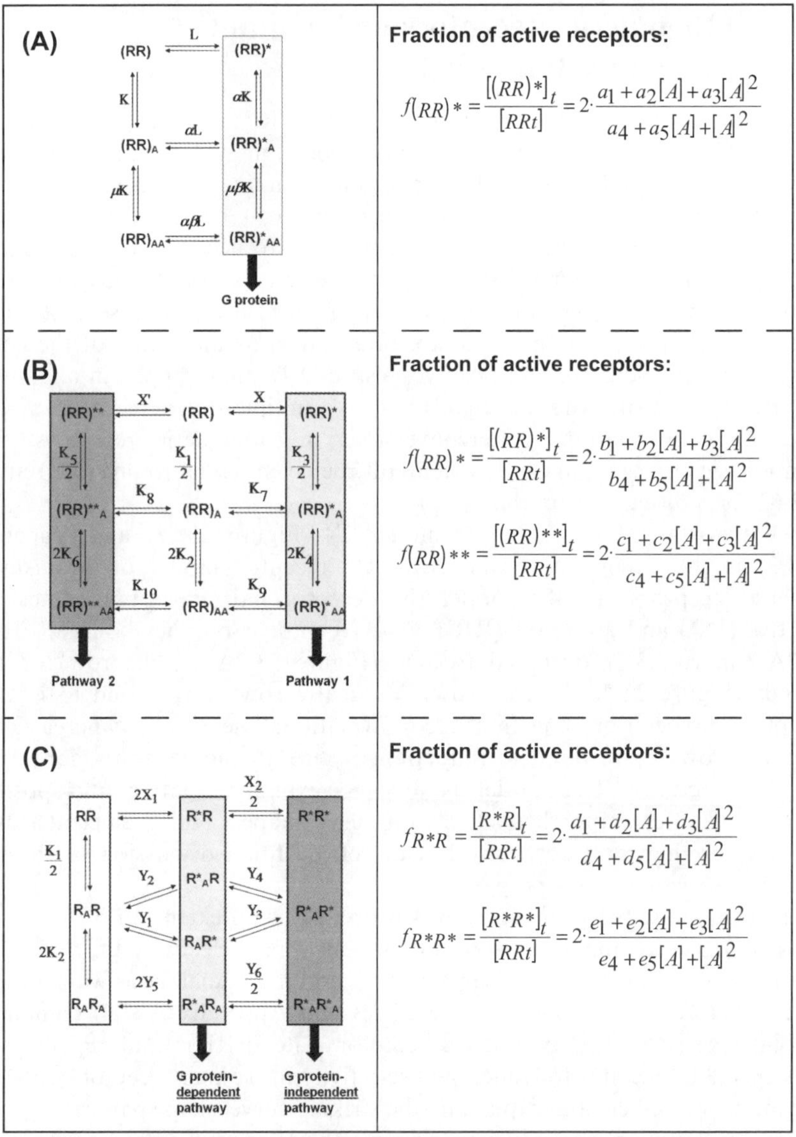

not be distinguished from each other. It seems, however, that the level of formalization of receptor signal transduction had not reached its true limit by the time these studies were published—we can ask ourselves whether there exists a limit for complexity in biological systems—and a dimer receptor model in which the protomers are explicitly differentiated was later proposed (see below).[68]

Figure 23.2 Dimeric models. (**A**) The two-state receptor dimer model of agonism: extension of the two-state receptor model (monomeric receptors) to dimeric receptors. The receptor is dimeric and there is only one signalling pathway, associated to the receptor active state (RR)* (originally a G protein-dependent pathway but could also be a G protein-independent pathway). (**B**) The three-state receptor dimer model of agonism: extension of the two-state receptor dimer model to account for functional selectivity. The receptor is dimeric and there are two signalling pathways arising from two interconvertible active conformations, (RR)* and (RR)** (originally both were proposed to be G protein-dependent pathways but one or both of them could be a G protein-independent pathway). (**C**) The asymmetric/symmetric three-state receptor dimer model of agonism. Adaptation of the three-state receptor dimer model to account for the possibility that functional selectivity depends on the stoichiometry of subunit activation. The receptor is dimeric and there are two signalling pathway arising from two interconvertible active conformations, R*R and R*R*. It was proposed that the asymmetrically activated dimer signals through a G protein-dependent pathway and the symmetrically activated dimer signals through a G protein-independent pathway. See Appendix A23.3, A23.4 and A.23.5, respectively, for definition of constants. The notations of the models A, B and C have been maintained as presented in previous studies (refs. 65, 67 and 68, respectively). In the models, horizontal arrows correspond to interconversions between receptor states and vertical arrows to ligand binding to receptor states. Note the conceptual equivalence between constants L, both X and X′, and both X_1 and X_2 in **A**, **B** and **C**, respectively. It is worth noting that the three models yield the same empirical equation, a ratio between two quadratic polynomials including five empirical parameters.

23.4.1 The Protomers within Activated Receptor Dimers can be Arranged in Either Symmetric or Asymmetric Conformations

Assuming that the receptor is dimeric when it interacts with an agonist, receptor activation prompts the question as to how the protomers are arranged within the activated dimer: are they in the same or in a different conformation? That is to say, is the activated dimer in a symmetric or an asymmetric state? Moreover, should symmetry play a role, might symmetry was associated with functional selectivity?

Classically, the function of GPCRs was associated to the activation of G proteins. Currently, it is known that GPCRs can signal through other pathways, amongst others, β-arrestins, tyrosine kinases and PDZ-domain containing proteins.[69] This versatility in receptor signalling raises some further questions. Can the symmetry/asymmetry of the protomers within a receptor dimer distinguish between different pathways? In other words, is the process known as functional selectivity related with the symmetry/asymmetry of dimeric receptors?

There are many data in the literature showing that the asymmetry of dimer receptor activation is associated with G protein-dependent pathways

Table 23.1 Receptors for which an asymmetric arrangement of their protomers within a dimer receptor structure have been proposed for G protein-dependent pathways.

Receptor	*GPCR class*	*Reference*
BLT1	A	70
Rhodopsin	A	15
Rhodopsin	A	80
5-HT2C	A	81
D2	A	8
LHR	A	22
V1a, V2	A	23
Oxytocin	A	23
D2	A	23
mGluRs	C	82
GABAB (heterodimer)	C	83
T1R taste (heterodimer)	C	84

(Table 23.1). Of particular interest is the study performed by Damian *et al.* on the conformational arrangements of the leukotriene B4 BLT1 receptor dimer subunits depending on whether the G protein was present or absent.[70] The authors found that, in the presence of the G protein, the two subunits within the activated BLT1 dimer display different conformations (asymmetry) whereas, in the absence of G protein, the two protomers display similar active conformations (symmetry).[70] Interestingly, the authors speculated with the possibility that, if activation of the G protein is associated with an asymmetry of the receptor dimer, then a symmetric dimer might be associated with G protein-independent (arrestin) signalling pathway; they concluded that, if confirmed, this would lead to the proposal that the signal transduction pathway of a receptor dimer may be controlled by the stoichiometry of subunit activation.[70] Interestingly, evidence in favour of a symmetric subunit arrangement for G protein-independent pathways was found for the binding of β-arrestin-1 to muscarinic M_3 receptor dimers, for which paired stimulation of the two receptor subunits was required.[71]

A mathematical model to account quantitatively for the functional consequences of symmetry in a GPCR dimer has been recently proposed, *i.e.* the asymmetric/symmetric three-state receptor dimer model (Figure 23.2C and Appendix A23.5).[68] The model includes three conformations of the receptor—an inactive RR, the asymmetrically activated R*R and the symmetrically activated R*R*. It is worth noting that the model includes the possibility of *cis* and *trans* activation through the equilibrium constants Y_2 and Y_1, respectively. Moreover, consistent with the experimental results outlined above, the authors assumed that the asymmetrically activated dimer governs the G protein-mediated pathway whereas the symmetrically activated dimer regulates the G protein-independent pathway. Both functional pathways are calculated as the fractional receptor populations of the corresponding asymmetric and symmetric species.[68]

23.4.2 Application of the Asymmetric/Symmetric Three-state Receptor Model to Particular Experimental Results

The model allowed[68] for the analysis of two remarkable experimental findings:

(i) dosage dependent switch from G protein-dependent to G protein-independent pathway;[72]
(ii) inverse agonists displaying agonist behavior.[8]

23.4.2.1 Dosage Dependent Switch from G Protein-dependent Pathway to G Protein-independent Pathway

Sun *et al.* showed that, at low concentration of an agonist, β_2-adrenergic receptors signal through $G\alpha_s$ to activate the mitogen-activated protein kinase pathway in mouse embryonic fibroblast cells. However, at high agonist concentrations, signals are also transduced through these receptors but *via* an additional pathway that is G protein-independent but tyrosine kinase Src-dependent. A biphasic curve was obtained; the first phase of the curve was associated to a G protein-dependent pathway whereas the second was supposed dependent on Src.[72]

The asymmetric/symmetric three-state receptor dimer model can describe these experimental results by assuming that the first phase results from the occupation of one ligand site in the asymmetrically activated dimer state and the second phase from the occupation of two sites in the symmetrically activated dimer state. Figure 23.3 shows a simulation under these premises. The solid grey curve corresponds to the G protein pathway, the dashed gray curve to the G protein-independent pathway and the black one to the sum of both pathways. The experimentally found dosage-dependent switch from G protein-coupled to G protein-independent signalling is well described by the simulation as the sum of two pathways; the G protein-dependent response appears at lower agonist concentrations and has a bell-shaped form whereas the G-protein-independent appears at higher agonist concentrations and has a sigmoid form. The G protein-independent signalling comes just when the G protein-dependent signalling initiates its decline; the top of the first bell-shaped curve provides the central plateau of the global curve.

Furthermore, the model provides a detailed description of the distribution of receptor species along ligand concentration (Figure 23.4). In the figure, three regions can be distinguished. The left region is constituted by unoccupied receptors, the central region contains receptors with one occupied site and the right region includes receptors with both sites occupied. With regard to the signalling pathways, the left region of the curve is mainly composed of the inactive RR, the central part of the curve includes the inactive R_AR and the G protein-dependent R^*_AR and the right part of the curve, the G protein-independent $R^*_AR^*_A$. Overall, the theoretical distribution of receptor species along ligand concentration provides a molecular explanation for the observed dosage dependent switch between the two signalling pathways.

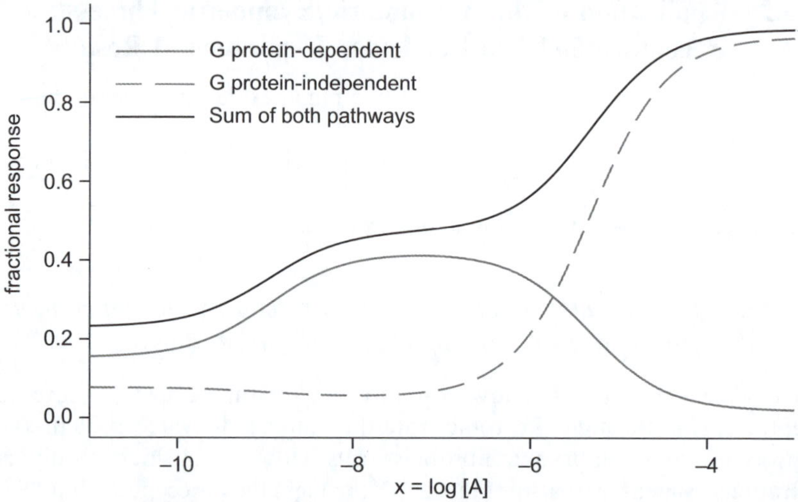

Figure 23.3 Simulation of the dosage dependent switch from a G protein-dependent pathway to a G protein-independent pathway. The G protein-dependent functional response is calculated as arising from the formation of the asymmetrically activated R*R dimer whereas the G protein-independent functional response is calculated as arising from the formation of the symmetrically activated R*R* dimer. The values of the constants for the simulation were: $X_1 = 10^{-1}$; $X_2 = 1$; $K_1 = 10^{-8.5}$; $K_2 = 10^{-4}$; $Y_1 = 10^{-3}$; $Y_2 = 0.8$; $Y_3 = 10^2$; $Y_5 = 0.8$; $Y_6 = 10^2$ (see Figure 23.2C and Appendix A23.5 for definition of constants).

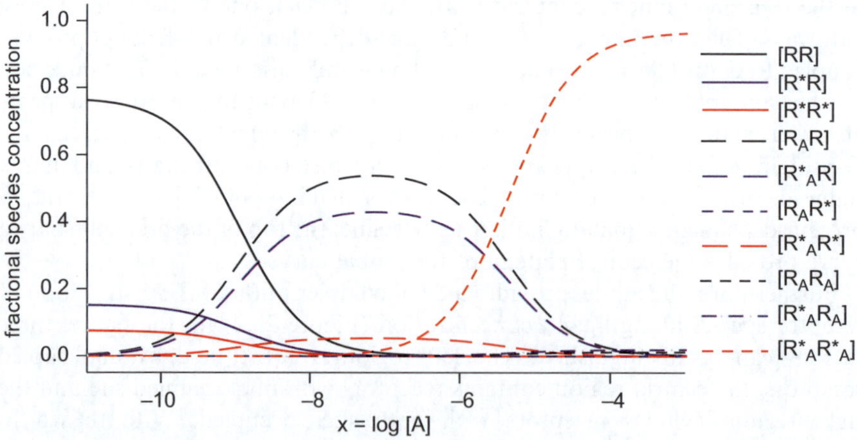

Figure 23.4 Simulation of the distribution of receptor species along ligand concentration for Figure 23.3. The values of the constants for the simulation were those used in Figure 23.3 (see Figure 23.2C and Appendix A23.5 for definition of constants). Colour code: black = inactive receptor; blue = asymmetrically activated receptor; red = symmetrically activated receptor.

Cooperativity is a central issue in oligomeric receptors. Cooperativity effects on signal transduction are reflected in the shapes of concentration–response curves, which are direct consequences of the values of the equilibrium constants present in the system. As discussed previously,[73] we distinguish between two forms of cooperativity:

- binding cooperativity (the binding of the first ligand facilitates or hampers the binding of the second); and
- induction cooperativity (the activation of the first dimer subunit facilitates or hampers the activation of the second).

In our simulation (Figure 23.3), the relation between the dissociation constants K_1 and K_2 implies a negative binding cooperativity between the binding sites of the inactive state (RR) of the receptor. In contrast, a positive induction cooperativity for protomer activation when the second subunit is occupied has been assumed (compare the induction constants Y_3 with Y_2 and Y_6 with Y_5). Importantly, Y_5 and Y_6, the induction constants for the activation of the asymmetric and symmetric receptor states in the fully occupied receptor, determine the maximum response of the system for both the G protein-dependent and the G protein–independent pathways (see efficacy definition in Appendix A23.5). Thus, modelling shows quantitatively that the cooperative allosteric interactions associated with both ligand binding and protomer activation together with the asymmetric/symmetric pathway specific-receptor states provide a plausible mechanistic explanation for the observed functional responses.

23.4.2.2 Inverse Agonists Displaying Agonist Behaviour

In a recent article, striking results were described for human D_2 receptor.[8] In a system consisting of a dimer receptor with one agonist molecule bound, the binding of an inverse agonist to the second protomer enhanced the G protein-dependent response whereas the binding of a second agonist decreased the response. The results are consistent with the asymmetric/symmetric three-state dimer receptor model—an asymmetrically activated dimer associated with G protein response and a symmetrically activated dimer associated with an independent G protein response. Let us suppose that the most abundant receptor species for a particular concentration of agonist A is $R^*{}_AR$ (an agonist binds preferably to active R* conformations). Subsequent addition of an inverse agonist B would imply the binding of molecule B to protomer R of $R^*{}_AR$ (an inverse agonist binds preferably to inactive R conformations). Formation of $R^*{}_AR_B$ leads to the stabilization of the asymmetrically activated dimer with the consequent increase of the G protein-dependent response. On the other hand, an increase of agonist A concentration would imply the stabilization of the symmetrically activated dimer ($R^*{}_AR^*{}_A$) with the consequent decrease of the G protein-dependent response.

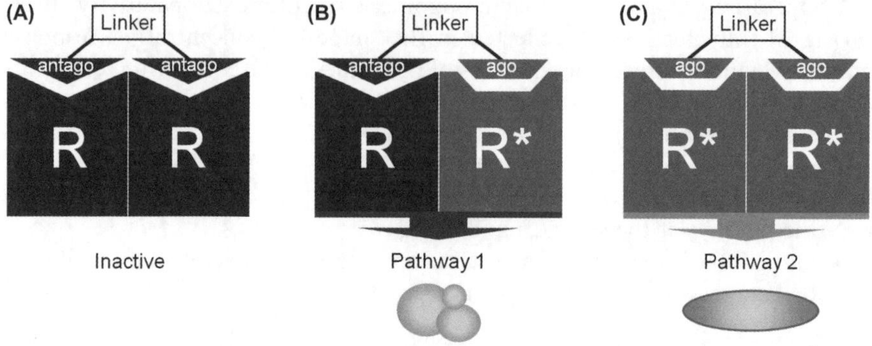

Figure 23.5 Proposal of bivalent ligand–receptor interactions for the asymmetric/symmetric three-state receptor dimer model. The pharmacological profiles (agonist, antagonist) of the pharmacophore pairs direct the dimeric receptor towards specific signalling pathways.

23.4.3 Functional Selectivity by Bivalent Ligands on Symmetric/Asymmetric Dimeric Receptors: Implications for Drug Discovery

GPCR dimerization offers new opportunities for drug discovery. One promising strategy is the development of bivalent ligands (two pharmacophores connected by a linker) with the aim of targeting both receptor protomers of the dimer by a single ligand molecule.[74] This approach has been applied both to homo- and heterodimeric receptors.[75–78] The asymmetric/symmetric three-state receptor model represents a homodimeric receptor. Although the model was constructed as the simplest proposal to account for the asymmetry/symmetry of receptor activated states, it contains structural and functional features which make it especially interesting for drug design. Of note, the model suggests that, in principle, dimeric ligands could be designed for specific signalling pathways:

- an antagonist–antagonist for the inactive pathway;
- an agonist–antagonist for the G protein-dependent pathway; and
- an agonist–agonist for the G protein-independent pathway (Figure 23.5).

If confirmed, this would have important consequences in drug discovery and therapeutics.

23.5 Conclusions

GPCRs are complex entities both in structural organization and functionality. The arrangements of these receptors in oligomeric states, which can be associated with specific signalling pathways, make GPCRs a challenging task in pharmacological research. Mathematical modelling has proved a useful tool for the description of pharmacological experiments. To this end, systematic and parsimonious construction of mathematical models can be instrumental in assessing the mechanisms involved in receptor binding and function.[60,73] Moreover, the combination of mathematical modelling with molecular modelling can synergistically strengthen both approaches providing complementary and deeper insights into these systems. In this regard, the recent crystallization of a CXCR4 chemokine GPCR in a dimeric state could be of key importance.[79] Drug design taking into account the presence of multiple signalling pathways, receptor oligomerization and the possibility that particular pathways can be associated with the stoichiometry of subunit activation within oligomers may revolutionize the processes of drug discovery and development.[50,68,70]

23.6 Appendix

A23.1 Two-state Receptor Model (Figure 23.1A)

Definition of the equilibrium constants of the model

$$X = \frac{[R^*]}{[R]} \quad Y = \frac{[R^*{}_A]}{[R_A]} \quad Z = \frac{[A][R]}{[R_A]} \quad T = \frac{[A][R^*]}{[R^*{}_A]}$$

Fraction of active receptors

$$f_{R^*} = \frac{[R^*]_t}{[Rt]} = \frac{T + [A]}{aT + b[A]}$$

where:

$$[R^*]_t = [R^*] + [R^*{}_A]$$

$$[Rt] = [R] + [R_A] + [R^*] + [R^*{}_A]$$

$$a = 1 + \frac{1}{X} \text{ and } b = 1 + \frac{T}{XZ}$$

Right asymptote: Efficacy

$$\lim_{[A]\to\infty} f_{R^*} = \frac{1}{b} = \frac{1}{1 + \frac{T}{XZ}}$$

Left asymptote: Basal activity

$$\underset{for[A]=0}{f_{R^*}} = \frac{1}{a} = \frac{1}{1 + \frac{1}{X}}$$

Note here and in the following that left and right asymptotes correspond to bottom and top asymptotes for a positive agonist, respectively, whereas the contrary happens for an inverse agonist.

A23.2 Three-state Receptor Model (Figure 23.1B)

Definition of the equilibrium constants of the model

$$L = \frac{[R]}{[R^*]} \quad M = \frac{[R]}{[R^{**}]}$$

$$K_A = \frac{[A][R]}{[R_A]} \quad K^*{}_A = \frac{[A][R^*]}{[R^*{}_A]} \quad K^{**}{}_A = \frac{[A][R^{**}]}{[R^{**}{}_A]}$$

Fraction of active receptors

$$f_{R^*} = \frac{[R^*]_t}{[Rt]} = \frac{K^*{}_A + [A]}{aK^*{}_A + b[A]}$$

$$f_{R^{**}} = \frac{[R^{**}]_t}{[Rt]} = \frac{K^{**}{}_A + [A]}{cK^{**}{}_A + d[A]}$$

where:

$$[R^*]_t = [R^*] + [R^*{}_A]$$

$$[R^{**}]_t = [R^{**}] + [R^{**}{}_A]$$

$$[Rt] = [R] + [R_A] + [R^*] + [R^*{}_A] + [R^{**}] + [R^{**}{}_A]$$

$$a = 1 + L + \frac{L}{M} \text{ and } b = 1 + \frac{LK^*{}_A}{K_A} + \frac{LK^*{}_A}{MK^{**}{}_A}$$

$$c = 1 + M + \frac{M}{L} \text{ and } d = 1 + \frac{MK^{**}{}_A}{K_A} + \frac{MK^{**}{}_A}{LK^*{}_A}$$

Right asymptote: Efficacy

$$\lim_{[A]\to\infty} f_{R^*} = \frac{1}{b} = \frac{1}{1 + \frac{LK^*{}_A}{K_A} + \frac{LK^*{}_A}{MK^{**}{}_A}}$$

$$\lim_{[A]\to\infty} f_{R^{**}} = \frac{1}{d} = \frac{1}{1 + \frac{MK^{**}{}_A}{K_A} + \frac{MK^{**}{}_A}{LK^*{}_A}}$$

Left asymptote: Basal activity

$$\underset{for[A]=0}{f_{R^*}} = \frac{1}{a} = \frac{M}{L + M + LM}$$

$$\underset{for[A]=0}{f_{R^{**}}} = \frac{1}{c} = \frac{L}{L + M + LM}$$

A23.3 Two-state Dimer Receptor Model (Figure 23.2A)

Definition of the equilibrium constants of the model

$$L = \frac{[(RR)^*]}{[(RR)]} \quad K = \frac{[(RR)_A]}{[A][(RR)]}$$

$$\alpha = \frac{[(RR)^*{}_A][(RR)]}{[(RR)^*][(RR)_A]} \quad \beta = \frac{[(RR)^*{}_{AA}][(RR)_A]}{[(RR)_{AA}][(RR)^*{}_A]} \quad \mu = \frac{[(RR)_{AA}][(RR)]}{[(RR)_A][(RR)_A]}$$

Fraction of active receptors

$$f_{(RR)^*} = \frac{[(RR)^*]_t}{[RR_t]} = 2 \cdot \frac{a_1 + a_2[A] + a_3[A]^2}{a_4 + a_5[A] + [A]^2}$$

where:

$$[(RR)^*]_t = [(RR)^*] + [(RR)^*{}_A] + [(RR)^*{}_{AA}]$$

$$[RRt] = [(RR)] + [(RR)_A] + [(RR)_{AA}] + [(RR)^*] + [(RR)^*{}_A] + [(RR)^*{}_{AA}]$$

$$a_1 = \frac{L}{K^2\mu(1+\alpha\beta L)}$$

$$a_2 = \frac{\alpha L}{K\mu(1+\alpha\beta L)}$$

$$a_3 = \frac{\alpha\beta L}{1+\alpha\beta L}$$

$$a_4 = \frac{1+L}{K^2\mu(1+\alpha\beta L)}$$

$$a_5 = \frac{1+\alpha L}{K\mu(1+\alpha\beta L)}$$

Right asymptote: Efficacy

$$\lim_{[A]\to\infty} f_{(RR)^*} = a_3 = \frac{\alpha\beta L}{1+\alpha\beta L}$$

Left asymptote: Basal activity

$$\underset{for[A]=0}{f_{(RR)^*}} = \frac{a_1}{a_4} = \frac{L}{1+L}$$

A23.4 Three-state Dimer Receptor Model (Figure 23.2B)

Definition of the equilibrium constants of the model

$$X = \frac{[(RR)^*]}{[(RR)]} \quad X' = \frac{[(RR)^{**}]}{[(RR)]}$$

$$\frac{K_1}{2} = \frac{[A][(RR)]}{[(RR)_A]} \quad 2K_2 = \frac{[A][(RR)_A]}{[(RR)_{AA}]} \quad \frac{K_3}{2} = \frac{[A][(RR)^*]}{[(RR)^*{}_A]}$$

$$2K_4 = \frac{[A][(RR)^*{}_A]}{[(RR)^*_{AA}]} \quad \frac{K_5}{2} = \frac{[A][(RR)^{**}]}{[(RR)^{**}{}_A]} \quad 2K_6 = \frac{[A][(RR)^{**}{}_A]}{[(RR)^{**}{}_{AA}]}$$

Fraction of active receptors

$$f_{(RR)^*} = \frac{[(RR)^*]_t}{[RR_t]} = 2 \cdot \frac{b_1 + b_2[A] + b_3[A]^2}{b_4 + b_5[A] + [A]^2}$$

$$f_{(RR)^{**}} = \frac{[(RR)^{**}]_t}{[RR_t]} = 2 \cdot \frac{c_1 + c_2[A] + c_3[A]^2}{c_4 + c_5[A] + [A]^2}$$

where:

$$[(RR)^*]_t = [(RR)^*] + [(RR)^*{}_A] + [(RR)^*_{AA}]$$

$$[(RR)^{**}]_t = [(RR)^{**}] + [(RR)^{**}{}_A] + [(RR)^{**}{}_{AA}]$$

$$\begin{aligned}[RR_t] = {} & [(RR)] + [(RR)_A] + [(RR)_{AA}] + [(RR)^*] + [(RR)^*{}_A] + [(RR)^*{}_{AA}] \\ & + [(RR)^{**}] + [(RR)^{**}{}_A] + [(RR)^{**}{}_{AA}]\end{aligned}$$

$$b_1 = \frac{X}{2\left(\frac{1}{K_1K_2} + \frac{X}{K_3K_4} + \frac{X'}{K_5K_6}\right)}$$

$$b_2 = \frac{X}{K_3\left(\frac{1}{K_1K_2} + \frac{X}{K_3K_4} + \frac{X'}{K_5K_6}\right)}$$

$$b_3 = \frac{X}{2K_3K_4\left(\frac{1}{K_1K_2} + \frac{X}{K_3K_4} + \frac{X'}{K_5K_6}\right)}$$

$$c_1 = \frac{X'}{2\left(\frac{1}{K_1K_2} + \frac{X}{K_3K_4} + \frac{X'}{K_5K_6}\right)}$$

$$c_2 = \frac{X'}{K_5\left(\frac{1}{K_1K_2} + \frac{X}{K_3K_4} + \frac{X'}{K_5K_6}\right)}$$

$$c_3 = \frac{X'}{2K_5K_6\left(\frac{1}{K_1K_2} + \frac{X}{K_3K_4} + \frac{X'}{K_5K_6}\right)}$$

$$b_4 = c_4 = \frac{1 + X + X'}{\frac{1}{K_1K_2} + \frac{X}{K_3K_4} + \frac{X'}{K_5K_6}}$$

$$b_5 = c_5 = \frac{2\left(\frac{1}{K_1} + \frac{X}{K_3} + \frac{X'}{K_5}\right)}{\frac{1}{K_1K_2} + \frac{X}{K_3K_4} + \frac{X'}{K_5K_6}}$$

Right asymptote: Efficacy

$$\lim_{[A]\to\infty} f_{(RR)^*} = 2b_3 = \frac{X}{K_3K_4\left(\frac{1}{K_1K_2} + \frac{X}{K_3K_4} + \frac{X'}{K_5K_6}\right)}$$

$$\lim_{[A]\to\infty} f_{(RR)^{**}} = 2c_3 = \frac{X'}{K_5K_6\left(\frac{1}{K_1K_2} + \frac{X}{K_3K_4} + \frac{X'}{K_5K_6}\right)}$$

Left asymptote: Basal activity

$$\underset{for[A]=0}{f_{(RR)^*}} = 2\frac{b_1}{b_4} = \frac{X}{1 + X + X'}$$

$$\underset{for[A]=0}{f_{(RR)^{**}}} = 2\frac{c_1}{c_4} = \frac{X'}{1 + X + X'}$$

A23.5 The Asymmetric/Symmetric Three-state Dimer Receptor Model (Figure 23.2C)

Definition of the equilibrium constants of the model

$$2X_1 = \frac{[R^*R]}{[RR]} \quad \frac{X_2}{2} = \frac{[R^*R^*]}{[R^*R]}$$

$$\frac{K_1}{2} = \frac{[A][RR]}{[R_AR]} \quad 2K_2 = \frac{[A][R_AR]}{[R_AR_A]} \quad Y_1 = \frac{[R_AR^*]}{[R_AR]} \quad Y_2 = \frac{[R^*{}_AR]}{[R_AR]}$$

$$Y_3 = \frac{[R^*{}_AR^*]}{[R_AR^*]} \quad Y_4 = \frac{[R^*{}_AR^*]}{[R^*{}_AR]} \quad 2Y_5 = \frac{[R^*{}_AR_A]}{[R_AR_A]} \quad \frac{Y_6}{2} = \frac{[R^*{}_AR^*{}_A]}{[R^*{}_AR_A]}$$

Fraction of active receptors

$$f_{R^*R} = \frac{[R^*R]_t}{[RR_t]} = 2 \cdot \frac{d_1 + d_2[A] + d_3[A]^2}{d_4 + d_5[A] + [A]^2}$$

$$f_{R^*R^*} = \frac{[R^*R^*]_t}{[RR_t]} = 2 \cdot \frac{e_1 + e_2[A] + e_3[A]^2}{e_4 + e_5[A] + [A]^2}$$

where:

$$[R^*R]_t = [R^*R] + [R^*{}_AR] + [R_AR^*] + [R^*{}_AR_A]$$

$$[R^*R^*]_t = [R^*R^*] + [R^*{}_AR^*] + [R^*{}_AR^*{}_A]$$

$$[RR_t] = [RR] + [R_AR] + [R_AR_A] + [R^*R] + [R^*{}_AR] + [R_AR^*] + [R^*{}_AR_A] + [R^*R^*] + [R^*{}_AR^*] + [R^*{}_AR^*{}_A]$$

$$d_1 = \frac{K_1K_2X_1}{1 + 2Y_5 + Y_5Y_6}$$

$$d_2 = \frac{K_2(Y_1 + Y_2)}{1 + 2Y_5 + Y_5Y_6}$$

$$d_3 = \frac{Y_5}{1 + 2Y_5 + Y_5Y_6}$$

$$e_1 = \frac{K_1K_2X_1X_2}{2(1 + 2Y_5 + Y_5Y_6)}$$

$$e_2 = \frac{K_2Y_1Y_3}{1 + 2Y_5 + Y_5Y_6}$$

$$e_3 = \frac{Y_5Y_6}{2(1 + 2Y_5 + Y_5Y_6)}$$

$$d_4 = e_4 = \frac{K_1K_2(1 + 2X_1 + X_1X_2)}{1 + 2Y_5 + Y_5Y_6}$$

$$d_5 = e_5 = \frac{2K_2(1 + Y_1 + Y_2 + Y_1Y_3)}{1 + 2Y_5 + Y_5Y_6}$$

Right asymptote: Efficacy

$$\lim_{[A]\to\infty} f_{R^*R} = 2d_3 = \frac{1}{1 + \frac{1}{2}\left(\frac{1}{Y_5} + Y_6\right)}$$

$$\lim_{[A]\to\infty} f_{R^*R^*} = 2e_3 = \frac{1}{1 + \frac{2}{Y_6} + \frac{1}{Y_5Y_6}}$$

Left asymptote: Basal activity

$$\underset{for[A]=0}{f_{R^*R}} = 2\frac{d_1}{d_4} = \frac{1}{1 + \frac{1}{2}\left(\frac{1}{X_1} + X_2\right)}$$

$$\underset{for[A]=0}{f_{R^*R^*}} = 2\frac{b_1}{b_4} = \frac{1}{1 + \frac{2}{X_2} + \frac{1}{X_1 X_2}}$$

Acknowledgements

This study was supported in part by the Spanish Ministerio de Ciencia e Innovación (SAF2007-65913) and Fundació La Marató de TV3 (Ref. 070530).

References

1. J. P. Overington, B. Al Lazikani and A. L. Hopkins, *Nat. Rev. Drug Discov.*, 2006, **5**, 993.
2. M. P. Bokoch, Y. Zou, S. G. Rasmussen, C. W. Liu, R. Nygaard, D. M. Rosenbaum, J. J. Fung, H. J. Choi, F. S. Thian, T. S. Kobilka, J. D. Puglisi, W. I. Weis, L. Pardo, R. S. Prosser, L. Mueller and B. K. Kobilka, *Nature*, 2010, **463**, 108.
3. D. T. Lodowski, K. Palczewski and M. Miyagi, *Biochemistry*, 2010, **49**, 9425.
4. V. P. Jaakola, J. R. Lane, J. Y. Lin, V. Katritch, A. P. Ijzerman and R. C. Stevens, *J. Biol. Chem.*, 2010, **285**, 13032.
5. J. A. Goncalves, S. Ahuja, S. Erfani, M. Eilers and S. O. Smith, *Prog. Nucl. Magn. Reson. Spectrosc.*, 2010, **57**, 159.
6. J. P. Vilardaga, M. Bunemann, T. N. Feinstein, N. Lambert, V. O. Nikolaev, S. Engelhardt, M. J. Lohse and C. Hoffmann, *Mol. Endocrinol.*, 2009, **23**, 590.
7. P. Rondard, C. Goudet, J. Kniazeff, J.-P. Pin and L. Prezeau, *Neuropharmacology*, 2010.
8. Y. Han, I. S. Moreira, E. Urizar, H. Weinstein and J. A. Javitch, *Nat. Chem. Biol.*, 2009, **5**, 688.
9. R. O. Dror, D. H. Arlow, D. W. Borhani, M. O. Jensen, S. Piana and D. E. Shaw, *Proc. Natl. Acad. Sci. U. S. A.*, 2009, **106**, 4689.
10. S. Bhattacharya and N. Vaidehi, *J. Am. Chem. Soc.*, 2010, **132**, 5205.
11. S. Ahuja and S. O. Smith, *Trends Pharmacol. Sci.*, 2009, **30**, 494.
12. R. Nygaard, T. M. Frimurer, B. Holst, M. M. Rosenkilde and T. W. Schwartz, *Trends Pharmacol. Sci.*, 2009, **30**, 249.
13. B. H. Meyer, J. M. Segura, K. L. Martinez, R. Hovius, N. George, K. Johnsson and H. Vogel, *Proc. Natl. Acad. Sci. U. S. A.*, 2006, **103**, 2138.
14. M. R. Whorton, M. P. Bokoch, S. G. Rasmussen, B. Huang, R. N. Zare, B. Kobilka and R. K. Sunahara, *Proc. Natl. Acad. Sci. U. S. A.*, 2007, **104**, 7682.

15. T. H. Bayburt, A. J. Leitz, G. Xie, D. D. Oprian and S. G. Sligar, *J. Biol. Chem.*, 2007, **282**, 14875.
16. J. F. White, J. Grodnitzky, J. M. Louis, L. B. Trinh, J. Shiloach, J. Gutierrez, J. K. Northup and R. Grisshammer, *Proc. Natl. Acad. Sci. U. S. A.*, 2007, **104**, 12199.
17. L. Arcemisbehere, T. Sen, L. Boudier, M. N. Balestre, G. Gaibelet, E. Detouillon, H. Orcel, C. Mendre, R. Rahmeh, S. Granier, C. Vives, F. Fieschi, M. Damian, T. Durroux, J. L. Baneres and B. Mouillac, *J. Biol. Chem.*, 2010, **285**, 6337.
18. S. Terrillon and M. Bouvier, *EMBO J.*, 2004, **5**, 30.
19. J. J. Carrillo, J. F. Lopez-Gimenez and G. Milligan, *Mol. Pharmacol.*, 2004, **66**, 1123.
20. B. Jastrzebska, D. Fotiadis, G. F. Jang, R. E. Stenkamp, A. Engel and K. Palczewski, *J. Biol. Chem.*, 2006, **281**, 11917.
21. J.-P. Pin, L. Comps-Agrar, D. Maurel, C. Monnier, M. L. Rives, E. Trinquet, J. Kniazeff, P. Rondard and L. Prezeau, *J. Physiol.*, 2009, **587**, 5337.
22. A. Rivero-Müller, Y. Y. Chou, I. Ji, S. Lajic, A. C. Hanyaloglu, K. Jonas, N. Rahman, T. H. Ji and I. Huhtaniemi, *Proc. Natl. Acad. Sci. U. S. A.*, 2010, **107**, 2319.
23. L. Albizu, M. Cottet, M. Kralikova, S. Stoev, R. Seyer, I. Brabet, T. Roux, H. Bazin, E. Bourrier, L. Lamarque, C. Breton, M. L. Rives, A. Newman, J. Javitch, E. Trinquet, M. Manning, J.-P. Pin, B. Mouillac and T. Durroux, *Nat. Chem. Biol.*, 2010, **6**, 587.
24. D. Fotiadis, Y. Liang, S. Filipek, D. A. Saperstein, A. Engel and K. Palczewski, *Nature*, 2003, **421**, 127.
25. G. Milligan, *Br. J. Pharmacol.*, 2008, **153**, S216.
26. P. Maurice, A. M. Daulat, R. Turecek, K. Ivankova-Susankova, F. Zamponi, M. Kamal, N. Clement, J. L. Guillaume, B. Bettler, C. Gales, P. Delagrange and R. Jockers, *EMBO J.*, 2010, **29**, 3636.
27. M. A. Ayoub, E. Trinquet, K. D. Pfleger and J.-P. Pin, *FASEB J.*, 2010, **24**, 3522.
28. J. D. Urban, W. P. Clarke, M. von Zastrow, D. E. Nichols, B. K. Kobilka, H. Weinstein, J. A. Javitch, B. L. Roth, A. Christopoulos, P. Sexton, K. Miller, M. Spedding and R. B. Mailman, *J. Pharmacol. Exp. Ther.*, 2006, **320**, 1.
29. E. J. Ariens, *Arch. Int. Pharmacodyn. Ther.*, 1954, **99**, 32.
30. W. P. Clarke and R. A. Bond, *Trends Pharmacol. Sci.*, 1998, **19**, 270.
31. S. Galandrin, G. Oligny-Longpre and M. Bouvier, *Trends Pharmacol. Sci.*, 2007, **28**, 423.
32. R. P. Stephenson, *Br. J. Pharmacol. Ther.*, 1956, **11**, 379.
33. R. F. Furchgott, in *Advances in Drug Research*, ed. N. J. Harper and A. B. Simmonds, Academic Press, New York, 1966, vol. 3, pp. 21–55.
34. J. Monod, J. Wyman and J. P. Changeux, *J. Mol. Biol.*, 1965, **12**, 88.
35. D. E. Koshland, Jr., G. Nemethy and D. Filmer, *Biochemistry*, 1966, **5**, 365.
36. A. Karlin, *J. Theor. Biol.*, 1967, **16**, 306.

37. P. Leff, *Trends Pharmacol. Sci.*, 1995, **16**, 89.
38. C. D. Thron, *Mol. Pharmacol.*, 1973, **9**, 1.
39. J. Giraldo, *FEBS Lett.*, 2004, **556**, 13.
40. J. Giraldo, J. Serra, D. Roche and X. Rovira, *Curr. Drug Targets*, 2007, **8**, 197.
41. J. Giraldo, *Drug Discov. Today*, 2010, **15**, 411.
42. P. Samama, S. Cotecchia, T. Costa and R. J. Lefkowitz, *J. Biol. Chem.*, 1993, **268**, 4625.
43. T. Kenakin, *Trends Pharmacol. Sci.*, 1995, **16**, 232.
44. B. K. Kobilka and X. Deupi, *Trends Pharmacol. Sci.*, 2007, **28**, 397.
45. J. G. Baker, I. P. Hall and S. J. Hill, *Mol. Pharmacol.*, 2003, **64**, 1357.
46. M. Azzi, P. G. Charest, S. Angers, G. Rousseau, T. Kohout, M. Bouvier and G. Pineyro, *Proc. Natl. Acad. Sci. U. S. A.*, 2003, **100**, 11406.
47. T. P. Kenakin, *Trends Pharmacol. Sci.*, 2009, **30**, 460.
48. P. Leff, C. Scaramellini, C. Law and K. McKechnie, *Trends Pharmacol. Sci*, 1997, **18**, 355.
49. S. Rajagopal, J. Kim, S. Ahn, S. Craig, C. M. Lam, N. P. Gerard, C. Gerard and R. J. Lefkowitz, *Proc. Natl. Acad. Sci. U. S. A.*, 2010, **107**, 628.
50. S. Rajagopal, K. Rajagopal and R. J. Lefkowitz, *Nat. Rev. Drug Discov.*, 2010, **9**, 373.
51. K. Palczewski, *Trends Biochem. Sci.*, 2010.
52. D. Colquhoun, in *A Symposium on Drug Receptors*, ed. H. P. Rang, University Park Press, Baltimore, 1973, pp. 149–182.
53. J. W. Wells, in *Receptor–Ligand interactions: A Practical Approach*, ed, E. C. Hulme, Oxford University Press, Oxford, 1992, pp. 289–395.
54. K. A. Wreggett and J. W. Wells, *J. Biol. Chem.*, 1995, **270**, 22488.
55. P. Chidiac, M. A. Green, A. B. Pawagi and J. W. Wells, *Biochemistry*, 1997, **36**, 7361.
56. D. Armstrong and P. G. Strange, *J. Biol. Chem.*, 2001, **276**, 22621.
57. A. Christopoulos and T. Kenakin, *Pharmacol. Rev.*, 2002, **54**, 323.
58. E. Urizar, L. Montanelli, T. Loy, M. Bonomi, S. Swillens, C. Gales, M. Bouvier, G. Smits, G. Vassart and S. Costagliola, *EMBO J.*, 2005, **24**, 1954.
59. L. Albizu, M. N. Balestre, C. Breton, J.-P. Pin, M. Manning, B. Mouillac, C. Barberis and T. Durroux, *Mol. Pharmacol.*, 2006, **70**, 1783.
60. X. Rovira, M. Vivo, J. Serra, D. Roche, P. G. Strange and J. Giraldo, *Br. J. Pharmacol.*, 2009, **156**, 28.
61. J. Y. Springael, E. Urizar, S. Costagliola, G. Vassart and M. Parmentier, *Pharmacol. Ther.*, 2007, **115**, 410.
62. J. Giraldo, N. M. Vivas, E. Vila and A. Badia, *Pharmacol. Ther.*, 2002, **95**, 21.
63. J. Giraldo, *Br. J. Pharmacol.*, 2008, **155**, 17.
64. R. Franco, V. Casadó, J. Mallol, S. Ferré, K. Fuxe, A. Cortés, F. Ciruela, C. Lluis and E. I. Canela, *Trends Biochem. Sci.*, 2005, **30**, 360.
65. R. Franco, V. Casadó, J. Mallol, C. Ferrada, S. Ferré, K. Fuxe, A. Cortés, F. Ciruela, C. Lluis and E. I. Canela, *Mol. Pharmacol.*, 2006, **69**, 1905.
66. J. Giraldo, *Trends Pharmacol. Sci.*, 2003, **24**, 63.

67. J. Brea, M. Castro, J. Giraldo, J. F. Lopez-Gimenez, J. F. Padin, F. Quintian, M. I. Cadavid, M. T. Vilaro, G. Mengod, K. A. Berg, W. P. Clarke, J. P. Vilardaga, G. Milligan and M. I. Loza, *Mol. Pharmacol.*, 2009, **75**, 1380.
68. X. Rovira, J.-P. Pin and J. Giraldo, *Trends Pharmacol. Sci.*, 2010, **31**, 15.
69. Y. Sun, D. McGarrigle and X. Y. Huang, *Mol. Biosyst.*, 2007, **3**, 849.
70. M. Damian, A. Martin, D. Mesnier, J.-P. Pin and J. L. Baneres, *EMBO J.*, 2006, **25**, 5693.
71. F. Novi, L. Stanasila, F. Giorgi, G. U. Corsini, S. Cotecchia and R. Maggio, *J. Biol. Chem.*, 2005, **280**, 19768.
72. Y. Sun, J. Huang, Y. Xiang, M. Bastepe, H. Juppner, B. K. Kobilka, J. J. Zhang and X. Y. Huang, *EMBO J.*, 2007, **26**, 53.
73. X. Rovira, D. Roche, J. Serra, J. Kniazeff, J.-P. Pin and J. Giraldo, *J. Pharmacol. Exp. Ther.*, 2008, **325**, 443.
74. I. Berque-Bestel, F. Lezoualc'h and R. Jockers, *Curr. Drug Discov. Technol.*, 2008, **5**, 312.
75. P. S. Portoghese, *Trends Pharmacol. Sci.*, 1989, **10**, 230.
76. F. Lezoualc'h, R. Jockers and I. Berque-Bestel, *Curr. Pharm. Des.*, 2009, **15**, 719.
77. S. Zhang, A. Yekkirala, Y. Tang and and P. S. Portoghese, *Bioorg. Med. Chem. Lett.*, 2009, **19**, 6978.
78. A. Soriano, R. Ventura, A. Molero, R. Hoen, V. Casado, A. Cortes, F. Fanelli, F. Albericio, C. Lluis, R. Franco and M. Royo, *J. Med. Chem.*, 2009, **52**, 5590.
79. B. Wu, E. Y. Chien, C. D. Mol, G. Fenalti, W. Liu, V. Katritch, R. Abagyan, A. Brooun, P. Wells, F. C. Bi, D. J. Hamel, P. Kuhn, T. M. Handel, V. Cherezov and R. C. Stevens, *Science*, 2010, **330**, 1066.
80. M. Filizola, S. X. Wang and H. Weinstein, *J. Comput. Aided Mol. Des.*, 2006, **20**, 405.
81. F. Mancia, Z. Assur, A. G. Herman, R. S. and W. A. Hendrickson, *EMBO Rep.*, 2008, **9**, 363.
82. V. Hlavackova, C. Goudet, J. Kniazeff, A. Zikova, D. Maurel, C. Vol, J. Trojanova, L. Prézeau, J.-P. Pin and J. Blahos, *EMBO J.*, 2005, **24**, 499.
83. J.-P. Pin, J. Kniazeff, V. Binet, J. Liu, D. Maurel, T. Galvez, B. Duthey, M. Havlickova, J. Blahos, L. Prézeau and P. Rondard, *Biochem. Pharmacol.*, 2004, **68**, 1565.
84. H. Xu, L. Staszewski, H. Tang, E. Adler, M. Zoller and X. Li, *Proc. Natl. Acad. Sci. U. S. A.*, 2004, **101**, 14258.

CHAPTER 24

Using Microfluidics, Real-time Imaging and Mathematical Modelling to study GPCR Signalling

ANDREJA JOVIC,[a] SHUICHI TAKAYAMA[ab] AND JENNIFER J. LINDERMAN*[c]

[a] Biomedical Engineering Department, University of Michigan, Ann Arbor, MI, USA; [b] Macromolecular and Materials Science Engineering, University of Michigan, Ann Arbor, MI, USA; [c] Department of Chemical Engineering, University of Michigan, Ann Arbor, MI, USA

24.1 Introduction

The combination of experimental and computational methods provides a powerful approach for understanding biological processes. Initial experiments can suggest hypotheses that can be captured by first-generation mathematical/computational models (henceforth simply called 'mathematical models' or 'models'), which can then be employed to generate new experimental approaches to test those models. Understanding signal transduction pathways, with their myriad of molecular interactions, feedback loops, and spatial and temporal variations in concentrations of key molecules, requires such an iterative process between experiments and modeling.[1]

RSC Drug Discovery Series No. 8
G Protein-Coupled Receptors: From Structure to Function
Edited by Jesús Giraldo and Jean-Philippe Pin

Published by the Royal Society of Chemistry, www.rsc.org

Models of signal transduction pathways that are mechanistic (reflecting actual physical processes that occur as opposed to empirical or statistical models) typically have many parameters. The concentrations of various molecules along the pathways, the forward and reverse rate constants for binding reactions, diffusivities of key molecules, and enzymatic activities are all physical parameters that can appear in such models.[2–5] Yet values for all of these parameters are not known with certainty. Assays can of course be developed to measure some of the parameters. Yet how well those values translate from the assay conditions (often *in vitro*) to the cellular environment during signalling is typically not known.

This uncertainty in parameter values, combined with uncertainty in experimental measurements, can make the comparison between model and experiment difficult: does, for example, a 30% difference between the two reflect an incorrect parameter value, a small error in the experimental measurement, or a poor understanding of the governing physical processes when the model was formulated? One approach that is helpful is to develop *qualitative* tests for the comparison. In other words, although the output of both the model and the experiment are quantitative (*e.g.* numbers of G proteins activated or concentration of intracellular free calcium), a comparison of qualitative trends (*e.g.* does a curve rise or fall, or shift right or left) may be most useful, at least in the initial stages of understanding a mechanism. Thus, it is helpful to develop new (and multiple) qualitative tests that can be used for the comparison between proposed mechanisms (as formulated in a model) and experimental data.

In this chapter, we focus on G protein signalling, in particular G protein-coupled receptor (GPCR) mediated calcium oscillations. Numerous models have been developed to describe these calcium oscillations.[6–12] Here we describe a novel approach, the modelling and experimental analysis of phase-locking, which provides a qualitative test that can be used to evaluate those models.

Phase-locking analysis involves application of periodic stimulation to an oscillatory pathway, which then becomes synchronized to the stimulation inputs. While the phenomenon of phase-locking is intrinsic to all non-linear oscillators,[13] phase-locking behaviour may be different for different oscillatory mechanisms.[14] We first describe proposed mechanisms for GPCR-induced calcium oscillations, and then introduce simulation and experimental methods to analyse phase-locking properties of the oscillations. Finally, we demonstrate phase-locking analysis as applied to a GPCR-mediated calcium signalling pathway to elucidate signalling mechanisms.

24.2 Models of GPCR-induced Calcium Signalling

Mathematical modelling of GPCR-initiated signal transduction pathways, like all modelling of biological processes, offers the opportunity for us to interpret data, analyse hypothesized mechanisms, run virtual (*in silico*) experiments and motivate new experiments. For example, different pathways or physical mechanisms may be suggested to explain the same biological data. Designing and performing experiments based on mathematical models that have been formulated for the proposed mechanisms can help distinguish between these mechanisms.

GPCR-induced intracellular calcium signalling has been the focus of many modelling studies, due in part to the discovery of calcium oscillations upon chemical stimulation of cells.[15–17] Another key observation was that increasing or decreasing the ligand concentration changes primarily the frequency and not the amplitude of calcium oscillations,[8] suggesting that critical information for signalling pathways is encoded in the frequency of calcium signals. These observations, along with estimation of some relevant kinetic parameters, led to the development of a variety of oscillatory calcium models whose mechanisms differ, often significantly, in terms of how oscillations are generated and calcium levels are set.[6–12] Most of these models are composed of ordinary differential equations based on mass action kinetics and assumptions of Michaelis–Menten kinetics, cooperativity (*e.g.* Hill coefficient), and similar. Upon solving these equations, one can track the dynamics of each of the cellular components [*e.g.* bound receptors, activated G proteins, inositol 1,4,5-trisphosphate (IP3) and intracellular free calcium concentration] over time.

For this chapter, we focus on two representative mathematical models of calcium oscillations with differing activation and recovery properties—one described by Chay *et al.*[7] and another by Politi *et al.*[11] (Figure 24.1). For the Chay *et al.* model (Figure 24.1 top left), calcium oscillations are produced by the switch-like (Hill coefficient of 4) activation of phospholipase C (PLC) by G proteins, which in turn produces IP3 and diacylglycerol (DAG). IP3 binds to IP3 receptors, opening calcium channels on intracellular stores and thus eliciting a calcium response. DAG initiates a negative feedback that subsequently reduces G protein activity, resulting in calcium oscillations. A basal level of G protein activity that results in basal IP3 production is included, ensuring that IP3 levels return to pre-stimulus levels. In this model, signalling is initiated by increasing the receptor contribution to the rate of G protein activation, indicated by the term 'stimulant' in Figure 24.1.

In contrast, the Politi *et al.* model (Figure 24.1 top right) assumes that activation of PLC by G proteins is a graded process (as opposed to switch-like). In other words, the rate of PLC activation is proportional to the number of activated G proteins. Production of IP3 initiates a feedback loop between the IP3 receptor (IP3R), calcium and PLC that leads to calcium oscillations. Furthermore, the model does not include a mechanism for basal IP3 production, highlighting a difference in recovery properties of the respective models in addition to the difference in their activation properties. In the Politi *et al.* model, signalling is initiated by increasing the rate at which phosphatidylinositol 4,5-bisphosphate (PIP2) is converted to IP3 (by activated PLC). Parameters for both models were derived from a combination of experimental data and estimation.

Despite having different mechanisms for generating calcium oscillations, the two models behave nearly identically when addressed with continuous stimulation; both models produce calcium oscillations upon application of continuous stimulation. The period of these oscillations decreases with increasing

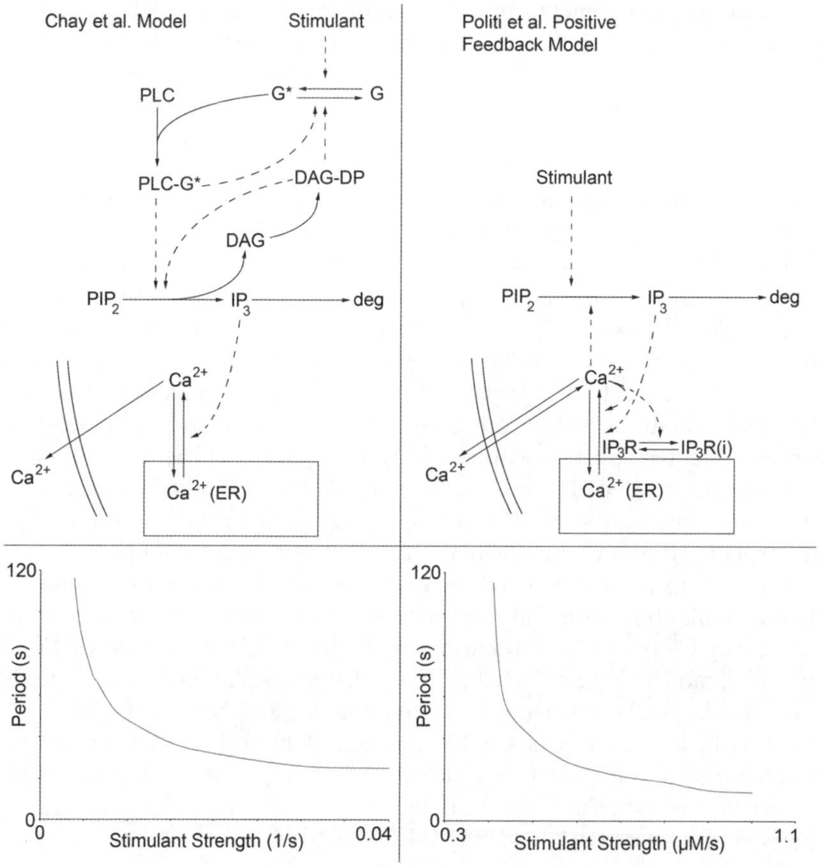

Figure 24.1 Schematics of the two mathematical models of oscillatory calcium signalling analysed in this chapter: the Chay *et al.* model[7] (top left) and the positive feedback Politi *et al.* model[11] (top right). Original published equations and parameters were used for both models. Under continuous stimulation, both models produce calcium oscillations whose period decreases with increases in stimulant strength (bottom figures). Despite significant differences in the mechanisms between these models, they exhibit the same behaviours under continuous stimulation. The stimulant strengths have different units, based upon the molecular species in the models that convey information about stimulant strength. For the Chay *et al.* model, the molecular species is the active agonist/receptor complex; for the Politi *et al.* model, the molecular species is the PLC/activated G-protein complex (PLC-G*). G = G protein; G* = activated G protein; PLC = phospholipase C; PIP2 = phosphatidylinositol 4,5-bisphosphate; IP3 = inositol 1,4,5-trisphosphate; DAG = diacylglycerol; DAG-DP = DAG dependent protein; deg = degradation; Ca^{2+} = calcium; ER = endoplasmic reticulum; IP3R = IP3 receptor; IP3R(i) = inactive IP3R.

stimulation strength (Figure 24.1 bottom left and 24.1 bottom right), demonstrating that this widely applied qualitative test is insufficient to distinguish between these (and many other) model mechanisms.

These and other results demonstrate that analysis of calcium transients elicited with continuous stimulation (*e.g.* the oscillations shown here) or simple step changes in stimulation (*e.g.* a sudden increase in ligand concentration) are insufficient to adequately probe the calcium signalling pathway. Are there additional tests available to distinguish between proposed model mechanisms for GPCR-induced calcium oscillations? As described below, phase-locking behaviour may provide one such test.

24.3 Phase-locking and Sub-threshold Calcium Responses

Phase-locking refers to the phenomenon whereby an oscillatory system synchronizes to a periodic input (Figure 24.2). Phase-locking is predicted to occur for all non-linear oscillators;[13] for example, it was explored in oscillatory electrical systems exposed to periodic electrical stimulation for the purpose of understanding how to better control such non-linear systems.[18] In a biological context, phase-locking has been observed in experiments with cardiac[19] and neuronal systems.[20] In these studies, periodic electrical stimulation was supplied and the corresponding cellular (electric) responses were measured.

Several studies have indicated that the simple observation of phase-locking alone does not provide insight into the mechanisms of oscillatory systems.[13] However, phase-locking properties (*e.g.* how phase-locking changes, or not, with different stimulation parameters) can provide insight into oscillatory mechanisms. Seminal studies of phase-locking in cardiac cell aggregates found that as the rest period between periodic electrical stimulation events was reduced, the cells' capacity to keep up with the stimulatory inputs was also reduced.[19] In other words, the number of cellular responses was less than the number of stimulation events, indicative of skipped beats. To quantify the cells' capacity to keep up with periodic stimulation, one can use the metric called the phase locking ratio (PLR), elsewhere referred to as the Winding number (Figure 24.2). One can define the phase locking ratio PLR as:

$$\text{PLR} = \frac{\text{number of system responses}}{\text{number of stimulation events}} \tag{24.1}$$

Theoretical studies have characterized the effects of periodic stimulation parameters on the PLR.[7] Periodic stimulation parameters include the stimulant concentration (C), the stimulation duration (D) and the rest period (R) (Figure 24.3). When the PLR is plotted against the value of a single periodic stimulation parameter (while holding the others constant), what has been termed a 'Devil's Staircase'[7] emerges (Figure 24.3).

Examining how the PLR changes with simulation parameters gives insight into activation and recovery properties of the oscillatory system. Depending upon the activation and recovery properties of the system, the staircase may increase or decrease; in other words, the PLR may increase or decrease with increasing C, D or R, based upon the activation and recovery properties of the

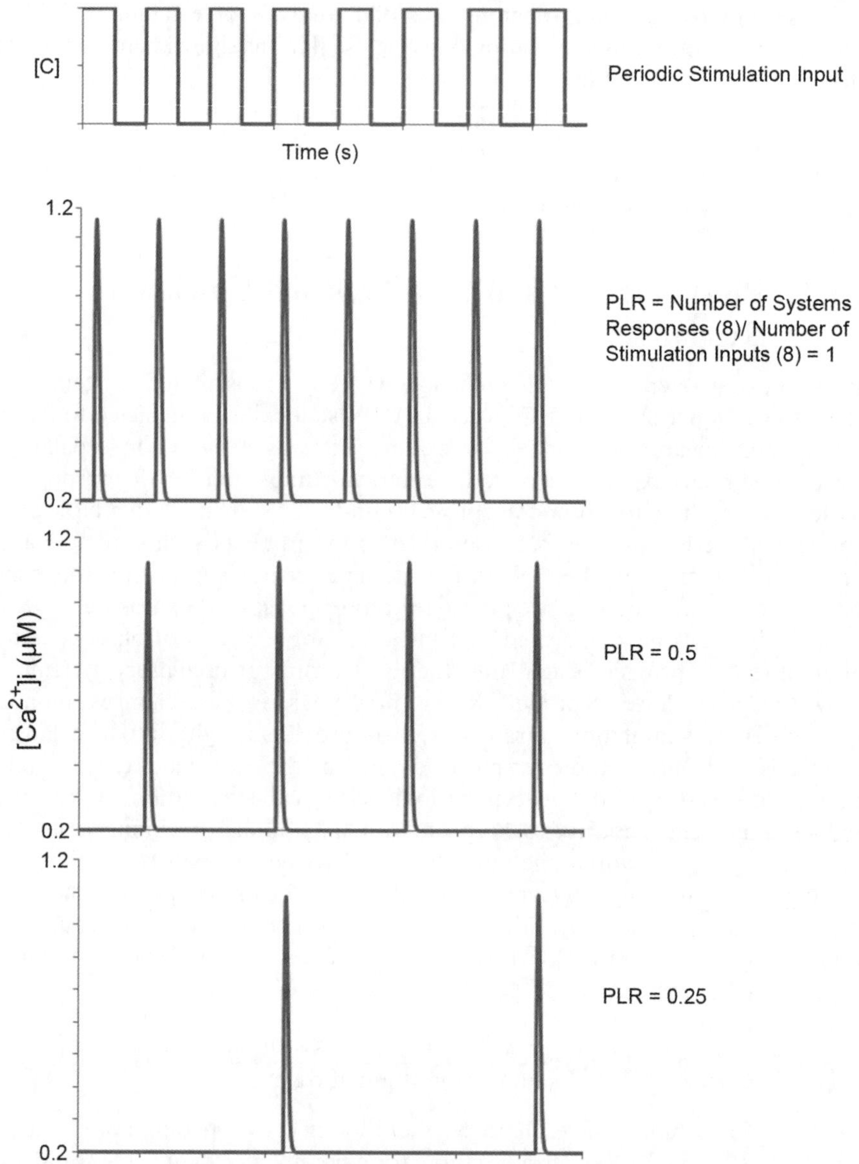

Figure 24.2 Phase-locking of calcium responses to a periodic stimulation input and calculation of the phase-locking ratio (PLR). Top trace shows the periodic stimulation input applied to a single cell, here a pulse of agonist with concentration [C]. For the three calcium traces depicted, the calcium responses are synchronized to the periodic stimulation input. A calcium response only results during a stimulation event; however, not every stimulation event necessarily elicits a calcium response, indicative of a loss of fidelity. To assess the degree of fidelity, the PLR is calculated by dividing the number of system responses by the number of stimulation events.

oscillatory system. In our experience with calcium oscillation models, increasing C or D typically results in an increase in the PLR, since both parameters increase the chances that a system response will occur. Thus, modifying these two stimulation parameters may not provide the insight needed to discriminate between various proposed oscillation mechanisms.

However, with calcium oscillation models, we find that when the PLR is plotted against the rest period R, the Devil's Staircase plot that results reveals recovery properties of the oscillatory system. Do cellular levels of key molecules (IP3, calcium) slowly or rapidly return to baseline values or perhaps overshoot baseline values and spend some time at reduced levels? These features affect the ability of the system to respond to the next stimulation pulse. Thus, periodic stimulation can provide insight into the recovery properties, a feat difficult to attain with continuous stimulation and conventional experimental techniques.

In calculating PLR, it is typical to count only 'full' responses; researchers have also observed 'sub-threshold' responses upon exposure of oscillatory systems to periodic stimulation[21] (Figure 24.3). Here 'sub-threshold' delineates that a system response did not reach full amplitude. This type of response represents a unique activation property that emerges only from periodic stimulation, and it may also be useful in comparing mechanisms.

24.4 Phase-locking Analysis of GPCR-induced Calcium Signalling in Two Models

Proposed GPCR-induced calcium oscillation models can be examined for the existence and properties of phase-locking.[14] Periodic stimulation in the form of ligand pulses can be used as a model input, and analysis proceeds as described in Figure 24.3. Here we focus on the Chay *et al.* and Politi *et al.* models. Because neither model explicitly includes receptor/ligand binding, periodic stimulation is implemented by periodic increases in the receptor contribution to the rate of G protein activation (Chay *et al.* model) or periodic increases in the rate at which PIP2 is converted to IP3 (Politi *et al.* model). We have also explicitly added the equations describing receptor/ligand/G protein dynamics to these models, and the phase-locking behaviour is similar to that shown here.[14] Phase-locking of the calcium responses is observed with both models, as expected for any non-linear oscillator[13] (Figure 24.4A). Under some periodic stimulation conditions, the calculated PLR for both models is less than one (Figure 24.4), indicative of skipped beats and consistent with observations in other experimental[19] and theoretical systems.[7] Note that the Politi *et al.* model, but not the Chay *et al.* model, shows sub-threshold calcium responses (Figure 24.4A). This result demonstrates that the activation properties of the two mathematical models are different, despite the similarity in signalling behaviour upon continuous stimulation seen in Figure 24.1.

To assess the phase-locking behaviours of the two models, each was exposed to periodic stimulation while the stimulant concentration (C), stimulation duration (D) or rest period (R) was varied. As seen in Figure 24.4B and 24.4C,

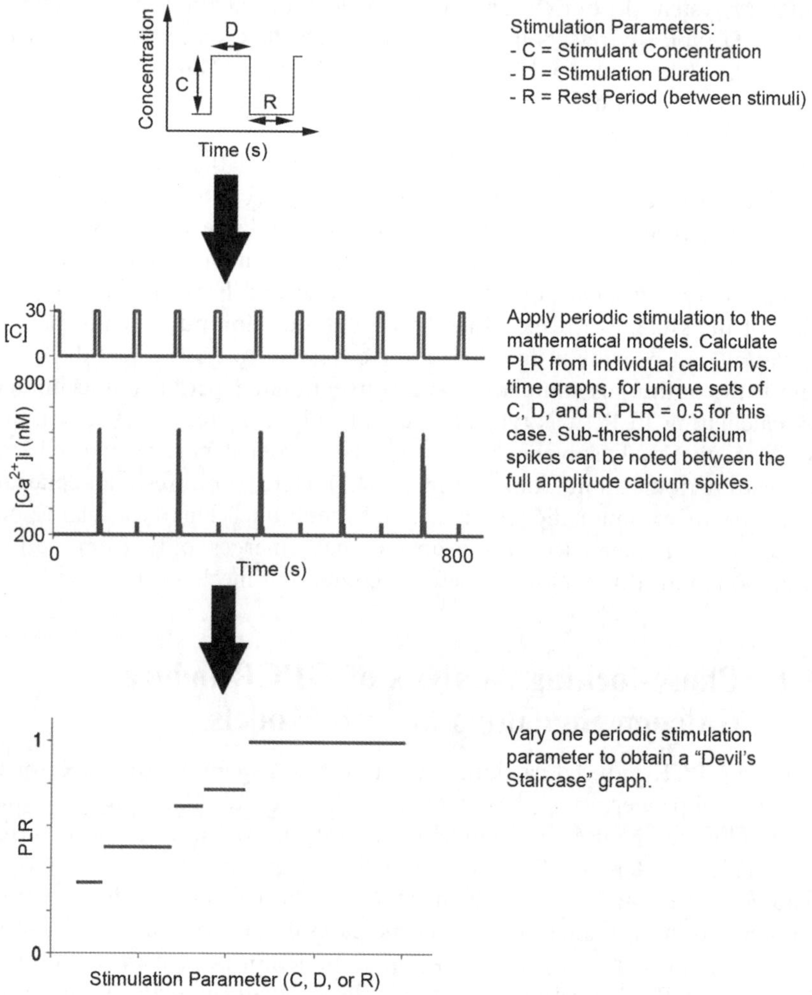

Figure 24.3 Procedure for assessing the effect of stimulation parameters on the phase-locking ratio (PLR) in mathematical models, to develop a set of discriminating markers for comparison to experiments. Periodic stimulation is applied to a single cell described by one of the mathematical models and the PLR is calculated. A single stimulation parameter (C, D or R) is varied and the PLR is calculated for every value. Plotting PLR *vs.* the stimulation parameter value results in a 'Devil's Staircase' graph. The relationship between the stimulation parameter and the PLR (increasing or decreasing) provides a discriminating feature for comparison to experiments.

the behaviour of both models when only C or D was varied is similar. However, plotting the phase-locking ratio against the rest period provides a discriminating marker: PLR increases as the rest period increases for the Chay *et al.* model, while the opposite trend occurs for the Politi *et al.* model

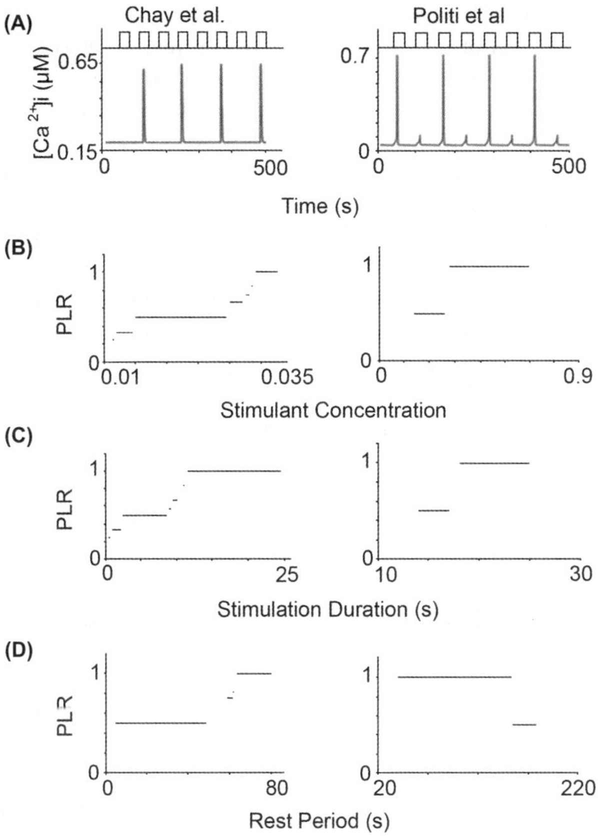

Figure 24.4 Unique signalling properties uncovered by phase-locking analysis of the Chay *et al.* and Politi *et al.* mathematical models. Original published equations and parameters were used for both models. (**A**) Periodic stimulation of the Chay *et al.* model leads to skipped beats with the absence of sub-threshold spikes. Periodic stimulation of the Politi *et al.* model also leads to skipped beats, but features sub-threshold spikes. (Chay *et al.* model: C = 0.03 1/s; Politi *et al.* model: C = 0.8 μM/s. For both: D = 30 s, R = 30 s). (**B**) PLR *vs.* C (Chay *et al.* model: D = 10 s, R = 50 s; Politi *et al.* model: D = 10 s, R = 50 s). (**C**) PLR *vs.* D (Chay *et al.* model: C = 0.03 1/s, R = 60 s; Politi *et al.* model: C = 0.8 ìM/s, R = 60 s). (**D**) PLR *vs.* R (Chay *et al.* model: C = 0.03 1/s, D = 10 s; Politi *et al.* model: C = 0.8 μM/s D = 10 s).

(Figure 24.4D). This difference suggests a difference in recovery properties for the two models. Importantly, note that the models can be easily distinguished *via* a qualitative comparison, *i.e.* one model predicts an increase in PLR with R and the other predicts a decrease, and that this behaviour should be readily apparent in experimental data.

Investigation of the respective model architectures reveals the mechanisms responsible for the differing activation and recovery properties. We can first examine the activation behaviour. The Chay *et al.* model does not exhibit sub-threshold calcium responses because of a switch-like activation of PLC by G proteins; the Hill coefficient of the reaction is four in this model. In contrast, the Politi *et al.* model exhibits sub-threshold calcium responses because G protein-mediated activation of PLC is graded. In terms of recovery properties, the Chay *et al.* model contains a basal G protein activity that results in basal IP3 production. Basal IP3 production promotes recovery to resting levels between stimulation events, and thus with longer rest periods, IP3 levels are higher and more likely to cross the threshold necessary to elicit a calcium response with a subsequent stimulation event. Ultimately, this results in an increase in PLR as R is increased. In contrast, the Politi *et al.* model has no such recovery mechanism; IP3 levels subside as the rest period increases and so calcium responses are less likely to occur with a subsequent stimulation event. As a result, one observes a reduction in PLR as R is increased.

These model observations provide a concrete set of discriminating markers that can be compared with experimental results. However, typical experiments assessing GPCR-induced calcium oscillations rely on continuous stimulation or on a single step change increase in ligand concentration. Thus experimental setups that are able to stimulate cells periodically and allow for real-time imaging of the resulting intracellular calcium responses must be developed.

24.5 Microfluidics to Enable Pulsatile Stimulation of Cells

The ability to carry out phase-locking analysis for the study of G protein dynamics is dependent upon the ability to reproducibly control cellular stimulation with high temporal resolution. Conventional techniques for creating dynamic cellular stimulation conditions are deficient in this respect; these methods typically involve adding known amounts of ligand to cells growing in culture dishes or on glass slides, effectively exposing cells only to step increases in stimulant concentration.[22] This approach is not amenable to the generation of reproducible periodic stimulation patterns necessary to study phase-locking. For example, it is difficult to exchange fluid rapidly because it requires a user to manually address cells repeatedly with stimulant, aspirate the stimulant away, and then reapply the stimulant, within a matter of tens of seconds. Perfusion chambers represent an advance in terms of reproducibility, but these setups lack versatility and scalability.[23]

The advent of microfluidic technology has overcome many of these limitations (Figure 24.5A). In terms of scalability, entire biochemical and genetic operations and manipulations are executed on a platform several square inches in area. The crux of this technology is the ability to harness the physical properties of liquids on the micron scale, enabling enhanced control over spatial and temporal facets of cellular stimulation, and thus versatility. While microfluidics initially garnered interest for studies of spatial dynamics,[24,25] recently it has become increasingly utilized for studies of temporal dynamics of

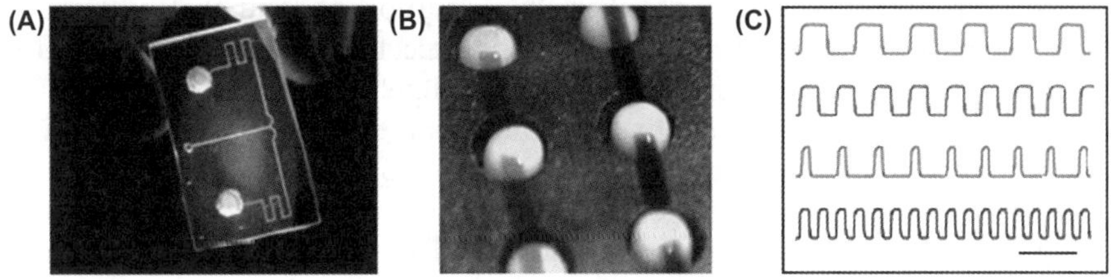

Figure 24.5 The microfluidic setup is able to control three stimulation parameters (C, D and R) (see Figure 24.3). (**A**) Image of the microfluidic setup filled with fluorescein solution. (**B**) Braille-actuated pumping controls fluid flow through channels. (**C**) Pulses generated by the microfluidic setup are reproducible and high temporal resolution is achievable. (Scale bar = 90 s).

cell signaling.[26] The appeal of this method is the rapid fluid exchange that can be achieved, as well as the low reagent consumption, portability and potential for high throughput analysis. For example, Hersen *et al.*[27] developed a microfluidic device capable of addressing cells with a chemical stimulant at frequencies ≤ 2 Hz without disrupting cell adhesion. G protein signalling occurs on the sub-second to seconds time scale, while G-protein-mediated calcium signalling is generally on the order of tens of seconds to minutes. Thus, microfluidic technology provides an optimal platform for investigations of these dynamics.

The ability to rapidly customize experimental setups for study of the dynamics of a particular GPCR ligand-receptor in a specific cell type is another major advantage of microfluidics over conventional techniques. Microfluidic devices can be designed and fabricated in ~24 hours. Most microfluidic devices are created by a rapid prototyping method pioneered by the Whitesides group.[28] Initially, microfluidic designs are created on a computer aided design (CAD) program and these designs are then converted into a transparency. The transparency design is transferred to a silicon wafer or glass slide through a photolithographic process, effectively creating a mould with positive relief features. The overwhelming choice of material for creating microfluidic devices is polydimethylsiloxane (PDMS); this silicon-based elastomer has favourable properties for cell culture and imaging.[29] PDMS is cast against the mould, and upon curing, the device is irreversibly sealed against a flat surface (usually glace or PDMS sheet) through plasma oxidation. Microfluidic devices can be customized for a particular cell type in a number of ways; for instance, for cell types that do not adhere well to PDMS or glass surfaces, the device surface can be coated with adhesion molecules such as laminin or fibronectin.[30] More elaborate manipulations can be implemented to support culture of cells that are difficult to grow *in vitro*.[31]

The method for pumping liquid in the microfluidic devices plays a tremendous role in the design and fabrication process. The most common pumping methods employed in microfluidics are gravity-driven and syringe-mediated pumping, of which the latter is more reliable. More elaborate pumping systems include 'Quake valves',[32] Braille actuation,[33,34] acoustics,[35] and most recently autonomous pumping regulation by embedded components.[36]

The reproducibility and versatility of generation of temporal patterns is in large part determined by the pumping mechanisms implemented for conveying fluids through the microfluidic channels. In our experiments, we use Braille-actuated pumping, which offers an excellent level of control, reproducibility and portability[33,34] (Figure 24.5B). With this approach, microfluidic channels are aligned on a Braille display that is connected to a computer *via* USB. Upon elevation, individual Braille pins are able to valve off the microfluidic channels that lie above, due to the elastomeric nature of the PDMS. In the appropriate sequence, consecutive Braille pin movements enable relatively unidirectional flow. Individual Braille pin movements are controlled by a computer program, such that the speed and direction of pumping can be regulated. This setup provides an optimal platform for conducting studies in which control over cellular stimulation parameters, such as stimulation duration and rest between stimulation events, is required. This is demonstrated in Figure 24.5C; one reservoir of the device was filled with fluorescein solution and the other with water. Upon alternating from which reservoir liquid was pumped, square-wave patterns of specific duration and rest period were reliably generated.[37]

24.6 Imaging of Signalling Dynamics in a Microfluidic Device

In order to monitor cellular signalling dynamics resulting from temporal patterns of stimulation, real-time imaging of appropriate readouts of cell signalling is needed. The advent of fluorescent reporters of cell signalling has enabled tracking of signalling behaviours of individual cells. In particular, green fluorescent protein (GFP) based readouts (or variants of GFP) have been employed to track the localization, translocation, appearance or degradation of intracellular components, representing 'passive' applications of these fluorescent constructs; in this context, 'passive' denotes that the component activity is not assessed.

Fluorescence resonance energy transfer (FRET) has been used in order to convey dynamic information about intracellular activity.[38] Although FRET probes for directly assessing G protein activity have been developed,[39] the dynamic range of these readouts has not reached a level that has led to widespread utilization. However, the following are some of the probes developed over the last 15 years that are able to track G protein-mediated signalling activities: Cameleon (for intracellular calcium);[40] Raichu-Ras (measures levels of activated Ras);[41] cGMP probe;[42] PKA probe; and a cAMP probe.[43]

In this chapter, the focus is on using G protein-mediated calcium signalling to elucidate molecular mechanisms. Since imaging probes for calcium are more developed compared with those for G proteins,[39] intracellular calcium is used as a readout to infer G protein signalling dynamics. While the fluorescent protein aequorin[44] has been utilized for assessing intracellular calcium levels in real-time, it is labour-intensive to introduce into cells, requiring micro-injection. Since high throughput analysis of cells is thus not feasible with this indicator, fluorescent dyes and genetically coded probes are preferred. Fura-2 and Fluo-4 are popular fluorescent dyes;[45,46] these probes can be easily introduced into cells and are commercially available. These dyes can leak out, rendering them infeasible for use in long-term experiments and it has been reported that the dyes can localize to various compartments of cells, complicating quantification of calcium levels. Genetically encoded FRET probes for calcium can be easily introduced into cell populations through transfection.[47] In addition, these probes are specifically designed to localize to a specific part of the cell and do not leak out, enabling long-term characterizations of calcium dynamics. For these reasons, we used the FRET probe YC3.60 developed by Nagai *et al.*,[48] which measures free calcium levels exclusively in the cytosol.

24.7 Experimental Observations of Phase-locking in GPCR-induced Calcium Signalling

Using the microfluidic and imaging techniques described above, one can measure GPCR-induced calcium oscillations and compare those data with the phase-locking properties of models to elucidate mechanisms. Using our Braille-actuated microfluidic platform, cells were periodically exposed to ligand and the resulting calcium signals were recorded in real-time using the FRET probe YC3.60.[14] The D and R values chosen for these studies were based upon the typical durations of single oscillatory calcium responses (tens of seconds) and typical oscillation periods (tens of seconds to minutes).

Figure 24.6 shows the calcium responses of three cell types exposed to periodic stimulation with different GPCR ligands: a HEK293 cell stimulated with carbachol (through the M3 receptor); a HeLa cell stimulated with histamine (through histamine receptors); and a HeyA8 cell stimulated with extracellular calcium (through the calcium-sensing receptor), respectively. Phase-locking was observed with each cell type, as calcium responses were synchronized to the periodic stimulation events. Furthermore, all three cells exhibited PLRs less than one, as predicted by the models under some periodic stimulation conditions. Thus phase-locking is a general result of periodic stimulation of GPCR systems and phase-locking analysis may be appropriate to analyse signalling mechanisms.

For the remainder of this chapter, we focus our phase-locking analysis on the M3 muscarinic pathway in HEK293 cells in order elucidate mechanisms of G protein (here, Gq) signalling. Periodic stimulation of HEK293 cells with carbachol resulted in phase-locking of the resulting calcium responses and sub-threshold spikes are present (Figure 24.6B). We then examined the phase-locking ratio as stimulant concentration (C), stimulation duration (D) or rest period (R) was varied. The PLR was measured for individual cells and then averaged. The population-averaged PLR can be used to compare between

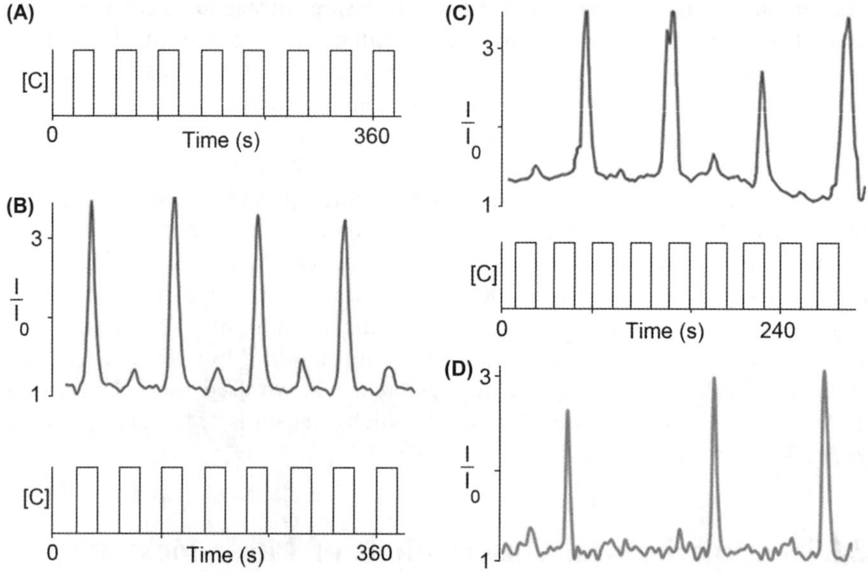

Figure 24.6 Observation of phase-locked calcium responses and sub-threshold spikes in three different cell types and with three different agonists. (**A**) Periodic stimulation pattern. (**B**) HEK293 cell periodically stimulated with carbachol (C = 10 nM, D = 24 s, R = 24 s). (**C**) HeLa cell periodically stimulated with histamine (C = 100 uM, D = 24 s, R = 24 s). (**D**) HeyA8 cell periodically stimulated with extracellular calcium (C = 2 mM, D = 16 s, R = 16 s). Here I/I0 signifies the FRET ratio of the calcium signal (I) normalized to the minimum FRET ratio (I0), as has been done previously for analysis of intracellular calcium responses.[48]

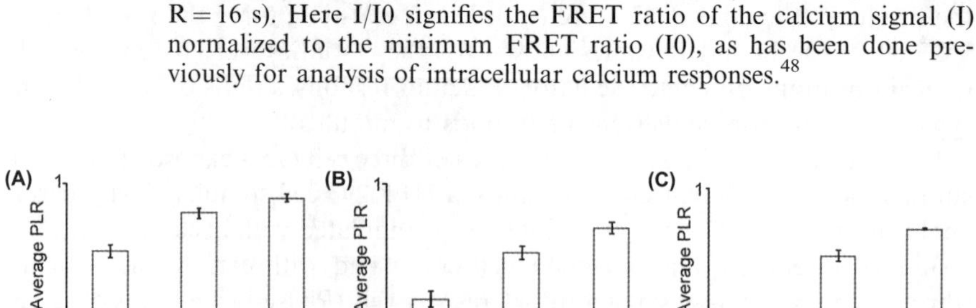

Figure 24.7 Experimentally measured average phase-locking ratios of cell populations *vs.* increasing C, D, or R. in HEK293 cells periodically stimulated with carbachol. (A) PLR *vs.* C (D = 24 s, R = 24 s). (B) PLR *vs.* D (C = 10 nM, R = 24 s). (C) PLR *vs.* R (C = 10 nM, D = 24 s). As each stimulation parameter was increased, there was a corresponding increase in the PLR, providing useful comparisons to mathematical model predictions. The results presented here are representative of at least 60 cells at each condition from three different experiments.[14]

different experimental conditions (Figure 24.7). Note that as C is increased, PLR increases; the same trend is observed when D is increased (Figure 24.7A and 24.7B). Thus increasing either C or D enhances the probability that a calcium response is going to result during a particular stimulation event. We also found that increases in R produced increases in PLR (Figure 24.7C), suggesting that a recovery mechanism allows a reset of the signalling system if there is enough time between stimulation events.

24.8 Comparing Model and Experimental Results

These experiments described above (Figure 24.7) provide data on the M3 system for comparison with model predictions on phase-locking. The models introduced above and the experimental data show increases in PLR with C or D, so this comparison does not allow the models to be distinguished. However, because sub-threshold oscillations are not seen with the Chay *et al.* model, and because PLR decreases with R rather than increases with the Politi *et al.* model, neither mathematical model is able to account for all of our experimental observations even at a qualitative level.

A comparison between the models offers insights into GPCR-induced calcium oscillations mechanisms, particularly the activation and recovery properties of the M3 receptor/cell system are studied. The presence of sub-threshold calcium responses (Figure 24.6B) suggests that the G protein activation of PLC is graded and not switch-like as modelled by Chay *et al.* The G protein activation properties thus appear more similar to those of the Politi *et al.* model. However, increasing the rest period R in experiments resulted in increases in PLR (Figure 24.7C). This result indicated that the recovery properties of the system were better described by the Chay *et al.* model, suggesting the existence of basal G protein activity driving basal IP3 production.

24.9 Model Revision

At least two approaches can be suggested to come up with a mechanistic model that is more consistent with the experimental data. First, we can learn from the comparison above and combine elements of each model to produce a model that does a better job of agreeing with the data. In order to undertake mechanism revision, elements from the respective calcium models that resulted in correct predictions of the activation and recovery properties of the system were combined.[14] Phase-locking analysis indicated that the Chay *et al.* model had the correct recovery properties, due to the inclusion of a basal G protein activity mechanism. Analysis also suggested that the Politi *et al.* model possessed the correct activation properties, due to the graded activation of PLC by G-proteins. Combining these two mechanistic elements, we created a revised model and then evaluated it using phase-locking analysis (Figure 24.8). Periodic stimulation resulted in the emergence of sub-threshold calcium responses, confirming an improvement in the activation properties of the new model (Figure 24.8A). As the rest period of the periodic stimulation increased, the phase-locking ratio correspondingly increased (Figure 24.8E), indicating that the recovery properties of the new system were accurate.

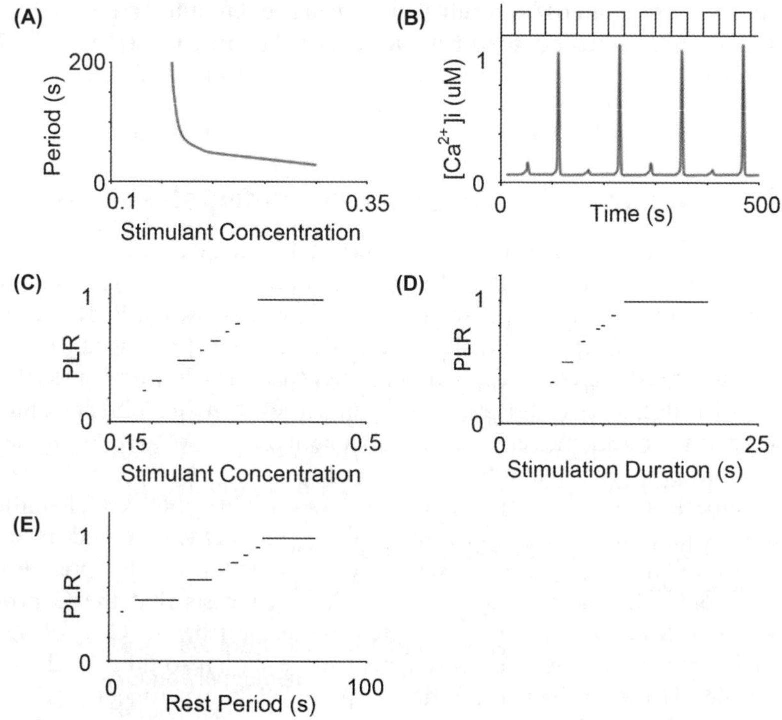

Figure 24.8 Behaviours of the revised model (with basal IP3 production = 0.3 μM/s) agree with experimental results of our phase-locking analysis. The model was revised by adding the basal IP3 production term to the rate equation describing IP3 production by PLC. (**A**) The revised model correctly predicts that under continuous stimulation, the calcium oscillation period decreases with increasing stimulant concentration (μM/s). (**B**) The model correctly predicts the presence of sub-threshold calcium spikes upon periodic stimulation (C = 0.3 μM/s, D = 10 s, R = 50 s). (**C,D,E**) Model predictions as stimulation concentration, stimulation duration, and rest period are varied: PLR *vs.* C (μM/s) (D = 10 s, R = 50 s). PLR *vs.* D (C = 0.3 μM/s, R = 50 s). PLR *vs.* R (C = 0.3 μM/s, D = 10 s).

The second approach takes account of the fact that the discrepancies between models and experiments could be the result of incorrect parameter values in the model. Uncertainty in model parameter values is a given, as assays to measure all parameter values *in vivo* simply are not available. The Politi *et al.* model has 17 independent parameters, while the Chay *et al.* model has ten—most of which were estimated and not measured directly. Rather than haphazardly attempting to explore all possible parameter space, a sampling algorithm can be used to effectively search a large parameter space to explore whether the differences between the models and experiments are the result of the model parameter values but not mechanisms. Latin Hypercube Sampling (LHS) is an algorithm that we have implemented for this purpose; it entails specifying a distribution (*e.g.* normal or uniform) and dividing it up into equal probability intervals from which to sample without replacement.[4,49,50] LHS is generally run for hundreds to thousands of iterations and allows for simultaneous variation of multiple model parameter values. In the context of this study, LHS was run for both the Chay *et al.* and Politi *et al.* models to observe whether any parameter set could result in the emergence of sub-threshold calcium responses. Parameters were sampled both from a normal distribution (with original parameters for the mean and with a standard deviation of 100%) and a uniform distribution (with a minimum of 0.1 times original parameter values and a maximum of ten times original parameter values). None of the parameter sets sampled were able to result in behaviour that agreed with our experimental results.

24.10 Future Directions

We have demonstrated in this chapter that phase-locking analysis can provide a useful tool for evaluating proposed mechanisms of GPCR-induced calcium oscillations. In particular, phase-locking can reveal aspects of the activation and recovery properties of oscillatory systems, particularly when the stimulation time and time between stimulation events are varied. Importantly, phase-locking analysis provides a qualitative tool: the ability of proposed models to reproduce experimental behaviour, *i.e.* the existence (or not) of sub-threshold oscillations and the behaviour (unchanged, increasing, decreasing) of the phase-locking ratio (PLR) as stimulation parameters are varied can be assessed. We demonstrated this concept here with two mathematical models of oscillatory calcium signalling (and with seven more models in ref. 14). As another qualitative test, Sneyd *et al.*[51] detailed an approach that involves subjecting cells to a pulse of IP3 and observing whether subsequent calcium oscillations increase or decrease in frequency. This qualitative readout of GPCR signalling can help determine whether IP3 oscillations are passive reflections of calcium oscillations or absolutely necessary to produce the calcium oscillations. A proposed mechanism that succeeds in passing qualitative tests can then be subjected to more quantitative tests and rigorous determination of physical parameter values. These tools complement existing genetic and chemical tools for deciphering GPCR signalling.

With advancements in imaging and synthesis, other probes should be available for phase-locking analysis and thereby provide even more discriminating markers for evaluation of mathematical models and elucidation of signalling pathways. Direct imaging of G protein dynamics has been achieved,[39] and with further development would be interesting to use for phase-locking analysis. Other relevant signalling components to image would be IP3,[52] and perhaps protein kinase C (PKC)[53] and the regulator of G protein signalling (RGS) proteins.[54]

Microfluidic technology will also allow other stimulation patterns to be tested. The phase-locking analysis described here relied on the ability to generate periodic (square-wave) stimulation. Other stimulation patterns such as saw-tooth patterns,[55] different step patterns[23,56] and sine waves[57] are examples of stimulation patterns that could be used in this context and might expand the number of discriminating markers available for evaluating models and elucidating mechanisms.

Collectively, the combination of microfluidics, real-time imaging, and mathematical modelling provides a means of effectively elucidating molecular mechanisms of signalling that has many future possibilities.

Acknowledgements

The authors would like to thank Bryan Howell, Michelle Cote, Susan M. Wade, Khamir Mehta, Atsushi Miyawaki and Richard R. Neubig for assistance with various aspects of the experimental or modelling work, and/or helpful comments.

This research was supported by the following sources: NIH Microfluidics in Biomedical Sciences Training Program (NIH NIBIB T32 EB005582), NIH ARRA Summer Supplement, NIH grants R01 HL084370-04, R01 CA136829-01, R33 HL092844, and the US Army Research Laboratories and the US Army Research Office (under contract number DAAD 19-03-1-0168).

References

1. J. J. Linderman, *J. Biol. Chem.*, 2009, **284**, 5427–5431.
2. P. Bisegna, G. Caruso, D. Andreucci, L. Shen, W. Gurevich, H. E. Hamm and E. DiBenedetto, *Biophys. J.*, 2008, **94**, 3363–3383.
3. S. J. Bornheimer, M. R. Maurya, M. G. Farquhar and S. Subramaniam, *Proc. Natl. Acad. Sci. U. S. A.*, 2004, **101**, 15899–15904.
4. T. L. Kinzer-Ursem and J. J. Linderman, *PloS Comput. Biol.*, 2007, **3** 84–94.
5. J. J. Saucerman, J. Zhang, J. C. Martin, L. X. Peng, A. E. Stenbit, R. Y. Tsien and A. D. McCulloch, *Proc. Natl. Acad. Sci. U. S. A.*, 2006, **103**, 12923–12928.
6. A. Atri, J. Amundson, D. Clapham and J. Sneyd, *Biophys. J.*, 1993, **65**, 1727–1739.
7. T. R. Chay, Y. S. Lee and Y. S. Fan, *J. Theor. Biol.*, 1995, **174**, 21–44.
8. K. S. R. Cuthbertson and T. R. Chay, *Cell Calcium*, 1991, **12**, 97–109.
9. G. Dupont and C. Erneux, *Cell Calcium*, 1997, **22**, 321–331.
10. Y. X. Li and J. Rinzel, *J. Theor. Biol.*, 1994, **166**, 461–473.
11. A. Politi, L. D. Gaspers, A. P. Thomas and T. Hofer, *Biophys. J.*, 2006, **90**, 3120–3133.
12. J. Sneyd, Modeling IP3-dependent calcium dynamics in non-excitable cells, in *Tutorials in Mathematical Biosciences II*, Springer-Verlag, Berlin, 2005, Lecture Notes in Mathematics, vol. 1867, pp. 631–634.
13. L. Glass and M. C. Mackey, *J. Math. Biol.*, 1979, **7**, 339–352.

14. A. Jovic, B. Howell, M. Cote, S. M. Wade, K. Mehta, A. Miyawaki, R. R. Neubig, J. J. Linderman and S. Takayama, *PLoS Comput. Biol.*, 2010, **6**(12), e1001040.
15. R. Jacob, J. E. Merritt, T. J. Hallam and T. J. Rink, *Nature*, 1988, **335** 40–45.
16. M. Prentki, M. C. Glennon, A. P. Thomas, R. L. Morris, F. M. Matschinsky and B. E. Corkey, *J. Biol. Chem.*, 1988, **263**, 11044–11047.
17. N. M. Woods, K. S. Cuthbertson and P. H. Cobbold, *Nature*, 1986, **319**, 600–602.
18. M. S. Baptista and I. L. Caldas, *Nonlinear Dyn.*, 1998, **17**, 119–139.
19. M. R. Guevara, L. Glass and A. Shrier, *Science*, 1981, **214**, 1350–1353.
20. R. F. Galán and N. N. Urban, *Phys. Rev. Lett.*, 2005, **94**, 158101.
21. R. Wessel, *Eigenmannia, Biophys. J.*, 1995, **69**, 1880–1890.
22. S. Paliwal, C. J. Wang and A. Levchenko, *HFSP J.*, 2008, **2**, 251–256.
23. K. R. King, S. Wang, A. Jayaraman, M. L. Yarmush and M. Toner, *Lab Chip*, 2008, **8**, 107–116.
24. A. Sawano, S. Takayama, M. Matsuda and A. Miyawaki, *Dev. Cell*, 2002, **3**, 245–257.
25. S. Takayama, E. Ostuni, P. LeDuc, K. Naruse, D. E. Ingber and G. M. Whitesides, *Nature*, 2001, **411**, 1016–1016.
26. A. Jovic, B. Howell and S. Takayama, *Microfluid. Nanofluid.*, 2009, **6**, 717–729.
27. P. Hersen, M. N. McClean, L. Mahadevan and S. Ramanathan, *Proc. Natl. Acad. Sci. U. S. A.*, 2008, **105**, 7165–7170.
28. G. M. Whitesides, E. Ostuni, S. Takayama, X. Y. Jiang and D. E. Ingber, *Annu. Rev. Biomed. Eng.*, 2001, **3**, 335–373.
29. J. C. McDonald, D. C. Duffy, J. R. Anderson, D. T. Chiu, H. Wu, O. J. Schueller and G. M. Whitesides, *Electrophoresis*, 2000, **21**, 27–40.
30. J. N. Lee, X. Jiang, D. Ryan and G. M. Whitesides, *Langmuir*, 2004, **20**, 11684–11691.
31. G. Mehta, M. J. Kiel, J. W. Lee, N. Kotov, J. J. Linderman and S. Takayama, *Adv. Funct. Mater.*, 2007, **17**, 2701–2709.
32. M. A. Unger, H. P. Chou, T. Thorsen, A. Scherer and S. R. Quake, *Science*, 2000, **288**, 113–116.
33. N. Futai, W. Gu, J. W. Song and S. Takayama, *Lab Chip*, 2006, **6**, 149–154.
34. W. Gu, X. Y. Zhu, N. Futai, B. S. Cho and S. Takayama, *Proc. Natl. Acad. Sci. U. S. A.*, 2004, **101**, 15861–15866.
35. S. M. Langelier, D. S. Chang, R. I. Zeitoun and M. A. Burns, *Proc. Natl. Acad. Sci. U. S. A.*, 2009, **106**, 12617–12622.
36. B. Mosadegh, C. H. Kuo, Y. C. Tung, Y. S. Torisawa, T. Bersano-Begey, H. Tavana and S. Takayama, *Nat. Phys.*, 2010, **6**, 433–437.
37. A. Jovic, S. M. Wade, A. Miyawaki, R. R. Neubig, J. J. Linderman, *et al.*, Hi-fi transmission of periodic signals amid cell-to-cell variability. *Mol. BioSyst.*, 2011, in press, DOI: 10.1039/c1mb05031a.
38. A. Miyawaki, *Dev. Cell*, 2003, **4**, 295–305.

39. M. Akgoz, I. Azpiazu, V. Kalyanaraman and N. Gautam, *FASEB J.*, 2004, **18**, C257–C257.
40. A. Miyawaki, J. Llopis, R. Heim, J. M. McCaffery, J. A. Adams, M. Ikura and R. Y. Tsien, *Nature*, 1997, **388**, 882–887.
41. N. Mochizuki, S. Yamashita, K. Kurokawa, Y. Ohba, T. Nagai, A. Miyawaski and M. Matsuda, *Nature*, 2001, **411**, 1065–1068.
42. Y. Nino, K. Hotta and K. Oka. *PLoS ONE*, 2010, **5**(2), e9164.
43. V. Nikolaev, M. Bunemann, A. Hannawacker and M. J. Lohse, *Naunyn Schmiedebergs Arch. Pharmacol.*, 2004, **369**, R46–R46.
44. J. R. Blinks, W. G. Wier, P. Hess and F. G. Prendergast, *Prog. Biophys. Mol. Biol.*, 1982, **40**, 1–114.
45. G. Grynkiewicz, M. Poenie and R. Y. Tsien, *J. Biol. Chem.*, 1985, **260**, 3440–3450.
46. K. R. Gee, K. A. Brown, W. N. U. Chen, J. Bishop-Stewart, D. Gray and I. Johnson, *Cell Calcium*, 2000, **27**, 97–106.
47. A. E. Palmer and R. Y. Tsien, *Nat. Protoc.*, 2006, **1**, 1057–1065.
48. T. Nagai, S. Yamada, T. Tominaga, M. Ichikawa and A. Miyawaki, *Proc. Natl. Acad. Sci. U. S. A.*, 2004, **101**, 10554–10559.
49. S. Marino, I. B. Hogue, C. J. Ray and D. E. Kirschner, *J. Theor. Biol.*, 2008, **254**, 178–196.
50. S. M. Blower and H. Dowlatabadi, *Int. Stat. Rev.*, 1994, **62**, 229–243.
51. J. Sneyd, K. Tsaneva-Atanasova, V. Reznikov, Y. Bai, M. J. Sanderson and D. I. Yule, *Proc. Natl. Acad. Sci. U. S. A.*, 2006, **103**, 1675–1680.
52. H. Shirakawa, M. Ito, M. Sato, Y. Umezawa and S. Miyazaki, *Biochem. Biophys. Res. Commun.*, 2006, **345**, 781–788.
53. J. Goedhart and T. W. J. Gadella, *J. R. Soc. Interface*, 2009, **6**, S27–S34.
54. X. Luo, S. Popov, A. K. Bera, T. M. Wilkie and S. Muallem, *Mol. Cell*, 2001, **7**, 651–660.
55. L. Chen, F. Azizi and C. H. Mastrangelo, *Lab Chip*, 2007, **7**, 850–855.
56. B. Kuczenski, W. C. Ruder, W. C. Messner and P. R. LeDuc, *PloS ONE*, 2009, **4**, 8.
57. J. Olofsson, H. Bridle, J. Sinclair, D. Granfeldt, E. Sahlin and O. Orwar, *Proc. Natl. Acad. Sci. U. S. A.*, 2005, **102**, 8097–8102.

Subject Index

Note: page numbers in *italic* refer to figures or tables